Proof and Computation

NATO ASI Series

Advanced Science Institutes Series

A series presenting the results of activities sponsored by the NATO Science Committee, which aims at the dissemination of advanced scientific and technological knowledge, with a view to strengthening links between scientific communities.

The Series is published by an international board of publishers in conjunction with the NATO Scientific Affairs Division

A Life Sciences **B Physics**	Plenum Publishing Corporation London and New York
C Mathematical and Physical Sciences **D Behavioural and Social Sciences** **E Applied Sciences**	Kluwer Academic Publishers Dordrecht, Boston and London
F Computer and Systems Sciences **G Ecological Sciences** **H Cell Biology** **I Global Environmental Change**	Springer-Verlag Berlin Heidelberg New York London Paris Tokyo Hong Kong Barcelona Budapest

PARTNERSHIP SUB-SERIES

1. **Disarmament Technologies**	Kluwer Academic Publishers
2. **Environment**	Springer-Verlag
3. **High Technology**	Kluwer Academic Publishers
4. **Science and Technology Policy**	Kluwer Academic Publishers
5. **Computer Networking**	Kluwer Academic Publishers

The Partnership Sub-Series incorporates activities undertaken in collaboration with NATO's Cooperation Partners, the countries of the CIS and Central and Eastern Europe, in Priority Areas of concern to those countries.

NATO-PCO DATABASE

The electronic index to the NATO ASI Series provides full bibliographical references (with keywords and/or abstracts) to about 50 000 contributions from international scientists published in all sections of the NATO ASI Series. Access to the NATO-PCO DATABASE compiled by the NATO Publication Coordination Office is possible in two ways:

- via online FILE 128 (NATO-PCO DATABASE) hosted by ESRIN, Via Galileo Galilei, I-00044 Frascati, Italy.

- via CD-ROM "NATO Science & Technology Disk" with user-friendly retrieval software in English, French and German (© WTV GmbH and DATAWARE Technologies Inc. 1992).

The CD-ROM can be ordered through any member of the Board of Publishers or through NATO-PCO, Overijse, Belgium.

Series F: Computer and Systems Sciences, Vol. 139

Proof and Computation

Edited by

Helmut Schwichtenberg

Ludwig-Maximilians-Universität München
Mathematisches Institut, Theresienstraße 39
D-80333 München, Germany

Springer

Published in cooperation with NATO Scientific Affairs Division

Proceedings of the NATO Advanced Study Institute on Proof and Computation, held in Marktoberdorf, Germany, July 20 – August 1, 1993

CR Subject Classification (1991): B.6, C.2, D.1–3, F.3–4, I.1–2

Additional material to this bood can be downloaded from http://extra.springer.com.

ISBN-13:978-3-642-79363-9 e-ISBN-13:978-3-642-79361-5
DOI: 10.1007/978-3-642-79361-5

CIP data applied for

Softcover reprint of the hardcover 1st edition 1995

Typesetting: Camera-ready by authors
Printed on acid-free paper
SAP: 10084496 45/3142 – 5 4 3 2 1 0

Preface

The Marktoberdorf Summer School 1993 was held under the title Proof and Computation. It was the 14th of a series of Advanced Study Institutes under the sponsorship of the NATO Scientific Affairs Division held in Marktoberdorf.

In recent years it has become obvious that logical methods and notions are of growing importance in many areas of computer science. The proofs-as-programs paradigm and the now well-established role of Prolog (e.g. as a tool to build expert systems) give a clear evidence of this fact. The logical notion of a formal proof in various constructive systems can be viewed as a very explicit way to describe a computation procedure. Many aspects of computation can so be captured in an illuminating way: a good example is the modelling of bounded resources in linear logic. But also conversely the development of logical systems (e.g. for verification purposes) has been influenced by accumulating knowledge on rewriting and unification techniques, in particular in a higher order context.

It was the purpose of the Summer School 1993 to study the impact of the concept of a formal proof to computation theory under several points of view:

- Direct use of the computational aspects of proofs (e.g. in logic programming or in type theory)
- New logical systems, specifically designed for computational applications (e.g. concurrency, bounded resources)
- Development of algebraic specifications and abstract data types as flexible tools to deal with the complexities of large-scale systems.

We again succeeded in bringing together leading researchers and a highly qualified group of young scientists for stimulating presentations and exchanges of new ideas. However, the Summer School would not have been possible without the help of the experienced scientific staff from Munich and the local Marktoberdorf crew. My sincere thanks go to all of them.

Munich, March 1995 H. Schwichtenberg

Table of Contents

Program Development by Proof Transformation

Ulrich Berger and Helmut Schwichtenberg
Mathematisches Institut der Universität München
Theresienstraße 39, D-80333 München

We begin by reviewing the natural deduction rules for the $\to\land\forall$–fragment of minimal logic. It is shown how intuitionistic and classical logic can be embedded. Recursion and induction is added to obtain a more realistic proof system. Simple types are added in order to make the language more expressive. We also consider two alternative methods to deal with the strong or constructive existential quantifier $\exists^*$. Finally we discuss the well–known notion of an extracted program of a derivation involving $\exists^*$, in order to set up a relation between the two alternatives.

Section 2 deals with the computational content of classical proofs. As is well–known a proof of a $\forall\exists$–theorem with a quantifier–free kernel — where $\exists$ is viewed as defined by $\neg\forall\neg$ — can be used as a program. We describe a "direct method" to use such a proof as a program, and compare it with Harvey Friedman's A–translation [3] followed by program extraction from the resulting constructive proof. It is shown that both algorithms coincide.

In section 3 Goad's method of pruning of proof trees is introduced. It is shown how a proof can be simplified after addition of some further assumptions. In a first step some subproofs are replaced by different ones using the additional assumptions. In a second step parts of the proof tree are pruned, i.e. cut out. Note that the first step involves searching for new proofs — using the new assumptions — of formulas in the proof tree. Hence we also have to discuss proof search in minimal logic.

Finally section 4 treats an example already considered by Goad in his thesis [5], the binpacking problem. The main difference to Goad's work is that he used a logic with the strong existential quantifier, whereas we work within the $\to\forall$–fragment. This example is particularly well–suited to demonstrate that the pruning method can be applied to adapt programs to particular situations, and moreover that pruning can change the functions computed by programs. In this sense this method is essentially different from program development by program transformation.

We would like to thank Michael Bopp and Karl–Heinz Niggl for their help in preparing these notes.

1. Proofs in minimal logic and arithmetic

Let us first fix our language $\mathcal{L}$. Let G be a set of ground types (e.g. nat and boole). *Types* (also called object types or simple types) are formed from G by the operations $\rho \to \sigma$ and $\rho \times \sigma$. For any type ρ let a countable infinite set of *variables* of type ρ be given. We denote variables of type ρ by $x^\rho, y^\rho, \ldots$. We also assume that a set C of *constants* denoted c^ρ is given, each of an arbitrary type ρ.

Terms and their types are defined inductively by

- x^ρ and c^ρ are terms of type ρ,
- If r is a term of type $\rho \to \sigma$ and s is a term of type ρ, the rs is a term of type σ.
- If r is a term of type σ, then $\lambda x^\rho\, r$ is a term of type $\rho \to \sigma$.
- If t_i is a term of type ρ_i for $i \in \{0, 1\}$, then $\langle t_0, t_1 \rangle$ is a term of type $\rho_0 \times \rho_1$.
- If t is a term of type $\rho_0 \times \rho_1$ and $i \in \{0, 1\}$, then $\pi_i(t)$ is a term of type ρ_i.

The set $\mathrm{FV}(r)$ of *free variables* of a term r is defined as usual. It is well known that any term has a unique normal form with respect to $\beta\eta$–conversion. In the sequel we will identify terms with the same $\beta\eta$-normal form.

We now fix our notion of a formula. We assume that a set $\mathcal{P}$ of *predicate symbols* P of arities $\rho_1, \ldots, \rho_m$ is given. 0–ary relation symbols are called propositional symbols. *Formulas* are defined inductively by

- If $t_1^{\rho_1}, \ldots, t_n^{\rho_n}$ are terms and $R \in \mathcal{P}$ is a relation symbol of arity $\rho_1, \ldots, \rho_n$, then $R(t_1, \ldots, t_n)$ is a formula.
- $\bot$ (to be read "falsity") is a formula.
- If A and B are formulas, then $A \to B$ is a formula.
- If A and B are formulas, then $A \wedge B$ is a formula.
- If A is a formula and x^ρ is a variable, then $\forall x^\rho\, A$ is a formula.

$R(t_1, \ldots, t_n)$ and $\bot$ are called *atomic formulas* or *atoms*. Note that $\bot$ is not a propositional symbol.

We use (also with indices)

$$
\begin{aligned}
r, s, t &\quad \text{for terms,} \\
x, y, z &\quad \text{for variables,} \\
a, b, c &\quad \text{for constants,} \\
P, Q, R &\quad \text{for relation symbols,} \\
f, g, h &\quad \text{for function symbols,} \\
A, B, C &\quad \text{for formulas.}
\end{aligned}
$$

Negation, disjunction and the existential quantifier are defined by

$$\neg A := A \to \bot,$$
$$A \vee B := \neg A \wedge \neg B \to \bot,$$
$$\exists x\, A := \neg \forall x\, \neg A.$$

A term t is called *closed*, if $\mathrm{FV}(t) = \emptyset$. We write $t[x_1, \ldots, x_n]$ to indicate that all variables free in t are in the list $x_1, \ldots, x_n$. Similar conventions apply to formulas.

As our deductive formalism we use the system of natural deduction introduced by Gerhard Gentzen in [4]. It consists of the following introduction and elimination rules for $\to$, $\wedge$ and $\forall$.

For any $\mathcal{L}$–formula A let countably many *assumption variables* of type A be given. We use u^A, v^A, w^A to denote assumption variables of type A.

The notions of a *derivation term* d^A in minimal logic and its set $\mathrm{FA}(d^A)$ of *free assumption variables* are defined inductively by

(A) u^A is a derivation term with $\mathrm{FA}(u^A) = \{u^A\}$.

($\to^+$) If d^B is a derivation term, then

$$(\lambda u^A\, d^B)^{A \to B}$$

is a derivation term with $\mathrm{FA}(\lambda u^A\, d^B) = \mathrm{FA}(d^B) \setminus \{u^A\}$.

($\to^-$) If $d^{A\to B}$ and e^A are derivation terms, then

$$(d^{A\to B} e^A)^B$$

is a derivation term with $\mathrm{FA}(d^{A\to B} e^A) = \mathrm{FA}(d^{A\to B}) \cup \mathrm{FA}(e^A)$.

($\wedge^+$) If d^B and e^B are derivation terms, then

$$\langle d^A, e^B \rangle^{A \wedge B}$$

is a derivation term with $\mathrm{FA}(\langle d^A, e^B \rangle^{A\wedge B}) := \mathrm{FA}(d^A) \cup \mathrm{FA}(e^B)$.

($\wedge^-$) If $d^{A_0 \wedge A_1}$ is a derivation term, then

$$\pi_i(d^{A_0\wedge A_1})^{A_i}$$

is a derivation term with $\mathrm{FA}(\pi_i(d^{A_0\wedge A_1})^{A_i}) := \mathrm{FA}(d^{A_0\wedge A_1})$ (for $i \in \{0, 1\}$).

($\forall^+$) If d^A is a derivation term and $x^\rho \notin \bigcup\{\mathrm{FV}(B) | u^B \in \mathrm{FA}(d^A)\}$, then

$$(\lambda x^\rho\, d^A)^{\forall x^\rho\, A}$$

is a derivation term with $\mathrm{FA}(\lambda x^\rho\, d^A) = \mathrm{FA}(d^A)$.

($\forall^-$) If $d^{\forall x^\rho A}$ is a derivation term and t^ρ is a term, then

$$(d^{\forall x^\rho A} t^\rho)^{A[t/x^\rho]}$$

is a derivation term with $\mathrm{FA}(d^{\forall x^\rho A} t^\rho) = \mathrm{FA}(d^{\forall x^\rho A})$.

Recall that we identify terms with the same $\beta\eta$-normal form. Hence e.g.

$$d^{R(r)\to B} e^{R(s)}$$

is a correct derivation provided r and s have the same $\beta\eta$-normal forms.

A derivation term d^A is called *closed*, if $\mathrm{FA}(d^A) = \emptyset$. We write

$$d^B[u_1^{A_1}, \ldots, u_n^{A_n}]$$

to indicate that the assumption variables free in d^B are in the list $u_1^{A_1}, \ldots, u_n^{A_n}$. We also use the notation $d{:}\, A$ instead of d^A.

Definition. *A formula A is called derivable from assumptions $A_1, \ldots, A_n$, is there is a derivation term $d^B[u_1^{A_1}, \ldots, u_n^{A_n}]$ with different assumption variables $u_1^{A_1}, \ldots, u_n^{A_n}$.*

Let S be a (finite or infinite) set of formulas. We write $S \vdash B$, if the formula B is derivable from finitely many assumptions $A_1, \ldots, A_n \in S$.

Examples for derivable formulas are

$$P \to (Q \to P),$$
$$(P \to Q \to R) \to (P \to Q) \to P \to R,$$
$$(\forall x.P(x) \to Q(x)) \to \forall x\, P(x) \to \forall x\, Q(x).$$

Definition. *For any derivation d we define its set* $\mathrm{FV}(d)$ *of free (object) variables by*

$$\begin{aligned}
\mathrm{FV}(u^A) &:= \mathrm{FV}(A).\\
\mathrm{FV}(\lambda u^A d^B) &:= \mathrm{FV}(A) \cup \mathrm{FV}(d^B).\\
\mathrm{FV}(d^{A\to B} e^A) &:= \mathrm{FV}(d^{A\to B}) \cup \mathrm{FV}(e^A).\\
\mathrm{FV}(\langle d^A, e^B\rangle) &:= \mathrm{FV}(d^A) \cup \mathrm{FV}(e^B).\\
\mathrm{FV}(\pi_i(d^{A\wedge B})) &:= \mathrm{FV}(d^{A\wedge B}).\\
\mathrm{FV}(\lambda x\, d^A) &= \mathrm{FV}(d^A) \setminus \{x\}.\\
\mathrm{FV}(d^{\forall x A} t) &:= \mathrm{FV}(d^{\forall x A}) \cup \mathrm{FV}(t).
\end{aligned}$$

Example: Let d be the derivation given by

$$\frac{u{:}\, R(x)}{R(x) \to R(x)} \to^+ u$$

Then we have

$$d = \lambda u^{R(x)}\, u^{R(x)},$$
$$\mathrm{FA}(d) = \emptyset,$$
$$\mathrm{FV}(d) = \{x\}.$$

For derivation terms d we have two kinds of substitution: we can substitute a derivation term f^A for a free assumption variable u^A, denoted $d[f/u]$, and we can substitute an object term t for a free object variable x, denoted $d[t/x]$.

Lemma. *(i) If $d\colon A$, $f\colon B$ are derivation terms and t is an object term, then $d[f/u^B]\colon A$ and $d[t/x]\colon A[t/x]$ are derivation terms.*
(ii) $\mathrm{FA}(d[f/u^B]) \subseteq (\mathrm{FA}(d) \setminus \{u^B\}) \cup \mathrm{FA}(f)$.
(iii) $\mathrm{FV}(d[f/u]) \subseteq \mathrm{FV}(d) \cup \mathrm{FV}(f)$.
(iv) $\mathrm{FA}(d[t/x]) = \{v^{A[t/x]} : v^A \in \mathrm{FA}(d)\}$.
(v) $\mathrm{FV}(d[t/x]) \subseteq (\mathrm{FV}(d) \setminus \{x\}) \cup \mathrm{FV}(t)$.

Proof by simultaneous induction (cf. [19], p. 30/31). □

According to this lemma β and η–reduction for derivations are well defined and preserve the derived formula and the free assumptions. Every derivation has a unique $\beta\eta$–normal form.

Derivation terms in *intuitionistic* and in *classical* logic are obtained by adding to the first (assumption–) clause of the definition

- in the case of intuitionistic logic: For any relation symbol $R \in \mathcal{P}$ we let

$$\mathrm{Efq}_R\colon \forall\vec{x}.\bot \to R(\vec{x})$$

  be a derivation term with $\mathrm{FA}(\mathrm{Efq}_R) = \emptyset$ (Ex–falso–quodlibet axiom).
- in the case of classical logic: For any relation symbol $R \in \mathcal{P}$ we let

$$\mathrm{Stab}_R\colon \forall\vec{x}.\neg\neg R(\vec{x}) \to R(\vec{x})$$

  be a derivation term with $\mathrm{FA}(\mathrm{Stab}_R) = \emptyset$ (Stability axiom).

Clearly $\mathrm{FV}(\mathrm{Stab}_R) := \mathrm{FV}(\mathrm{Efq}_R) := \emptyset$.

We write $S \vdash A$ ($S \vdash_I A$, $S \vdash_C A$), if there is a derivation term d^A in minimal (intuitionistic, classical) logic such that for any $u^B \in \mathrm{FA}(d)$ we have $B \in S$.

Here are some more interesting examples: The Peirce formula

$$((P \to Q) \to P) \to P$$

is derivable in classical, but not in minimal logic. The Mints formula

$$((((P \to Q) \to P) \to P) \to Q) \to Q$$

is derivable in minimal logic. However, its variant

$$((((P \to Q) \to P) \to P) \to R) \to R$$

is derivable in classical but not in minimal logic.

By obvious reasons the stability axiom is also called the *principle of indirect proof* for the relation symbol R. From our stability axioms we can derive the principle of indirect proof for arbitrary formulas (in our $\to \wedge\forall$–language).

Stability Lemma. *From stability assumptions* Stab_R *for any relation symbol* R *occurring in a formula* A *we can derive* $\neg\neg A \to A$.

Proof by induction on A. □

Similarly from our ex–falso–quodlibet axioms we can derive ex–falso–quodlibet for arbitrary formulas (again in our $\to \wedge\forall$–language).

Ex–falso–quodlibet Lemma. *From assumptions* Efq_R *for any relation symbol* R *occurring in a formula* A *we can derive* $\bot \to A$ *in intuitionistic logic.*

Proof by induction on A. □

From $\neg\neg A \to A$ one can clearly derive $\bot \to A$. Hence any formula derivable in intuitionistic logic is also derivable in classical logic.

We now extend our language $\mathcal{L}$ by a strong existential quantifier written $\exists^*$ (as opposed to $\exists$ defined by $\neg\forall\neg$). There are two approaches to deal with formulas containing $\exists^*$ in a constructive setting, e.g. in minimal or intuitionistic logic.

- (Weyl [20]) A formula containing $\exists^*$ is considered not to be an entity the deduction system can deal with: some "realizing terms" are required to turn it into a "judgement". E.g. r realizes $\exists^* x\, A$ is a judgement, which can be translated into $A[r/x]$.
- (Heyting) The logic is extended by axioms expressing the intended meaning of the strong existential quantifier.

We will treat both approaches here. At first sight, Weyl's point of view is more convincing. However, Heyting's is more prominent in the literature, and we also need it to properly discuss Friedman's A–translation.

Let us first describe Heyting's approach. Here we extend our notion of an $\mathcal{L}$–formula by adding a clause

- If A is a formula, then $\exists^* x^\rho\, A$ is a formula.

In the inductive definition of derivation terms d^A in minimal logic and their sets $\mathrm{FA}(d^A)$ of free assumptions we have to add two more clauses:

($\exists^{*+}$) If A is a formula and x is a variable of type ρ, then

$$\exists^{*+}_{x,A} \colon \forall x. A \to \exists^* x\, A$$

is a derivation term and $\mathrm{FA}(\exists^{*+}_{x,A}) = \emptyset$. $\forall C$ denotes the universal closure of C.

($\exists^{*-}$) If A, B are formulas and x is a variable of type ρ such that $x \notin \mathrm{FV}(B)$, then

$$\exists^{*-}_{x,A,B} \colon \forall.(\exists^* x^\rho\, A) \to (\forall x^\rho.A \to B) \to B$$

is a derivation term with $\mathrm{FA}(\exists^{*-}_{x,A,B}) = \emptyset$.

Clearly $\mathrm{FV}(\exists^{*+}) := \mathrm{FV}(\exists^{*-}) := \emptyset$.

For these new derivation terms we have the following conversion rule:

$$\exists^{*-}\vec{t}\left(\exists^{*+}\vec{t}t\, d^{A[t/x]}\right) e \mapsto etd. \qquad (\exists^*)$$

It can be shown that any derivation term has a unique normal form with respect to $\beta\eta\exists^*$–conversion.

Similarly one could also introduce a *strong disjunction* into our natural deduction calculus, written $\vee^*$ (as opposed to $\vee$ defined by $A \vee B := \neg(\neg A \wedge \neg B)$).

Note that one can easily extend the Ex–falso–quodlibet Lemma to the present situation and prove $\bot \to A$ for an arbitrary formula A. In the cases $\exists^* x\, A$ and $A \vee^* B$ just use the corresponding introduction axiom.

Let us now describe Weyl's approach. We restrict ourselves to formulas without $\vee^*$, since later in an arithmetical system we will define $A \vee^* B$ by

$$A \vee^* B := \exists^* p.(p = \mathrm{true} \to A) \wedge (p = \mathrm{false} \to B),$$

where p is a variable of type boole. To every formula A and terms $\vec{r} = r_1^{\rho_1}, \ldots, r_m^{\rho_m}$ we associate a *judgement*

$$\vec{r} \,\mathrm{mr}\, A$$

(to be read $\vec{r}$ modified realizes A), which will be a formula not containing $\exists^*$. The list of types $\rho_1, \ldots, \rho_m = \tau(A)$ is defined as follows.

$$\tau(R(\vec{t})) := \varepsilon,$$

where ε denotes the empty list, and if $\tau(A) = \vec{\rho}$ and $\tau(B) = \sigma_1, \ldots, \sigma_n$ we let

$$\begin{aligned}
\tau(A \to B) &:= \vec{\rho} \to \sigma_1, \ldots, \vec{\rho} \to \sigma_n,\\
\tau(A \wedge B) &:= \vec{\rho}, \vec{\sigma},\\
\tau(\forall x^\rho\, B) &:= \rho \to \sigma_1, \ldots, \rho \to \sigma_n,\\
\tau(\exists^* x^\rho\, B) &:= \rho, \vec{\sigma}.
\end{aligned}$$

Instead of $\vec{\rho} \to \sigma_1, \ldots, \vec{\rho} \to \sigma_n$ we will sometimes write $\vec{\rho} \to \vec{\sigma}$. To give some examples, let n, m, k be of type nat. Then

$$\begin{aligned}
\tau(\forall n \exists^* m\, R(n,m)) &= \mathrm{nat} \to \mathrm{nat},\\
\tau(\forall n \exists^* m \exists^* k\, R(n,m,k) &= (\mathrm{nat} \to \mathrm{nat}), (\mathrm{nat} \to \mathrm{nat}),\\
\tau(\forall n \exists^* m\, R(n,m) \to \exists^* k\, Q(k)) &= (\mathrm{nat} \to \mathrm{nat}) \to \mathrm{nat}.
\end{aligned}$$

Note that $\tau(A) = \varepsilon$ iff A is a *Harrop formula* (i.e. contains $\exists^*$ in premises of $\to$ only).

Now judgements $\vec{r}^{\tau(A)} \operatorname{mr} A$ are defined by

$$\begin{aligned}
\varepsilon \operatorname{mr} R(\vec{t}) &:= R(\vec{t}),\\
r_1, \dots, r_n \operatorname{mr} (A \to B) &:= \forall \vec{x}. \vec{x} \operatorname{mr} A \to r_1\vec{x}, \dots, r_n\vec{x} \operatorname{mr} B,\\
\vec{r}, \vec{s} \operatorname{mr} (A \wedge B) &:= \vec{r} \operatorname{mr} A \wedge \vec{s} \operatorname{mr} B,\\
r_1, \dots, r_n \operatorname{mr} \forall x^\rho\, B &:= \forall x^\rho\, r_1 x, \dots, r_n x \operatorname{mr} B,\\
r, \vec{s} \operatorname{mr} \exists^* x^\rho\, B &:= \vec{s} \operatorname{mr} B[r/x].
\end{aligned}$$

Note that for Harrop formulas A we have $\varepsilon \operatorname{mr} A \equiv A$ iff A does not contain $\exists^*$.

Let us now set up a relation between the implicit and the explicit approach to deal with the existential quantifier.

Definition. *Assume that to any assumption variable u^B we have assigned a list $\vec{x}_u^{\tau(B)} = x_{u1}^{\rho_1}, \dots, x_{un}^{\rho_n}$ of distinct variables, where $\rho_1, \dots, \rho_n = \tau(B)$. Relative to this assignment we define for any derivation d^A its extracted terms* $\operatorname{ets}(d^A)$, *by induction on d^A. If $\tau(A) = \sigma_1, \dots, \sigma_k$, then* $\operatorname{ets}(d^A)$ *will be a list $r_1^{\sigma_1}, \dots, r_k^{\sigma_k}$.*

$$\begin{aligned}
\operatorname{ets}(u^A) &= \vec{x}_u^{\tau(A)},\\
\operatorname{ets}(\lambda u^A\, d^B) &= \lambda \vec{x}_u^{\tau(A)} \operatorname{ets}(d),\\
\operatorname{ets}(d^{A\to B} e^A) &= \operatorname{ets}(d)\operatorname{ets}(e),\\
\operatorname{ets}(\langle d^A, e^B\rangle) &= \operatorname{ets}(d^A), \operatorname{ets}(e^B),\\
\operatorname{ets}(\pi_0(d^{A\wedge B})) &= \textit{the head of } \operatorname{ets}(d^{A\wedge B}) \textit{ of same length as } \tau(A),\\
\operatorname{ets}(\pi_1(d^{A\wedge B})) &= \textit{the tail of } \operatorname{ets}(d^{A\wedge B}) \textit{ of same length as } \tau(B),\\
\operatorname{ets}(\lambda x^\rho\, d^A) &= \lambda x^\rho \operatorname{ets}(d),\\
\operatorname{ets}(d^{\forall x^\rho A} t^\rho) &= \operatorname{ets}(d)t,\\
\operatorname{ets}(\exists^{*+}_{x,A}) &= \lambda \vec{x} \lambda x \lambda \vec{y}. x, \vec{y},\\
\operatorname{ets}(\exists^{*-}_{x,A,B}) &= \lambda \vec{x} \lambda x \lambda \vec{y} \lambda z_1 \dots \lambda z_n . z_1 x \vec{y}, \dots, z_n x \vec{y}.
\end{aligned}$$

Note that if $\operatorname{ets}(d) = r_1, \dots, r_k$ and $\operatorname{ets}(e) = \vec{s}$, then $\operatorname{ets}(d)\operatorname{ets}(e) = r_1\vec{s}, \dots, r_k\vec{s}$ and $\lambda\vec{x} \operatorname{ets}(d) = \lambda\vec{x}\, r_1, \dots, \lambda\vec{x}\, r_k$. In the last clause the (omitted) types are

$$x^\rho, \quad (\vec{y})^{\tau(A)} \quad \text{and} \quad z_j^{\rho\to\tau(A)\to\sigma_j},$$

where $\tau(B) = \sigma_1, \dots, \sigma_n$.

The following can be proved easily.

Lemma. $\operatorname{FV}(\operatorname{ets}(d)) \subseteq \operatorname{FV}(d) \cup \{\vec{x}_u^{\tau(A)} : u^A \in \operatorname{FA}(d)\}$.

Lemma. $\operatorname{ets}(d[t/x]) = \operatorname{ets}(d)[t/x]$ *and* $\operatorname{ets}(d[e/u]) = \operatorname{ets}(d)[\operatorname{ets}(e)/\vec{x}_u]$, *hence*

$$\begin{aligned}
d =_{\beta\exists^*} e \quad &\Longrightarrow \quad \operatorname{ets}(d) =_\beta \operatorname{ets}(e)\\
d =_\eta e \quad &\Longrightarrow \quad \operatorname{ets}(d) =_\eta \operatorname{ets}(e)
\end{aligned}$$

Hence we can safely identify derivations with the same $\beta\eta\exists^*$–normal forms.

Soundness Theorem. *Assume that to any assumption variable u^B we have assigned a list $\vec{x}_u^{\tau(B)}$ and a new assumption variable $\tilde{u}\colon \vec{x}_u^{\tau(B)} \operatorname{mr} B$. Relative to this assignment we can find for any derivation $d\colon A$ a derivation*

$$\mu(d)\colon \mathrm{ets}(d) \operatorname{mr} A$$

with $\mathrm{FA}(\mu(d)) \subseteq \{\tilde{u}\colon \vec{x}_u \operatorname{mr} B | u\colon B \in \mathrm{FA}(d)\}$.

Proof by induction on d.

Case $u_i\colon A_i$. Then $\tilde{u}_i\colon \vec{x}_{u_i} \operatorname{mr} A_i$.

Case $\lambda u^B d^A$. Let $\tau(B) = \vec{\rho}$ and $\tau(A) = \sigma_1, \dots, \sigma_m$. Then we have $\tau(B \to A) = \vec{\rho} \to \sigma_1, \dots, \vec{\rho} \to \sigma_m$. We look for a derivation of

$$\lambda \vec{x}^{\vec{\rho}}\, \mathrm{ets}(d^A) \operatorname{mr} (B \to A),$$

i.e. of

$$\forall \vec{x}^{\vec{\rho}}. \vec{x} \operatorname{mr} B \to \mathrm{ets}(d^A) \operatorname{mr} A,$$

since $(\lambda \vec{x}^{\vec{\rho}}\, \mathrm{ets}(d^A))\vec{x} =_\beta \mathrm{ets}(d^A)$ and terms with the same β–normal form are identified. Hence we can take

$$\mu(\lambda u^B\, d^A) := \lambda \vec{x} \lambda \tilde{u}^{\vec{x} \operatorname{mr} B} \mu(d).$$

Case $d^{A \to B} e^A$. By IH $\mu(d)\colon \mathrm{ets}(d) \operatorname{mr} (A \to B)$, $\mu(e)\colon \mathrm{ets}(e) \operatorname{mr} A$. By definition the first of these means

$$\mu(d)\colon \forall \vec{x}. \vec{x} \operatorname{mr} A \to \mathrm{ets}(d)\vec{x} \operatorname{mr} B.$$

Hence $\mu(d)\mathrm{ets}(e)\mu(e)\colon \mathrm{ets}(d)\mathrm{ets}(e) \operatorname{mr} B$, and since $\mathrm{ets}(de) = \mathrm{ets}(d)\mathrm{ets}(e)$ we can take $\mu(de) := \mu(d)\mathrm{ets}(e)\mu(e)$.

Cases $\langle d^A, e^B \rangle$, $\pi_i(d^{A \wedge B})$. Easy.

Case $\lambda x^\rho\, d^A$. We have to find a derivation of $\lambda x^\rho \mathrm{ets}(d) \operatorname{mr} \forall x^\rho A$, i.e. of (since we identify terms with the same β–normal forms) $\forall x^\rho\, \mathrm{ets}(d) \operatorname{mr} A$. By IH we have $\mu(d)\colon \mathrm{ets}(d) \operatorname{mr} A$ with free assumptions not involving x. Hence we can take

$$\mu(\lambda x^\rho\, d) := \lambda x^\rho\, \mu(d).$$

Case $d^{\forall x^\rho A} t^\rho$. By IH we have

$$\mu(d)\colon \mathrm{ets}(d) \operatorname{mr} \forall x\, A,$$

i.e.

$$\mu(d)\colon \forall x\, \mathrm{ets}(d)x \operatorname{mr} A.$$

Hence $\mu(d)t\colon \mathrm{ets}(d)t \operatorname{mr} A[t/x]$ and we can take $\mu(dt) := \mu(d)t$.

Case $\exists^{*+}_{x,A}$. We look for a derivation $\mu(\exists^{*+}_{x,A})$ of

$$(\lambda\vec{x}\lambda x\lambda\vec{y}.x,\vec{y})\,\mathrm{mr}\,\forall\vec{x}\forall x^\rho.A \to \exists^* x^\rho\, A$$
$$\forall\vec{x}\forall x^\rho(\lambda\vec{y}.x,\vec{y})\,\mathrm{mr}\,(A \to \exists^* x^\rho\, A)$$
$$\forall\vec{x}\forall x^\rho\forall\vec{y}.\vec{y}\,\mathrm{mr}\,A \to x,\vec{y}\,\mathrm{mr}\,\exists^* x^\rho\, A$$
$$\forall\vec{x}\forall x^\rho\forall\vec{y}.\vec{y}\,\mathrm{mr}\,A \to \vec{y}\,\mathrm{mr}\,A$$

Hence we can take $\mu(\exists^{*+}_{x,A}) := \lambda\vec{x}\lambda x\lambda\vec{y}\lambda u\, u$ where $u\colon \vec{y}\,\mathrm{mr}\,A$.

Case $\exists^{*-}_{x,A,B}$. Recall that

$$\exists^{*-}_{x,A,B}\colon \forall\vec{x}.\exists^* x\, A \to (\forall x.A \to B) \to B$$

with $x \notin \mathrm{FV}(B)$. We look for a derivation $\mu(\exists^{*-}_{x,A,B})$ of

$$(\lambda\vec{x}\lambda x\lambda\vec{y}\lambda z_1\ldots\lambda z_n.z_1x\vec{y},\ldots,z_nx\vec{y})\,\mathrm{mr}\,\forall\vec{x}.\exists^* x\, A \to (\forall x.A \to B) \to B$$
$$\forall\vec{x}(\lambda x\lambda\vec{y}\lambda z_1\ldots\lambda z_n.z_1x\vec{y},\ldots,z_nx\vec{y})\,\mathrm{mr}\,(\exists^* x\, A \to (\forall x.A \to B) \to B)$$
$$\forall\vec{x}\forall x\forall\vec{y}.x,\vec{y}\,\mathrm{mr}\,\exists^* x\, A \to (\lambda z_1\ldots\lambda z_n.z_1x\vec{y},\ldots,z_nx\vec{y})\,\mathrm{mr}\,((\forall x.A \to B) \to B)$$
$$\forall\vec{x}\forall x\forall\vec{y}.\vec{y}\,\mathrm{mr}\,A \to \forall\vec{z}.\vec{z}\,\mathrm{mr}\,(\forall x.A \to B) \to z_1x\vec{y},\ldots,z_nx\vec{y}\,\mathrm{mr}\,B$$

Now since $\vec{z}\,\mathrm{mr}\,\forall x.A \to B$ is the same as

$$\forall x\forall\vec{y}.\vec{y}\,\mathrm{mr}\,A \to z_1x\vec{y},\ldots,z_nx\vec{y}\,\mathrm{mr}\,B$$

we can easily derive this formula by

$$\mu(\exists^{*-}_{x,A,B}) := \lambda\vec{x}\lambda x\lambda\vec{y}\lambda u\lambda\vec{z}\lambda v.vx\vec{y}u,$$

where $u\colon \vec{y}\,\mathrm{mr}\,A$ and $v\colon \forall x\forall\vec{y}.\vec{y}\,\mathrm{mr}\,A \to z_1x\vec{y},\ldots,z_nx\vec{y}\,\mathrm{mr}\,B$. □

Remark: Mints has shown already in [8] that for any $d\colon \exists^* x\, A$ with $\mathrm{FA}(d) = \mathrm{FV}(d) = \emptyset$ the extracted terms $\mathrm{ets}(d)$ reduce into the terms that can be read off from the long normal form of $d\colon \exists^* x\, A$. This has later been generalized by Stein in his thesis [16].

Let us now extend these considerations to arithmetic. We allow constants for primitive recursive functionals of arbitrary types (i.e. terms of Gödel's T), identifying terms with the same normal form (w.r.t. the usual conversion rules for Gödel's T). It is assumed that at least the ground types nat of natural numbers and boole of booleans are present. We restrict ourselves to decidable atomic formulas; it is convenient to represent them by boolean terms, i.e. in the form $\mathrm{atom}(t^{\mathrm{boole}})$ where atom is a distinguished relation symbol. We could equally well take equations $r = s$ with r, s terms of type nat as the only boolean formulas. We do not need $\perp$ as an extra atomic formula, since it can be defined by $\perp := \mathrm{atom}(\mathrm{false})$. Let us use n, m

as variables of type nat and p, q as variables of type boole. Our *induction schemata* are the universal closures of

$$A[0/n] \to (\forall n.A \to A[n+1/n]) \to \forall n\, A,$$
$$A[\mathrm{true}/p] \to A[\mathrm{false}/p] \to \forall p\, A.$$

We can use boolean induction (i.e. case analysis) to prove stability

$$\neg\neg\mathrm{atom}(p) \to \mathrm{atom}(p)$$

and from this we can as before conclude the stability $\neg\neg A \to A$ of formulas A built from atoms by $\to$, $\wedge$ and $\forall$. As already remarked, strong disjunction $\vee^*$ can be defined by means of the strong existential quantifier $\exists^*$.

Let us now carry out this program. First we extend the notion of a term by adding the clauses

- The constants 0^{nat}, $S^{\mathrm{nat}\to\mathrm{nat}}$, $\mathrm{true}^{\mathrm{boole}}$ and $\mathrm{false}^{\mathrm{boole}}$ are terms.
- For each type ρ the constants

$$R_{\mathrm{nat},\rho}\colon \rho \to (\mathrm{nat} \to \rho \to \rho) \to \mathrm{nat} \to \rho,$$
$$R_{\mathrm{boole},\rho}\colon \rho \to \rho \to \mathrm{boole} \to \rho$$

  are terms.

We add the following conversion rules (writing $t+1$ for St).

$$R_{\mathrm{nat},\rho} rs0 \to_0 r,$$
$$R_{\mathrm{nat},\rho} rs(t+1) \to_0 st(R_{\mathrm{nat},\rho} rst)$$

and

$$R_{\mathrm{boole},\rho} rs\, \mathrm{true} \to_0 r,$$
$$R_{\mathrm{boole},\rho} rs\, \mathrm{false} \to_0 s.$$

Formulas are built from atomic formulas by means of $\to$, $\wedge$, $\forall$ and $\exists^*$. We let $\bot := \mathrm{atom}(\mathrm{false})$, $\neg A := A \to \bot$ and (writing p for $\mathrm{atom}(p)$) $A \vee^* B := \exists^* p.(p \to A) \wedge (\neg p \to B)$.

We also extend the notion of a derivation term by constants for the truth axiom and induction axioms. Hence derivation terms in arithmetic are obtained by adding the clauses

(T) $\mathrm{T}\colon \mathrm{atom}(\mathrm{true})$ is a derivation term with $\mathrm{FA}(\mathrm{T}) = \emptyset$.

(Ind) For any formula $\forall n\, A$

$$\mathrm{Ind}_{n,A}\colon \forall.A[0/n] \to (\forall n.A \to A[n+1/n]) \to \forall n\, A$$

is a derivation term with $\mathrm{FA}(\mathrm{Ind}_{n,A}) = \emptyset$. Similarly, for any formula $\forall p\, A$

$$\mathrm{Ind}_{p,A}\colon \forall.A[\mathrm{true}/p] \to A[\mathrm{false}/p] \to \forall p\, A$$

is a derivation term with $\mathrm{FA}(\mathrm{Ind}_{p,A}) = \emptyset$.

Clearly $\mathrm{FV}(\mathrm{T}) := FV(\mathrm{Ind}_{n,A}) := \mathrm{FV}(\mathrm{Ind}_{p,A}) = \emptyset$.

In the sequel $\vdash$ refers to derivability in the arithmetical system determined by the derivation terms just described.

$\mathrm{Stab}_{\mathrm{atom}}\colon \forall p.\neg\neg\mathrm{atom}(p) \to \mathrm{atom}(p)$ can now be proved easily by boolean induction, using the truth axiom in the case true. As above we can conclude $\vdash \neg\neg A \to A$ for formulas A without $\exists^*$, i.e. built with $\to$, $\wedge$ and $\forall$. We also have $\vdash \bot \to A$ for any A. Furthermore we can derive the usual axioms for $\vee^*$.

The notion of extracted terms can straightforwardly be extended to this situation, and the Soundness Theorem carries over easily. In the case of $\mathrm{Ind}_{n,A}$ we have to prove

$$\mathrm{ets}(\mathrm{Ind}_{n,A})\,\mathrm{mr}\,\forall\vec{x}.A[0/n] \to (\forall n.A \to A[n+1/n]) \to \forall n\, A,$$

i.e.

$$\begin{aligned}\forall\vec{x}\forall\vec{y}\forall\vec{f}\forall n.\vec{y}\,\mathrm{mr}\,A[0/n] &\to (\forall n\forall\vec{y}_1.\vec{y}_1\,\mathrm{mr}\,A \to \vec{f}n\vec{y}_1\,\mathrm{mr}\,A[n+1/n])\\ &\to \mathrm{ets}(\mathrm{Ind}_{n,A})\vec{x}\vec{y}\vec{f}n\,\mathrm{mr}\,A.\end{aligned}$$

Hence we let $\mathrm{ets}(\mathrm{Ind}_{n,A}) := \lambda\vec{x}.R_1,\ldots,R_k$ where k is the length of $\tau(A) = \rho_1,\ldots,\rho_k$ and $R_1,\ldots,R_k$ are simultaneous primitive recursion operators of type

$$R_i\colon \vec{\rho} \to (\mathrm{nat} \to \vec{\rho} \to \vec{\rho}) \to \mathrm{nat} \to \rho_i$$

satisfying

$$\begin{aligned}R_i\vec{y}\vec{f}0 &= y_i,\\ R_i\vec{y}\vec{f}(z+1) &= f_i z(R_1\vec{y}\vec{f}z)\ldots(R_k\vec{y}\vec{f}z),\end{aligned}$$

where $=$ denotes equality of $\beta\eta R$–normal forms. Using these equations we can then prove the above claim easily (recall that terms with the same normal form are identified). The operators $R_1,\ldots,R_n$ can be defined easily from the recursion constant $R_{\mathrm{nat},\rho_1\times\ldots\times\rho_k}$. We could equally well have introduced them as constants and added the equations above as conversion rules.

Boolean induction, i.e. case analysis is treated similarly. We let

$$\mathrm{ets}(\mathrm{Ind}_{p,A}) := \lambda\vec{x}.R_1,\ldots,R_k,$$

where now $R_1,\ldots,R_k$ are simultaneous primitive recursion (or case splitting) operators of type $R_i\colon \vec{\rho} \to \vec{\rho} \to \mathrm{boole} \to \rho_i$ satisfying

$$\begin{aligned}R_i\vec{y}\vec{z}\,\mathrm{true} &= y_i,\\ R_i\vec{y}\vec{z}\,\mathrm{false} &= z_i.\end{aligned}$$

The lemmata stating that ets commutes with substitution and reduction remain valid since the conversion rules for induction and recursion fit together.

The following remarks and definitions will be helpful later. Let us call a formula A *decidable* if there is a term t_A such that $\vdash A \leftrightarrow \mathrm{atom}(t_A)$.

1. Every quantifier–free formula is decidable. Let $\supset := \lambda p \lambda q.Rq\,\mathrm{true}\,p$ and $\& := \lambda p\lambda q.R(R\,\mathrm{true}\,\mathrm{false}\,q)\,\mathrm{false}\,p$. Clearly

$$\forall p, q.(\mathrm{atom}(p) \to \mathrm{atom}(q)) \leftrightarrow \mathrm{atom}(\supset pq)$$
$$\forall p, q.(\mathrm{atom}(p) \wedge \mathrm{atom}(q)) \leftrightarrow \mathrm{atom}(\& pq)$$

are provable. Hence we let

$$t_{\mathrm{atom}(r)} := r,$$
$$t_{A\to B} := \supset t_A t_B$$
$$t_{A\wedge B} := \& t_A t_B.$$

2. We can do case splitting according to decidable formulas A, i.e. for every formula $B[\vec{x}]$ we can prove

$$\mathrm{Cases}_{A,B}\colon (A \to B) \to (\neg A \to B) \to B.$$

The derivation $\mathrm{Cases}_{A,B}$ is given by

$$\lambda u_1, u_2.\mathrm{Ind}\,\vec{x}(\lambda u_3\lambda u_4.u_3\mathrm{T})(\lambda u_5\lambda u_6.u_6\neg\mathrm{F})t_A(\lambda u_7.u_1(d_1u_7))(\lambda u_8.u_2(d_2u_8))$$

where $d_1^{\mathrm{atom}(t_A)\to A}$, and $d_2^{\neg\mathrm{atom}(t_A)\to\neg A}$ are derivations which exist according to 1 and the axioms and assumption variables with indices are (writing t for $\mathrm{atom}(t)$)

$$\mathrm{Ind}_{p,(p\to B)\to(\neg p\to B)\to B},$$

$u_1^{A\to B}$, $u_2^{\neg A\to B}$, $u_3^{\mathrm{true}\to B}$, $u_4^{\neg\mathrm{true}\to B}$, $u_5^{\mathrm{false}\to B}$, $u_6^{\neg\mathrm{false}\to B}$, $u_7^{t_A}$ and $u_8^{\neg t_A}$. The extracted terms of $\mathrm{Cases}_{A,B}$ are given by

$$\mathrm{ets}(\mathrm{Cases}_{A,B}) = \lambda\vec{y},\vec{z}.\mathbf{if}\,t_A\vec{y}\vec{z} =_\eta \mathbf{if}\,t_A$$

where $\mathbf{if} := \vec{R}(\lambda\vec{y}_1,\vec{z}_1.\vec{y}_1)(\lambda\vec{y}_1,\vec{z}_1.\vec{z}_1)$ and $\vec{y},\vec{z},\vec{y}_1,\vec{z}_1$ are lists of variables of types $\vec{\rho} := \tau(B)$. Clearly

$$\mathbf{if}\,\mathrm{true}\,\vec{r}\vec{s} =_{\beta R} \vec{r}, \quad \mathbf{if}\,\mathrm{false}\,\vec{r}\vec{s} =_{\beta R} \vec{s}.$$

For better readability we use for $\mathbf{if}\,t_A\vec{r}\vec{s}$ the notation

$$\mathbf{if}\ \ A\ \ \mathbf{then}\ \ \vec{r}\ \ \mathbf{else}\ \ \vec{s}\ \ \mathbf{fi}.$$

2. Classical proofs as programs

As is well known a proof of a $\forall\exists$–theorem with a quantifier–free kernel — where $\exists$ is viewed as defined by $\neg\forall\neg$ — can be used as a program. We describe a "direct method" to use such a proof as a program, and compare it with Harvey Friedman's A–translation [3] followed by the well–known program extraction from constructive proofs.

First note that a classical proof of $\forall x \exists y\, A$ generally does *not* yield a program to compute y from x. The reason for this is that there might be a universal quantifier $\forall z$ right after $\exists y$, i.e. after $\neg\forall y \neg$, and this makes it possible that an assumption

$$\forall y \neg \forall z\, B$$

is instantiated with a non–constant term containing critical variables which are bound later by $\forall z$.

It is well known that this is not just a technical difficulty: if T denotes Kleene's T–predicate, then

$$\forall n \exists m \forall k. T(n,n,k) \rightarrow T(n,n,m)$$

is trivially provable even in minimal logic (with $\exists m$ defined as $\neg\forall m\neg$, i.e. in classical logic), but there is no computable function f satisfying

$$\forall n, k. T(n,n,k) \rightarrow T(n,n,f(n)),$$

for then $\exists k\, T(n,n,k)$ would be decidable: it would be true if and only if $T(n,n,f(n))$ holds.

Hence in the rest of this section we will only consider formulas of the form $\forall x \exists y\, A$ with A quantifier–free.

We now describe a "direct method" (cf. [13]) to extract the computational content from a classical proof.

By a Π–*formula* we mean a formula built without the strong existential quantifier $\exists^*$, which has no (universal) quantifier in premises of implications. For instance any Horn formula $\forall \vec{x}. P_1(\vec{x}) \rightarrow \cdots \rightarrow P_n(\vec{x}) \rightarrow Q(\vec{x})$ is a Π–formula, but

$$\forall x, y, u. (\forall z. xz = yz) \rightarrow ux = uy$$

is not. Clearly every Π–formula is equivalent (in minimal logic) to a conjunction of formulas $\forall C$ where C is quantifier–free and without $\wedge$. So from now on we will assume that Π–formulas are of this form.

A derivation d is called a *refutation of* Π*–assumptions* if d derives a closed false atom from assumptions $\mathrm{FA}(d) = \{v_1 : \forall C_1, \ldots, v_n : \forall C_n\}$ where each C_i is quantifier–free.

Now let d be a refutation of Π–assumptions. We may assume $\mathrm{FV}(d) = \emptyset$ (if not, substitute arbitrary closed terms for the free variables in d). Next we can normalize d. Let $d\downarrow$ be the result. Again $d\downarrow$ is a refutation of Π–assumptions. We then can read off from $d\downarrow$ a list $|d|$ of closed terms called the "first instance" of $d\downarrow$ (cf. [13]) such that one of the Π–assumptions is false at $|d|$. To make this notion easier to understand let us restrict the general situation slightly. A closed quantifier–free formula B is *true* respectively *false* if t_B normalizes to the boolean constant true respectively false. A closed Π–formula $\forall \vec{x}\, C$ is true iff for all closed terms $\vec{t}$ the formula $C[\vec{t}/\vec{x}]$ is true.

Let $d\colon \bot$ be a normal derivation with $\mathrm{FV}(d) = \emptyset$ of $\bot$ from assumptions

$$u\colon \forall \vec{y}.B_1[\vec{y}] \to \ldots \to B_m[\vec{y}] \to \bot,$$
$$v_1\colon \forall C_1, \ldots, v_n\colon \forall C_n$$

where $\forall C_1, \ldots, \forall C_n$ are true closed Π–formulas. We define a list $|d|$ of closed terms, called the *first instance* of d, such that $B_1[|d|], \ldots, B_m[|d|]$ are true. $|d|$ is defined by induction on d. Since d is normal and $\mathrm{FV}(d) = \emptyset$ it does not contain axioms (exept the truth axiom, which is a closed Π–formulas and hence may be assumed to be among the Π–assumptions $\forall C_i$). To see this recall that the normal form of any closed term of type nat is of the form $S(S(S\ldots(S0)\ldots))$ and of any closed term of type boole is either true or false; hence all induction axioms unfold. Therefore d is of the form

$$w\vec{s}d_1 \ldots d_k,$$

where $\vec{s}$ are closed terms and $d_1, \ldots, d_k$ are derivations of closed quantifier–free formulas. We distinguish two cases.

1. $d_1, \ldots, d_k$ derive only true formulas (which can be decided, since the formulas are quantifier–free and closed). Then w cannot be one of the v_i since all $\forall C_i$ are true. Hence $d = u\vec{s}d_1 \ldots d_k$ and the d_i derive $B_i[\vec{s}]$. So let $|d| := \vec{s}$.
2. There is a minimal i such that d_i derives a false formula, $A_1 \to \cdots \to A_{n_i} \to \bot$ say. Then $A_1 \ldots, A_{n_i}$ are true. Without loss of generality we may assume that $d_i = \lambda w_1^{A_1} \ldots \lambda w_{n_i}^{A_{n_i}} e$ where $e\colon \bot$ contains assumptions among

$$u\colon \forall \vec{y}.B_1[\vec{y}] \to \ldots \to B_m[\vec{y}] \to \bot,$$
$$v_1\colon \forall C_1, \ldots, v_n\colon \forall C_n,$$
$$w_1\colon A_1, \ldots, w_{n_i}\colon A_{n_i}.$$

Therefore we can recursively define $|d| := |e|$.

Hence from a proof d of $\forall \vec{x} \exists \vec{y} \bigwedge_i B_i[\vec{x}, \vec{y}]$ from true Π–assumptions with B_i quantifier–free we can obtain the following algorithm to compute for any $\vec{r}$ an $\vec{s}$ such that all $B_i[\vec{r}, \vec{s}]$ hold. First instanciate d with $\vec{r}$, i.e. form $d\vec{r}\colon \exists \vec{y} \bigwedge_i B_i[\vec{r}, \vec{y}]$. Since $\exists$ is $\neg\forall\neg$, we have $d\vec{r}u\colon \bot$ with $u\colon \forall \vec{y}.B_1[\vec{r}, \vec{y}] \to \ldots \to$

$B_m[\vec{r}, \vec{y}] \to \bot$ a new Π–assumption (of a false formula!). Now normalize $d\vec{r}u$. From its normal form $(d\vec{r}u)\!\downarrow$, which is a refutation from Π–assumptions, we can read off the first instance $|(d\vec{r}u)\!\downarrow\!|$ of $(d\vec{r}u)\!\downarrow$. These are closed terms $\vec{s}$ such that all $B_i[\vec{r}, \vec{s}]$ are true.

It might seem that instead of the method described, which chooses the branch to follow by checking whether some quantifier–free formulas are false or true, one could alternatively look for an occurrence of the false Π–assumption u in the proof whose arguments do not contain u any more. However, in our general case where Π–assumptions (and not just Horn formulas) are allowed these arguments may contain free assumption variables bound later (by $\to^+$) in the proof, and so we cannot conclude that all arguments of u derive true formulas. If, however, we restrict attention to the special case where we only allow Horn–formulas as assumptions, then this phenomenon cannot happen and we have a variant of the direct method.

Some comments are to be made here.

1. In principle, of course, we could replace any quantifier–free formula C by the atomic formula atom(t_C). However, this introduces quite a lot of somewhat artificial boolean functions, which makes the proof much harder to read and work with.
2. If we are prepared to apply some preparatory "pruning"–step to our derivation then we may assume that such additional assumptions v are always true. For if we had a $v\colon A$ with A false which is bound later in the proof yielding $A \to C$, then we could replace the whole subproof above this occurrence of $A \to C$ by a derivation of $A \to C$ using ex–falso–quodlibet. If we do this preparatory step first, then the variant of the direct method can be applied to the general case of Π–assumptions as well.

We now describe Friedman's A–translation from [3]. Let A be an arbitrary but fixed formula. The A–translation B^A of a formula B is obtained by replacing any atomic subformula P of B by $(P \to A) \to A$.

Note that any derivation d of some formula B from assumptions $C_1, \ldots,$ C_n becomes after the A–translation a derivation of B^A from $C_1^A, , \ldots, C_n^A$. To see this recall that our logical rules are those of minimal logic and hence give no extra treatment to falsity. Also the axiom schemes (exept the truth axiom, which can be viewed as a Π–assumption) remain instances of the same axiom scheme after the A–translation. E.g. boolean induction

$$B[\text{true}/p] \to B[\text{false}/p] \to \forall p\, B$$

is translated into

$$B^A[\text{true}/p] \to B^A[\text{false}/p] \to \forall p\, B^A,$$

which again is an instance of boolean induction.

Let us look at what happens with Π–assumptions under the A–translation. As before we may assume that all formulas considered do not contain $\wedge$.

Lemma 1. *For any quantifier–free formula C we can find a derivation $d\colon C \to C^A$.*

Proof by induction on C. Let $C \equiv B_1 \to \ldots \to B_m \to R$ with an atom R. We must derive

$$(\vec{B} \to R) \to \vec{B}^A \to (R \to A) \to A.$$

So assume

$$\begin{aligned}
&\tilde{u}\colon \vec{B} \to R,\\
&\tilde{v}_i\colon \vec{B}_i^A,\\
&w\colon R \to A.
\end{aligned}$$

We must show A.

Case $u_i^-\colon \neg B_i$ for some i. Let $B_i \equiv \vec{C}_i \to P_i$ with atoms P_i. Then

$$\tilde{v}_i\colon \vec{C_i}^A \to (P_i \to A) \to A$$

and we have

$$\begin{aligned}
&e_{ij}[u_i^-]\colon\equiv \mathrm{Stab}_{C_{ij}} \lambda v^{\neg C_{ij}}.u_i^- \lambda \vec{u}^{\vec{C}_i}.\mathrm{Efq}_{P_i}(vu_j)\colon C_{ij},\\
&e_i[u_i^-]\colon\equiv \lambda w_i^{P_i}.u_i^- \lambda \vec{u}^{\vec{C}_i} w_i\colon \neg P_i.
\end{aligned}$$

By IH we have $d_{C_{ij}}\colon C_{ij} \to C_{ij}^A$. Hence

$$\tilde{v}_i(d_{C_{i1}} e_{i1}) \ldots (d_{C_{in_i}} e_{in_i})(\lambda w_i^{P_i}.\mathrm{Efq}_A(e_i w_i))\colon A.$$

Case $u_i^+\colon B_i$ for all i. Then

$$w(\tilde{u} u_1^+ \ldots u_m^+)\colon A$$

The extracted terms for this derivation are

$$\begin{aligned}
d^{\mathrm{ets}} \equiv \lambda \vec{x}_1, \ldots, \vec{x}_m, \vec{z}.\ &\textbf{if}\quad \neg B_1 \quad \textbf{then}\quad \vec{x}_1 d^{\mathrm{ets}}_{C_{11}} \ldots d^{\mathrm{ets}}_{C_{1n_1}} \vec{0} \quad \textbf{else}\\
&\ldots\\
&\textbf{if}\quad \neg B_m \quad \textbf{then}\quad \vec{x}_m d^{\mathrm{ets}}_{C_{m1}} \ldots d^{\mathrm{ets}}_{C_{mn_m}} \vec{0} \quad \textbf{else}\\
&\vec{z}\quad \textbf{fi} \ldots \textbf{fi},
\end{aligned}$$

where $\vec{x}_i, \vec{z}$ are the lists of variables associated with $\tilde{v}_i\colon B_i^A, w\colon R \to A$. □

Here we have used case splitting according to the quantifier free formulas B_i which is admissible by a remark above.

If we want to use the A–translation to extract the computational content from a classical proof we have to choose a particular A involving the strong existential quantifier.

Lemma 2. *Let $B_i[\vec{x}, \vec{y}]$ be quantifier–free formulas and*

$$A[\vec{x}] := \exists^* \vec{y} \bigwedge_i B_i[\vec{x}, \vec{y}].$$

Then we can find a derivation of $(\forall \vec{y}. B_1[\vec{x}, \vec{y}] \to \ldots \to B_m[\vec{x}, \vec{y}] \to \bot)^{A[\vec{x}]}$.

Proof. Let $\vec{y}$ be given and assume $\tilde{v}_i : \vec{B}_i^A$ and $w : \bot \to A$. We must show A.

Case $u_i^- : \neg B_i$ for some i. Let $B_i \equiv \vec{C}_i \to P_i$ with atoms P_i. Then

$$\tilde{v}_i : \vec{C}_i^{\,A} \to (P_i \to A) \to A$$

and we have

$$e_{ij}[u_i^-] :\equiv \mathrm{Stab}_{C_{ij}} \lambda v^{\neg C_{ij}}. u_i^- \lambda \vec{u}^{\vec{C}_i}. \mathrm{Efq}_{P_i}(v u_j) : C_{ij},$$
$$e_i[u_i^-] :\equiv \lambda w_i^{P_i}. u_i^- \lambda \vec{u}^{\vec{C}_i} w_i : \neg P_i.$$

Using $d_{C_{ij}} : C_{ij} \to C_{ij}^A$ from Lemma 1 we obtain

$$\tilde{v}_i(d_{C_{i1}} e_{i1}) \ldots (d_{C_{in_i}} e_{in_i})(\lambda w_i^{P_i}. \mathrm{Efq}_A(e_i w_i)) : A.$$

Case $u_i^+ : B_i$ for all i. Then

$$\exists^+ \vec{y} \langle u_1^+, \ldots, u_m^+ \rangle : A$$

The extracted terms for this derivation are

$$\begin{aligned} d^{\mathrm{ets}} \equiv \lambda \vec{y}, \vec{x}_1, \ldots, \vec{x}_m, \vec{z}.\ & \mathbf{if} \quad \neg B_1 \quad \mathbf{then} \quad \vec{x}_1 d^{\mathrm{ets}}_{C_{11}} \ldots d^{\mathrm{ets}}_{C_{1n_1}} \vec{0} \quad \mathbf{else} \\ & \ldots \\ & \mathbf{if} \quad \neg B_m \quad \mathbf{then} \quad \vec{x}_m d^{\mathrm{ets}}_{C_{m1}} \ldots d^{\mathrm{ets}}_{C_{mn_m}} \vec{0} \quad \mathbf{else} \\ & \vec{y} \quad \mathbf{fi} \ldots \mathbf{fi}, \end{aligned}$$

where $\vec{x}_i, \vec{z}$ are the lists of variables associated with $\tilde{v}_i : B_i^A, w : \bot \to A$. □

Theorem (Friedman). *For any derivation*

$$d[u : \forall \vec{y}. B_1[\vec{x}, \vec{y}] \to \ldots \to B_m[\vec{x}, \vec{y}] \to \bot, v_1 : \forall C_1, \ldots, v_n : \forall C_n] : \bot$$

with B_i, C_j quantifier–free we can find a derivation

$$d^{\mathrm{tr}}[v_1 : \forall C_1, \ldots, v_n : \forall C_n] : \exists^* \vec{y} \bigwedge_i B_i[\vec{x}, \vec{y}].$$

Proof. Let $A[\vec{x}] := \exists^* \vec{y} \bigwedge_i B_i[\vec{x}, \vec{y}]$ and consider the $A[\vec{x}]$–translation

$$d^{A[\vec{x}]}[u' : (\forall \vec{y}. B_1[\vec{x}, \vec{y}] \to \ldots \to B_m[\vec{x}, \vec{y}] \to \bot)^{A[\vec{x}]}, v_1' : (\forall C_1)^{A[\vec{x}]}, \ldots v_n' : (\forall C_n)^{A[\vec{x}]}]$$
$$: (\bot \to A[\vec{x}]) \to A[\vec{x}]$$

of d, obtained by just changing some formulas. By Lemma 1 we have

$$d_{v_i}[v_i\colon \forall C_i]\colon (\forall C_i)^{A[\vec{x}]}.$$

By Lemma 2 (now using the particular choice of $A[\vec{x}]$) the $A[\vec{x}]$–translation of the assumption u is provable without assumptions:

$$d_u\colon (\forall \vec{y}.B_1[\vec{x},\vec{y}] \to \ldots \to B_m[\vec{x},\vec{y}] \to \bot)^{A[\vec{x}]}$$

Substituting $d_{v_i}[v_i\colon \forall C_i]$ for v_i' and d_u for u' we obtain

$$d^{\mathrm{tr}}[v_1\colon \forall C_1,\ldots,v_n\colon \forall C_n]\colon \exists^* \vec{y} \bigwedge_i B_i[\vec{x},\vec{y}],$$

where

$$d^{\mathrm{tr}} \equiv d^{A[\vec{x}]}[d_u, d_{v_1},\ldots,d_{v_n}]\mathrm{Efq}_{A[\vec{x}]}. \qquad \square$$

Having obtained a proof d^{tr} of an existential formula $\exists^* \vec{y} \bigwedge_i B_i[\vec{x},\vec{y}]$ we can then apply the general method of extracting terms to this proof. It yields

$$(d^{\mathrm{tr}})^{\mathrm{ets}} \equiv (d^{A[\vec{x}]})^{\mathrm{ets}}[d_u^{\mathrm{ets}}, d_{v_1}^{\mathrm{ets}},\ldots,d_{v_n}^{\mathrm{ets}}]\vec{0}, \tag{1}$$

since extracting terms commutes with substitution.

Note that there are many ways to use Friedman's method to extract a term from a given arithmetical proof of a weak existential formula.

1. The most straightforward way is to prove the formula without any assumptions. This means that we are not allowed to use lemmata, and hence that such a proof tends to be rather long, and difficult to produce.
2. The next straightforward way is to pack all Π–Lemmata used in the proof (and proved explicitely in 1 above) into purely generalized atoms $\forall \vec{x}\, \mathrm{atom}(t)$. However, this means that we have to introduce rather complex boolean terms, which will later show up in the proof as tests for case distinctions.

We now prove that the value of the extracted terms $(d^{\mathrm{tr}})^{\mathrm{ets}}$ when instanciated with a list $\vec{r}$ of closed terms is in fact the same as the result of the direct method described above.

So consider again the situation of Friedman's Theorem, i.e. a derivation

$$d[u\colon \forall \vec{y}.B_1[\vec{x},\vec{y}] \to \ldots \to B_m[\vec{x},\vec{y}] \to \bot, v_1\colon \forall C_1,\ldots,v_n\colon \forall C_n]\colon \bot$$

with B_i, C_j quantifier–free. We just observed that the program $(d^{\mathrm{tr}})^{\mathrm{ets}}$ extracted from the translated derivation has the form (1) above. Let us try to understand how this program works. First, $(d^{A[\vec{x}]})^{\mathrm{ets}}$ closely follows the structure of d. The reason is that $d^{A[\vec{x}]}$ differs from d only with respect to the formulas affixed, and when forming the extracted terms this affects only the types and the arities of the lists of object variables associated with assumption variables.

In order to comprehend $d_{v_i}^{\text{ets}}$ and d_u^{ets} let us have a second look at the proofs of Lemma 1 and 2. First note that $d_{v_i}[v_i\colon \forall C_i]\colon (\forall C_i)^{A[\vec{x}]}$ is obtained from $d_i\colon C_i \to C_i^{A[\vec{x}]}$ constructed in the proof of Lemma 1 by

$$d_{v_i} \equiv \lambda \vec{y}_i . d_i(v_i \vec{y}_i).$$

Since v_i has type $\forall C_i$, which is a Harrop formula, we have $d_{v_i}^{\text{ets}} \equiv \lambda \vec{y}_i\, d_i^{\text{ets}}$. Now from the proof of Lemma 1 we obtain

$$\begin{aligned} d_i^{\text{ets}} \equiv \lambda \vec{x}_1, \ldots, \vec{x}_m, \vec{z}.\ & \textbf{if} \quad \neg B_1 \quad \textbf{then} \quad \vec{x}_1 d_{C_{11}}^{\text{ets}} \ldots d_{C_{1n_1}}^{\text{ets}} \vec{0} \quad \textbf{else} \\ & \ldots \\ & \textbf{if} \quad \neg B_m \quad \textbf{then} \quad \vec{x}_m d_{C_{m1}}^{\text{ets}} \ldots d_{C_{mn_m}}^{\text{ets}} \vec{0} \quad \textbf{else} \\ & \vec{z} \quad \textbf{fi} \ldots \textbf{fi}, \end{aligned} \tag{2}$$

where $C_i \equiv B_1 \to \ldots \to B_m \to R$ with $B_i \equiv \vec{C}_i \to P_i$ and $\vec{x}_i, \vec{z}$ are the lists of variables associated with $\vec{v}_i, w$. Furthermore, $d_{C_{ij}}^{\text{ets}}$ are the extracted terms of derivations $d_{C_{ij}}\colon C_{ij} \to C_{ij}^{A[\vec{x}]}$ constructed by previous applications of Lemma 1. Similarly

$$\begin{aligned} d_u^{\text{ets}} \equiv \lambda \vec{y}, \vec{x}_1, \ldots, \vec{x}_m, \vec{z}.\ & \textbf{if} \quad \neg B_1 \quad \textbf{then} \quad \vec{x}_1 d_{C_{11}}^{\text{ets}} \ldots d_{C_{1n_1}}^{\text{ets}} \vec{0} \quad \textbf{else} \\ & \ldots \\ & \textbf{if} \quad \neg B_m \quad \textbf{then} \quad \vec{x}_m d_{C_{m1}}^{\text{ets}} \ldots d_{C_{mn_m}}^{\text{ets}} \vec{0} \quad \textbf{else} \\ & \vec{y} \quad \textbf{fi} \ldots \textbf{fi}, \end{aligned} \tag{3}$$

where $B_i \equiv \vec{C}_i \to P_i$ and $\vec{x}_i, \vec{z}$ are the lists of variables associated with $\vec{v}_i, w$. Furthermore, $d_{C_{ij}}^{\text{ets}}$ are the extracted terms of derivations $d_{C_{ij}}\colon C_{ij} \to C_{ij}^{A[\vec{x}]}$ constructed by previous applications of Lemma 1.

This analysis makes it possible to prove that module $\beta\eta R$ the extracted terms when instanciated with a list $\vec{r}$ of closed terms are in fact the same as the result of the direct method described above to read off the first instance provided by the instanciated derivation

$$\bar{d}[\bar{u}\colon \forall \vec{y}. B_1[\vec{r}, \vec{y}] \to \ldots \to B_m[\vec{r}, \vec{y}] \to \bot, v_1\colon \forall C_1, \ldots, v_n\colon \forall C_n]\colon \bot$$

Below we will show the following

Claim. *For any normal derivation*

$$e[\bar{u}\colon \forall \vec{y}. B_1[\vec{r}, \vec{y}] \to \ldots \to B_m[\vec{r}, \vec{y}] \to \bot, v_1\colon \forall C_1, \ldots, v_n\colon \forall C_n]\colon \bot$$

with $\mathrm{FV}(e) = \emptyset$ *we have*

$$|e| =_{\beta\eta R} (e^{A[\vec{r}]})^{\text{ets}}[d_u^{\text{ets}}[\vec{r}/\vec{x}], d_{v_1}^{\text{ets}}, \ldots, d_{v_n}^{\text{ets}}]\vec{0}.$$

We then obtain that the instanciation of the extracted terms (1) with $\vec{r}$ for $\vec{x}$, i.e.

$$(d^{A[\vec{x}]})^{\mathrm{ets}}[d_u^{\mathrm{ets}}, d_{v_1}^{\mathrm{ets}}, \ldots, d_{v_n}^{\mathrm{ets}}]\vec{0}[\vec{r}/\vec{x}] \equiv (d[\vec{r}/\vec{x}]^{A[\vec{r}]})^{\mathrm{ets}}[d_u^{\mathrm{ets}}[\vec{r}/\vec{x}], d_{v_1}^{\mathrm{ets}}, \ldots, d_{v_n}^{\mathrm{ets}}]\vec{0}$$

has as its value the list of closed terms which is the first instance of the instanciated derivation $d[\vec{r}/\vec{x}]$, i.e. $|d[\vec{r}/\vec{x}]{\downarrow}|$. For by the claim we have

$$\begin{aligned} |d[\vec{r}/\vec{x}]{\downarrow}| &=_{\beta\eta R} ((d[\vec{r}/\vec{x}]{\downarrow})^{A[\vec{r}]})^{\mathrm{ets}}[d_u^{\mathrm{ets}}[\vec{r}/\vec{x}], d_{v_1}^{\mathrm{ets}}, \ldots, d_{v_n}^{\mathrm{ets}}]\vec{0} \\ &=_{\beta\eta R} ((d[\vec{r}/\vec{x}]^{A[\vec{r}]})^{\mathrm{ets}}){\downarrow}[d_u^{\mathrm{ets}}[\vec{r}/\vec{x}], d_{v_1}^{\mathrm{ets}}, \ldots, d_{v_n}^{\mathrm{ets}}]\vec{0} \\ &=_{\beta\eta R} (d[\vec{r}/\vec{x}]^{A[\vec{r}]})^{\mathrm{ets}}[d_u^{\mathrm{ets}}[\vec{r}/\vec{x}], d_{v_1}^{\mathrm{ets}}, \ldots, d_{v_n}^{\mathrm{ets}}]\vec{0}, \end{aligned}$$

since normalization (i.e. $\beta\eta\exists^* R$–conversion) commutes with $A[\vec{r}]$–translation and the formation of extracted terms.

It remains to prove the claim. We use induction on e. Since e is normal, it must be of the form $e = w\vec{s}e_1 \ldots e_k$ with $w \in \{\bar{u}, v_1, \ldots, v_n\}$.

Case 1. $e_1, \ldots, e_k$ derive only true formulas. Then $w = \bar{u}$, $k = m$ and the e_i derive $B_i[\vec{r}, \vec{s}]$. By definition $|e| := \vec{s}$. Furthermore, letting $\vec{d}^{\mathrm{ets}} := d_{v_1}^{\mathrm{ets}}, \ldots, d_{v_n}^{\mathrm{ets}}$

$$\begin{aligned} &(e^{A[\vec{r}]})^{\mathrm{ets}}[d_u^{\mathrm{ets}}[\vec{r}/\vec{x}], \vec{d}^{\mathrm{ets}}]\vec{0} \\ &\equiv d_u^{\mathrm{ets}}[\vec{r}/\vec{x}]\vec{s}(e_1^{A[\vec{r}]})^{\mathrm{ets}}[d_u^{\mathrm{ets}}[\vec{r}/\vec{x}], \vec{d}^{\mathrm{ets}}] \ldots (e_m^{A[\vec{r}]})^{\mathrm{ets}}[d_u^{\mathrm{ets}}[\vec{r}/\vec{x}], \vec{d}^{\mathrm{ets}}]\vec{0} \\ &=_{\beta R} \vec{s} \end{aligned}$$

by the form (3) of d_u^{ets}, since all $B_i[\vec{r}, \vec{s}] \equiv (\vec{B}_i \to P_i)[\vec{r}, \vec{s}]$ are true.

Case 2. There is a minimal i such that e_i derives a false formula, $D_{i1}[\vec{s}] \to \cdots \to D_{in_i}[\vec{s}] \to \bot$ say. Then $D_{i1}[\vec{s}], \ldots, D_{in_i}[\vec{s}]$ are true. Without loss of generality we may assume that $e_i = \lambda w_1^{D_{i1}[\vec{s}]} \ldots \lambda w_m^{D_{in_i}[\vec{s}]} f$ where $f\colon \bot$ contains assumptions among

$$\begin{aligned} &\bar{u}\colon \forall\vec{y}.B_1[\vec{r}, \vec{y}] \to \ldots \to B_m[\vec{r}, \vec{y}] \to \bot, \\ &v_1\colon \forall C_1, \ldots, v_n\colon \forall C_n, \\ &w_1\colon D_{i1}, \ldots, w_m\colon D_{in_i}. \end{aligned}$$

Hence by definition $|e| := |f|$. Furthermore using the notation $d_{w_j}[w_j]\colon D_{ij}^{A[\vec{r}]}$ for the derivation obtained by applying Lemma 1 to D_{ij} we have, letting again $\vec{d}^{\mathrm{ets}} := d_{v_1}^{\mathrm{ets}}, \ldots, d_{v_n}^{\mathrm{ets}}$

$$\begin{aligned} &(e^{A[\vec{r}]})^{\mathrm{ets}}[d_u^{\mathrm{ets}}[\vec{r}], \vec{d}^{\mathrm{ets}}]\vec{0} \\ &\equiv \begin{cases} d_u^{\mathrm{ets}}[\vec{r}]\vec{s}(e_1^{A[\vec{r}]})^{\mathrm{ets}}[d_u^{\mathrm{ets}}[\vec{r}], \vec{d}^{\mathrm{ets}}] \ldots (e_k^{A[\vec{r}]})^{\mathrm{ets}}[d_u^{\mathrm{ets}}[\vec{r}], \vec{d}^{\mathrm{ets}}]\vec{0} & \text{if } w = \bar{u} \\ d_{v_i}^{\mathrm{ets}}\vec{s}(e_1^{A[\vec{r}]})^{\mathrm{ets}}[d_u^{\mathrm{ets}}[\vec{r}], \vec{d}^{\mathrm{ets}}] \ldots (e_k^{A[\vec{r}]})^{\mathrm{ets}}[d_u^{\mathrm{ets}}[\vec{r}], \vec{d}^{\mathrm{ets}}]\vec{0} & \text{if } w = v_i \end{cases} \\ &=_{\beta R} (e_i^{A[\vec{r}]})^{\mathrm{ets}}[d_u^{\mathrm{ets}}[\vec{r}], \vec{d}^{\mathrm{ets}}]d_{w_1}^{\mathrm{ets}} \ldots d_{w_{n_i}}^{\mathrm{ets}}\vec{0} \quad \text{by (3) and (2), respectively} \\ &=_{\beta} (f^{A[\vec{r}]})^{\mathrm{ets}}[d_u^{\mathrm{ets}}[\vec{r}], \vec{d}^{\mathrm{ets}}, d_{w_1}^{\mathrm{ets}}, \ldots, d_{w_{n_i}}^{\mathrm{ets}}]\vec{0}, \end{aligned}$$

so the claim follows from the IH. □

In applications it will be important to produce extracted terms with as few as possible case distinctions, and also that the case distinctions should be over as simple as possible boolean terms. The following example will show that such improvements are indeed necessary.

Let $f\colon \text{nat} \to \text{nat}$ be an unbounded function with $f(0) = 0$. Then we can prove

$$\forall n \exists m. f(m) \leq n < f(m+1).$$

If e.g. $f(m) = m^2$, then this formula expresses the existence of an integer square root $m := [\sqrt{n}]$ for any n. More formally we can prove

$$\forall n \exists m. \neg n < f(m) \wedge n < f(m+1) \tag{1}$$

from the assumptions

$$v_1\colon \forall n \neg n < f(0), \qquad v_2\colon \forall n\, n < f(g(n)).$$

Here $<^{\text{nat}\to\text{nat}\to\text{boole}}$ is the characteristic function of the natural ordering of the natural numbers and $r < s$ denotes $\text{atom}(<rs)$. We expressed $f(m) \leq n$ by $\neg n < f(m)$ and $f(0) = 0$ by $\forall n \neg n < f(0)$ to keep the formal proof as simple as possible. In order to have Π–assumptions we had to express the unboundedness of f by a witnessing function g. First note that the Π–assumptions v_1 and v_2 are not closed as required previously. But this is no problem since A–translation and program extraction clearly also work in this case: we just get a program containing f and g as parameters.

Now let us prove (1). Let n be given and assume

$$u\colon \forall m. \neg n < f(m) \to n < f(m+1) \to \bot.$$

We have to show $\bot$. From v_1 and u we inductively get $\forall m \neg n < f(m)$. For $m := g(n)$ this yields a contradiction to v_2.

The derivation term corresponding to this proof is

$$d := \text{Ind}_{m,\neg n<f(m)} n(v_1 n)u(g(n))(v_2 n)\colon \bot.$$

Now let

$$A := \exists^* m. \neg n < f(m) \wedge n < f(m+1).$$

The program extracted from d is

$$(d^{\text{tr}})^{\text{ets}} \equiv (d^A)^{\text{ets}}[d_u^{\text{ets}}, d_{v_1}^{\text{ets}}, d_{v_2}^{\text{ets}}]0\colon \text{nat}.$$

$(d^A)^{\text{ets}}$ is the same as d exept that $\text{Ind}_{m,\neg n<f(m)}$ has to be replaced by

$$\lambda n\, R_{\text{nat},(\text{nat}\to\text{nat})\to\text{nat}\to\text{nat}}$$

(since $\tau((\neg n < f(m))^A) = \tau(((n < f(m) \to A) \to A) \to (\bot \to A) \to A) =$ $(\text{nat} \to \text{nat}) \to \text{nat} \to \text{nat})$, and the assumption variables u, v_1, v_2 in d have to be replaced by (unary lists of) object variables

$$x_u\colon \text{nat} \to [(\text{nat} \to \text{nat}) \to \text{nat} \to \text{nat}] \to (\text{nat} \to \text{nat}) \to \text{nat} \to \text{nat},$$
$$x_{v_1}\colon \text{nat} \to (\text{nat} \to \text{nat}) \to \text{nat} \to \text{nat},$$
$$x_{v_2}\colon \text{nat} \to \text{nat} \to \text{nat}.$$

The subprograms $d_u^{\text{ets}}, d_{v_1}^{\text{ets}}, d_{v_2}^{\text{ets}}$ (of the same types as x_u, x_{v_1}, x_{v_2}) are given by

$$\begin{aligned} d_u^{\text{ets}} &\equiv \lambda m, x_1, x_2, k.\ \textbf{if}\quad n < f(m) \quad \textbf{then} \quad x_1\text{id}0 \quad \textbf{else} \\ &\qquad\qquad\qquad\quad \textbf{if}\quad \neg n < f(m+1) \quad \textbf{then} \quad x_2 0 \quad \textbf{else} \\ &\qquad\qquad\qquad\quad m \quad \textbf{fifi}, \\ d_{v_1}^{\text{ets}} &\equiv \lambda n, x_1, k.\ \textbf{if}\quad \neg n < f(0) \quad \textbf{then} \quad x_1 0 \quad \textbf{else} \\ &\qquad\qquad\qquad\quad k \quad \textbf{fi}, \\ d_{v_2}^{\text{ets}} &\equiv \lambda n, k\, k. \end{aligned}$$

Hence the normal form of $(d^{\text{tr}})^{\text{ets}}$ is

$$\begin{aligned} &R(\lambda x_1, k.\ \textbf{if}\quad \neg n < f(0) \quad \textbf{then} \quad x_1 0 \quad \textbf{else} \quad k \quad \textbf{fi}) \\ &\quad (\lambda m, w_1, w_2, k.\ \textbf{if}\quad n < f(m) \quad \textbf{then} \quad w_1\text{id}0 \quad \textbf{else} \\ &\qquad\qquad\qquad\qquad \textbf{if}\quad \neg n < f(m+1) \quad \textbf{then} \quad w_2 0 \quad \textbf{else} \\ &\qquad\qquad\qquad\qquad m \quad \textbf{fifi}) \\ &\quad (g(n)) \\ &\quad (\lambda k\, k) \\ &\quad 0 \end{aligned}$$

Informally, $(d^{\text{tr}})^{\text{ets}} = H(g(n), \lambda k\, k, 0)$ where $H\colon \text{nat} \to (\text{nat} \to \text{nat}) \to \text{nat} \to \text{nat}$ is such that

$$\begin{aligned} H(0, x_1, k) &= \textbf{if}\quad \neg n < f(0) \quad \textbf{then} \quad x_1 0 \quad \textbf{else} \quad k \quad \textbf{fi} \\ H(m+1, x_1, k) &= \textbf{if}\quad n < f(m) \quad \textbf{then} \quad H(m, \text{id}, 0) \quad \textbf{else} \\ &\quad\ \ \textbf{if}\quad \neg n < f(m+1) \quad \textbf{then} \quad x_1 0 \quad \textbf{else} \\ &\quad\ \ m \quad \textbf{fifi} \end{aligned}$$

This program is correct, but it is unnecessarily complicated. However, one can define a refined A–translation which simplifies the type of the auxiliary function H as well as its if–then–else structure. The type reduction can be achieved by *not* replacing *all* atoms P by $(P \to A) \to A$; this of course requires that atom is not the only relation symbol. Furthermore, to reduce case splitting one can construct "better" proofs of $C \to C^A$ for quantifier-free C without $\wedge$. In our example these refinements produce the following simplified progam:

$$h(0) = 0,$$
$$h(m+1) = \textbf{if} \quad n < f(m) \quad \textbf{then} \quad h(m) \quad \textbf{else} \quad m \quad \textbf{fi}.$$

However, space does not permit us to carry the details of the refinements here.

3. Pruning proof trees

It is shown how a proof can be simplified after addition of some further assumptions. In a first step some subproofs are replaced by different ones which use the additional assumptions. In a second step parts of the proof tree are pruned, i.e. cut out. Note that the first step involves searching for new proofs — using the new assumptions — of formulas in the proof tree. Hence we also have to discuss methods for proof search in minimal logic.

Let us describe both pruning steps in more detail.

For the first pruning step let a derivation d be given. For simplicity let us suppose that d contains no free assumption variables, i.e. $\mathrm{FA}(d) = \emptyset$. Assume also that some additional assumptions $v_1 \colon B_1, \ldots, v_m \colon B_m$ are given. We then go through all nodes of this proof. At any node, we have

- the formula A derived at this node, and
- the assumptions $u_1 \colon A_1, \ldots, u_n \colon A_n$ active at this node; they are bound further down in the proof by $\to^+$, with u_1 bound above u_2 bound above u_3 and so on.

Then we search for a new proof of A, using the additional assumptions

$$v_1 \colon B_1, \ldots, v_m \colon B_m$$

and a subset $\{u_1 \colon A_1, \ldots, u_i \colon A_i\}$ of the active assumptions with i minimal. The reason for giving preference to cutting out assumptions bound further down (i.e. more towards the root) in the proof is that the second pruning step described below then has a chance to prune out greater parts of the proof.

Clearly some systematic way to search for proofs in minimal logic is necessary here. In our implementation of the pruning method we have chosen a variant of Dale Miller's higher order logic programming framework in what he calls the $\mathcal{L}_\lambda$–fragment of our language $\mathcal{L}$; see [7]. It is based on an unification algorithm also given by Miller in [7] for $\mathcal{L}_\lambda$–terms (they are called higher–order patterns in Nipkow [10]). The reason for the restriction to the $\mathcal{L}_\lambda$–fragment is that the unification problem is decidable in this case. It should be noted that any first–order formula is in $\mathcal{L}_\lambda$. — Other interesting approaches to proof search in minimal logic can be found in [2, 6, 12, 14, 15].

We now describe the second pruning step. Let a derivation of a formula A be given. Assume that it contains a proper subderivation e of the same formula A. Assume also that between the endformula A of e and the global endformula A no $\to^+$ rule appears which binds an assumption variable that actually occurs free in e. More precisely, we assume that the derivation term is of the form

$$d^A[e^A/u^A].$$

We then "prune" the proof tree and replace it by its subtree e. This yields a correct derivation, since the subderivation contains no free object variables or assumption variables which have been bound in the original derivation. More precisely, we replace

$$d^A[e^A/u^A] \qquad \text{by} \qquad e^A.$$

Note that this pruning process is not unique: there may be many different occurrences of such subderivations of A. More formally, we may have

$$d^A[e^A/u^A] \equiv d^A[e'^A/v^A].$$

Note also that this is a generalization of Prawitz's immediate simplifications from [11].

The second pruning step then consists in applying the procedure just described to any subderivation of the given derivation.

Example: Proof by cases.

$$\dfrac{\dfrac{(A \to B) \to (\neg A \to B) \to B \qquad \dfrac{\begin{array}{c} u\colon A \\ | \\ B \end{array}}{A \to B}\,(\to^+ u)}{(\neg A \to B) \to B} \qquad \dfrac{\begin{array}{c} v\colon \neg A \\ | \\ B \end{array}}{\neg A \to B}\,(\to^+ v)}{B}$$

More precisely, we have the derivation term

$$\left[(\mathrm{Cases}_{A,B})^{(A\to B)\to(\neg A\to B)\to B}(\lambda u^A\, d_1^B)^{A\to B}(\lambda v^{\neg A}\, d_2^B)^{\neg A\to B}\right]^B.$$

Now assume that e.g. $\neg A$ can be proved from some additional assumptions. Let $e^{\neg A}$ be such a derivation term. Then we can replace the assumption $v\colon \neg A$ by this derivation $e^{\neg A}$:

$$\dfrac{\dfrac{(A \to B) \to (\neg A \to B) \to B \qquad \dfrac{\begin{array}{c} u\colon A \\ | \\ B \end{array}}{A \to B}\,(\to^+ u)}{(\neg A \to B) \to B} \qquad \dfrac{\begin{array}{c} | \\ \neg A \\ | \\ B \end{array}}{\neg A \to B}\,(\to^+ v)}{B}$$

or written as a derivation term

$$\left[(\mathrm{Cases}_{A,B})^{(A\to B)\to(\neg A\to B)\to B}(\lambda u^A\, d_1^B)^{A\to B}(\lambda v^{\neg A}\, d_2^B[e^{\neg A}/v^{\neg A}])^{\neg A\to B}\right]^B.$$

Now since $v^{\neg A}$ does not occur free in $e^{\neg A}$, i.e. $v^{\neg A} \notin \mathrm{FA}(e^{\neg A})$, we can write this derivation term as, letting $C := (A \to B) \to (\neg A \to B) \to B$,

$$\left[(\mathrm{Cases}_{A,B})^C (\lambda u^A\, d_1^B)^{A\to B} (\lambda v^{\neg A}\, w^B)^{\neg A \to B}\right]^B [d_2^B[e^{\neg A}/v^{\neg A}]/w^B]$$

and then prune it to obtain

$$d_2^B[e^{\neg A}/v^{\neg A}].$$

Note that by this second pruning step a β–normal derivation may be turned into a non–β–normal one. Moreover it can happen that after forming the β–normal form again the resulting derivation once more has a subderivation reducible by the second pruning step. In fact, the following example (due to Jan Ekman) shows that one may get back to the original derivation, and hence one has non–termination. The example only uses minimal implicational logic.

$$((p \to q) \to p) \to (p \to p \to q) \to q$$

(which corresponds to $\neg(p \leftrightarrow \neg p)$ if q is viewed as falsity) is derivable by

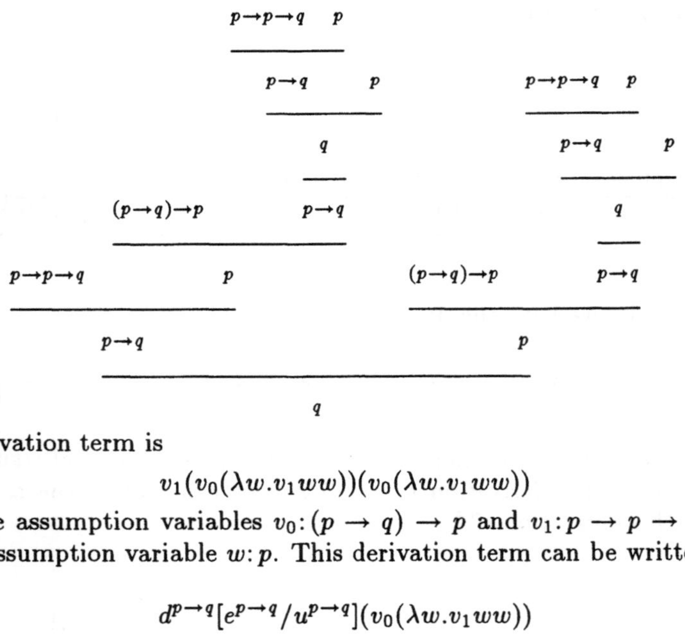

The derivation term is

$$v_1(v_0(\lambda w.v_1 ww))(v_0(\lambda w.v_1 ww))$$

with free assumption variables $v_0\colon (p \to q) \to p$ and $v_1\colon p \to p \to q$, and a bound assumption variable $w\colon p$. This derivation term can be written in the form

$$d^{p\to q}[e^{p\to q}/u^{p\to q}](v_0(\lambda w.v_1 ww))$$

where

$$d^{p\to q} := v_1[v_0 u^{p\to q}],$$
$$e^{p\to q} := \lambda w.v_1 ww.$$

Replacing the subderivation $d^{p\to q}[e^{p\to q}/u^{p\to q}]$ by $e^{p\to q}$ yields

$$(\lambda w.v_1 ww)(v_0(\lambda w.v_1 ww)),$$

which β–converts into the original derivation. We therefore restrict the second pruning step to cases where no new β–redexes are created.

4. An example of program development by pruning

The ideas discussed in the previous section are applied to an example already treated by Goad in his thesis [5], the binpacking problem. The main difference to Goad's work is that he used a logic with the strong existential quantifier, whereas we work within the $\to\forall$–fragment. This example is particularly well–suited to demonstrate that the pruning method can be applied to adapt programs to particular situations, and moreover that pruning can change the functions computed by programs. In this sense this method is essentially different from program development by program transformation.

The one–dimensional *binpacking problem* can be described as follows. We are given

- p one–dimensional blocks whose size is given by natural numbers $i_0, \ldots, i_{p-1}$, and
- q one–dimensional bins whose size is given by natural numbers $j_0, \ldots, j_{q-1}$.

The problem is to decide whether it is possible to pack the blocks into the bins, and if so, to give a solution.

We will treat the obvious first–fit algorithm to solve this problem: First try to pack i_0 into j_0. If it fits, try to pack the remaining blocks $i_1, \ldots, i_{p-1}$ into the modified bins $j_0 - i_0, j_1, \ldots, j_{q-1}$. If they do not fit, try to pack i_0 into the next bin j_1, and so on.

Now consider the particular situation where we know more about the input data, e.g. some order relations between the blocks to be packed and/or between the bins. Then it might be possible to adapt the general algorithm to this particular situation, and it is at least conceivable that the new algorithm which works for our particular situation is even extensionally different from the original one. We will see that this actually can happen, and that one can apply general proof–theoretic machinery to obtain the new, specialized algorithm from the general one.

To apply proof–theoretic methods we first have to construct a proof to work with. This proof should correspond as closely as possible to the general binpacking algorithm described above.

First note that it will be useful to refine the binpacking problem slightly: We also give as an input a natural number m and then look for a solution under the additional condition that the first block i_0 is to be packed in one of the first m bins $j_0, \ldots, j_{m-1}$. Let us call this variant the *bounded binpacking problem*.

Now a proof must be a proof of something, so we must start by giving a formal specification of the bounded binpacking problem. As is well known, there are many choices at this point. In particular, we have to fix a language and the basic properties of the functions and predicates in that language.

We use the two ground types nat for natural numbers and seq for finite sequences of natural numbers. i, j, k, m, n are variables for natural numbers, and a, b, c, x are variables for finite sequences of natural numbers. The empty sequence is denoted by `nil`, and `i::x` denotes the result of constructing a new sequence by adding the number `i` in front of the sequence `x`. The intended domains D_{nat} and D_{seq} are the obvious flat domains. Hence we have undefined or bottom objects denoted by **undefined-nat** and **undefined-seq**.

To formulate the specification we use the predicates **blegal** (for bounded legal) and **unsolvable**. The intended meaning of **(blegal a x b c m)** is that the sequence a is a defined answer to the bounded binpacking problem given by the blocks x= $i_0, \ldots, i_{p-1}$, the bins b= $j_0, \ldots, j_{q-1}$ and the bound m. This is to mean that a is a sequence a= $k_0, \ldots, k_{p-1}$, where k_n is the index of the bin the block i_n is to be packed into. Here the side condition that the first block i_0 is to be packed into a bin with index < m must be observed. So we require

$$\sum_{k_n=r} i_n \leq j_r \qquad \text{and} \qquad k_0 < m.$$

The answer **a=undefined-seq** is to mean that there is no solution.

The reason for the appearance of c is only technical: representing the bins by two sequences b and c allows to avoid the subtraction function. b remains fixed throughout, and c is initialized by a sequence of zeros of the same length as b.

The recursive definition of **blegal** is as follows.

```
blegal(k::a)(i::x)b c(m+1)
  =  blegal(k::a)(i::x)b c m
     v (k<=m &
        (ref c k)+i<=(ref b k) &
        blegal a x b(incr c k i)(len b))
blegal(k::a)(i::x)b c 0 = false
blegal nil(i::x)b c m   = false
blegal(k::a)nil b c m   = false
blegal nil nil b c m    = true
```

Here we have used some auxiliary functions: **(ref b k)** is the **k**–th member of the sequence b, and **(incr c k i)** is the sequence obtained from c by increasing its **k**–th element by **i**. Both functions should return **undefined-seq** if **k** is greater than or equal to the length **(len b)** (or **(len c)**).

To actually carry out proofs involving **blegal** it will be convenient to use the last four clauses in the definition of **blegal** as rewrite rules, and to encode the meaning of the essential first clause into two lemmata treated as "global assumptions" in the proof:

```
(add-global-assumption
 'lemma1 (pf "all a^,x,b,c,m.blegal a^ x b c m
                                    -> blegal a^ x b c(m+1)"))

(add-global-assumption
 'lemma2 (pf "all k,a^,i,x,b,c,m.k<=m
                           -> (ref c k)+i<=(ref b k)
                           -> blegal a^ x b(incr c k i)(len b)
                           -> blegal(add k a^)(add i x)b c(m+1)"))
```

(`pf` means parse–formula here, and `(add i x)` is a prefix notation for `i::x`).

Now we could define `unsolvable` in terms of `blegal` by letting `(unsolvable x b c m)` denote `all a.(blegal a x b c m) -> F`, and then take as our specification or goal formula

```
(define goal-formula
 (pf "all x,m,b,c.
      (all a^.(delta-seq a^ -> (blegal a^ x b c m))
           -> ((delta-seq a^ -> F) -> (unsolvable x b c m))
           -> F)
      -> F"))
```

i.e.

$$\forall x, m, b, c \exists \hat{a}. (\delta \hat{a} \to \mathtt{blegal}\, \hat{a} x b c m) \wedge (\neg \delta \hat{a} \to \mathtt{unsolvable}\, x b c m).$$

(The reason for using **â** instead of **a** here is to tell the machine that the value `undefined-seq` is allowed. `(delta-seq â)` means that **â** is a defined sequence.)

However, we run into a problem here. This goal formula is not of the form $\forall x \exists y A$ with quantifier–free A, since `(unsolvable x b c m)` by our definition contains a universal quantifier. Therefore we cannot guarantee that the general machinery for extracting programs from proofs will work here (cf. the example $\forall n \exists m \forall k. Tnnk \to Tnnm$ in Section 2).

Our solution to this problem is to observe that the inner universal is essentially bounded. Hence we can define `unsolvable` recursively, just as we have done for `blegal`. The intended meaning for `(unsolvable x b c m)` is that there is no possibility to pack the blocks `x` in such a way into the bins `b-c` that the first block is packed into a bin with index `<m`. The recursive definition of `(unsolvable x b c m)` is as follows.

```
unsolvable(i::x)b c(m+1)
 =  unsolvable(i::x)b c m
```

```
    & ((ref c m)+i<=(ref b m)
        -> unsolvable x b(incr c m i)(len b))
unsolvable(i::x)b c 0 = true
unsolvable nil b c m  = false
```

Now with this definition of **unsolvable** the goal formula has the right form $\forall x \exists y\, A$ with quantifier–free A.

To actually carry out proofs involving **unsolvable** it will again be convenient to use the last two clauses in the definition as rewrite rules, and to encode the meaning of the essential first clause into a lemma:

```
(add-global-assumption
 'lemma3 (pf "all i,x,b,c,m.
   unsolvable(add i x)b c m
   -> ((ref c m)+i<=(ref b m)
        -> unsolvable x b(incr c m i)(len b))
   -> unsolvable(add i x)b c(m+1)"))
```

In the proof of our goal we will make use of two more lemmata expressing properties of defined/undefined sequences.

```
(add-global-assumption
 'lemma4 (pf "all a^,m.delta-seq(add m a^) -> delta-seq a^"))

(add-global-assumption
 'lemma5 (pf "all a^,m.delta-seq a^ -> delta-seq(add m a^)"))
```

Let us now carry out the proof of our goal formula informally. The proof is by induction on the sequence **x**, and an auxiliary induction on **m**.

i::x, **m+1**. Choose **â** by hypothesis of the auxiliary induction.

Case **(delta-seq â)**. Take â and use Lemma 1.

Case **(delta-seq â) -> F**.

Subcase **(ref c m)+i<=(ref b m)**. Choose **â1** by hypothesis of the (main) induction for the modified bins, i.e. for **x (len b) b (incr c m i)**. Take **m::â1**. To prove that this is a solution we need Lemmata 2 and 4 in the case **delta-seq m::â1**, and Lemmata 3 and 5 in the case **delta-seq m::â1 -> F**.

Subcase **(ref c m)+i<=(ref b m) -> F**. Unsolvable by Lemma 3.

i::x, **0**. Unsolvable.
nil, **m**. Take **nil**.

We have carried out this proof in the interactive prover MINLOG developed in Munich. In Appendix 1 the formal representation of this proof is shown. The two applications of induction and also the two uses of case distinctions can be seen clearly in the type–free λ–term representing the proof.

Note that the axiom scheme `if-at` taking care of case distinctions has the general form

$$(P(\vec{r}) \to A) \to (\neg P(\vec{r}) \to A) \to A,$$

where the atom $P(\vec{r})$ is given by a boolean term. Hence this axiom can be proved easily by boolean induction. In the particular case $A \equiv \bot$ — and only this case is used in our derivation — the axiom reads

$$(P(\vec{r}) \to \bot) \to ((P(\vec{r}) \to \bot) \to \bot) \to \bot,$$

and hence is easily derivable, even in minimal logic. However, we have kept it in the proof, since it helps to make the structure of the case distinctions in the proof more visible.

Let us now aim at demonstrating the pruning technique in this example. The first step consists in specializing the general problem to a situation where the length of the sequences of blocks and bins is fixed, say 2 for both, and `m` = 1. Then we are to pack the blocks `i0::i1::nil` into the bins `j0::j1::nil` with the side condition that the first block `i0` has to go into the first bin `j0`. If we specialize the general proof to this particular situation and normalize it, then both inductions unfold completely and we obtain a proof as shown in Appendix 2.

Note that this proof has the following structure.

Case **u26**: $i_0 \leq j_0$.

Subcase **u27**: $i_0 + i_1 \leq j_0$. Take `0::0::nil`. To prove that this is a solution we need **u26**, **u27**, Lemmata 1 and 2 in the case **delta-seq 0::0::nil** *and Lemma 3 in the case* **delta-seq 0::0::nil -> F**.

Subcase **u31**: $i_0 + i_1 \not\leq j_0$.

Subsubcase **u32**: $i_1 \leq j_1$. Take `0::1::nil`. To prove that this is a solution we need **u26**, **u32** and Lemma 2 in the case **delta-seq 0::1::nil** *and* **u31** *and Lemmata 3 and 5 in the case* **delta-seq 0::1::nil -> F**.

Subsubcase **u38**: $i_1 \not\leq j_1$. Unsolvable by **u26**, **u31**, **u38** and Lemmata 2, 3 and 4.

Case **u44**: $i_0 \not\leq j_0$. Unsolvable by **u44** and Lemma 3.

Here for the first time we can observe that we have a proof produced by a machine and not by a human: It contains in its emphasized parts arguments a human would avoid to carry out, namely to start from some obviously false assumption (**u29:delta-seq 0::0::nil -> F** in the first instance and **u34:delta-seq 0::1::nil -> F** in the second instance) and then carry out a somewhat elaborate argument to arrive at some conclusion (as it turns out, atomic formulas in both instances). We could use the first pruning step to remove these and other unneccessary parts of the proof. However, such an application of pruning does not yet show the effect we want to demonstrate, since the algorithm represented by the proof has not changed.

To achieve this desired effect, we first have to provide some additional assumptions on the input data. So let us assume that the second block is less than or equal to the first one, and that the first bin is less than or equal to the second one, i.e. that

$$\texttt{v1}\colon i_1 \leq i_0 \qquad \text{and} \qquad \texttt{v2}\colon j_0 \leq j_1$$

holds. Then we can replace quite a number of subproofs by different ones — this is the first pruning step of Section 2 —, where we always aim at replacing the use of assumption variables bound later in the proof (e.g. case assumptions) by the use of our additional assumptions **v1** and **v2** and — if necessary — some other assumption variables. Let us look at these replacements individually. The result is shown in Appendix 3.

(1) Here the machine replaced the derivation using Lemmata 3 and 5 of an atom from the false assumption `u29:delta-seq 0::0::nil -> F` — which rewrites into `T -> F` — by a trivial derivation using ex–falso–quodlibet and the truth–axiom `T`.

(2) The case assumption `u32`: $i_1 \leq j_1$ is replaced by a simple derivation (involving a global assumption **trans** expressing transitivity of $\leq$) of $i_1 \leq j_1$ from the additional assumptions `v1`: $i_1 \leq i_0$ and `v2`: $j_0 \leq j_1$, together with the outer case assumption `u26`: $i_0 \leq j_0$.

(3) Similar to (1): The lengthy derivation using Lemmata 3 and 5 of an atom from the false assumption `u34:delta-seq 0::1::nil -> F` — which rewrites into `T -> F` — is replaced by a trivial derivation using ex–falso–quodlibet and the truth–axiom `T`. Note that an occurrence of the case assumption `u31`: $i_0+i_1 \not\leq j_0$ present in the original proof has disappeared.

(4) Here the derivation `((lemma4 undefined-seq) 0) u39` of $\perp$ has been replaced by `u39`: $\perp$. We have used here that `delta-seq undefined-seq` rewrites into $\perp$ and `0::undefined-seq` rewrites into `undefined-seq`,

(5) Here the subproof `(lambda (u43) ((efq false) (u38 u43)))` of $i_1 \not\leq j_1$ is replaced by `u38`, which derives the same formula.

(6) As in (5), we could replace the subproof `(lambda (u42) ((efq false) (u31 u42)))` of $i_0 + i_1 \not\leq j_0$ by `u31`: $i_0 + i_1 \not\leq j_0$. However, here we can go one step further and replace the subproof by a simple derivation (involving a global assumption `<=-+-ax` expressing compatibility of $\leq$ with $+$ and again **trans** expressing transitivity of $\leq$) of $i_0+i_1 \not\leq j_0$ from the additional assumptions `v1`: $i_1 \leq i_0$ and `v2`: $j_0 \leq j_1$, together with the outer case assumption `u38`: $i_1 \not\leq j_1$.

(7) Similar to (5).

Now we come to the essential second pruning step. It consists in going through every node of the derivation (from the bottom upwards) and checking whether the formula A at this node also appears somewhere further up in the proof in such a way that between the two occurrences of A no $\to^{+}$–rule appears which binds an assumption variable that actually occurs free further up in the proof tree.

In the subcase **u31**: $i_0 + i_1 \not\leq j_0$ all occurrences of the case assumption **u31** have been pruned away. Hence the proof of $\bot$ provided by this subderivation is independent of **u31**. We can remove the case axiom and replace the subderivation of $\bot$ starting with **((if-at 'false)** $i_0 + i_1 \leq j_0$**)** by the proof of that case starting with **((if-at 'false)** $i_1 \leq j_1$**)**.

Moreover, the proof of $\bot$ starting with this case axiom **((if-at 'false)** $i_1 \leq j_1$**)** has again the property that its first subsubcase **u32**: $i_1 \leq j_1$ does not use its case assumption **u32**: it has been pruned away in step (2). So again we can remove the case axiom and replace the subderivation of $\bot$ starting with **((if-at 'false)** $i_1 \leq j_1$**)** by the proof of that case. The resulting derivation is shown in Appendix 4.

Note that we have indeed changed the program provided by our classical proof in an extensional way. In the general situation — without additional assumptions — our proof yielded **0::0::nil** as an answer for the particular arguments i_0, i_1 and j_0, j_1 satisfying $i_0 + i_1 \leq j_0$, $i_1 \leq i_0$ and $j_0 \leq j_1$ (see Appendix 2). After the pruning procedure, the resulting proof yields **0::1::nil** as an answer for these particular arguments satisfying the same conditions (see Appendix 4).

Appendix 1

Proof of the general goal formula for the one–dimensional binpacking problem.

```
#(() (all x (all m (all b (all c (imp (all a^ (imp (imp (atom
(delta-seq a^)) (atom ((((blegal a^) x) b) c) m))) (imp (imp
(imp (atom (delta-seq a^)) (atom false)) (atom (((unsolvable
x) b) c) m))) (atom false)))) (atom false)))))) (lambda (x)
(((seq-ind-at '(all x (all m (all b (all c (imp (all a^ (imp
(imp (atom (delta-seq a^)) (atom ((((blegal a^) x) b) c) m)))
(imp (imp (imp (atom (delta-seq a^)) (atom false)) (atom
(((unsolvable x) b) c) m))) (atom false)))) (atom false))))))
(lambda (m) (lambda (b) (lambda (c) (lambda (u22) (((u22 nil)
(lambda (u23) truth-axiom-symbol)) (lambda (u24) ((efq
(((unsolvable nil) b) c) m)) (u24 truth-axiom-symbol)))))))))
(lambda (x) (lambda (n10) (lambda (u11) ((((lambda (x) (lambda
(i) (lambda (IV) ((((natind-at '(all m (all b (all c (imp (all
a^ (imp (imp (atom (delta-seq a^)) (atom ((((blegal a^)
((add i) x)) b) c) m))) (imp (imp (imp (atom (delta-seq a^))
(atom false)) (atom (((unsolvable ((add i) x)) b) c) m)))
(atom false)))) (atom false)))))) i) x) (lambda (b) (lambda (c)
(lambda (u19) (((u19 undefined-seq) (lambda (u20) ((efq
((((blegal undefined-seq) ((add i) x)) b) c) 0)) u20)))
(lambda (u21) truth-axiom-symbol)))))) (lambda (m) (lambda
(NIV) (lambda (b) (lambda (c) (lambda (u1) (((NIV b) c)
(lambda (a^) (lambda (u2) (lambda (u3) ((((if-at '(atom false))
(delta-seq a^)) (lambda (u4) (((u1 a^) (lambda (u5)
((((((lemma1 a^) ((add i) x)) b) c) m) (u2 u4)))) (lambda (u6)
((efq (((unsolvable ((add i) x)) b) c) ((+-nat m) 1)))
(u6 u4)))))) (lambda (u7) ((((if-at '(atom false)) (<=-nat
((+-nat ((ref c) m)) i) ((ref b) m))) (lambda (u8) ((((IV
(len b)) b) (((incr c) m) i)) (lambda (a^1) (lambda (u9)
(lambda (u10) (((u1 ((add m) a^1)) (lambda (u11)
(((((((((lemma2 m) a^1) i) x) b) c) m) truth-axiom-symbol) u8)
(u9 (((lemma4 a^1) m) u11))))) (lambda (u12) ((((((lemma3 i)
x) b) c) m) (u3 u7)) (lambda (u13) (u10 (lambda (u14) (u12
(((lemma5 a^1) m) u14)))))))))))))) (lambda (u15) (((u1 a^)
(lambda (u16) ((efq ((((blegal a^) ((add i) x)) b) c)
((+-nat m) 1))) (u7 u16)))) (lambda (u17) ((((((lemma3 i) x)
b) c) m) (u3 u17)) (lambda (u18) ((efq ((((unsolvable x) b)
(((incr c) m) i)) (len b))) (u15 u18)))))))))))))))))))))) x)
n10) u11))))) x)))
```

Appendix 2

This is the result of specializing the general problem to a situation where the length of the sequences of blocks and bins is fixed, say 2 for both, and m = 1. We are to pack the blocks `i0::i1::nil` into the bins `j0::j1::nil` with the side condition that the first block `i0` has to go into the first bin `j0`. If we specialize the general proof to this particular situation and normalize it, then both inductions unfold completely and we obtain the following type-free proof term.

```
(lambda (u25)
 ((((if-at '(atom false)) (<=-nat i0 j0))
   (lambda (u26)
    ((((if-at '(atom false)) (<=-nat ((+-nat i0) i1) j0))
      (lambda (u27)
       (((u25 ((add 0) ((add 0) nil)))
         (lambda (u28)
          (((((((((lemma2 0) ((add 0) nil)) i0)
                 ((add i1) nil))
                ((add j0) ((add j1) nil)))
               ((add 0) ((add 0) nil)))
              0)
             truth-axiom-symbol)
            u26)
           (((((lemma1 ((add 0) nil)) ((add i1) nil))
               ((add j0) ((add j1) nil)))
              ((add i0) ((add 0) nil)))
             1)
            (((((((((lemma2 0) nil) i1) nil)
                    ((add j0) ((add j1) nil)))
                 ((add i0) ((add 0) nil)))
                0)
               truth-axiom-symbol)
              u27)
             truth-axiom-symbol)))))
       (lambda (u29)
(1)     ((((((lemma3 i0) ((add i1) nil)) ((add j0)
 |                                              ((add j1) nil)))
 |           ((add 0) ((add 0) nil)))
 |          0)
 |         truth-axiom-symbol)
 |        (lambda (u30)
 |         ((efq false)
 |          (u29 (((lemma5 ((add 0) nil)) 0)
```

```
(1)                   truth-axiom-symbol)))))))))
         (lambda (u31)
          ((((if-at '(atom false)) (<=-nat i1 j1))
            (lambda (u32)
             (((u25 ((add 0) ((add 1) nil)))
               (lambda (u33)
                ((((((((((lemma2 0) ((add 1) nil)) i0)
                       ((add i1) nil))
                      ((add j0) ((add j1) nil)))
                     ((add 0) ((add 0) nil)))
                    0)
                   truth-axiom-symbol)
                  u26)
                 (((((((((lemma2 1) nil) i1) nil)
                        ((add j0) ((add j1) nil)))
                      ((add i0) ((add 0) nil)))
                     1)
                    truth-axiom-symbol)
(2)                u32)
                  truth-axiom-symbol))))
              (lambda (u34)
(3)            (((((((lemma3 i0) ((add i1) nil))
 |                  ((add j0) ((add j1) nil)))
 |                 ((add 0) ((add 0) nil)))
 |               0)
 |              truth-axiom-symbol)
 |             (lambda (u35)
 |              (((((((lemma3 i1) nil)
 |                    ((add j0) ((add j1) nil)))
 |                   ((add i0) ((add 0) nil)))
 |                  1)
 |                 (((((((lemma3 i1) nil)
 |                       ((add j0) ((add j1) nil)))
 |                      ((add i0) ((add 0) nil)))
 |                    0)
 |                   truth-axiom-symbol)
 |                  (lambda (u36) ((efq false) (u31 u36)))))
 |               (lambda (u37)
 |                ((efq false)
 |                 (u34
 |                  (((lemma5 ((add 1) nil)) 0)
(3)                  (((lemma5 nil) 1)
                      truth-axiom-symbol)))))))))))
           (lambda (u38)
```

```
        (((u25 undefined-seq)
          (lambda (u39)
           (((((((((lemma2 0) undefined-seq) i0)
                  ((add i1) nil))
                 ((add j0) ((add j1) nil)))
                ((add 0) ((add 0) nil)))
               0)
              truth-axiom-symbol)
             u26)
            ((efq
              (((((blegal undefined-seq) ((add i1) nil))
                 ((add j0) ((add j1) nil)))
                ((add i0) ((add 0) nil)))
               2))
(4)          (((lemma4 undefined-seq) 0) u39)))))
         (lambda (u40)
          ((((((lemma3 i0) ((add i1) nil))
               ((add j0) ((add j1) nil)))
              ((add 0) ((add 0) nil)))
             0)
            truth-axiom-symbol)
           (lambda (u41)
            ((((((lemma3 i1) nil) ((add j0) ((add j1) nil)))
                ((add i0) ((add 0) nil)))
               1)
              ((((((lemma3 i1) nil) ((add j0) ((add j1) nil)))
                  ((add i0) ((add 0) nil)))
                 0)
                truth-axiom-symbol)
(6)            (lambda (u42) ((efq false) (u31 u42)))))
(5)          (lambda (u43) ((efq false) (u38 u43))))))))))))
  (lambda (u44)
   (((u25 undefined-seq)
     (efq
      (((((blegal undefined-seq) ((add i0) ((add i1) nil)))
         ((add j0) ((add j1) nil)))
        ((add 0) ((add 0) nil)))
       1)))
    (lambda (u46)
     ((((((lemma3 i0) ((add i1) nil))
          ((add j0) ((add j1) nil)))
         ((add 0) ((add 0) nil)))
        0) truth-axiom-symbol)
(7)   (lambda (u47) ((efq false) (u44 u47)))))))))
```

Appendix 3

Here we assume that we have the additional assumptions

$$\mathtt{v1}\colon i_1 \leq i_0 \qquad \text{and} \qquad \mathtt{v2}\colon j_0 \leq j_1.$$

Then we can replace quite a number of subproofs by different ones — this is the first pruning step of Section 2 —, where we always aim at replacing the use of assumption variables bound later in the proof (e.g. case assumptions) by the use of our additional assumptions **v1** and **v2** and — if necessary — some other assumption variables. The resulting proof term looks as follows.

```
(lambda (u25)
 ((((if-at '(atom false)) (<=-nat i0 j0))
   (lambda (u26)
    ((((if-at '(atom false)) (<=-nat ((+-nat i0) i1) j0))
      (lambda (u27)
       (((u25 ((add 0) ((add 0) nil)))
         (lambda (u28)
          (((((((((lemma2 0) ((add 0) nil)) i0)
                 ((add i1) nil))
                ((add j0) ((add j1) nil)))
               ((add 0) ((add 0) nil)))
              0)
             truth-axiom-symbol)
            u26)
           ((((((lemma1 ((add 0) nil)) ((add i1) nil))
                ((add j0) ((add j1) nil)))
               ((add i0) ((add 0) nil)))
              1)
             (((((((((lemma2 0) nil) i1) nil)
                  ((add j0) ((add j1) nil)))
                 ((add i0) ((add 0) nil)))
                0)
               truth-axiom-symbol)
              u27)
             truth-axiom-symbol)))))
        (lambda (u29)
(1)      ((efq
 |         ((and-strict true)
 |          ((imp-strict ((leq-strict-nat i0) j1)) false)))
(1)        (u29 truth-axiom-symbol))))))
     (lambda (u31)
      ((((if-at '(atom false)) (<=-nat i1 j1))
```

```
        (lambda (u32)
         (((u25 ((add 0) ((add 1) nil)))
           (lambda (u33)
            (((((((((lemma2 0) ((add 1) nil)) i0)
                   ((add i1) nil))
                  ((add j0) ((add j1) nil)))
                 ((add 0) ((add 0) nil)))
                0)
               truth-axiom-symbol)
              u26)
             (((((((((lemma2 1) nil) i1) nil)
                   ((add j0) ((add j1) nil)))
                  ((add i0) ((add 0) nil)))
                 1)
                truth-axiom-symbol)
(2)            (((((trans i1) j0) j1)
 |               (((((trans i1) i0) j0) v1) u26)) v2))
(2)           truth-axiom-symbol))))
          (lambda (u34)
(3)        ((efq
 |           ((and-strict true)
 |            ((imp-strict ((leq-strict-nat i0) j1)) false)))
(3)         (u34 truth-axiom-symbol))))))
       (lambda (u38)
        (((u25 undefined-seq)
          (lambda (u39)
           (((((((((lemma2 0) undefined-seq) i0)
                  ((add i1) nil))
                 ((add j0) ((add j1) nil)))
                ((add 0) ((add 0) nil)))
               0)
              truth-axiom-symbol)
             u26)
            ((efq
              (((((blegal undefined-seq) ((add i1) nil))
                 ((add j0) ((add j1) nil)))
                ((add i0) ((add 0) nil)))
               2))
(4)          u39))))
         (lambda (u40)
          ((((((lemma3 i0) ((add i1) nil))
               ((add j0) ((add j1) nil)))
              ((add 0) ((add 0) nil)))
             0)
```

```
               truth-axiom-symbol)
              (lambda (u41)
               (((((((lemma3 i1) nil) ((add j0) ((add j1) nil)))
                   ((add i0) ((add 0) nil)))
                  1)
                 (((((((lemma3 i1) nil) ((add j0) ((add j1) nil)))
                     ((add i0) ((add 0) nil)))
                    0)
                   truth-axiom-symbol)
(6)               (lambda (u48)
 |                 (u38
 |                  ((((trans i1) j0) j1)
 |                    ((((trans i1) i0) j0) v1)
 |                     ((((<=-+-ax i0) i1) j0) u48)))
(6)                 v2)))))
               u38)))))))))))
  (lambda (u44)
   (((u25 undefined-seq)
     (efq
      (((((blegal undefined-seq) ((add i0) ((add i1) nil)))
          ((add j0) ((add j1) nil)))
         ((add 0) ((add 0) nil)))
        1)))
    (lambda (u46)
     (((((((lemma3 i0) ((add i1) nil))
          ((add j0) ((add j1) nil)))
         ((add 0) ((add 0) nil)))
        0)
       truth-axiom-symbol)
(7)   u44))))))
```

Appendix 4

This is the result of applying the second pruning step to the derivation shown in Appendix 3.

```
(lambda (u25)
 ((((if-at '(atom false)) (<=-nat i0 j0))
   (lambda (u26)
    (((u25 ((add 0) ((add 1) nil)))
      (lambda (u33)
       (((((((((lemma2 0) ((add 1) nil)) i0) ((add i1) nil))
             ((add j0) ((add j1) nil)))
            ((add 0) ((add 0) nil)))
           0)
          truth-axiom-symbol)
         u26)
        (((((((((lemma2 1) nil) i1) nil)
               ((add j0) ((add j1) nil)))
             ((add i0) ((add 0) nil)))
            1)
           truth-axiom-symbol)
          (((((trans i1) j0) j1)
            (((((trans i1) i0) j0) v1) u26)) v2))
         truth-axiom-symbol))))
     (lambda (u34)
      ((efq ((and-strict true)
             ((imp-strict ((leq-strict-nat i0) j1)) false)))
       (u34 truth-axiom-symbol))))))
  (lambda (u44)
   (((u25 undefined-seq)
     (efq
      (((((blegal undefined-seq) ((add i0) ((add i1) nil)))
         ((add j0) ((add j1) nil)))
        ((add 0) ((add 0) nil)))
       1)))
    (lambda (u46)
     ((((((lemma3 i0) ((add i1) nil))
         ((add j0) ((add j1) nil)))
        ((add 0) ((add 0) nil)))
       0)
      truth-axiom-symbol)
     u44))))))
```

References

[1] Ulrich Berger. Program extraction from normalization proofs. In M. Bezem and J.F. Groote, editors, *Typed Lambda Calculi and Applications*, pages 91–106. Springer Lecture Notes in Computer Science Vol. 664, 1993.

[2] Roy Dyckhoff. Contraction–free sequent calculi for intuitionistic logic. *Journal of Symbolic Logic*, 57:793–807, 1992.

[3] Harvey Friedman. Classically and intuitionistically provably recursive functions. In Dana S. Scott and Gert H. Müller, editors, *Higher Set Theory*, pages 21–28. Springer Lecture Notes in Mathematics, Volume 699, 1978.

[4] Gerhard Gentzen. Untersuchungen über das logische Schließen. *Mathematische Zeitschrift*, 39:176–210, 405–431, 1934.

[5] Christopher Alan Goad. *Computational uses of the manipulation of formal proofs.* PhD thesis, Stanford University, August 1980. Stanford Department of Computer Science Report No. STAN–CS–80–819.

[6] Jörg Hudelmaier. Bounds for cut elimination in intuitionistic propositional logic. *Archive for Mathematical Logic*, 31:331–354, 1992.

[7] Dale Miller. A logic programming language with lambda–abstraction, function variables and simple unification. *Journal of Logic and Computation*, 2(4):497–536, 1991.

[8] G.E. Mints. On e–theorems (in russian). *Zapiski*, 40:110–118, 1974.

[9] Chetan Murthy. Extracting constructive content from classical proofs. Technical Report 90–1151, Dep.of Comp.Science, Cornell Univ., Ithaca, New York, 1990. PhD thesis.

[10] Tobias Nipkow. Functional unification of higher–order patterns. In *Proc. 8th IEEE Symp. Logic in Computer Science*, pages 64–74, 1993.

[11] Dag Prawitz. *Natural Deduction*, volume 3 of *Acta Universitatis Stockholmiensis. Stockholm Studies in Philosophy.* Almqvist & Wiksell, Stockholm, 1965.

[12] Dan Sahlin, Torkel Franzen, and Seif Haridi. An intuitionistic predicate logic theorem prover. *Journal of Logic and Computation*, 2(6):619–656, 1992.

[13] Helmut Schwichtenberg. Proofs as programs. In Peter Aczel, Harold Simmons, and Stanley S. Wainer, editors, *Proof Theory. A selection of*

papers from the Leeds Proof Theory Programme 1990, pages 81–113. Cambridge University Press, 1992.

[14] N. Shankar. Proof search in intuitionistic sequent calculus, 1991.

[15] Wilfried Sieg and Richard Scheines. Searching for proofs (in sentential logic). In Leslie Burkholder, editor, *Philosophy and the computer*, pages 137–159, Boulder, San Francisco, Oxford, 1992. Westview Press.

[16] Martin Stein. *Interpretationen der Heyting-Arithmetik endlicher Typen.* PhD thesis, Universität Münster, Fachbereich Mathematik, 1976.

[17] Anne S. Troelstra, editor. *Metamathematical Investigations of Intuitionistic Arithmetic and Analysis*, volume 344 of *Lecture Notes in Mathematics*. Springer, Berlin, 1973.

[18] Anne S. Troelstra and Dirk van Dalen. *Constructivism in Mathematics. An Introduction*, volume 121, 123 of *Studies in Logic and the Foundations of Mathematics.* North–Holland, Amsterdam, 1988.

[19] Anton Wallner. Komplexe Existenzbeweise in der Arithmetik. Master's thesis, Mathematisches Institut der Universität München, 1993.

[20] Hermann Weyl. Über die neue Grundlagenkrise der Mathematik. *Mathematische Zeitschrift*, 10, 1921.

Concurrent Processes and Petri Nets

Wilfried Brauer and Robert Gold

Institut für Informatik, Technische Universität München

Arcisstr. 21, D-80333 München

Abstract. Petri nets as models for distributed systems and concurrent processes are briefly introduced, by means of two examples: a distributed control and a systolic stack. Different ideas to define semantics that distinguish concurrency from nondeterminism and that model causality and temporal succession are sketched. Methods of behaviour preserving refinements (of actions/transitions and states/places) as well as related modular composition techniques are described and appropriate semantic notions are looked for. Petri nets are also studied from the viewpoint of data flow theory, and several new Petri net semantics (also useful for coloured nets) are obtained. Also an algebraic extension to Petri nets in the sense of process algebra is briefly described - the Petri box calculus. Much of the research reported here was carried out in the ESPRIT Basic Research Actions DEMON and CALIBAN and in the subproject SEMAFOR of the Sonderforschungsbereich 342 of the Deutsche Forschungsgemeinschaft at TU München.

1. Introduction

1.1 General remarks

In the real world there are no completely independent sequential processes like those which are abstractly dealt with in traditional informatics. There are always several concurrent processes interacting with each other. Processes are caused by activities of processors combined in a system. Such systems are usually non-sequential, i.e. there is, in general, no fixed linear order in which processors have to be active, and processors usually are positioned at different locations in space. Moreover systems ususally are not isolated but interact with their environments. To apply informatics for solving problems in the real world therefore amounts to modelling distributed, reactive systems. The behaviours of the processors and the ways of interaction may be rather varied. Therefore as a paradigm for distributed, reactive systems one should better consider a group of (intelligent) persons together with some resources which they need (see [Bra 93], [BB 92]); the group may be constituted by the necessity to organize access to a scarce resource or by the goal of doing a difficult job by collaboration. This paradigm is helpful not only for understanding how complicated such systems might be, but also for getting ideas on how to construct useful systems.

On a rather abstract level such a system can be described by the actions which the persons might execute, the states the system may attain, and the mutual dependencies between states and actions. Therefore traditional system descriptions use at least three different description mechanisms: for the

- structure of the system, i.e. the components and their interconnections (structure graphs)
- state space (all sorts of data models)
- dynamic behaviour (usually the sequences of actions or states generated or accepted by some sort of automata, transition diagrams, flow graphs or the like).

Traditionally the notion of state means the global state - even if this state is considered as structured (e.g. as a state vector). In distributed systems it is in general not possible to discern global states - only local states in (small) components can be related to individual actions.

A natural requirement on a formalism for modeling the structure and the operational behaviour of distributed systems therefore is that it can represent local states, (local) actions and the (local) interrelations between the holdings of states and the executions of actions (which implies that the structure and the dynamic behaviour descriptions should be very closely related).

The only formalism fulfilling these requirements is that of Petri nets. The structure of a Petri net always is a bipartite directed graph which may have labels on the nodes and the arcs: one type of nodes (usually drawn as circles) represents the local states the other one (usually depicted by rectangles) the actions, arcs lead from states to actions (signifying that the holding of the states is a necessary condition for action executions) and actions to states (signifying that the executions of actions change the states). The holding of a state is depicted by a marking of the respective state node (i.e. a token is placed into the circle).

The change of the marking of a state node can also be interpreted as sending a message to the actor depending on the respective local state. This leads to a more general concept: state changes are communicated by message passings, and the transport of objects (form one location, person, etc.) to another is considered as a state change. To distinguish different messages and different object types a number of formal tools from the field of data modeling is available.

Obviously a Petri net model of a distributed system also explicitely models the concurrency inherent in the system: actions which are not in conflict can be executed concurrently; to check for conflict it is not necessary to consider global states but only the local states in relation to the actions in question.

The mathematical foundations of Petri nets allows for quite a number of sophisticated tools for analysing the behaviour of systems modeled by Petri nets and gives methods at hand to construct net models in a hierarchical and modular way (by top-down refinement using techniques as in sequential program development: macro substitution or procedure call, or by bottom-up composition of modules using a number of operators, in particular for different ways of parallel composition) - see the volumes of the subseries "Advances in Petri Nets" of the Springer lecture Notes in Computer Science (LNCS) and other volumes of the LNCS series (in particular also the proceedings of the conference series "Concur 'XY" — the latest being LNCS 715 — and the proceedings of the conference series "Applications and Theory of Petri Nets 19XY" — the latest being LNCS 691).

The diversity of application systems modelled and the complexity of these systems has led to a rather large number of extensions and variations of the basic notion of Petri nets. Many of these variants are of little use (apart from the purpose they were invented for, perhaps) because they divert so much from the core concepts, that most of the formal tools are not applicable for them. The most elaborate extension which is also mostly applied is the class of coloured Petri nets (see [Jen 92]), for which also computer-aided design and analysis tools are available. A more recent concept to combine the power of algebraic specifications with Petri nets is presented by Reisig [Rei 91].

There is one drawback with Petri nets: their structure is fixed and the effect of an action execution cannot change dynamically. There are several proposals to extend Petri nets by mechanisms for reconfigurability or for changing action effects, but these sometimes lead out of the class of Petri nets (see [Kie 90]) or introduce other difficulties (like noncompositionality - see [Val 81]). It seems that it is not yet clear which forms of reconfiguration and modification of nets are really necessary and useful for applications. High level nets (like algebraic nets, coloured nets, predicate/transition nets) offer already a number of possibilities for data-driven structure change in nets. But even the combination of these concepts with the recursion concept of Kiehn's Petri net systems [Kie 90] and the selfmodification conept of Valk [Val 81] does not yield the same power of flexibility as Milner's π-calculus [Mil 93]. Therefore it is a great challenge to search for application oriented restricted concepts for such dynamics on the level of Petri net extensions.

1.2 Two examples

We will look at two examples of distributed systems from the literature (which are not treated by the use of Petri nets there): one concerning distributed control, one concerning the use of asynchronous operations.

1.2.1 Distributed control

A. Bergeron in [Ber 93] gives an automata-theoretic approach to the notions of controllability and observability of discrete event processes and of methods for constructing control systems for such processes. Here is her example: Suppose three rooms are connected with doors as indicated in Figure 1.0.

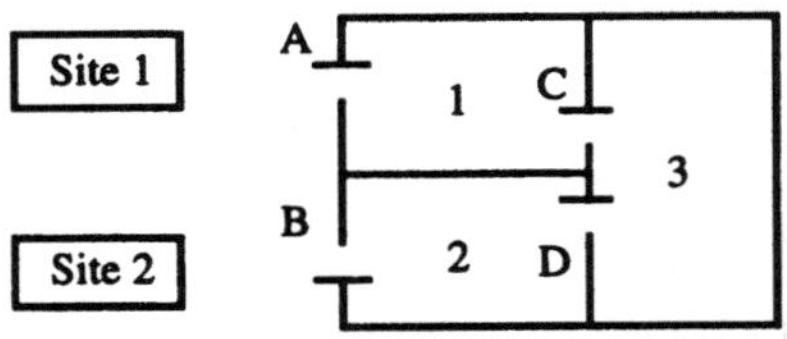

Figure 1.0

"Each door, labelled A, B, C or D can allow objects to move in either direction. Observable and/or controllable events will be, for example, of the form A_{in} or C_{out} meaning that an object just passed through a door in a given direction (fixing arbitrarily the "in" direction as left to right). We are to construct two controllers, one at Site 1 and one at Site 2, that should prevent two objects from being in the same room simultaneously. The controller at Site 1 can observe events at doors A, B and C, and the controller at Site 2 events at doors, A, B and D. Six events are controllable, ie. all except A_{out} and B_{out}. Is it possible to distribute these controllable events among the two controllers such that they can enforce the constraint, while allowing maximum freedom to objects?"

We can approach the problem as follows: At each door place two controllers: one who allows the action X_{in}, the other allows $X_{out} (X = A, B, C, D)$. Obviously the following restrictions have to be obeyed:

(a) None of the following pairs of actions (X_{in}, X_{in}), (X_{out}, X_{out}), (X_{in}, X_{out}) may occur concurrently, where $X = A, B, C, D$ (Only one object at a time passes through the door).

(b) None of the following pairs of actions may occur concurrently: $(A_{in}, C_{out}), (B_{in}, D_{out}), (C_{in}, D_{in})$. (No two objects enter a room concurrently).

In order to control the restrictions the controllers have to communicate by sending a (1-bit) message to each relevant partner, if they let an object pass. They give the way free only if they have got the appropriate messages from their partners. We can model each controller as a transition of a Petri net with an appropriate number of input places (for messages from partners) and output places (for messages to partners). This gives the Petri net in Figure 1.1.

It is easy to check that there will never be two tokens on a place, which means that this net can perform the control task.

To get the controllers for the sites 1 and 2, we only need to join the controllers for the doors A, B and C as well as those for A, B and D together with the communication links between them. The two controllers still have to communicate about the situation in room 3. We see immediately that the controller at site 1 (resp. 2) does not need to know anything about the events at door B (resp. A) if the other controller works properly.

If we do not allow communication of the controllers at sites 1 and 2, then each of them must be more restrictive, e.g. from site 1, C_{in} must be forbidden as soon as B_{in} occurs etc., i.e. we have the additional communication depicted in Figure 1.2.

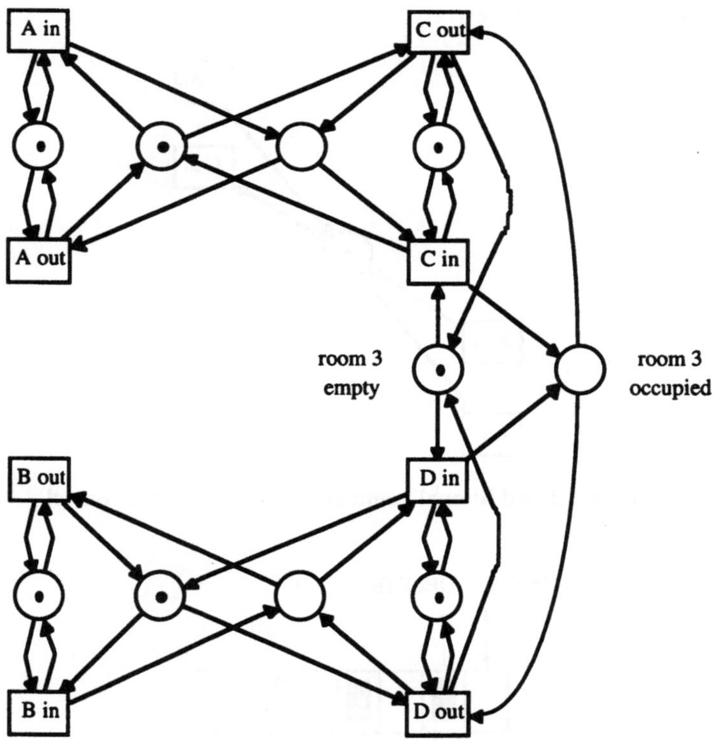

Figure 1.1: The net of local controllers

1.2.2 An asynchronous pushdown storage system

Already Petri in his dissertation [Pet 62] has made clear that an automaton model with an infinite storage device (like a Turing machine or a pushdown automaton) cannot be implemented without using asynchronously operating elements (see also [Bra 91]). Already in 1973 Furtek [Fur 73] gave a construction of an asynchronously operating pushdown store system using Petri nets (see also [Rei 93]). We consider here a construction of a so-called systolic stack which was developed without ideas from Petri net theory. The stack is described in [GL 82] as follows (with some small modifications): The stack is a linear array of cells (potentially infinite to the right), each cell directly communicating with its left and right neighbors. Cells can be in one of two states, *occupied*, when they hold a value (stack element), or *empty* (see Figure 1.3, where the occupied cells are shown containing a black rectangle). Insertions take place at the leftmost cell of the stack, and deletions take place from the cell next to the leftmost if the leftmost cell is empty, otherwise from the leftmost cell. Thus when the stack is ready to be operated on, the leftmost cell is empty (unless the stack is full) and the one next to it is full (unless the stack is empty). An insertion or deletion simply updates the leftmost, or next to leftmost cell, respectively.

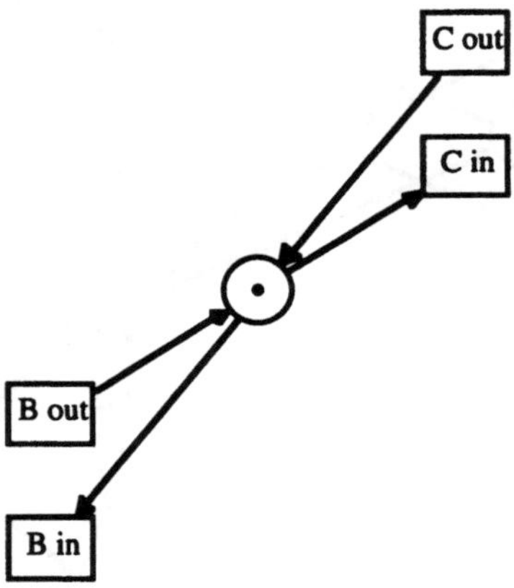

Figure 1.2: Additional communications for Site 1 control

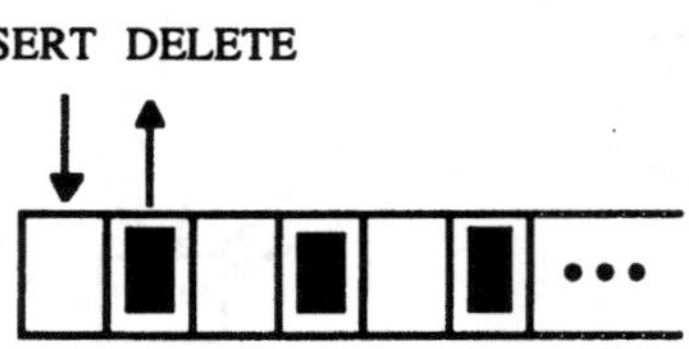

Figure 1.3: The systolic stack

In order to support a sequence of insertions and deletions, the stack obviously needs some way to move the entries around. We introduce two operations, called L and R operations, whose function is to redistribute the stack entries among the cells so as to make further operations possible. The operations are visualized as the application of two rewrite rules, as shown in Figure 1.4. The first rule states that if we ever have two occupied cells to the left of an empty cell, then in the R operation the element in the rightmost of the two occupied cells will move one over to the right into the empty cell. The second rule is the dual of the first, with the roles of empty and occupied interchanged. In other words, if we have two consecutive empty cells followed by an occupied cell, then during the L operation the element in the occupied cell will move one left into the rightmost of the two empty cells.

An implementation can be described as a potentially infinite Petri net composed of two sequences of one-transition-nets $L_1, L_2, L_3, \ldots$ and $R_0, R_1, R_2, \ldots$, where all L_i are isomorphic and all $R_i, i \geq 1$ are isomorphic: see Figure 1.5.

A place d_i is marked by a token, if the cell i of the stack contains a data item; the place e_i is just complementary to d_i: it is marked if and only if d_i is empty (this is necessary since in a Petri net (positive) actions cannot be based on the (negative) fact of absence of information). The net controlling the stack is obtained by taking the nets $R_0, R_1, L_1, R_2, L_2, \ldots$ and identifying places with identical labels, see Figure 1.6.

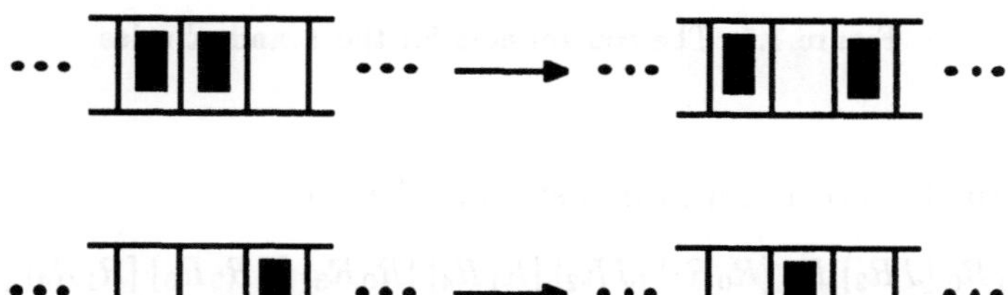

Figure 1.4: The re-write rules for the stack

In the start configuration (when the stack is empty) all e_i have to be marked. Insertion of a data element to the stack can only occur if e_0 is marked and d_0 not, and by insertion of the data d_0 gets marked and e_0 unmarked. It is easy to see that at no time any two places d_i and e_i will be marked. If we, in addition, assume that each transition (L_i, R_i) will act as soon as it is activated (by seeing the appropriate marking on the connected places) and if each action (of L_i or R_i) takes one unit of time then three units of time after an insertion (or deletion) another insertion (or deletion) can take place - irrespective of the size of the stack.

Example: If $[x, y, z]$ denotes a "step", i.e. the concurrent execution of x, y and z, and if we denote the insert action by J, then a sequence of 5 insertions

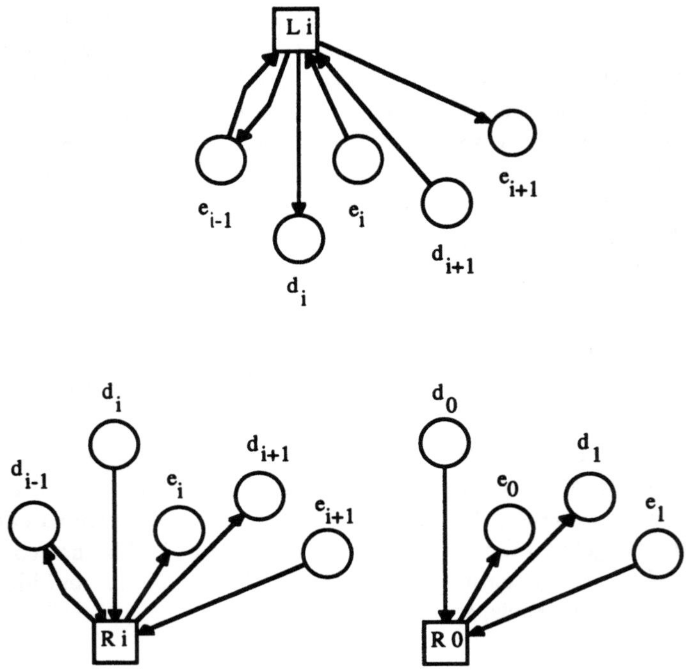

Figure 1.5: The control nets for the L and R rules

leads to the following sequence of steps in the net

$$JR_0\ JR_1R_0\ [JR_2]\ R_1\ [R_0R_3]\ [JR_2]\ [R_1R_4]\ [R_0R_3]\ [JR_2R_5]\ [R_1R_4]\ [R_0R_3R_6]$$

If we want to have a stack of n cells we only need to take transitions $R_0, R_1, \ldots, R_{n-1}$ and $L_1, \ldots, L_{n-1}$. One nice feature of this stack is also that all its cells can be used for data storage.

1.3 Remarks on construction and uses of Petri nets

The two examples show how to construct Petri nets in a bottom-up way by connecting components asynchronously, i.e. by identifying places - those places can be considered as shared variables or communication channels.

The nets are used to control actions of discrete event systems. If we model such systems also by Petri nets - e.g. the stack by the net in Figure 1.7, where the transitions l_i, r_i each move one data item to the left or right respectively and the places are the stack cells - then the control is exercised by synchronization of the corresponding transitions (L_i with l_i and R_i with r_i) - on the net level this amounts simply to the merging of transitions (in the stack example we can also identify the cells with the places d_i). We then obtain a net in which different sorts of tokens (messages and data) are used: a so-called high-level Petri net (see 1.1).

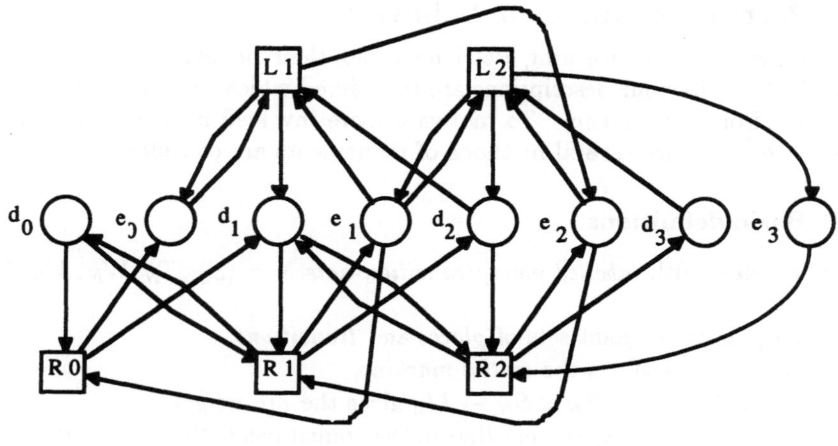

Figure 1.6: A net for a stack with 3 cells

In a similar way as for control, nets may be also used to test other systems in order to check whether they behave properly or in order to learn about their behaviour. This gives a method to check equivalence of nets following the testing theory of De Nicola and Hennessy [DNH 84]; this is particularly interesting when timed nets are considered (see [Vog 93]).

For sequential systems a much advocated and very successful design method is stepwise refinement. In the case of distributed systems with concurrent processes, refinement is rather complicated: if an action (a transition) is refined by a concurrent system (a Petri net), the action may be activated before the internal processes have terminated, actions inside and outside the refined part may occur concurrently (if they look the same for an external observer, he may get problems in understanding the behaviour of the system). Therefore much research has been devoted to refinement of nets (see [BGV 90], [Vog 92], [JM 92], [BDE 93]).

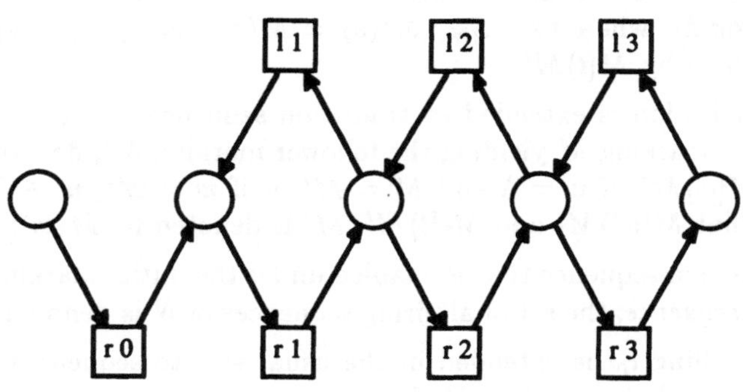

Figure 1.7: The stack without control

2. Petri nets and their behaviour

Basic definitions concerning Petri nets and their behaviour are given. In particular behaviour descriptions are considered which distinguish nondeterminism from concurrency. To motivate more involved notions of behaviour and of equivalence several methods of refinements are considered.

2.1 Basic definitions

We will deal with *labelled place/transition-nets* $N = (S_N, T_N, W_N, M_N, l_N)$, where

- S_N, T_N are disjoint sets of *places* and *transitions*,
- $M_N : S_N \to \mathbb{N}_0$ is the *initial marking*,
- $W_N : S_N \times T_N \cup T_N \times S_N \to \mathbb{N}_0$ gives the *arc weights*,
- $l_N : T_N \to \Sigma \cup \{\lambda\}$ is a *labelling* of the transitions with labels from a fixed infinite *alphabet* Σ or with the empty string $\lambda \notin \Sigma$. The elements of Σ are *actions*, which may be represented by several different transitions; transitions with label λ stand for *internal* actions; these transitions perform *silent moves*.

The sets S_N and T_N need not be finite. We call a net N *finite* if S_N and T_N are finite. From W_N we deduce the set of arcs $F_N = \{(x,y) \mid W_N(x,y) \neq 0\}$. Often unlabelled nets are considered. In these cases the labelling l_N can be assumed to be the identity; mostly we will simply omit this labelling. We will use the following definitions:

- ${}^\bullet x := \{y \in S_N \cup T_N \mid (y,x) \in F_N\}$, $x^\bullet := \{y \in S_N \cup T_N \mid (x,y) \in F_N\}$ for $x \in S_N \cup T_N$ are called the *pre-*, *postset* resp. of x; for subsets X of $S \cup T$ we put ${}^\bullet X = \cup_{x \in X} {}^\bullet x$ and $X^\bullet = \cup_{x \in X} x^\bullet$;
- N is called *T-restricted* if $\forall t \in T_N : {}^\bullet t \neq \emptyset, t^\bullet \neq \emptyset$;
- $M : S_N \to \mathbb{N}_0$ is called a *marking*;
- $t \in T_N$ is *enabled* under a marking M if $\forall s \in S_N : W_N(s,t) \leq M(s)$; this is denoted by $M[t\rangle$;
- the firing of a transition t that is enabled under M yields the follower marking M' where $\forall s \in S_N : M'(s) := M(s) - W_N(s,t) + W_N(t,s)$; this is denoted by $M[t\rangle M'$;
- this definition is extended to transition sequences: $w \in T_N^*$ is enabled under a marking M yielding the follower marking M', denoted by $M[w\rangle$ and $M[w\rangle M'$, if $w = \lambda$ and $M = M'$ or if $w = w't$, $w' \in T_N^*$, $t \in T_N$ such that $M[w'\rangle M_1$ and $M_1[t\rangle M'$; M' is denoted by $M(w)$;
- a transition sequence that is enabled under the initial marking is called a *firing sequence*; the set of all firing sequences of N is denoted by $FS(N)$;
- the labelling l_N is extended in the usual way to sequences $w \in T^*$ by $l_N(\lambda) = \lambda$, $l_N(wt) = l_N(w) l_N(t)$;

- $a \in \Sigma \cup \lambda$ is *image enabled* under a marking M if $\exists w \in T_N^* : l_N(w) = a \wedge M[w\rangle$; this is denoted by $M[a\rangle\rangle$; if w yields the follower marking M' we write $M[a\rangle\rangle M'$; observe that w is a sequence of transitions which are labelled by λ with one exception (if $a \neq \lambda$), which is labelled a.
- $l_N(FS(N))$ is called the *language* of N;
- $[M\rangle := \{M' : S_N \to \mathbb{N}_0 \mid \exists w \in T_N^* : M[w\rangle M'\}$ is called the set of *markings reachable* from the marking M;
- $[M_N\rangle$ is called the set of *reachable markings*;
- $t \in T_N$ is *live* if $\forall M \in [M_N\rangle\ \exists M' \in [M\rangle : M'[t\rangle$;
- N is *live* if all $t \in T_N$ are live;
- $s \in S_N$ is *m-bounded* if $\forall M \in [M_N\rangle : M(s) \leq m$;
- $s \in S_N$ is *bounded* if there exists $m \in \mathbb{N}_0$ such that s is m-bounded;
- $s \in S_N$ is *safe* if s is 1-bounded;
- N is *(m-)bounded* if all $s \in S_N$ are (m-)bounded;
- N is *safe* if N is 1-bounded;
- $t \in T_N$ is *dead* if $\forall M \in [M_N\rangle : \neg M[t\rangle$.

Sometimes we will need the *restriction* of a relation. Let $r \subseteq X \times Y$ be a relation and $X' \subseteq X, Y' \subseteq Y$. The relation $r \cap (X' \times Y') \subseteq X \times Y$ is denoted by $r|_{X' \times Y'}$. If r is a function, then the function $r \cap (X' \times Y)$ is denoted by $r|_{X'}$. For a sequence $w \in A^*$ over some set A and an element a of A we denote the *number of occurrences* of a in w by $\#(w, a)$. For the *disjoint union* of two sets we use the symbol $\dot\cup$, i.e. $X = Y \dot\cup Z$ iff $X = Y \cup Z$ and $Y \cap Z = \emptyset$.

A function $X \to \mathbb{N}_0$ is called a *multiset* over X where X is a set. The firing rule defined above can be extended to multisets of transitions. Let N be a net, let $s \in S_N, t \in T_N$ and u a multiset over T_N.

- $u^-(s) := \sum_{t \in T_N} u(t) W_N(s,t)$, $u^+(s) := \sum_{t \in T_N} u(t) W_N(t,s)$, $\Delta u(s) := u^+(s) - u^-(s)$;
- u is *enabled* under a marking M if $\forall s \in S_N$: $u^-(s) \leq M(s)$; this is denoted by $M[u\rangle$; u is called a *step*;
- the multiset $t + t$ has value 2 at t and 0 everywhere else; for $t_1 \neq t_2$ the multiset $t_1 + t_2$ has value 1 at t_1 and t_2 and 0 everywhere else;
- a transition t is *self-concurrent* if $\exists M \in [M_N\rangle : M[t + t\rangle$.

In general we do not distinguish between isomorphic nets, partial orders (see below) etc.

2.2 Behaviour of Petri nets

In the previous section we have already introduced liveness, boundedness and safeness. These are very basic notions of behaviour, where two nets have the same behaviour if either both are live or both are not live etc. The notions of liveness and safeness for nets (which are used since the 1970's) differ from the notions of liveness and safety properties introduced by Lamport in [Lam 77].

Very often the behaviour of a net is described by its language, i.e. by all sequences of labelled transition firings starting at an initial marking. There are two aspects in which this semantics can be seen as unsatisfactory:

- It is an interleaving semantics, i.e. it does not distinguish between the concurrent execution of actions and their execution in arbitrary order. Instead one could take a partial order semantics like the set of (images of) partial words or of (event structures of) processes.
- It is a linear time semantics (like processes and partial words), i.e. it contains no information at which stage of an execution which choices have been made. Instead one could work with failures semantics which contains some information on the choices – this semantics has been developed for TCSP [BHR 84], [BR 84]. Or one could model the branching behaviour completely by taking a class of bisimilar transition systems as semantics; this has most often been studied for CCS [Mil 83], [Par 81]. Both are interleaving semantics.

In the following we define the above mentioned behaviour notions and the related equivalences. These are those equivalences of nets we will need in this paper. For a more detailed discussion of equivalences of Petri nets see [PRS 92].

Definition 2.1

Two nets N_1, N_2 are called *language-equivalent* if

$$l_{N_1}(FS(N_1)) = l_{N_2}(FS(N_2))$$

Another well-known notion of behaviour are processes; they are intended to model the causality of a system run.

Definition 2.2

A *causal net* is a triple $N = (B, E, F)$, where

- B, E are finite, disjoint sets of *conditions* and *events*,
- $F \subseteq B \times E \cup E \times B$ is the *flow relation* (by F^+ we denote the transitive closure of F, by $^\bullet x$ and $x^\bullet$ we denote the pre-, postset resp. of $x \in B \cup E$, i.e. $^\bullet x := \{y \in B \cup E \mid (y, x) \in F\}$, $x^\bullet := \{y \in B \cup E \mid (x, y) \in F\}$),
- $\forall x, y \in B \cup E : (x, y) \in F^+ \Rightarrow (y, x) \notin F^+$ (acyclicity),
- $\forall b \in B : |^\bullet b| \leq 1, |b^\bullet| \leq 1$ (no branching of places).

A *process* $\pi = (B, E, F, p)$ of a net $N = (S_N, T_N, W_N, M_N, l_N)$ *enabled* under a marking M consists of a causal net (B, E, F) and a labelling $p : B \cup E \rightarrow S_N \cup T_N$ such that

- $p(B) \subseteq S_N$, $p(E) \subseteq T_N$,
- $\forall s \in S_N : M(s) = |p^{-1}(s) \cap \{b \in B|\ {}^\bullet b = \emptyset\}|$ (minimal conditions correspond to the marking M),
- $\forall e \in E\ \forall s \in S_N : W_N(s, p(e)) = |p^{-1}(s) \cap {}^\bullet e|,\ W_N(p(e), s) = |p^{-1}(s) \cap e^\bullet|$ (transition environments are respected).

We write $M[\pi\rangle$ in this situation. The marking M' reached after the firing of π is defined by $\forall s \in S_N : M'(s) = |p^{-1}(s) \cap \{b \in B|\ b^\bullet = \emptyset\}|$ (denoted by $M[\pi\rangle M'$). A process of a net N enabled under the initial marking is simply called *process of N*. The *initial process* of a net N, denoted by $\pi_0(N)$, is the unique process of N with an empty set of events.

For the following we need some definitions about partial orders.

Definition 2.3

Let $(A, <)$ be a finite strict partial order, i.e. let $<$ be a transitive, irreflexive relation on a finite set A, and let $\beta : A \to X$ be a function into some set X, called a labelling. Then $(A, <, \beta)$ is called a *strict partial order labelled over X*. For a strict partial order $(A, <)$, for $k, l \in A$ and $B \subseteq A$ we define, as usual, $l \leq k :\Leftrightarrow l < k \vee l = k$ and $\downarrow B := \{l \in A|\ \exists k \in B : l \leq k\}$. We call B *downward-closed*, if $B = \downarrow B$. Let $(A_i, <_i, \beta_i)$, $i = 1, 2$, be strict partial orders labelled over X. They are said to be *isomorphic*, if there exists a bijection $\alpha : A_1 \to A_2$ that is *label* and *order preserving*, i.e. $\beta_1(k) = \beta_2(\alpha(k))$ and $k <_1 l \Leftrightarrow \alpha(k) <_2 \alpha(l)$ for all $k, l \in A_1$. α is called an *isomorphism*. $(A_1, <_1, \beta_1)$ is called *prefix* of $(A_2, <_2, \beta_2)$ if $A_1 \subseteq A_2$, $A_1 = \downarrow_2 A_1$, $<_1 = <_2|_{A_1 \times A_1}$, $\beta_1 = \beta_2|_{A_1}$.

A process $\pi = (B, E, F, p)$ of a net N enabled under some marking M defines a strict partial order labelled over $S_N \cup T_N$, namely $(B \cup E, F^+, p)$.

Definition 2.4

Let N be a net and $\pi_i = (B_i, E_i, F_i, p_i)$, $i = 1, 2$, be processes of N enabled under some marking M. Then π_1 is a *prefix* of π_2 if $(B_1 \cup E_1, F_1^+, p_1)$ is a prefix of $(B_2 \cup E_2, F_2^+, p_2)$. We also call π_2 an *extension* of π_1.

From a process we derive a strict partial order labelled over Σ
- by supressing all conditions and those events labelled by λ-transitions and
- by changing the labels of the remaining events from transitions of N to their labels.

Definition 2.5

Let $\pi = (B, E, F, p)$ be a process of a net N enabled under some marking M. The *event structure* of π is the strict partial order $ev(\pi) := (A, <, \beta)$ labelled over Σ, where $A := \{e \in E|\ l_N(p(e)) \neq \lambda\}$, $< := F^+|_{A \times A}$, $\beta := l_N \circ p|_A$. We say that $ev(\pi)$ is *enabled* under M yielding the follower marking M', denoted by $M[ev(\pi)\rangle\rangle$, and $M[ev(\pi)\rangle\rangle M'$, if π is a process in N that is enabled under M yielding the follower marking M'. Two nets N_1, N_2 are called *process-equivalent* if $\{ev(\pi) | \ \pi \text{ is a process of } N_1\} = \{ev(\pi) | \ \pi \text{ is a process of } N_2\}$.

A third notion of behaviour, which is a less well-known partial order semantics of nets, is based on so-called partial words [Gra 81]. Intuitively a partial word can be seen as the causal ordering of the events in a system run together with a partial observation of this run; consequently also causally independent events can be ordered.

Definition 2.6

Let N be a net, let $p = (A, <, \beta)$ be a strict partial order labelled over T_N, $C \subseteq A$, M be a marking of N. The multiset βC over T_N is defined by $\beta C(t) := |\beta^{-1}(t) \cap C|$. p is said to be *enabled* (under M_N) if $\forall B, C \subseteq A, B \cap C = \emptyset$: If all elements of C are pairwise unordered, and B and $B \cup C$ are downward-closed, then

$$\forall s \in S_N : (\beta C)^-(s) \leq M_N(s) + \sum_{l \in B} \Delta\beta(l)(s).$$

Then p is called a *partial word* of N. The *image* of p, $(A', <', \beta')$, is the strict partial order labelled over Σ defined by $A' := \{a \in A | \ l_N(\beta(a)) \neq \lambda\}$, $<' := < |_{A' \times A'}$, $\beta' := l_N \circ \beta|_{A'}$.

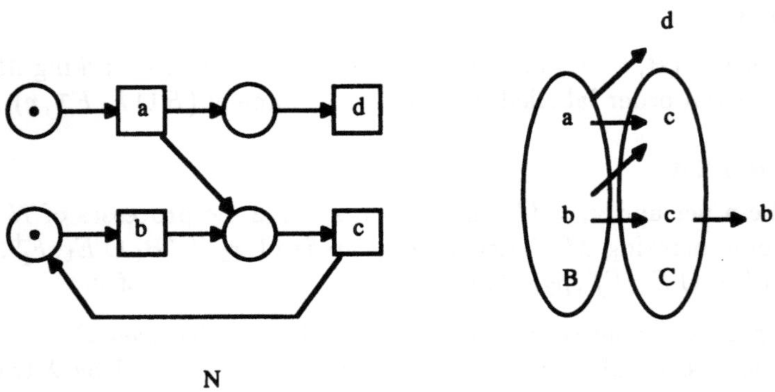

Figure 2.1

Figure 2.1 shows a net N and a partial word enabled in N, where some sets B and C are indicated. The unordered elements of C can be fired independently, if we respect the given precedences, i.e. fire the elements of B first. The precedence relation in this partial word does not describe causality in the sense of processes: the first occurrence of b is an immediate predecessor of both occurrences of c, but b produces only one token. It is easy to see that in general the set of partial words is closed under making partial words more sequential.

The two ways to describe the behaviour of nets by a partial order labelled with transitions (by event structures and by partial words) have been compared by A. Kiehn.

Theorem 2.1

(i) Every linearization of an event structure of a process of a net N is a firing sequence of N; for each firing sequence w of a safe net N there is a process π uniquely, such that w is a linearization of the event structure of π.

(ii) If π is a process of a net N, then $ev(\pi)$ is a partial word of N.

(iii) [Kie 88] For each net $\{p|p$ is a least sequential partial word of $N\} \subseteq \{ev(\pi)|\pi$ is a process of $N\}$. If N is safe, then equality holds.

The proof of (iii) given by A. Kiehn is rather complicated; a simpler proof which uses a well-known mathematical result (the so-called marriage theorem) is given in [Vog 92].

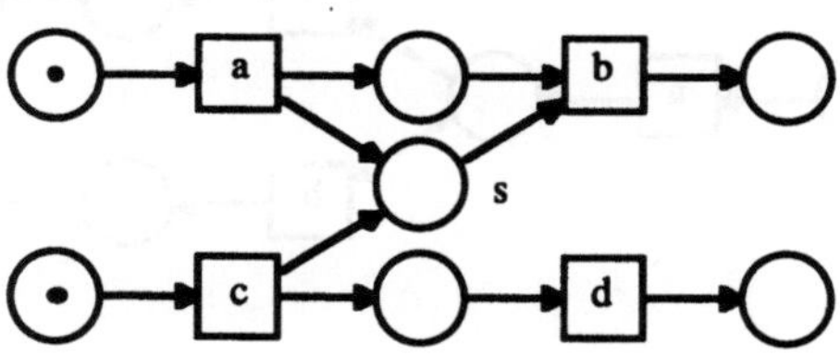

Figure 2.2

In a non-safe net a firing sequence needs not always determine a unique process; in particular - as Figure 2.2 shows - not for every process π is $ev(\pi)$ a least sequential partial word: After firing transitions a and c in the net of Figure 2.2 the transition b can either take from place s the token produced by a or that produced by c - this defines two different processes π_1, π_2 such that $ev(\pi_1)$ is strictly less sequential than $ev(\pi_2)$.
The following translation of failures semantics to nets can be found in [Pom 86]. In [Vog 89] it has been shown to be just the right semantics for the modular construction of nets, if one uses a certain well-known synchronization operator and is interested in nets that cannot deadlock. There is also a variant that takes divergence into account. Divergence or internal looping is an infinite firing sequence of silent moves that might occur at some reachable marking.

Definition 2.7

The ***(interleaving) failures semantics*** of a net N is $\mathcal{F}(N) = \{(v, X) \mid v \in \Sigma^*, X \subseteq \Sigma, \exists w \in FS(N) : l_N(w) = v \wedge \forall a \in X : \neg M(w)[a\rangle\}$. Two nets N_1, N_2 are ***$\mathcal{F}$-equivalent*** if $\mathcal{F}(N_1) = \mathcal{F}(N_2)$.

Figure 2.3a shows two nets which are not failures equivalent: Only the second net can perform a such that no b is possible, it can refuse b after a. Formally, we have $(a, \{b\}) \in \mathcal{F}(N_2) - \mathcal{F}(N_1)$.

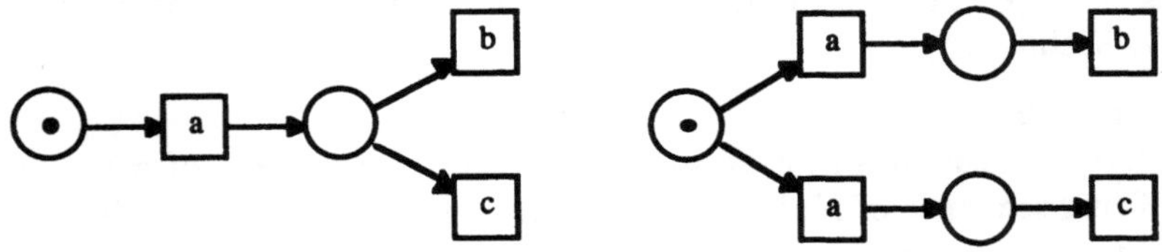

Figure 2.3a

On the other hand, the nets of Figure 2.3b are not distinguished by failures semantics, since only the second net can perform a such that bc is not possible. This is because of the fact that failures semantics only has a 'look-ahead' of 1. These nets are distinguished by bisimilarity.

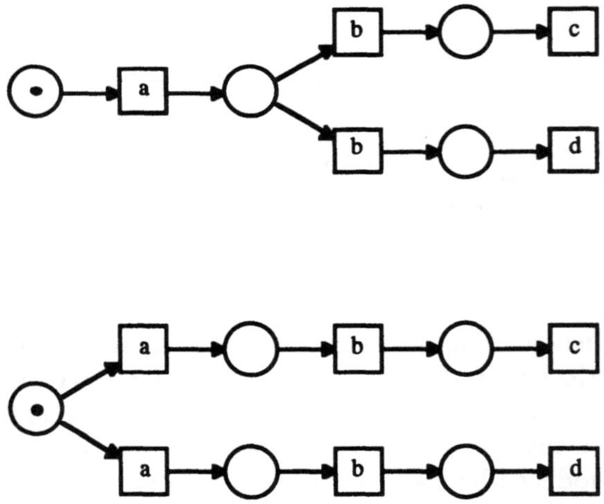

Figure 2.3b

Definition 2.8

Two nets N_1, N_2 are *(interleaving) bisimilar* if there is a *bisimulation*, i.e. a relation $\rho \in [M_{N_1}\rangle \times [M_{N_2}\rangle$ with

- $(M_{N_1}, M_{N_2}) \in \rho$,
- for all $M_1 \in [M_{N_1}\rangle$ and all $M_2 \in [M_{N_2}\rangle$ such that $(M_1, M_2) \in \rho$ we have: if there is $a \in \Sigma \cup \{\lambda\}$ and $M_1' \in [M_{N_1}\rangle$ such that $M_1[a\rangle\rangle M_1'$ then for some M_2' we have $M_2[a\rangle\rangle M_2'$ and $(M_1', M_2') \in \rho$,
  if there is $a \in \Sigma \cup \{\lambda\}$ and $M_2' \in [M_{N_2}\rangle$ such that $M_2[a\rangle\rangle M_2'$ then for some M_1' we have $M_1[a\rangle\rangle M_1'$ and $(M_1', M_2') \in \rho$.

2.3 Behaviour preserving refinements

2.3.1 A basic technique for the refinement of transitions

We define a refinement technique, which is widely used in the literature. We give the definition of the refinement nets, i.e. the nets that substitute transitions, and a description, how this substitution is done. A *refinement net D* is a net, which contains a special place, called the *idle place*, denoted by $idle_D$. We assume that the arcs from and to the idle place have weight 1. The set of places of D is $S_D \dot{\cup} \{idle_D\}$. This is a slight abuse of notation, since here S_D is not the set of all places of D; the additional place represents the interface to a *host* net, i.e. the net into which D is inserted and 'really' belongs to the host. The idle place is removed, when D is inserted into a host. The transitions in the postset of the idle place are called *initial transitions*, those in the preset of the idle place are called *final transitions* (Figure 2.4). (It is also possible that a transition is initial and final at the same time, also that there are no initial or no final transitions.)
The *refined net* is constructed by removing a transition t from the host and inserting the refinement net (without the idle place). To the preset of each initial transition the preset of t is added and to the postset of each final transition the postset of t is added (Figure 2.5).

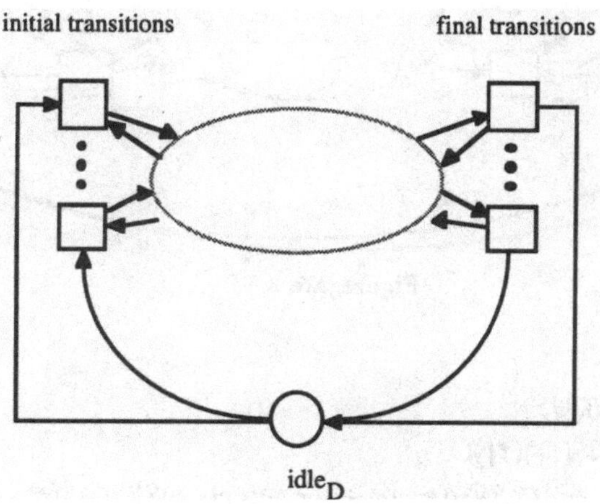

Figure 2.4: A refinement net D

Definition 2.9

Let N be a net, $t \in T_N$ and D a refinement net. We assume that $S_N \cup T_N$ and $S_D \cup T_D$ are disjoint. Refining t by D yields the refined net, denoted by $N[t/D]$, which is defined by:

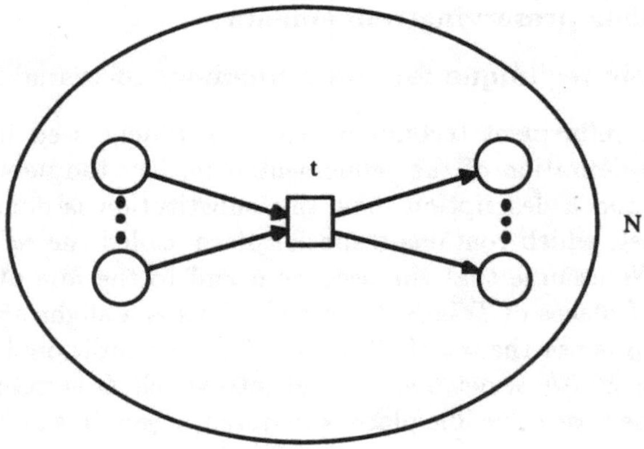

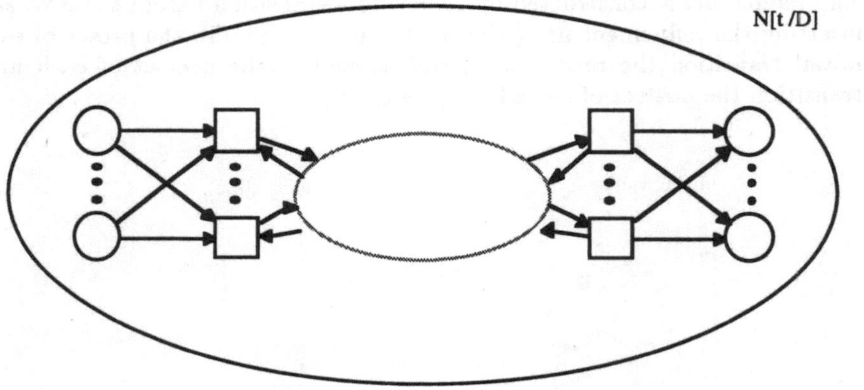

Figure 2.5

$$S_{N[t/D]} = S_N \dot{\cup} S_D$$
$$T_{N[t/D]} = (T_N - \{t\}) \dot{\cup} T_D$$
$$W_{N[t/D]}(x,y) = \begin{cases} W_N(x,y), & \text{if } x,y \in S_N \cup (T_N - \{t\}) \\ W_D(x,y), & \text{if } x,y \in S_D \cup T_D \\ W_N(x,t), & \text{if } x \in S_N \text{ and } y \text{ is an initial transition} \\ W_N(t,y), & \text{if } y \in S_N \text{ and } x \text{ is a final transition} \\ 0, & \text{otherwise} \end{cases}$$
$$M_{N[t/D]}(s) = \begin{cases} M_N(s), & \text{if } s \in S_N \\ M_D(s), & \text{if } s \in S_D \end{cases}$$
$$l_{N[t/D]}(x) = \begin{cases} l_N(x), & \text{if } x \in T_N - \{t\} \\ l_D(x), & \text{if } x \in T_D \end{cases}$$

The marking of the idle place is not relevant for the refinement technique. But it will be relevant, when we study the behaviour of the refinement net in order to determine the behaviour of the refined net. This refinement technique obviously also applies to unlabelled nets.

2.3.2 Refinement nets with one initial and one final transition

The first attempts on behaviour preserving refinement of transitions are based on the idea that the firing of a transition has a 'beginning' and an 'end'. Therefore transitions can be split in one initial transition, which models the 'beginning', an inner part and one final transition, which models the 'end'. In this approach, the refinement net does not have an initial conflict. But of course we could embed this choice between an internal initial and an internal final transition.

In [Val 79] a refinement net[†] with one initial transition ($t_{initial}$) and one final transition (t_{final}) with $t_{initial} \neq t_{final}$ is called *block*. In order to study the behaviour of a block the idle place is initially marked with one token (Figure 2.6).

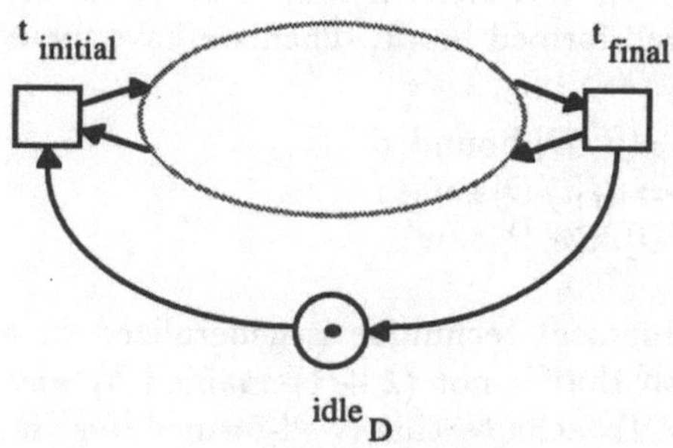

Figure 2.6

For the refinement by an arbitrary block we cannot expect behaviour preservation. Therefore we only use well-formed blocks.

Definition 2.10

A block D is called *well-formed*, if

- D is live
- $\forall M \in [M_D\rangle : (M(idle_D) > 0 \Rightarrow M = M_D)$
- $\forall t \in T_D : (M_D[t\rangle \Rightarrow t = t_{initial})$.

Well-formed blocks can simulate a transition: The only transition that can activate D is the initial transition. Thus a simulation can only start, if the

[†] Throughout the chapters 2.3.2 and 2.3.3 all nets are supposed to be unlabelled and finite.

environment — which is represented by the idle place — allows this. If D is activated, always a transition sequence can fire that enables the final transition, since D is live. Hence every simulation can be finished. The firing of the final transition marks the idle place, i.e. produces the output for the environment. Now we have reached the initial marking again, i.e. a well-formed block does not have a 'memory', just like a procedure call does not depend on earlier calls. During the simulation of a transition t by D the idle place is empty. Therefore it is not possible to start a second simulation of t before the first is finished. Only when the simulation has come to an end, a new simulation can be initiated by the initial transition. The environment of t in a host net N has to ensure this property, therefore the transition that is refined must not be 2-enabled by any reachable marking of N. A transition $t \in T_N$ is ***k*-enabled** ($k \in \mathbb{N}_0$) by a marking M, if

$$\forall s \in S_N : k \cdot W_N(s,t) \leq M(s).$$

Then the following results concerning behaviour preservation are obtained.

Theorem 2.2 ([Val 79])
Let N be a net, $t \in T_N$ a transition that is not 2-enabled by any reachable marking and D a well-formed block. Then we have for $N[t/D]$ as defined in 2.9:

(i) N bounded $\Leftrightarrow$ $N[t/D]$ bounded
(ii) N and D safe $\Leftrightarrow$ $N[t/D]$ safe
(iii) N live $\Leftrightarrow$ $N[t/D]$ live.

In [SM 83] this refinement technique is generalized such that it is possible to refine a transition that is not $(k+1)$-enabled by any reachable marking even if $k+1 > 2$. To achieve this, well-formedness must be strenghtened (to k-well-behavedness). 1-well-behavedness is a slightly relaxed version of well-formedness, but is sufficient for behaviour preservation, if the transition is not 2-enabled by any reachable marking.

In [SM 83] it is suggested to use the same refinement technique also for the refinement of places. The place s that should be refined is split as shown in Figure 2.7. Then the transition t is refined. The results gained for the refinement of transitions can be transfered to this case.

2.3.3 An alternative technique for the refinement of transitions

Let us discuss the basic refinement technique defined in 2.3.1. Let N be a net and t a transition in N that is refined by some refinement net D. In the refined net every place s in the preset of t is connected with every initial transition of D with the weight of the arc between s and t. Symmetrically final transitions and places in the postset of t are connected. Thus the firing of an initial transition absorbs the same number of tokens from each place in

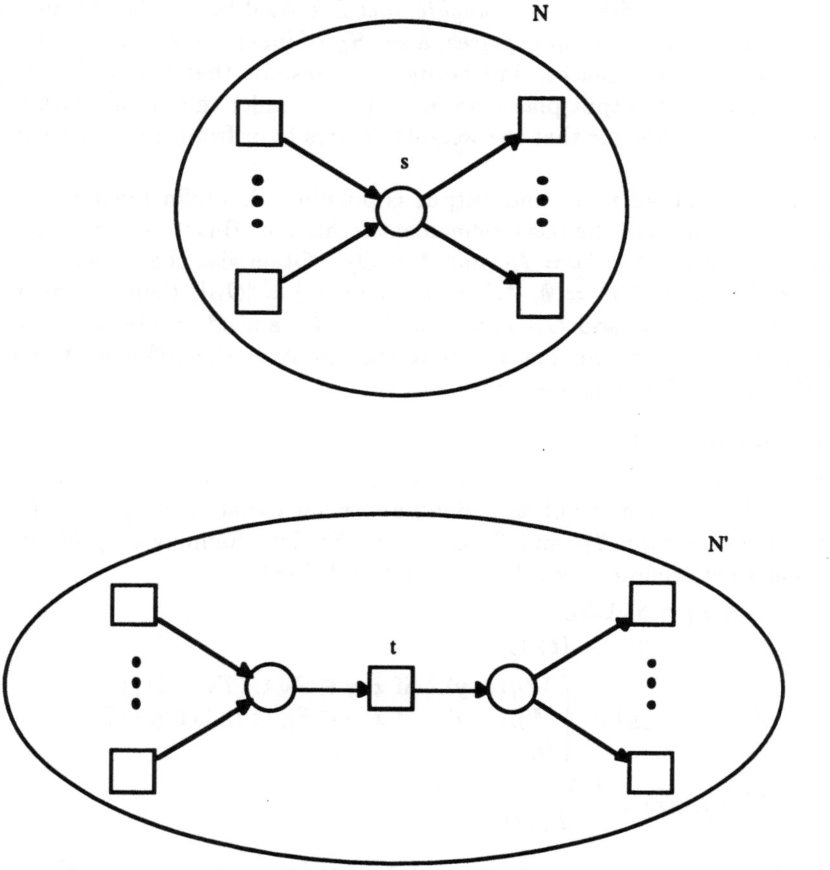

Figure 2.7

the preset of t as the firing of t. These tokens cannot be distributed between two initial transitions or between an initial transition and a transition not belonging to D. Symmetrically the firing of a final transition produces the same number of tokens on each place in the postset of t as the firing of t. Especially, it is not possible that some output is produced and used by the environment before the rest is produced. We say that ***distributed input*** and ***distributed output*** is not allowed.

In [Vog 87] a refinement technique is proposed, where distributed input and distributed output is possible (Figure 2.8). A transition t is removed and a daughter-net D is inserted by identifying all places in the preset of t with some special places of D, called the input places, and all places in the postset of t with some other places of D, called the output places. Thus several initial transitions (the postset of the input places) and several final transitions (the preset of the output places) are allowed.

Hence the set of places of a *daughter-net* D should be the disjoint union of a non-empty set I_D of input places, a set S_D of inner places, and a non-empty set O_D of output places. Furthermore we assume that ${}^\bullet I_D = \emptyset = O_D^\bullet$ and that input and output places are not marked under the initial marking. For simplicity we assume that the weights on arcs to or from the transition t that is refined are 1.
Since distributed input and output is possible, this refinement technique is more general than the basic refinement technique. But it can only be applied to a transition t, if ${}^\bullet t = I_D$ and $t^\bullet = O_D$. Otherwise the refinement is not possible. (If ${}^\bullet t \cap t^\bullet = \emptyset$, $|{}^\bullet t| = |I_D|$ and $|t^\bullet| = |O_D|$, then we can rename the places in I_D and O_D such that ${}^\bullet t = I_D$ and $t^\bullet = O_D$ and apply the refinement technique; observe that the result of the refinement might be different for different renamings.)

Definition 2.11
Let N be a net, $t \in T_N$ such that $\forall s \in S_N : W_N(s,t) \leq 1, W_N(t,s) \leq 1$ and let D be a daughter-net as defined above such that ${}^\bullet t = I_D, t^\bullet = O_D$. We assume that $S_N \cup T_N$ and $S_D \cup T_D$ are disjoint. Refining t by D yields the refined net, denoted by $N[t/D]$, which is defined by:

$$S_{N[t/D]} = S_N \dot{\cup} S_D$$

$$T_{N[t/D]} = (T_N - \{t\}) \dot{\cup} T_D$$

$$W_{N[t/D]}(x,y) = \begin{cases} W_N(x,y), & \text{if } x,y \in S_N \cup (T_N - \{t\}) \\ W_D(x,y), & \text{if } x,y \in S_D \cup I_D \cup O_D \cup T_D \\ 0, & \text{otherwise} \end{cases}$$

$$M_{N[t/D]}(s) = \begin{cases} M_N(s), & \text{if } s \in S_N \\ M_D(s), & \text{if } s \in S_D \end{cases}$$

In the previous chapter, blocks were designed such that they simulate a transition. As a result we got behaviour preservation. Now we look at things the other way round. We give the desired property, namely N and $N[t/D]$ should have equivalent behaviour, and try to determine, which daughter-nets ensure this. These daughter-nets are called modules.
The main problem is to define an appropriate notion of behaviour. The following definition requires a close correspondence of the reachable markings in N and $N[t/D]$. This seems reasonable, since properties like liveness or boundedness depend on the reachable markings.

Definition 2.12 ([Vog 87])
Let D be a daughter-net. D is called a *module* if, for all nets N and all $t \in T_N$ with $\forall s \in S_N : W_N(s,t) \leq 1, W_N(t,s) \leq 1$ and ${}^\bullet t = I_D, t^\bullet = O_D$, the nets N and $N[t/D]$ have the same behaviour, i.e.

$$\forall M \in [M_{N[t/D]}\rangle \ \exists M' \in [M_N\rangle,\ w \in T_D^* :$$

$$M[w\rangle M'' \wedge M''|_{S_N} = M' \wedge$$

$$(M'[t\rangle \Rightarrow \exists w'' \in T_D^* : M''[w''\rangle \ \wedge \ M''(w'')|_{S_N} = M'(t))$$

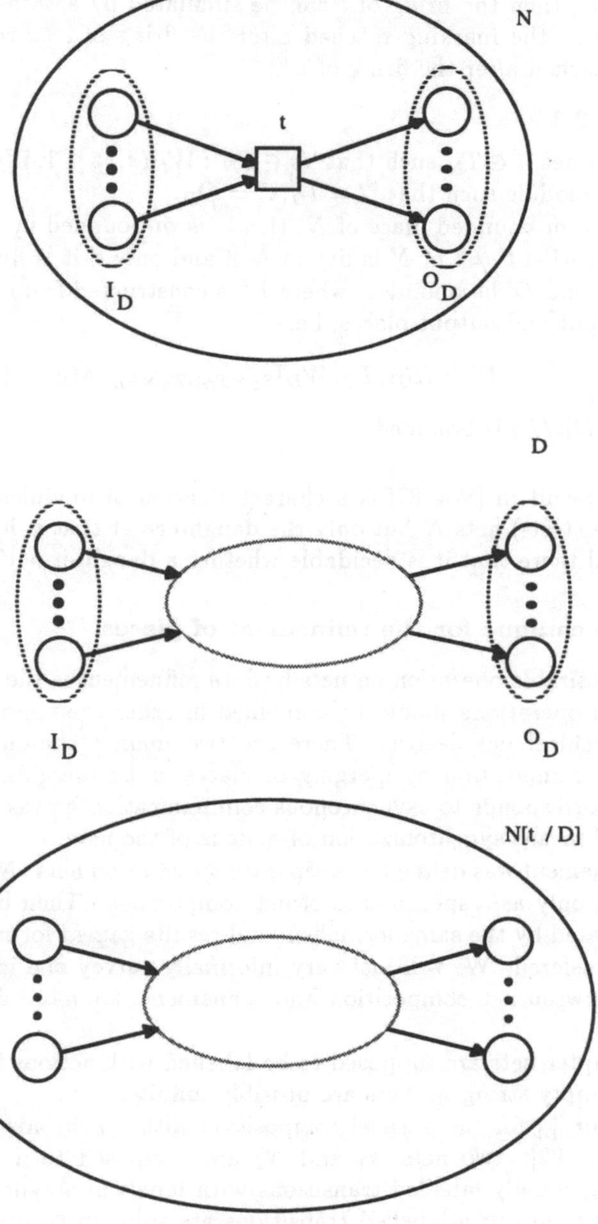

Figure 2.8

This behaviour notion means the following. Each reachable marking M in the refined net corresponds to a reachable marking M' in N after normalization

of M by a firing sequence of D, and if t is enabled under the corresponding marking M', then the firing of t can be simulated by a transition sequence w'' in D, i.e. the marking reached after the firing of t corresponds to the marking reached after the firing of w''.

Theorem 2.3

Let N be a net, $t \in T_N$ such that $\forall s \in S_N : W_N(s,t) \leq 1, W_N(t,s) \leq 1$ and let D be a module such that ${}^\bullet t = I_D, t^\bullet = O_D$.

- If s is a m-bounded place of N, then s is m-bounded in $N[t/D]$.
- A transition $t' \neq t$ of N is live in N if and only if it is live in $N[t/D]$.
- Let N and D' be bounded, where D' is constructed from D by removing the input and output places, i.e.

$$D' = (S_D, T_D, W_D|_{S_D \times T_D \cup T_D \times S_D}, M_D|_{S_D}).$$

  Then $N[t/D]$ is bounded.

The main result in [Vog 87] is a characterization of modules that does not involve the (test-) nets N but only the daughter-net that is in question, it is also proved there that it is decidable whether a daughter-net is a module.

2.3.4 A technique for the refinement of places

Another desirable operation on nets besides refinement is the composition of nets. Both operations should be combined in order to support the modular and hierarchical net design. There are two main techniques used in the literature: composition by merging of places or by merging of transitions. The first corresponds to asynchronous communication between the two nets, the second means synchronization of actions of the nets.

So far refinement was defined as a separate operator on nets. Now we consider refinement only as a special case of net composition. Then both operations can be treated by the same formalism and results gained for net composition can be transfered. We will just very informally survey one idea on the connection between net composition and refinement, for more details see [Vog 92].

In this chapter nets are supposed to be labelled with actions from a set Σ or with the empty string λ. Nets are possibly infinite.

An operator $\|_A$ for the parallel composition with synchronization is defined e.g. in [Vog 92]: two nets N_1 and N_2 are composed to a net $N_1 \|_A N_2$ by merging equally labelled transitions with labels in a synchronization set $A \subseteq \Sigma$. For this all a-labelled transitions are split up (if $a \in A$) such that each a-labelled transition from one net can be combined with every a-labelled transition from the other net. This operator corresponds to the parallel composition with synchronization of TCSP; it means that the nets are put in parallel such that actions from A are synchronized.

For the modular construction of nets we consider two nets N_1 and N_2 as equivalent, if exchanging them in any context preserves the behaviour. A simple notion of behaviour is the existence or non-existence of deadlocks: a net N can ***deadlock*** if there is $M \in [M_N\rangle$ such that $\forall a \in \Sigma : \neg M[a\rangle\rangle$ (i.e. no action is image enabled under M). But there may be λ-transitions enabled under M. Now nets N_1, N_2 are ***deadlock-equivalent*** if for all nets N and $A \subseteq \Sigma : N \parallel_A N_1$ cannot deadlock if and only if $N \parallel_A N_2$ cannot deadlock. Deadlock-equivalence is an external equivalence (see [Bau 88]), since the nets must be tested in all environments N. But deadlock-equivalence can be characterized by an internal equivalence that involves only the nets N_1, N_2 and no other nets.

Theorem 2.4 ([Vog 89], [Vog 92])
Let N_1, N_2 be nets such that $\Sigma - (l_{N_1}(T_{N_1}) \cup l_{N_2}(T_{N_2})) \neq \emptyset$. Then N_1 is deadlock-equivalent to N_2 if and only if N_1 is $\mathcal{F}$-equivalent to N_2.

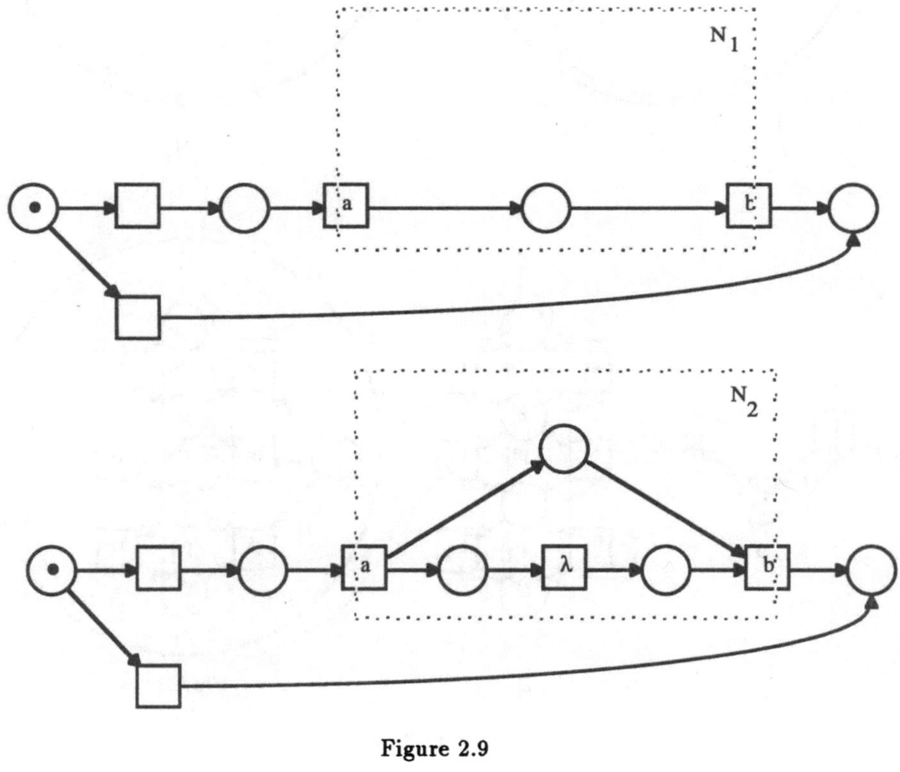

Figure 2.9

Now we transfer the definitions and results from net composition to refinement. In order to describe the refinement of places via parallel composition we fix an atomic net N_p, i.e. a net with only one place p, with transition set

T_p and the identity as labelling. We call a net N an ***N_p-host***, if it contains N_p as an induced subnet - except for the transition lablelling, which might be different in N. Now we can consider an N_p-host N as being constructed from some net N_1 and N_p by parallel composition $\|_{T_p}$ and a relabelling of the transitions adjacent to p; combining the net N_1 and some net D in the same way, we obtain the refined net $N[p/D]$.

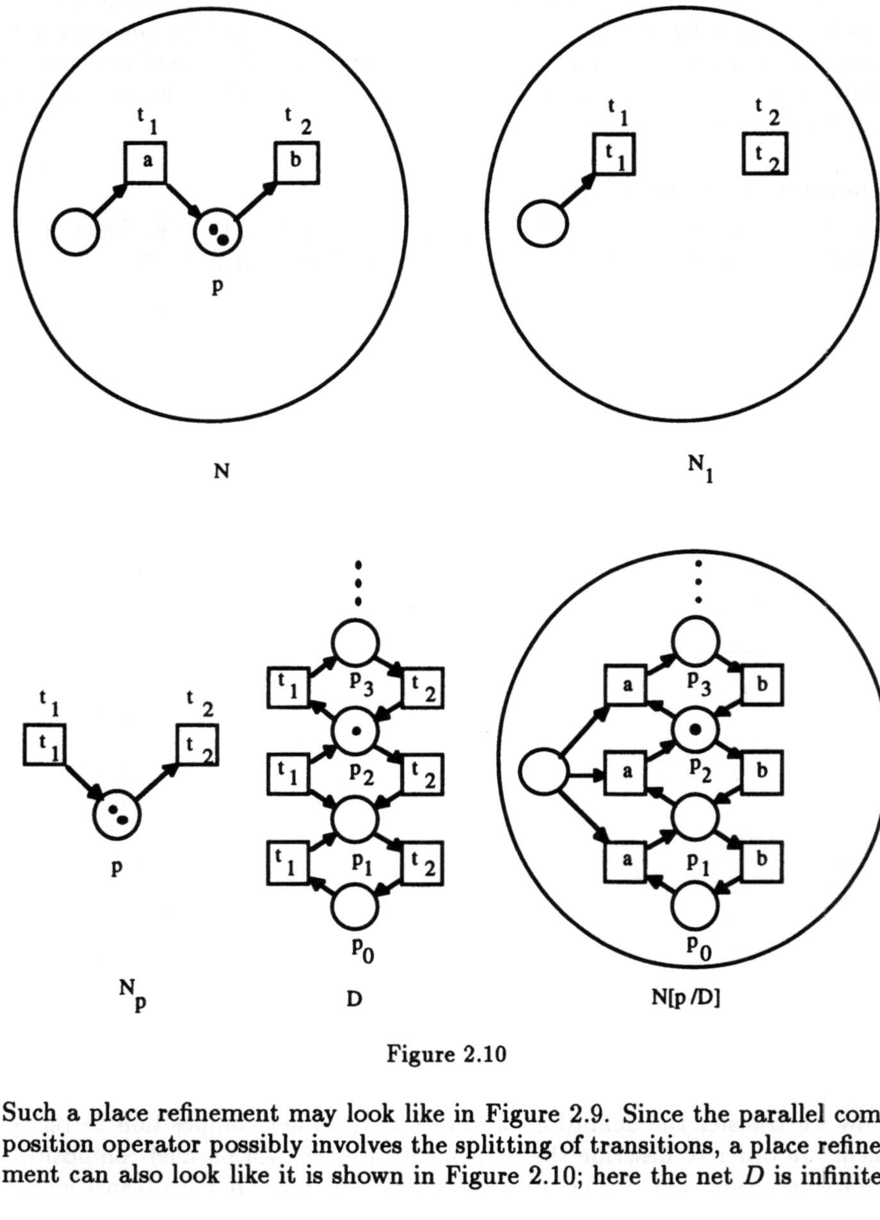

Figure 2.10

Such a place refinement may look like in Figure 2.9. Since the parallel composition operator possibly involves the splitting of transitions, a place refinement can also look like it is shown in Figure 2.10; here the net D is infinite, and the place p is replaced by an infinite sequence $p_0, p_1, p_2, \ldots$ such that one token on p_i corresponds to i tokens on p.

A net D is called *N_p-deadlock-module* if for all N_p-hosts N: N cannot deadlock $\Leftrightarrow N[p/D]$ cannot deadlock. In contrast to the definition of deadlock-equivalence we use as test nets only those nets N that contain the place p and a special surrounding of p.

N_p-deadlock-modules are defined in the same style as modules in [Vog 87]. The desired property, behaviour preservation, is demanded for every host net N.

Theorem 2.5 ([Vog 92])

A net D is an N_p-deadlock-module if and only if N_p and D are $\mathcal{F}$-equivalent. Moreover we have in this case that $\mathcal{F}(N) = \mathcal{F}(N[p/D])$ and that N and $N[p/D]$ are even bisimilar.

The last part of this theorem is due to the fact that an N_p-deadlock-module is in some sense deterministic, and studying these deterministic nets is the specific adaption of the general approach of net composition to the special case of place refinement. Thus for place refinement we have only required a very simple sort of behaviour preservation, but automatically gained behaviour preservation in a very strong sense. Since D must be deterministic, it is possible to decide, whether for a given finite D the nets N_p and D are $\mathcal{F}$-equivalent. Hence with Theorem 2.5 it is decidable, whether D is an N_p-deadlock-module. In contrast to this result it is in general undecidable, whether two nets are $\mathcal{F}$-equivalent, even if the nets are finite.

Let us remark that in [Vog 92] for the refinement of transitions the dual approach is taken. For this a composition operator for nets by merging of places is defined. Then we have the same technique that was used in [Vog 87] but now in the framework of modular construction of nets. Therefore failures semantics is used, and the conditions for modules are formulated in terms of this semantics. Next it is studied in which sense modules for transition refinement may be considered as deterministic.

For both approaches also a variation is considered that takes divergence into account, and similar results for both net composition techniques are obtained.

2.4. Congruences with respect to action refinement

2.4.1 Motivation

The topic of this part is action refinement: In this section all transitions are labelled by actions, and when we refine a net we replace every transition with some given label by a copy of the same refinement net. E.g. we want to be able to replace every action *exchange_x_y* by a sequence *move_x_to_h*; *move_y_to_x*; *move_h_to_y*. Furthermore, our interest is semantic equivalence

of nets, and we require that equivalent nets must have at least the same language.
In this situation we cannot expect that action refinement is e.g. language preserving: If we replace *exchange_x_y* by the sequence *move_x_to_h*; *move_y_to_x*; *move_h_to_y*, then the refined net can perform totally different actions than the unrefined net. Hence these two nets cannot possibly have the same language. Thus we have quite a different starting point compared with the last section where the refined and the unrefined net were expected to be equivalent, i.e. to have the same relevant properties or to have the same behaviour in some other sense.

Instead we are interested in the preservation of semantic equivalence: Refining equivalent nets in the same way should give equivalent refined nets. In other words, we look for semantic equivalences that are congruences with respect to refinement.

What makes the issue of action refinement so interesting (especially to the Petri net community) is its connection to partial order semantics. In [Pra 86] and [CDP 87] it is suggested that partial order semantics is useful when considering action refinement, see also [Lam 86]. In [CDP 87] the following example is discussed:

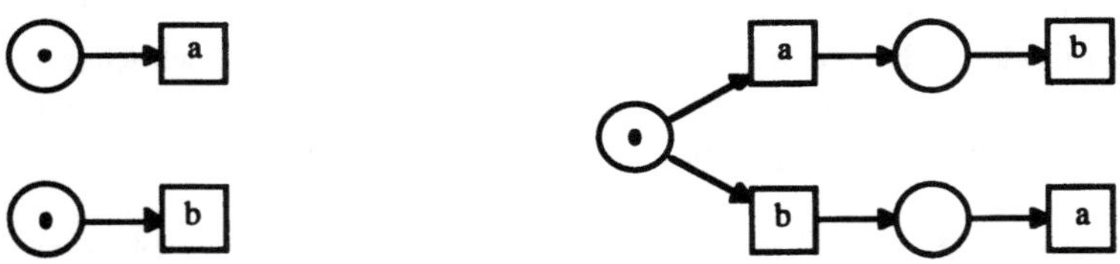

Figure 2.11

The nets of Figure 2.11 are interleaving bisimilar, but if we refine a to a sequence a_1a_2 the first net can perform a_1ba_2 while the second cannot, i.e. the refined nets do not even have the same language.
Hence the two nets should be distinguished, which can be done by using some partial order semantics. It is argued in [CDP 87] that partial order semantics gives a congruence for refinement, indeed. We will present precise results below.

2.4.2 Refinement technique and refinement nets

Before we can look for a congruence w.r.t. refinement we have to decide what refinement technique and what refinement nets we want to use.
In the two papers [BDKP 89], [Vog 90], which we will mainly refer to, distributed input or output are forbidden. This can be explained by the following examples, see [Vog 90].

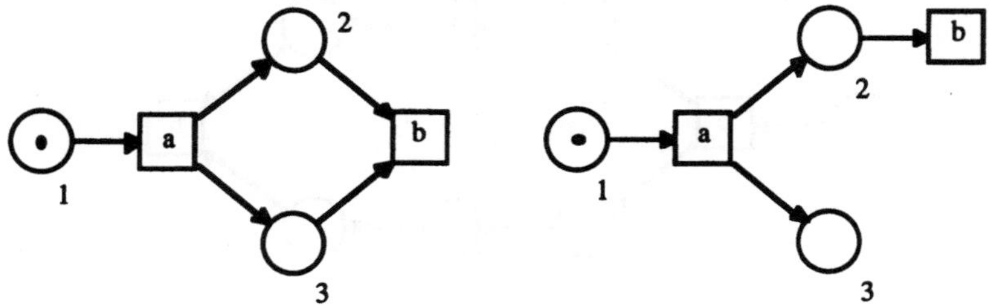

Figure 2.12

Figure 2.12 shows two nets which are sequential and conflict-free. Since they have the same language they also have the same event structures of processes and are (interleaving) bisimilar, too, and they certainly should be semantically equivalent. Figure 2.13 shows a net that in view of the results in Section 2.3.3 one might consider as a suitable refinement net with distributed output.

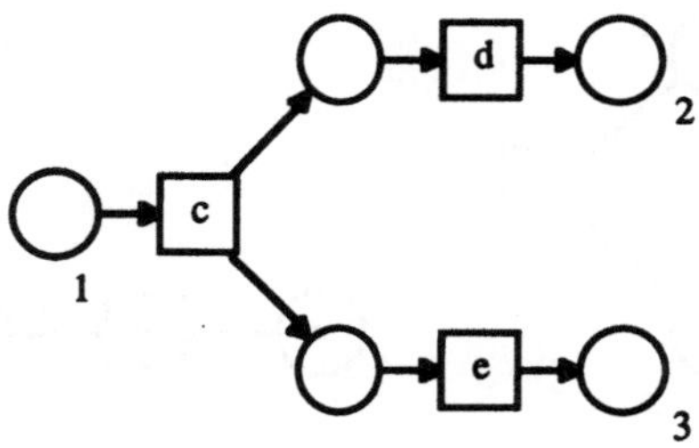

Figure 2.13

Refining the *a*-labelled transitions with this net transforms the nets of Figure 2.12 to the nets in Figure 2.14. The second of these refined nets can perform *cdbe* while the first one cannot, hence there cannot be an acceptable semantic equivalence that is a congruence w.r.t. this sort of refinement.
One might feel that in the context of this section we should use refinements that preserve behaviour w.r.t. the markings, e.g. in the sense of Definition 2.12. Then we can also use the results of the previous section regarding behaviour preservation. This would especially forbid distributed input. But we can also give an argument against distributed input that is valid in our context alone. In fact, it is symmetrical to the above example.
The nets of Figure 2.15 should certainly be semantically equivalent. But if we refine the *b*-labelled transitions with the net shown in Figure 2.16 we get the nets of Figure 2.17; the second can perform *cad*, while the first cannot.
As a consequence of these observations we use the basic refinement technique of Section 2.3.1 in order to define action refinement:

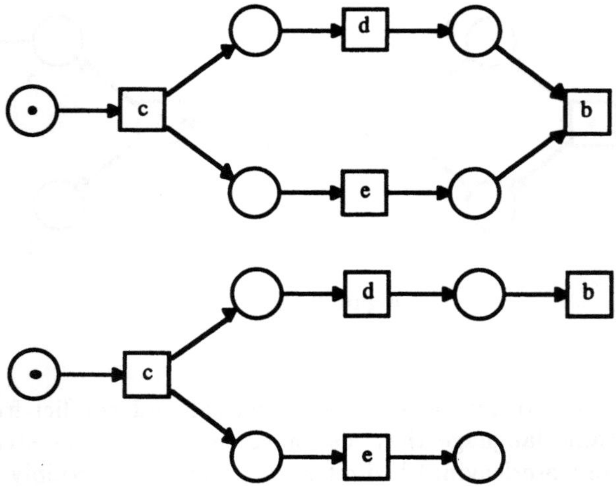

Figure 2.14

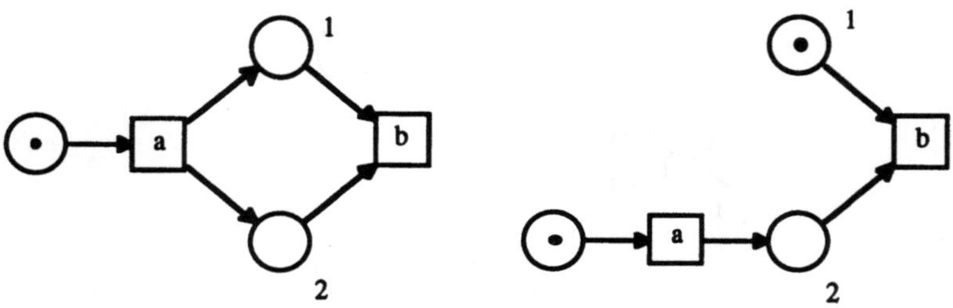

Figure 2.15

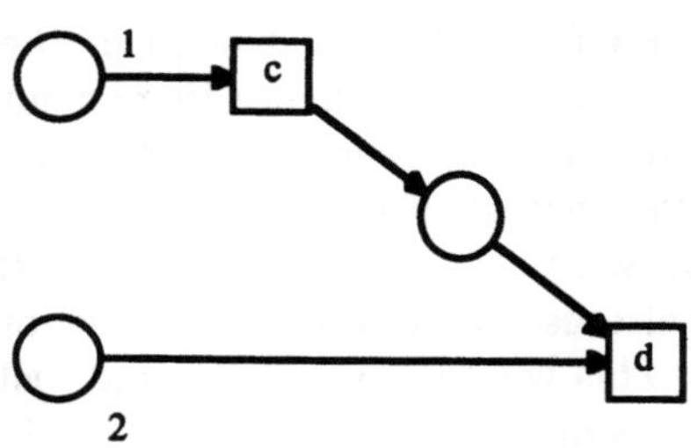

Figure 2.16

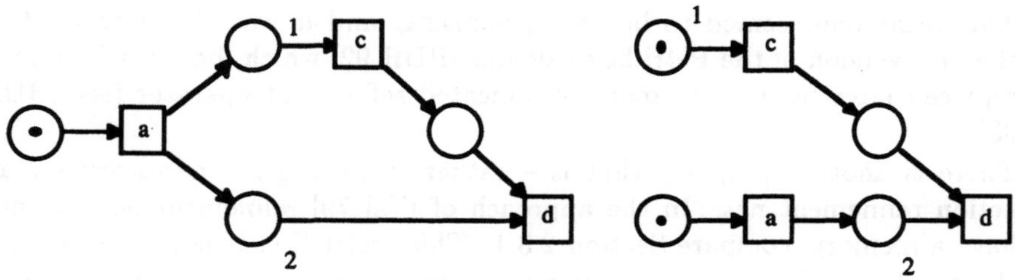

Figure 2.17

Definition 2.13

An *action refinement* r is a pair (a, D) where $a \in \Sigma$ and D is a refinement net as in Section 2.3.1. For a net N the refined net $r(N)$ is obtained by replacing every a-labelled transition by a copy of D as explained in Definition 2.9.

In [GG 90] a different technique for action refinement is presented, which allows initial and final concurrency while avoiding distributed input or output. We will explain this technique by way of the above example. If we refine the second net of Figure 2.12 by (a, D) with D as in Figure 2.13, then the postset of the a-labelled transition is replaced by the cartesian product of this postset and the set $\{2, 3\}$ of output places of D, and analogously for the preset. The transitions of N and D are connected to these new places in the natural way as shown in Figure 2.18.

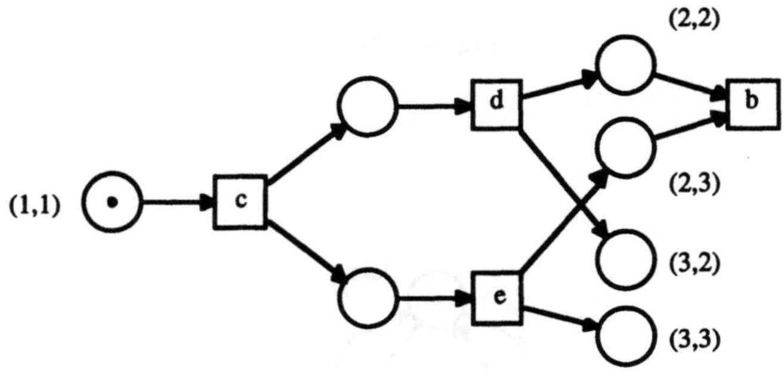

Figure 2.18

We see that this net cannot perform the undesirable sequence *cdbe*; here we do not have a distributed output in the sense that part of the output can be used by the environment while some other part has not been produced, yet. But this technique allows final (and initial) concurrency since the refinement net D finishes its simulation by firing concurrently d and e.

This technique seemed to be very promising, and it was therefore used in the first version of the Petri box calculus [BDH 92] which, however had been replaced later by a much more complicated refinement operator (see [BDE 93]).
There is another property that is – rather surprisingly – necessary for an action refinement net. In the approach of [Val 79] refinement nets do not have a memory, compare Section 2.3.1. This restriction is not necessary for the results of [Val 79], but it turns out to be necessary in the context of action refinement. The nets shown in Figure 2.19 should be semantically equivalent by the same reasoning as above. Figure 2.20 shows a refinement net D describing the alternation of actions a_1 and a_2, i.e. D 'remembers' the parity of the number of simulations it has performed. Refining a by D yields the nets of Figure 2.21, where the first can only perform a_1ba_2 while the second can only perform a_1ba_1. Again we conclude that we cannot allow such a refinement net.

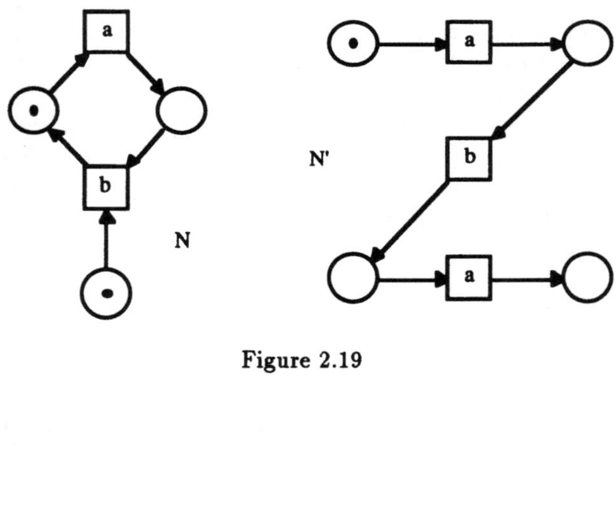

Figure 2.19

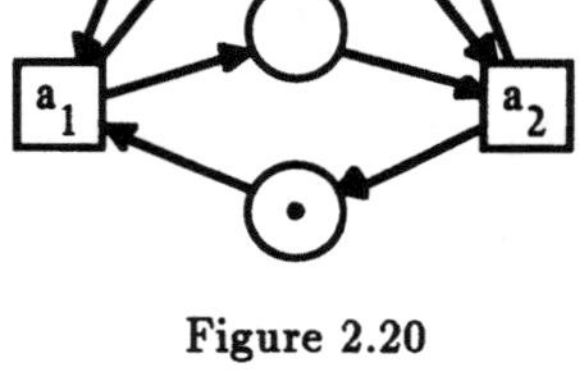

Figure 2.20

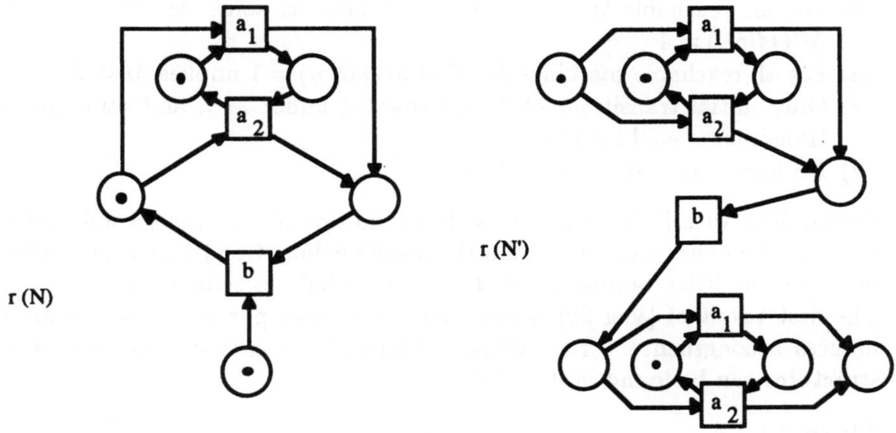

Figure 2.21

Also we cannot allow that a non-initial transition is initially enabled. Consider the refinement net of Figure 2.22. If we refine the nets of Figure 2.19 with this refinement net, then the first can only perform one *d* while the second can perform *d* twice.

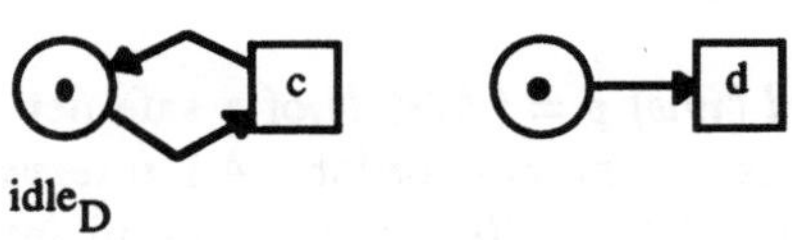

Figure 2.22

Consequently, the refinement nets in this section will have a considerable similarity with the well-formed blocks of [Val 79].

2.4.3 Interval word semantics

In this section we will mainly report the results of [Vog 90], (see also [Vog 92]), and the nets in this subsection may also be infinite. We will present two linear time and a failures semantics for safe nets which induce congruences w.r.t. action refinement. Especially, partial order semantics is shown to be useful and, to a certain degree, even necessary for this purpose.

Definition 2.14

A *V-refinement net* D is a safe refinement net with the following properties:
i) $M_D(idle_D) = 1$

ii) For all reachable markings M of D there is some $M' \in [M\rangle$ with $M'(idle_D) = 1$
iii) For all reachable markings M of D $M(idle_D) = 1$ implies $M = M_D$.
iv) Only initial transitions of D are enabled under M_D, and some initial transition is enabled under M_D.
v) No final transition is λ-labelled.

Comparison with [GG 90] suggests that condition ii) and the second part of iv) might be unnecessary to obtain the results below. Condition v) is a rather technical condition necessary for the result on failures semantics.
The first result of [Vog 90] shows that linear time partial order semantics induces a congruence w.r.t. action refinement. A similar result for event structures can be found in [GG 89].

Theorem 2.6

For safe nets process equivalence is a congruence w.r.t. action refinement with V-refinement nets.

The main aim of [Vog 90] is to determine the coarsest congruence w.r.t. refinement contained in failures equivalence. Crucial for this purpose is the notion interval order.

Definition 2.15

A partial order $(A, <)$ is an *interval order* if there are closed intervals $I(x) \subseteq \mathbb{R}$, $x \in A$, such that $x < y$ if and only if for all $z \in I(x)$, $z' \in I(y)$ we have $z < z'$.
An *interval partial word* (*ipw*) $p = (A, <, l)$ of a safe net N is a partial word of N such that $(A, <)$ is an interval order. An *interval partial word with termination set* (*ipwt*) $p = (A, <, l, ter)$ consists of an *ipw* $(A, <, l)$ and a set $ter \subseteq A$ containing at least all non-maximal elements of A.
The marking reached after an *ipwt* p is

$$M_N + \sum_{x \in ter} W_N(l(x), .) - \sum_{x \in A} W_N(., l(x)).$$

The *image* of p is $(A', <', l', ter')$, where $A' = \{x \in A \mid l_N \circ l(x) \neq \lambda\}$, $ter' = ter \cap A'$, and $<'$ and l' are the restrictions of $<$ and $l_N \circ l$ to A'.

The idea of an *ipwt* is the following: The elements of A are occurrences of transitions of N. In an interval observation of a system run we observe in which order such occurrences begin and end. Thus each occurrence corresponds to a time interval. If two such intervals overlap, then we can directly observe that the corresponding transition occurrences are concurrent. If one interval is as a whole earlier than the other, then the corresponding transition occurrence preceeds the other. This way the partial order of an *ipwt* can be

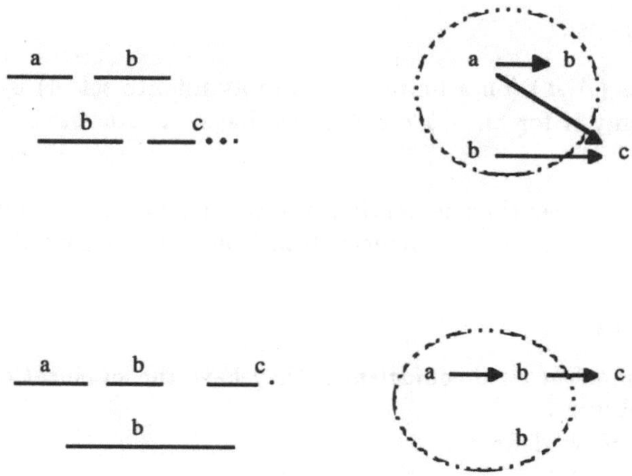

Figure 2.23

observed. Figure 2.23 shows two examples of intervals and the corresponding interval orders (more precisely the images of some *ipwt*'s are shown).

We can add a termination set to an interval partial word to indicate that the corresponding transition occurrences have already terminated. Such a set is indicated in Figure 2.23 by a dotted line; thus the two observations have stopped in the middle of the time interval belonging to c. Naturally, if $x \in A$ is not maximal, then something has begun after the occurrence of $l(x)$, hence x must belong to the termination set.

In an interval observation we observe the beginning and ending of transition occurrences. One might wonder whether we can describe such an observation in the following way without using partial order semantics: We split every action into a sequence of two and take a usual sequential observation of the resulting system, i.e. an element of its language. (This approach is taken in [Hen 87] and [AH 88]). But both the above interval observations correspond to the sequence $a_1 b_1 a_2 b_1 b_2 c_1 b_2$ in the splitted system. Hence with an interval observation we observe not only that action b occurs concurrently to itself, we also observe which of these occurrences ends first.

Interval orders can be characterized by the forbidden substructure depicted in Figure 2.24 as shown by the following result of [Fis 70]:

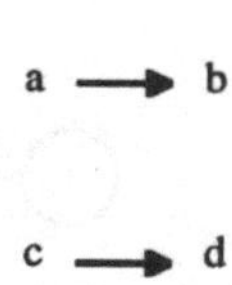

Figure 2.24

Theorem 2.7

A partial order $(A, <)$ (on a finite or countably infinite set A) is an interval order if and only if for all $a, b, c, d \in A$ we have: $a < b$ and $c < d$ implies $a \leq d$ or $c \leq b$.

We can define a linear time semantical equivalence based on interval partial words and a sort of failures semantics based on interval partial words with termination set.

Definition 2.16

Safe nets are ***interval word equivalent*** if they have the same set of images of interval partial words.
For a safe net N we define

$$\begin{aligned} \mathit{ref}\mathcal{F}(N) = \{(p, X) \mid\ & p \text{ is the image of an } \mathit{ipwt}\ q \text{ of } N, \\ & \text{and for the marking } M \text{ reached} \\ & \text{after } q \text{ and every } a \in X : \neg M[a\rangle\}. \end{aligned}$$

Two safe nets N_1, N_2 are ***ref*** $\mathcal{F}$***-equivalent*** if $\mathit{ref}\,\mathcal{F}(N_1) = \mathit{ref}\,\mathcal{F}(N_2)$.

In [Vog 90] the following results are shown:

Theorem 2.8

For safe nets interval word equivalence is a congruence w.r.t. action refinement with V-refinement nets.

Theorem 2.9

For safe nets *ref* $\mathcal{F}$-equivalence is the coarsest congruence w.r.t. action refinement with V-refinement nets contained in failures equivalence.

In fact, the proofs and results of [Vog 90] also show that interval word equivalence is the coarsest congruence w.r.t. action refinement contained in language equivalence. Hence here we have results that show that some partial order semantics is not only useful for action refinement, but in fact the power of partial order semantics is necessary, too.

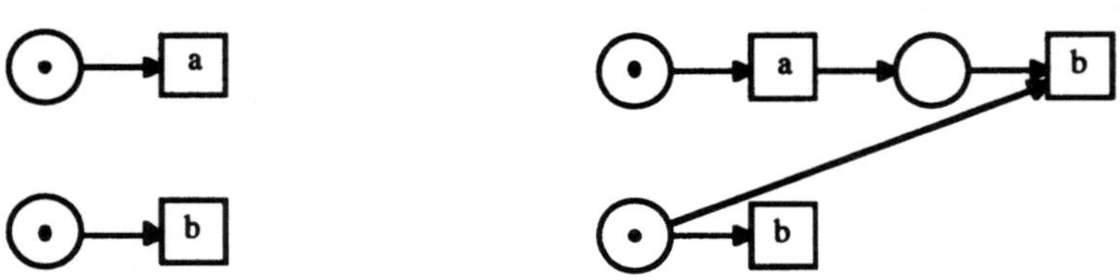

Figure 2.25

Let us consider some examples to see that interval word semantics is less discriminating than process semantics. Figure 2.25 shows two nets which are interval word equivalent, but not process equivalent: Only the second net can perform *a* and then some *b* that is necessarily after *a*, but this cannot be observed by an interval observation.

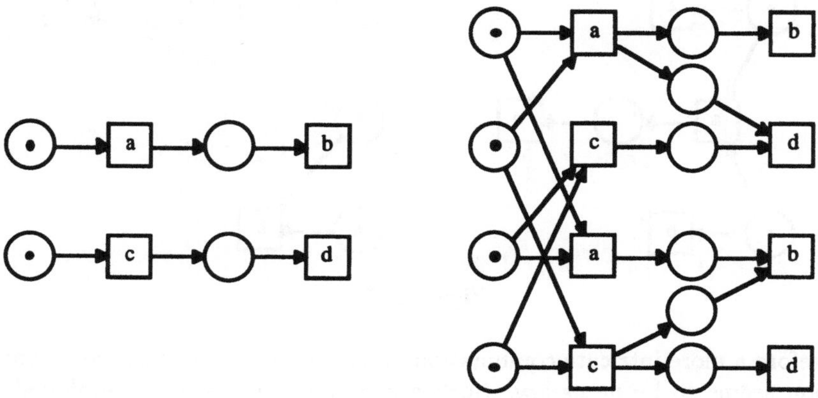

Figure 2.26

Also Figure 2.26 shows two nets which are interval word, but not process equivalent. The image partial word of Figure 2.24 is the event structure of a process of the first, but not of the second net. Such an image partial word cannot be observed by an interval observation, compare Figure 2.23 and Figure 2.24.

An equivalence which is based on a duration of transition occurrences can already be found in [GV 87]. It has been shown in [Gla 90] for event structures without silent moves that ST-bisimilarity is a congruence for action refinement. In the same paper also ST-traces are introduced, and it is shown that we can base a congruence on these, too. It should be remarked that ST-traces are a (more or less) sequential representation of *ipwt*'s. Interval orders are also studied in the context of action refinement for a process algebra in [NEL 89].

2.4.4 History preserving bisimulation

We have seen in the last subsection that partial order semantics can be very useful when looking for a congruence w.r.t. action refinement. Thus when looking for a bisimulation type congruence it is natural to consider a combination of partial order semantics and bisimulation. The most straightforward combination is pomset bisimulation [BC 87], where one changes Definition 2.8 by considering $M[ev(\pi)\rangle\rangle M'$, π a process, instead of $M[a\rangle\rangle M'$. Somewhat surprisingly, pomset bisimilarity is not a congruence, see [Dev 88], [GG 89]: The two nets of Figure 2.27 (see [Dev 88]) are pomset bisimilar, but if we refine a to a_1a_2 then only the first net can perform a_1 in such a way that no b is possible afterwards.

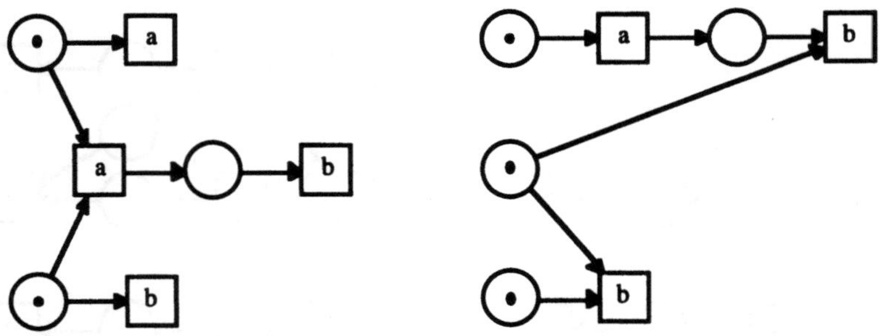

Figure 2.27

Therefore a more intricate combination of partial order semantics and bisimulation seems to be necessary. Such a combination has been developed in [RT 88] under the name of BS-bisimulation, and has also been studied in [DDM 89] for a semantics for CCS; in [GG 89] it is given the name history preserving bisimulation – a name we will stick to – and shown to induce a congruence w.r.t. action refinement for event structures without silent moves. If we translate this notion to Petri nets we get the fully concurrent bisimulation of [BDKP 89], and we will report now the results of this paper. (All nets in [BDKP 89] are finite and T-restricted.)
In pomset bisimulation, system states are described by markings as usual. In order to transform one state to another, a partial order of actions is performed, which is generated from a process. For the bisimulation it is required that such a state transformation in one system is matched by a transformation in the other system involving an isomorphic partial order of actions.
In history preserving bisimulation, a system state is described by the process that leads to this state. When a bisimulation relates two states the processes that lead to these states must have isomorphic event structures. Naturally, a state transformation is an extension of one process π to another π'. For the bisimulation we do not only require that the part we have added to π can be matched in the second system: We require that the isomorphism from the event structure of π can be extended to the event structure of π'; without this requirement the results would fail as shown in [BDKP 89].

Definition 2.17

Nets N_1 and N_2 are ***history preserving bisimilar***, denoted $N_1 \approx_h N_2$, if there exists a ***history preserving bisimulation***, i.e. a set

$\mathcal{B} \subseteq \{(\pi_1, \pi_2, f) \mid \pi_1$ is a process of N_1, π_2 a process of N_2, $f : ev(\pi_1) \rightarrow ev(\pi_2)$ is an isomorphism$\}$

such that:

i) $(\pi_0(N_1), \pi_0(N_2), \emptyset) \in \mathcal{B}$ (i.e. the initial processes are related.)

ii) For all $(\pi_1, \pi_2, f) \in \mathcal{B}$:

a) If the process π_1' of N_1 is an extension of π_1, then there exists an extension π_2' of π_2 and an extension f' of f such that $(\pi_1', \pi_2', f') \in \mathcal{B}$

b) vice versa

As observed in [BDKP 89] it is — in this definition — sufficient to consider the case that π_1' has one event more than π_1. In [BDKP 89] two sorts of refinement nets are studied:

Definition 2.18

A refinement net D is a *simple refinement net* if

i) $M_D(idle_D) = 1$ and for all $s \in S_D - \{idle_D\} : M_D(s) = 0$

ii) For all reachable markings M of D $M(idle_D) > 0$ implies $M = M_D$.

iii) No transition of T_D is dead.

If we compare this definition with Definition 2.14 we note that i) is a rather severe restriction. It seems that this severeness is not necessary for the results below, but motivated by the proof technique. Condition iii) looks like a counterpart to 2.14 ii), thus it is also probably not necessary.

As we will see below, an even simpler sort of refinement net is needed, which is based on state machine nets. Such a refinement net introduces no concurrency.

Definition 2.19

A simple refinement net D is an *SM-refinement net*, if for all $t \in T_D$ we have $|{}^\bullet t| = 1 = |t^\bullet|$ and $W_D({}^\bullet t, t) = 1 = W_D(t, t^\bullet)$.

The first congruence result of [BDKP 89] applies to nets where the refined transitions are not self-concurrent and where there are no internal (λ-labelled) transitions immediately after them.

Theorem 2.10

Let $r = (a, D)$ be an action refinement with a simple refinement net D. Let N_1, N_2 be two nets such that for every a-labelled transition t of N_1 or N_2 we have

- t is not self-concurrent
- $t' \in t^{\bullet\bullet}$ implies that t' is not λ-labelled.

Then $N_1 \approx_h N_2$ implies $r(N_1) \approx_h r(N_2)$.

Let us exhibit two examples that show the necessity of the two restrictions on the nets N_1 and N_2. The two nets of Figure 2.28 are obviously history preserving bisimilar, but the second net has a self-concurrent a-transition.

Figure 2.28

If we refine the a-actions by the net of Figure 2.29 the results are not even process equivalent: Figure 2.29 also shows the event structure of a process that only the second refined net can exhibit.

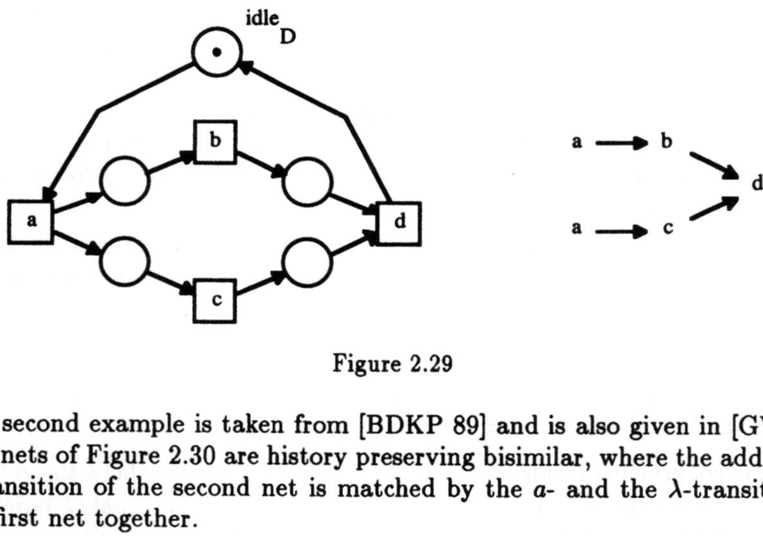

Figure 2.29

The second example is taken from [BDKP 89] and is also given in [GW 89]: The nets of Figure 2.30 are history preserving bisimilar, where the additional a-transition of the second net is matched by the a- and the λ-transition of the first net together.
If we refine all a-actions by the sequence cd, then the second net can perform c such that no b is possible, while the first cannot.
As a consequence of the first example one might restrict attention to safe nets, as it is done in [Vog 90] and also advocated in [GG 90]. One can also restrict the refinement nets: The following is again a result of [BDKP 89].

Theorem 2.11

Let $r = (a, D)$ be an action refinement with an SM-refinement net D. Let N_1, N_2 be two nets such that for every a-labelled transition t of N_1 or N_2 we have

– $t' \in t^{\bullet\bullet}$ implies that t' is not λ-labelled.

Then $N_1 \approx_h N_2$ implies $r(N_1) \approx_h r(N_2)$.

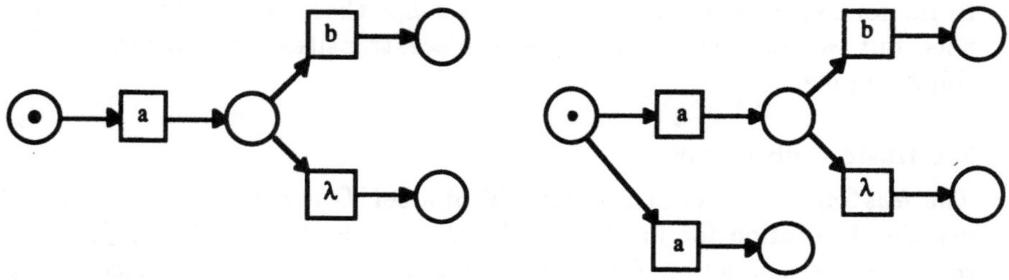

Figure 2.30

The second example above indicates that history preserving bisimulation might not be the right notion for the refinement of nets with silent moves, see below. One might also wonder whether history preserving bisimulation is necessary to obtain a congruence w.r.t. action refinement which respects bisimilarity. For more results see [Vog 92] and [JM 92].

3. Data flow semantics for Petri nets

Already at the end of the sixties J. Dennis developed his ideas on data flow schemata and architectures which were influenced by Petri (see [Den 70]). And already 1973 G. Kahn started the theory of data flow networks (see [Kah 73]) which led to an interesting theory and a methodology for developing parallel programs and distributed systems (see e.g. [Bro 90]). In spite of the close relationship between Petri nets and data flow graphs only very recently people have started to relate again these two approaches in a more formal way (in particular [BS 92] gives a Petri net semantics using the stream concept and the technique of scheduling from data flow theory). In a recent dissertation at TU München ([Gol 93]) a study of coloured Petri nets from the point of view of data flow theory has been made. In this approach a Petri net is regarded as a module that transforms input data into output data. For that the set of places of a net is divided into input places (places with empty preset), output places (places with empty postset) and internal places. The internal places and the transitions are hidden from the outside and thus from observers. The aim is to examine the functional dependence of the markings produced on the output places — after possibly infinite runs of the net — from the marking on the input places.

One problem with semantics of dataflow nets is the appropriate description of non-determinism, since if we allow non-deterministic functions for the nodes of dataflow nets, the merge-anomaly occurs ([Kel 78, BA 81]). It can be avoided by looking at a non-deterministic function as a set of deterministic functions and at the net as a set of deterministic dataflow nets ([Bro 88, BS 92]). An other possibility is to consider traces in the sense of [Jon 89].

In the following we will describe how to adapt these two approaches to Petri nets, but we will restrict ourselves to place/transition-nets in order to keep things simple.

3.1 Basic definitions

The ***least upper bound*** of a subset M of a set D is denoted by *lub* M. We use the λ-notation for functions: $f : S \to D$ is denoted by $\lambda s.f(s)$. The restriction of a function $f : S \to D$ to a subset $S' \subseteq S$ is denoted by $f|_{S'}$. *fix* f is the ***least fixed point*** of the function $f : D \to D$.

We extend the natural numbers by a top element ω: $\mathbb{N}_0^\omega = \mathbb{N}_0 \cup \{\omega\}$. In the usual way the order $\leq$ on $\mathbb{N}_0$ and the operators $\cdot, +, -$ on $\mathbb{N}_0$ are extended.

We allow in multisets infinitely many copies of an element. This means that in this chapter a ***multiset over a set*** D is a a function $m : D \to \mathbb{N}_0^\omega$. The set of multisets over a set D is denoted by $\mathcal{M}(D)$. We define $\leq, +, -$ on multisets as componentwise extensions from the order, operators resp. on $\mathbb{N}_0^\omega$. $(\mathcal{M}(D), \leq)$ forms a complete partial order. A multiset m is called ***finite***, if $\forall d \in D : m(d) \in \mathbb{N}_0$. We write $\mathcal{M}_f(D)$ for the set of all finite multisets over D. For $d \in D$ the multiset containing only one copy of d is denoted by $\mathbf{d}$.

By $STREAM(D)$ we denote the set of finite or infinite sequences (***streams***) over the set D, i.e. $STREAM(D) = D^* \cup D^\omega$. ϵ is the empty sequence. By $l(v) \in \mathbb{N}_0^\omega$ we denote the length of the stream v. For $i \in \mathbb{N}$ with $i \leq l(v)$ the i-th element of v is denoted by v_i. The concatenation of two streams u and v is denoted by uv. We define a partial order $\sqsubseteq$ (***prefix***) on $STREAM(D)$ by $u \sqsubseteq w$, if it exists a stream v such that $uv = w$. For $d \in D$ the stream of length 1 consisting only of the element d is denoted by $\mathbf{d}$. The projection of a stream $v \in STREAM(D)$ to a subset D' of D is denoted by $v\restriction_{D'}$.

The function *multiset* : $STREAM(D) \to \mathcal{M}(D)$ assigns to each stream v the multiset $\lambda d.$(number of occurrences of d in v).

In this chapter we will deal with ***unlabelled place/transition-nets*** $N = (S_N, T_N, W_N, M_N)$. We assume that S_N, T_N are finite, that the preset of every transition is non-empty and that no place is isolated, i.e. $\forall s \in S_N : {}^\bullet s \cup s^\bullet \neq \emptyset$.

Let S' be a subset of S. A finite multiset over S' is called ***marking of*** S' (as in Chapter 2.1), an element of $\mathcal{M}(S') = S' \to \mathbb{N}_0^\omega$ an ***environment of*** S'. Note that ***steps*** as defined in Chapter 2.1 are finite multisets over T_N, i.e. in $\mathcal{M}_f(T_N)$, and that μ^-, μ^+ for a step μ are environments of S_N, i.e. in $\mathcal{M}(S_N)$. A step μ is ***activated*** under a marking M, denoted by $M[\mu\rangle$, if $\mu^- \leq M$. Firing of a step μ yields the ***follower marking*** $M' = M - \mu^- + \mu^+$, denoted by $M[\mu\rangle M'$. A sequence $v \in STREAM(\mathcal{M}_f(T_N))$ is called ***step sequence***, if

$$\forall j \in \mathbb{N}, j \leq l(v)\ \exists \text{ marking } M_j : M_{j-1}[v_j\rangle M_j$$

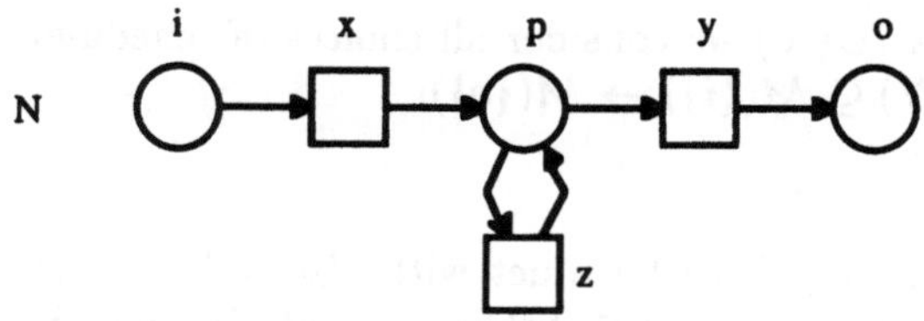

Figure 3.1

where M_0 is the initial marking. A step sequence $w \in STREAM(T_N)$ is called ***firing sequence.***

We devide the set of places of a net N into ***input places*** $IN_N = \{s \in S_N \mid {}^\bullet s = \emptyset\}$, ***output places*** $OUT_N = \{s \in S_N \mid s^\bullet = \emptyset\}$ and ***internal places*** $S_N - (IN_N \cup OUT_N)$. Since there are no isolated places, IN_N and OUT_N are disjoint.

3.2 The semantics IO

In this approach we make nets deterministic by giving schedulers for sets of conflicting transitions. Since schedulers are assigned statically, we introduce the notion of structural conflicts. Two transitions are said to be in ***structural conflict***, if their presets are not disjoint. The equivalence classes of the transitive closure of this conflict relation are called ***conflict regions***. Conflict regions are considered e.g. in [Tau 88] for the implementation of Petri nets. The set of conflict regions of a net N is denoted by U_N.

We give a ***scheduler*** $e \in STREAM(u)$ for each conflict region u. A scheduler fixes the order of transition firings in the conflict region. This solves the conflicts in the region. Transitions in different conflict regions are allowed to fire concurrently.

Let us consider the net N (Figure 3.1). We give schedulers e_1, e_2 for the conflict regions $\{x\}$, $\{y, z\}$ resp. For the computation of the output on o for an input (environment) a of i we introduce a function $F_N(a, e_1, e_2) : \mathcal{M}(S_N) \to \mathcal{M}(S_N)$. (We will need such a function for each choice of a, e_1 and e_2.) This function associates to each environment d of S_N the environment μ^+ of S_N where μ is the maximal step that is 'consistent' with the schedulers e_1, e_2 and activated under the environment d plus the data on i described by a. Since we want to fix the behaviour according to the schedulers e_1, e_2, we demand that this step has to be 'consistent' with the schedulers in the sense that there are prefixes lub_1, lub_2 of the schedulers e_1, e_2 resp. with $\mu = multiset(lub_1) + multiset(lub_2)$. The output that we are interested in will then be the least fixed point of $F_N(a, e_1, e_2)$ restricted to the output place o. Since $F_N(a, e_1, e_2)$ is a continuous function in the complete partial order $\mathcal{M}(S_N)$, the least fixed point is well defined. We compute the least fixed point for each environment a of the input place i and thus get a function with functionality $\mathcal{M}(\{i\}) \to \mathcal{M}(\{o\})$ as description of the behaviour corresponding to the schedulers e_1, e_2.

For the semantics $IO(N)$ we consider all choices of schedulers and get a set of functions $IO(N) \subseteq \mathcal{M}(\{i\}) \rightarrow \mathcal{M}(\{o\})$.

Definition 3.1

Let $N = (S_N, T_N, W_N, M_N)$ be a net with $U_N = \{u_1, \ldots, u_n\}$. We define the functions $F_N(a, e_1, \ldots, e_n) \in \mathcal{M}(S_N) \rightarrow \mathcal{M}(S_N)$ for all $a \in \mathcal{M}(IN_N)$, $e_j \in STREAM(u_j)$, $j = 1, \ldots, n$, by

$$F_N(a, e_1, \ldots, e_n)(d) = (\sum_{j=1}^{n} multiset(lub_j))^+ \text{ for all } d \in \mathcal{M}(S_N)$$

where $lub_j = lub\{e \sqsubseteq e_j \mid multiset(e)^- \leq M_N + a_S + d\} \in STREAM(u_j)$ for all $j = 1, \ldots, n$ and $a_S = \lambda s. \begin{cases} a(s), & \text{if } s \in IN_N \\ 0, & \text{otherwise} \end{cases} \in \mathcal{M}(S_N)$. The semantics $IO(N)$ is defined by

$$\begin{aligned} IO(N) = \{f \in \mathcal{M}(IN_N) \rightarrow \mathcal{M}(OUT_N) \mid \\ \exists e_j \in STREAM(u_j),\ j = 1, \ldots, n : \\ f = \lambda a.[fix\ F_N(a, e_1, \ldots, e_n)]|_{OUT_N} \} \end{aligned}$$

It is possible to show that the functions in the semantics of a net are continuous. This means that we can restrict these functions from environments of the input places to markings of the input places without loss of information. We allow finite and infinite schedulers since we want to model arbitrary behaviours.

Let us now consider the function $f \in IO(N)$ (for the net N in Figure 3.1) induced by the schedulers $e_1 = x^\omega$ and $e_2 = (zy)^\omega$ (i.e. infinitely manz x's, (zy)'s resp.). We compute $fix\ F_N(a, e_1, e_2)$ for some environment a of i by the iteration $d_1, d_2, \ldots$, where $d_j = [F_N(a, e_1, e_2)]^j(0,0,0)$ for $j \in \mathbb{N}$. Let $a = 2$.

For the computation of d_1 we have to consider $M_N + a_S + (0,0,0) = (2,0,0)$. We get $lub_1 = xx$, $lub_2 = \epsilon$ and $d_1 = (x + x)^+ = (0,2,0)$. For the computation of $d_2 = F_N(a, e_1, e_2)(d_1)$ we have to consider $M_N + a_S + d_1 = (2,2,0)$. This is *not* the follower marking of the step $x+x$. It can be interpreted as the history after firing the step $x+x$: we had $(2,0,0)$ at the beginning and added $(0,2,0)$. We get $lub_1 = xx$, $lub_2 = zy$ and $d_2 = (x + x + z + y)^+ = (0,3,1)$. $M_N + a_S + d_2 = (2,3,1)$ is the history after firing the steps $x + x$ and $z + y$: $(2,0,0)$ at the beginning, $(0,2,0)$ added by $x+x$, $(0,1,1)$ added by $z+y$. We have $d_3 = F_N(a, e_1, e_2)(d_2) = (x+x+z+y+z)^+ = (0,4,1)$ where $lub_1 = xx$, $lub_2 = zyz$.

The least fixed point of $F_N(a, e_1, e_2)$ is $d_4 = (0,4,2)$ and we get $f(2) = 2$. We interpreted the iteration process by a step sequence $(x + x)(z + y)(z)(y)$ that

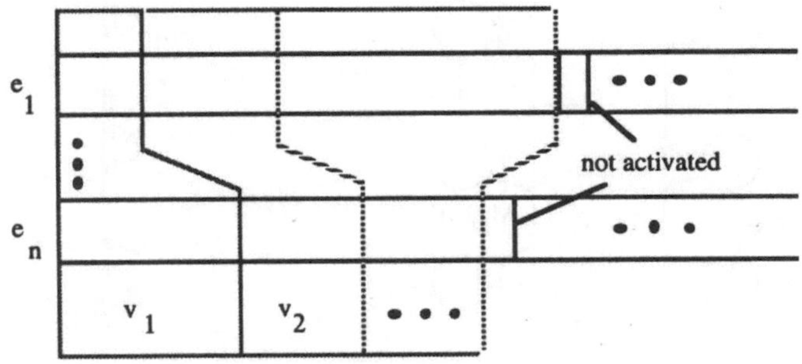

Figure 3.2: sketch of a scheduled step sequence

is activated under the marking $M_N + a_S = (2,0,0)$. This close relationship to step sequences gives us a strong consistency result w.r.t. the step sequence semantics.

To establish this relationship we need the notion of *scheduled step sequences*. Informally such a step sequence works off the schedulers as long as possible. In Figure 3.2 a step sequence v (beginning with steps v_1, v_2), which is scheduled w.r.t. schedulers $e_1, \ldots, e_n$, is indicated. It 'covers' prefixes of the schedulers. The first transitions following these prefixes are not activated under the corresponding markings. Note that these prefixes can also be infinite.

Theorem 3.1

Let N be a net with $U_N = \{u_1, \ldots, u_n\}$. Let the function $f \in IO(N)$ be induced by schedulers $e_1, \ldots, e_n$. Let a be a marking of the input places of N. For all scheduled step sequences v w.r.t. $e_1, \ldots, e_n$ in N_a (the net resulting from N by adding $a(s)$ tokens to the initial marking of s for all input places s)

$$f(a) = \sum_{k=1}^{\omega} v_k^+|_{OUT_N}$$

If we merge the functions in $IO(N)$ we obtain the semantics $SIO(N)$ that associates to each marking a of the input places of N a set of markings of the output places of N: $SIO(N)(a) = \{f(a) \mid f \in IO(N)\}$. With Theorem 3.1 we get

Theorem 3.2

Let N be a net. For all markings a of the input places of N

$$SIO(N)(a) = \{multiset(w)^+|_{OUT_N} \mid w \text{ is a firing sequence in } N_a\}$$

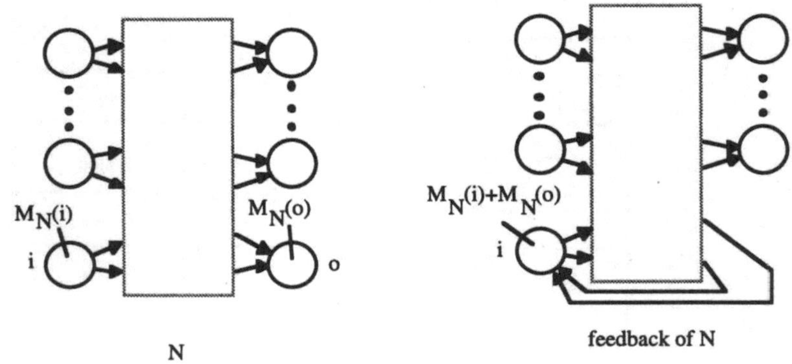

Figure 3.3: feedback

This theorem means that the outputs of all firing sequences are represented in $IO(N)$. Therefore IO is consistent with the firing sequence semantics in a very strong sense.

The semantics IO is compositional with respect to ***parallel composition*** without synchronization (simply placing two nets side by side), ***feedback***, i.e. merging of an input place and an output place (Figure 3.3), and ***output merging***, i.e. merging of two output places. With these operators asynchronous communication can be modelled. In the terminology of Chapter 2:

Theorem 3.3

The equivalence on nets induced by the semantics IO is a congruence w.r.t. parallel composition, feedback and output merging.

Nets with at most one conflict region and without internal places are called ***atomic nets***.

Theorem 3.4

Every net is composable by parallel composition, feedback and output merging from atomic nets.

In [Gol 93] some variants of IO, which describe properties like maximality or fairness, are examined. It turns out that for extended free choice nets these semantics simplify to some extent.

3.3 The semantics $\mathcal{T}$

One disadvantage of IO is that it is not fully abstract w.r.t. SIO. Fully abstract semantics are interesting, since such a semantics adds the right amount of information to a non-compositional basic semantics in order to achieve compositionality.

In the terminology of Chapter 2 *full abstractness* of a semantics Sem w.r.t. a semantics Tem means that the equivalence on nets induced by Sem is the coarsest congruence w.r.t. the above mentioned net operators contained in the equivalence induced by Tem.

We will look for a semantics that is fully abstract with respect to a slightly modified semantics ST, where $ST(N)$ associates to each marking of the input places the set $ST(N)(a) = \{a' \in \mathcal{M}(OUT_N) \mid \exists a \in SIO(N)(a) : a' \leq a\}$.

In order to get a fully abstract semantics we have a look at semantics for dataflow nets. It is known that the operational behaviour (histories) of dataflow networks is not compositional with respect to feedback (Brock-Ackerman-anomaly [BA 81]) and that the trace semantics ([Jon 89]) is fully abstract with respect to the operational behaviour. As a consequence of that we introduce traces for Petri nets.

A pair of a input or output place and a natural number of tokens is called communication event. We define communication firing sequences by extending firing sequences by communication events. From these firings sequences we get traces by abtracting from the (internal) transition occurrences.

Definition 3.2

Let $N = (S_N, T_N, W_N, M_N)$ be a net. The set of *communication events* of N is

$$CE_N = \{(s,n) \mid s \in IN_N \cup OUT_N,\ n \in I\!N_0\}$$

Let $\rightarrow\ \subseteq \mathcal{M}_f(S_N) \times \mathcal{M}_f(S_N) \times (CE_N \cup T_N)$ be the relation defined in Figure 3.4. We write $M \rightarrow^{ev} M'$ for a triple $(M, M', ev) \in \rightarrow$. A sequence $cs \in STREAM(CE_N \cup T_N)$ is called *communication firing sequence*, if $\forall j \in I\!N, j \leq l(cs)\ \exists$ marking M_j : $M_{j-1} \rightarrow^{cs_j} M_j$, where M_0 is the initial marking. We define the semantics $\mathcal{T}(N)$ as the set of *traces of* N:

$$\mathcal{T}(N) = \{\tau \in STREAM(CE_N) \mid \exists \text{ communication firing sequence } cs \text{ in } N : \tau = cs|_{CE_N}\}$$

Now we have the following result.

Theorem 3.5

The semantics $\mathcal{T}$ is compositional w.r.t. the net operators parallel composition, feedback and output merging and even fully abstract w.r.t. the semantics ST.

The semantics $\mathcal{T}(N)$ can be computed from $IO(N)$ (for each net N). This means that the equivalence induced by IO is contained in the equivalence induced by $\mathcal{T}$.

for all $M, M' \in \mathcal{M}_f(S)$, $ev \in CE_N \cup T_N$:

$$(M, M', ev) \in \rightarrow \;\Leftrightarrow\; (ev \in T_N \wedge M[ev\rangle M')$$

(firing of the transition ev)

$$\vee\; (ev = (i,n) \in CE_N \wedge i \in IN_N \wedge M' = \lambda s. \begin{cases} M(i) + n, & \text{if } s = i \\ M(s), & \text{otherwise} \end{cases})$$

(input of n tokens on i)

$$\vee\; (ev = (o,n) \in CE_N \wedge o \in OUT_N \wedge n \leq M(o) \wedge M' = \lambda s. \begin{cases} M(o) - n, & \text{if } s = o \\ M(s), & \text{otherwise} \end{cases})$$

(output of n tokens from o)

Figure 3.4: definition of $\rightarrow$

3.4 Deadlocking

In the case of synchronous communication the deadlocking behaviour can be decribed by failures semantics (see Theorem 2.4). In fact the failures semantics is fully abstract w.r.t. a semantics that detects deadlocks but nothing else. In the asynchronous case deadlocking was examined e.g. in [BKPR 91]. There a general paradigm for concurrent systems with asynchronous communication is presented. Dataflow nets and Petri nets can be considered as special cases thereof. As basic semantics for the description of deadlocks a semantics is chosen, which gives for every initial state of the system a set of sequences of states, which begin with the initial state and are possibly completed with a token for deadlock. In [BKPR 91] it is argumented that a fully abstract semantics does not need failure sets.

In our framework we can even simplify this approach by choosing a basic semantics, which only registers deadlocks, but not states of the system.

Definition 3.3

Let $N = (S_N, T_N, W_N, M_N)$ be a net. A firing sequence w is called *deadlocking*, if it is finite and if after firing w no transition is enabled. We define the semantics $DLOCK(N)$ that associates to each marking a of the input places of N a truth value:

$$(DLOCK(N)(a) = true) \Leftrightarrow (\exists \text{ deadlocking firing sequence in the net } N_a)$$

where N_a is the net resulting from N by adding $a(s)$ tokens to the initial marking of s for all input places s.

In order to get a semantics that is fully abstract w.r.t. $DLOCK$ we restrict the trace semantics $\mathcal{T}$ to deadlocking traces.

Definition 3.4
Let $N = (S_N, T_N, W_N, M_N)$ be a net. A finite communication firing sequence cs is called ***deadlocking***, if

$$\left(\neg\exists t \in T_N : M[t\rangle\right) \wedge \left(\forall s \in OUT_N : M(s) = 0\right)$$

where M is the marking reached after the firing of cs. We define the semantics

$$\mathcal{T}_{DLOCK}(N) = \{\tau \in CE_N^* \mid \exists \text{ deadlocking communication firing sequence } cs \text{ in } N : \tau = cs|_{CE_N}\}$$

Theorem 3.6
The semantics $\mathcal{T}_{DLOCK}$ is compositional w.r.t. the net operators parallel composition, feedback and output merging and even fully abstract w.r.t. the semantics $DLOCK$.

This result shows that a semantics, which is fully abstract w.r.t. deadlocks, neither needs failure sets nor knowledge about internal states.

4. The Petri box algebra

It was an old goal to develop a Petri net based programming language (see [Bra 87]) - most people used Petri nets only as a modelling and analysing tool. One rather early step was the dissertation [Sen 83] which describes the development of a graphical programming language based on Petri nets (see also [Sen 82]). Difficulties concerned not only the problems of implementation of nets (see [Tau 88]) but mainly the fact that Petri nets have not an inherently modular structure. The process algebra approach started by Hoare and Milner has had great influence on research on modular constructions for Petri nets (see [Bra 91], [BGV 90], [Vog 92]).

In the ESPRIT BRA project DEMON a basic Petri net programming notation was developed, which consists of two parts: the box algebra (which owes much to CCS) and the semantic domain of Petri boxes (special Petri nets and operators on them) such that the Petri boxes constitute a compositional semantics for the box expressions.

Petri boxes are place/transition-nets equipped with special interfaces, which allow asynchronous and synchronous communication. The entry and exit interfaces (Figure 4.1) consist of places labelled by e (entry), x (exit) resp., the communication interface is formed by transitions labelled by multisets of actions. The empty labelling of transitions (empty multiset of actions) denotes silent transitions, which cannot communicate with others. Variables

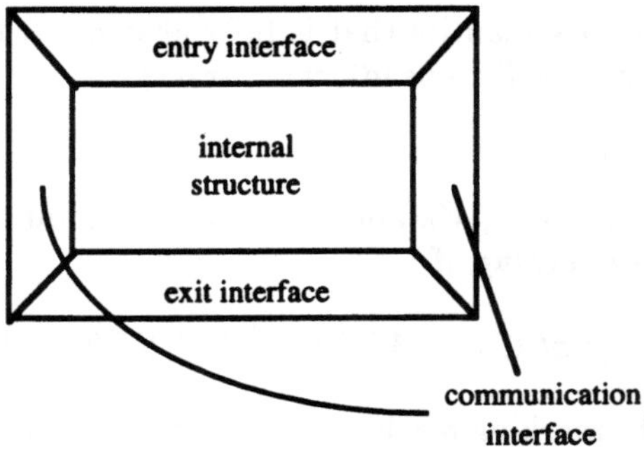

Figure 4.1: box scheme

are also allowed as labels of transitions. The presets, postsets resp. of the entry, exit places resp. have to be empty. Thus these places correspond to input, output places resp. in the approach of [Gol 93]. Boxes carry no initial marking, but natural candidates for initially marked places are the entry places.

Some basic boxes are shown in Figure 4.2 †.

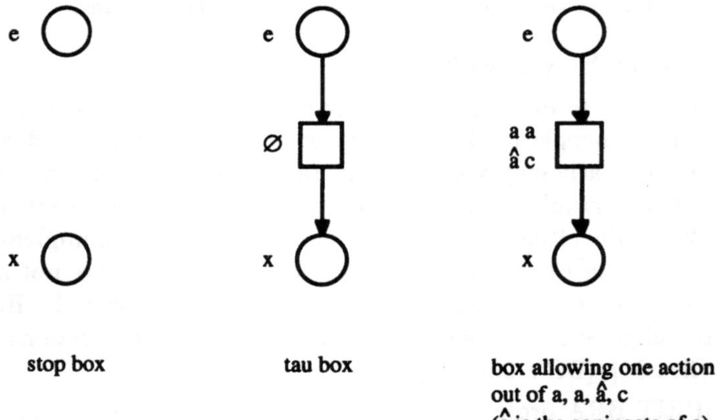

Figure 4.2: basic boxes

Sequential composition is done by multiplication of the set of exit places of the first component with the set of entry places of the second component. This differs from the approach taken in [Gol 93], since there sequential composition can be modelled by parallel composition followed by feedback (Figure 4.3). Note that the composition of N_1 and N_2 in this example would in [Gol 93] be

† Actually, the stop box is derivable e.g. from an a-box by restricting the action a.

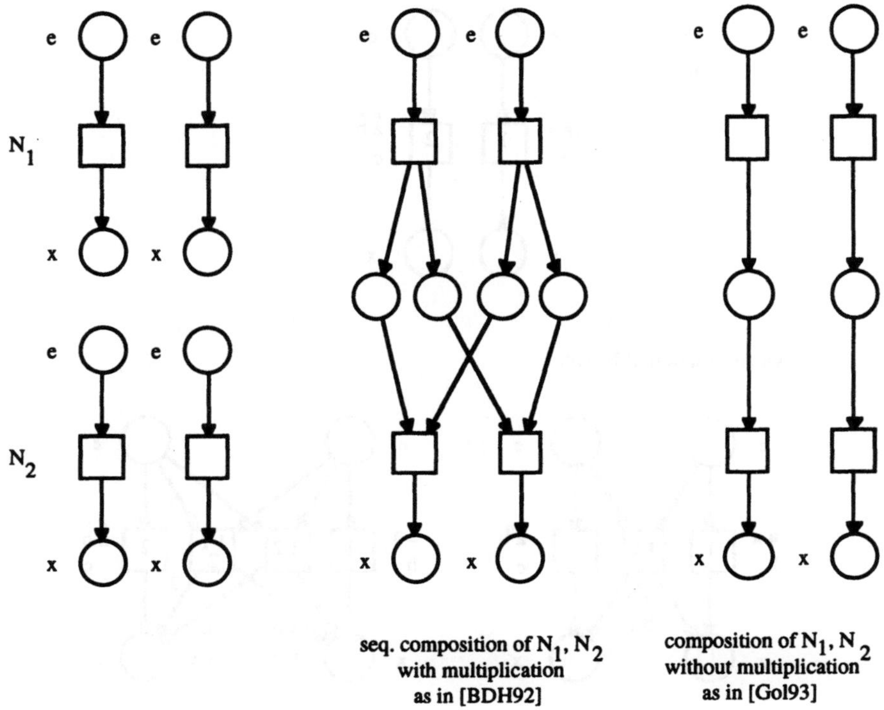

Figure 4.3

considered as the application of two times an operator composing an input and an output place. Therefore the second component may start even if the first component has not finished completely.

Parallel composition and synchronisation is done by two separate operators || and sy. The operator || (named concurrent composition in [BDH 92]) simply places two boxes side by side. For the synchronisation over an action name (say a) choose a transition, whose labelling contains a and a transition, whose labelling contains the conjugate $\hat{a}$. Then merge copies of both transitions and label the resulting transition by the (multiset)sum of the labels of the components, but omit one instance of a and $\hat{a}$ (Figure 4.4). This synchronisation procedure is iterated until no pairs of transitions with labels containing a and $\hat{a}$, which have not already merged, exist.

Note that copies of the transitions are merged. The original transitions can, but need not be deleted by the restriction operator. In comparison to that the parallel composition operator of [Vog 89] described in Chapter 2.3.4 (multiplies and) merges the transitions themselves.

It is possible that infinitely many transitions are created, e.g. when synchronising the basic box that allows actions a and $\hat{a}$. Multilabelling (i.e. labelling with multisets of actions) of transitions has been introduced in order to allow

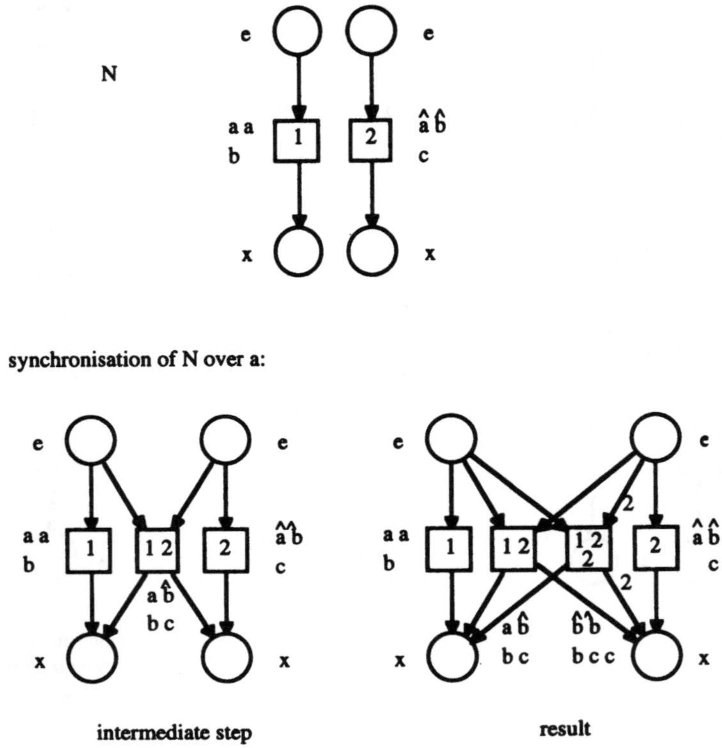

Figure 4.4

synchronisation over multiple actions, which is useful in some examples as pointed out in [BDH 92]. But also a more technical reason exists: multisets of actions are necessary to achieve commutativity of synchronisation.

A number of other operators on boxes are defined in [BDH 92]: choice (like CCS's +), restriction, scoping, relabelling, refinement, recursion, iteration. For new versions of refinement and recursion see [BDE 93].

The refinement technique allows distributed input and distributed output by multiplying the preset of the transition, which should be refined, with the set of entry places of the refinement net, and symmetrically the postset with the exit places. In the new version ([BDE 93]) special care is taken by an elaborate naming strategy that simultaneous refinements and refinements of transitions with side loops are possible (Figure 4.5).

Recursion is defined by successively refining variable-labelled transitions. This differs very much from other approaches, e.g. [Gol 88, Tau 89].

Many of the operators can be synthesized from the refinement operator. For example the iteration of boxes B_1, B_2, B_3 is obtained by simultaneously refining the transitions labelled by the variables X, Y, Z in box B_* (Figure

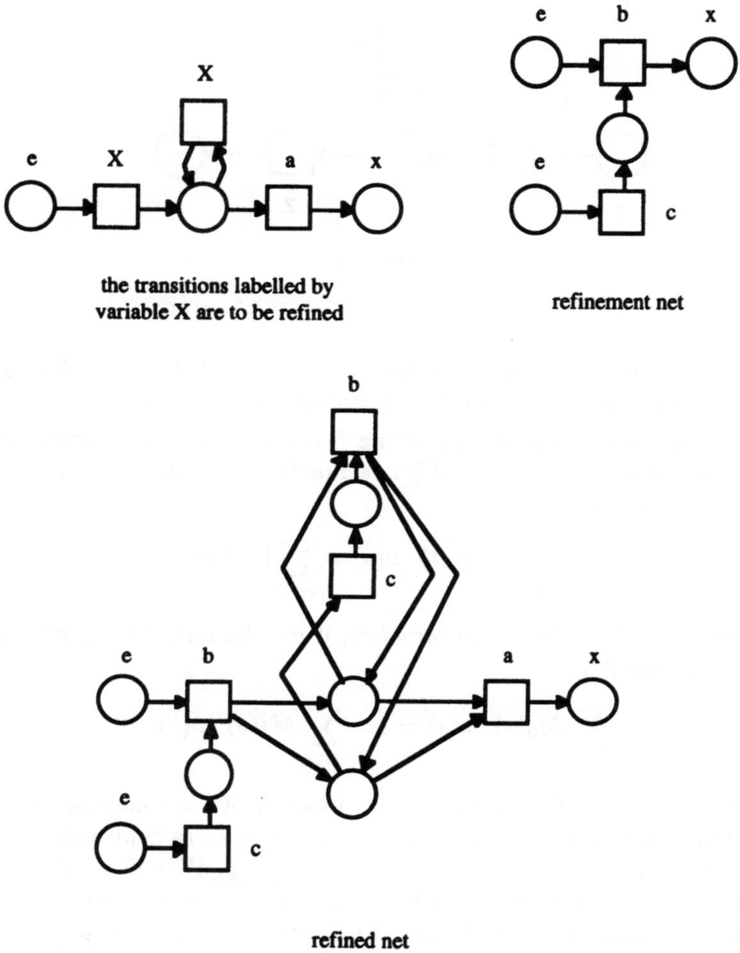

Figure 4.5

4.6) by B_1, B_2, B_3 resp.

Besides structural equations some behavioural properties have been obtained, e.g.

Theorem 4.1

Every box composed from basic boxes by sequential composition, choice, concurrent composition, synchronisation, restriction, scoping, relabelling and iteration is a safe net, if we initially mark every entry place and leave the other places unmarked. Iteration is only allowed, if the box has exactly one entry place or exactly one exit place.

The proof can be done by showing that boxes are covered by S-components.

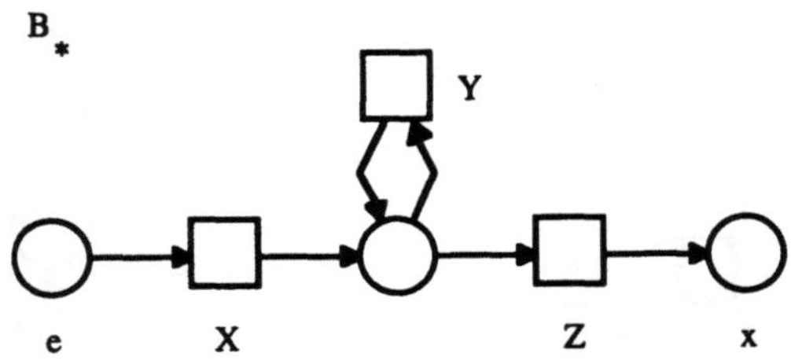

Figure 4.6

A similar result for the refinement operator is given in [Dev 93]. The central property in this paper is S-invariant synthesis for finite nets.

First recall some definitions concerning S-invariants. An *S-invariant* of a place/transition-net $N = (S, T, W)$ is a function ν from S into integers or real numbers such that

$$\forall t \in T : \sum_{s \in S} W(s,t) \cdot \nu(s) = \sum_{s \in S} W(t,s) \cdot \nu(s)$$

An S-invariant ν is called *1-conservative*, if for the initial marking M_0 and the final marking M_f:

$$\sum_{s \in S} M_0(s) \cdot \nu(s) = 1 = \sum_{s \in S} M_f(s) \cdot \nu(s)$$

In the box calculus only the entry, exit places resp. are marked under the initial, final marking resp. *S-components* can be seen as S-invariants ν with $\forall s \in S : \nu(s) \in \{0,1\}$ and $\forall t \in T : \nu(t) := \sum_{s \in S} W(s,t) \cdot \nu(s) \in \{0,1\}$. If we restrict ourselves to nets where each transition has non-empty pre- and postset, this corresponds to the usual definition. By the places s with $\nu(s) = 1$ and the transitions t with $\nu(t) = 1$ a subnet $N' = (S', T', W')$ of N is induced with $T' = {}^\bullet S' \cup S'^\bullet$ and $\forall t \in T' : |{}^\bullet t \cap S'| \leq 1 \ \wedge \ |t^\bullet \cap S'| \leq 1$ (pre- and postsets are taken with respect to N). A net N is called *S-covered (T-covered) by S-components*, if each $s \in S$ ($t \in T$) belongs to some 1-conservative S-component ν of N, i.e. $\nu(s) = 1$ ($\nu(t) = 1$).

Now let N be a finite net that contains transitions labelled by variables X_i, $i \in I$, and let N_i, $i \in I$, be finite nets. Define $\tilde{N}$ as the net obtained by refining all transitions labelled by X_i with N_i.

Theorem 4.2

An S-invariant ν of N and 1-conservative S-invariants ν_t of N_i for each transition t labelled by some variable X_i induce an S-invariant of the net $\tilde{N}$.

This theorem implies the following important property.

Theorem 4.3
If N and N_i, $i \in I$, are S-covered by S-components, then $\tilde{N}$ is 2-safe. If N and N_i, $i \in I$, are T-covered by S-components, then $\tilde{N}$ is free of self-concurrency (from the initial marking).

S-invariant analysis can also be done for infinite nets, but special care has to been taken in this case (see [Dev 92]).
In a similar fashion recursion has been examined. As a result 2-safeness of nets constructed by the recursion operator is obtained (see [Dev 92]).

References

[AH 88] L. Aceto and M. Hennesy: Towards action-refinement in process algebras. Technical Report 3/88, Dept. Comp. Sci. Univ. of Sussex, Brighton, (1988).

[BA 81] J.D. Brock and W.B. Ackerman: Scenarios: A model of non-determinate computation. In: J. Diaz and I. Ramos (eds.), Proc. of the International Colloquium on Formalization of Programming Concepts, Peniscola 1981, Lect. Notes Comp. Sci. 107 (1981) 252-259.

[Bau 88] B. Baumgarten: On internal and external characterizations of PT-net building block behaviour. In: G. Rozenberg (ed.), Advances in Petri Nets 1988, Lect. Notes Comp. Sci. 340 (1988) 44–61.

[BB 92] W. Brauer and U. Brauer: Wissenschaftliche Herausforderung für die Informatik: Änderungen von Forschungszielen und Denkgewohnheiten. In: W. Langenheder, G. Müller, B. Schinzel (eds.), Informatik cui bono? Springer (1992) 11-19.

[BC 87] G. Boudol and I. Castellani: On the semantics of concurrency: Partial orders and transition systems. In: H. Ehrig et al. (eds.), TAPSOFT 87, Vol. I, Lect. Notes Comp. Sci. 249 (1987) 123–137.

[BDE 93] E. Best, R. Devillers and J. Esparza: General refinement and recursion for the box calculus. In: P. Engalbert, A. Finkel, K.W. Wagner (eds.), STACS 93, Lect. Notes Comp. Sci. 665 (1993) 130-140.

[BDH 92] E. Best, R. Devillers and J.G. Hall: The box calculus: a new causal algebra with multi-label communication. In: G. Rozenberg (ed.), Advances in Petri Nets 1992, Lect. Notes Comp. Sci. 609 (1992) 21-69.

[BDKP 89] E. Best, R. Devillers, A. Kiehn, and L. Pomello: Concurrent bisimulations in Petri nets. Acta Informatica 28 (1991) 231-264.

[Ber 93] A. Bergeron: A unified approach to control problems in discrete event processes. R.A.I.R.O Informatique Theorique 27/6 (1993) 555-573.

[BGV 90] W. Brauer, R. Gold and W. Vogler: A survey of behaviour and equivalence preserving refinements of Petri nets. In: G. Rozenberg (ed.), Advances in Petri Nets 1990, Lect. Notes Comp. Sci. 483 (1990) 1-46.

[BHR 84] S.D. Brookes, C.A.R. Hoare, and A.W. Roscoe: A theory of communicating sequential processes. J. ACM 31 (1984) 560–599.

[BKPR 91] F.S. de Boer, J.N. Kok, C. Palamidessi and J.J.M.M. Rutten: The failure of failures in a paradigm for asynchronous communication. In: J.C.M. Baeten, J.F. Groote (eds.), CONCUR 91, Proc. of the 2nd International Conference on Concurrency Theory, Amsterdam 1991, Lect. Notes Comp. Sci. 527 (1991) 111-126.

[BR 84] S.D. Brookes and A.W. Roscoe: An improved failures model for communicating processes. In: S.D. Brookes, A.W. Roscoe, and G. Winskel (eds.), Seminar on Concurrency, Lect. Notes Comp. Sci. 197 (1984) 281–305.

[Bra 87] W. Brauer: Carl Adam Petri and informatics. In: G. Rozenberg, K. Voss and H. Genrich (eds.), Concurrency and Nets, Springer (1987) 13-21.

[Bra 91] W. Brauer: Formal approaches to concurrency. In: F.L. Bauer (ed.), Logic, Algebra and Computation, NATO ASI Series, Vol. F 79, Springer(1991) 335-358.

[Bra 93] W. Brauer: Distributed action systems In: F.L. Bauer (ed.), Logic, Algebra and Computation, NATO ASI Series, Vol. F 79, Springer (1993) 1-30.

[Bro 88] M. Broy: Nondeterministic data flow programs: how to avoid the merge anomaly. Science of Computer Programming 10 (1988) 65-85.

[Bro 90] M. Broy: Functional specification of time sensitive communicating systems. In: J.W. de Bakker, W.P. de Roever, G. Rozenberg (eds.), Stepwise Refinement of Distributed Systems, Lect. Notes Comp. Sci. 430 (1990) 153-179.

[BS 92] M. Broy and T. Streicher: Modular functional modelling of Petri nets with individual tokens. In: G. Rozenberg (ed.), Advances in Petri Nets 1992, Lect. Notes Comp. Sci. 609 (1992) 70-88.

[CDP 87] L. Castellano, G. De Michelis, and L. Pomello: Concurrency vs. interleaving: An instructive example. Bull. EATCS 31 (1987) 12–15.

[DDM 89] P. Degano, R. De Nicola, and U. Montanari: Partial orderings descriptions and observations of nondeterministic concurrent processes. In: J.W. de Bakker et al. (eds.), Proc. REX School / Workshop Linear Time, Branching Time and Partial Order in Logic and Models of Concurrency, Noordwijkerhout 1988, Lect. Notes Comp. Sci. 354 (1989) 438-466.

[Den 70] J.B. Dennis: Modular, asynchronous control structures for a high performance processor. In: ACM Conf. Rec. Concurrent Systems and Parallel Computation (1970) 55-80.

[Dev 88] R. Devillers: On the definition of a bisimulation notion based on partial words. Petri Net Newsletter 29 (1988) 16-19.

[Dev 92] R. Devillers: S-invariant analysis of Petri boxes. Report LIT-273, Université Libre de Bruxelles, Laboratoire d'Informatique Théorique (1992).

[Dev 93] R. Devillers: Construction of S-invariants and S-components for refined Petri nets. In: M.A. Marsan (ed.), Proceedings of the 14th International Conference on Application and Theory of Petri Nets, Chicago 1993, Lect. Notes Comp. Sci. 691 (1993) 242-261.

[DNH 84] R. De Nicola, M.C.B. Hennessy: Testing equivalence for processes. Theoret. Comp. Sci. 34 (1984) 83-133.

[Fis 70] P.C. Fishburn: Intransitive indifference with unequal indifference intervals. J. Math. Psych. 7 (1970) 144-149.

[Fur 73] F.C. Furtek: Asynchronous pushdown stacks. Computation Structures Group Memo 86, Project MAC MIT, (1973)

[GG 89] R.J. v. Glabbeek and U. Goltz: Equivalence notions for concurrent systems and refinement of actions. In: A. Kreczmar and G. Mirkowska (eds.), MFCS 89, Lect. Notes Comp. Sci. 379 (1989) 237-248.

[GG 90] R.J. v. Glabbeek and U. Goltz: Refinement of actions in causality based models. In: J.W. de Bakker, W.P. de Roever, G. Rozenberg (eds.), Proc. REX Workshop on Stepwise Refinement of Distrib. Systems 1989, Lect. Notes Comp. Sci. 430 (1990) 267-300.

[GL 82] L.J. Guibas and F.M. Liang: Systolic stacks, queues and counters. 1982 Conference on Advanced Research in VLSI, M.I.T.

[Gla 90] R.J. v. Glabbeek: The refinement theorem for ST-bisimulation semantics. In: M. Broy and C.B. Jones (eds.), Proc. IFIP Working Conference on Programming Concepts and Methods, Sea of Galilee, Israel, (1990).

[Gol 88] U. Goltz: On representing CCS programs by finite Petri nets. In: M.P. Chytil et al (eds.), MFCS 88, Lect. Notes Comp. Sci. 324 (1988) 339-350.

[Gol 92] R. Gold: Dataflow semantics for Petri nets. In: I.M. Havel and V. Koubek (eds.), MFCS 92, Lect. Notes Comp. Sci. 629 (1992) 275-283.

[Gol 93] R. Gold: Datenflußsemantiken für Petrinetze. Dissertation, TU München, Fakultät für Informatik (1993).

[Gra 81] J. Grabowski: On partial languages. Fundamenta Informaticae IV.2 (1981) 428-498.

[GV 87] R.J. v. Glabbeek and F. Vaandrager: Petri net models for algebraic theories of concurrency. In: J.W. de Bakker et al. (eds.), PARLE Vol. II, Lect. Notes Comp. Sci. 259 (1987) 224-242.

[Hen 87] M. Hennessy: Axiomatising finite concurrent processes. Technical Report 4/87, Dept. Comp. Sci. Univ. of Sussex, Brighton, (1987).

[Jen 92] K. Jensen: Coloured Petri nets. EATCS Monographs on Theoretical Computer Science, Vol. II, III to appear in 1994, 1995.

[JM 92] L. Jategaonkar and A.R. Meyer: Testing equivalence for Petri nets with action refinement. In: W.R. Cleaveland (ed.), CONCUR 92, Lect. Notes Comp. Sci. 630 (1992) 17-31.

[Jon 89] B. Jonsson: A fully abstract trace model for dataflow networks In: Proc. of the 16th Annual ACM Symposium on Principles of Programming Languages, Austin 1989 (1989) 155-165.

[Kah 73] G. Kahn: A preliminary theory for parallel programs. IRIA Rapport de Rechereche 6 (1973).

[Kel 78] R.M. Keller: Denotational models for parallel programs with indeterminate operators In: E.J. Neuhold (ed.), Formal Description of Programming Concepts, North-Holland (1978) 337-366.

[Kie 88] A. Kiehn: On the interrelationship between synchronized and nonsynchronized behaviour of Petri nets. J. Inf. Process. Cybern. EIK 24 (1988) 3-18.

[Kie 90] A. Kiehn: Petri net systems and their closure properties. In: G. Rozenberg (ed.), Advances in Petri Nets 1989, Lect. Notes Comp. Sci. 424 (1990) 306-328.

[Lam 77] L. Lamport: Proving the correctness of multiprocess programs. IEEE Transactions on Software Engineering. No. 2 (1977) 125-143.

[Lam 86] L. Lamport: On interprocess communication I. Distributed Comp. 1 (1986) 77-85.

[Mil 83] R. Milner: Calculi for synchrony and asynchrony. Theor. Comput. Sci. 25 (1983) 267-310.

[Mil 93] R. Milner: The polyadic π-calculus: a tutorial. In: F.L. Bauer (ed.), Logic, Algebra and Computation, NATO ASI Series, Vol. F 79, Springer (1993) 202-246.

[NEL 89] M. Nielsen, U. Engberg, and K. Larsen: Partial order semantics for concurrency. In: J.W. de Bakker et al. (eds.), Proc. REX School / Workshop Linear Time, Branching Time and Partial Order in Logic and Models of Concurrency, Noordwijkerhout 1988, Lect. Notes Comp. Sci. 354 (1989) 523-548.

[Par 81] D. Park: Concurrency and automata on infinite sequences. In: P. Deussen (ed.), Proc. 5th GI Conf. on Theoretical Comp. Sci., Lect. Notes Comp. Sci. 104 (1981) 167-183.

[Pet 62] C.A. Petri: Kommunikation mit Automaten. Schriften des Rhein.-Westf. Inst. f. Instr. Mathem. a. d. Univ. Bonn, Nr. 2, Bonn, (1962).

[Pom 86] L. Pomello: Some equivalence notions for concurrent systems - an overview. In: G. Rozenberg (ed.), Advances in Petri Nets 1985, Lect. Notes Comp. Sci. 222 (1986) 381-400.

[PRS 92] L. Pomello, G. Rozenberg and C. Simone: A survey of equivalence notions for net based systems. In: G. Rozenberg (ed.), Advances of Petri Nets 1992, Lect. Notes Comp. Sci. 609 (1992) 410-472.

[RT 88] A. Rabinovich and B.A. Trakhtenbrot: Behaviour structures and nets. Fundamenta Informaticae 11 (1988) 357–404.

[Rei 91] W. Reisig: Petri nets and algebraic specifications. Theor. Comp. Sci. 80 (1991) 1-34.

[Rei 93] W. Reisig: Petri nets: Fundamentals, essentials, consequences. EATCS Bulletin Nr. 49 (1993) 185-191.

[Sen 82] H.E. Sengler: Programmieren mit graphischen Mitteln: Die Sprache GRADE und ihre Implementation. In: H. Wössner (ed.), Programmiersprachen und Programmentwicklung, Informatik-Fachberichte 53 (1982) 67-92.

[Sen 83] H.E. Sengler: Ein Modell des Verstehens von Programmen und seine Anwendung beim Entwurf der Programmiersprache GRADE. Dissertation, Universität Hamburg, FB Informatik (1983).

[SM 83] I. Suzuki and T. Murata: A method for stepwise refinement and abstraction of Petri nets. J. Comp. Sys. Sci. 27 (1983) 51-76.

[Tau 88] D. Taubner: On the implementation of Petri nets. In: G. Rozenberg (ed.), Advances in Petri Nets 1988, Lect. Notes Comp. Sci. 340 (1988) 418-439.

[Tau 89] D. Taubner: Finite representations of CCS and TCSP programs by automata and Petri nets. Lect. Notes Comp. Sci. 369, Springer (1989).

[Val 79] R. Valette: Analysis of Petri nets by stepwise refinement. J. Comp. Sys. Sci. 18 (1979) 35-46.

[Val 81] R. Valk: Generalizations of Petri nets. In: J. Gruska and M. Chytil (eds.), MFCS 81, Lect. Notes Comp. Sci. 118 (1981) 140-155.

[Vog 87] W. Vogler: Behaviour preserving refinements of Petri nets. In: G. Tinhofer and G. Schmidt (eds.), Graph-Theoretic Concepts in Computer Science, Proc. WG 86, Bernried, Lect. Notes Comp. Sci. 246 (1987) 82-93.

[Vog 89] W. Vogler: Failures semantics and deadlocking of modular Petri nets. Acta Informatica 26 (1989) 333-348.

[Vog 90] W. Vogler: Failures semantics based on interval semiwords is a congruence for refinement. In: C. Choffrut and T. Lengauer (eds.), STACS 90, Lect. Notes Comp. Sci. 415 (1990) 285–297; also in Distributed Computing 4 (1991) 139-162.

[Vog 92] W. Vogler: Modular construction and partial order semantics of Petri nets. Lect. Notes Comp. Sci. 625, Springer-Verlag (1992).

[Vog 93] W. Vogler: Timed testing of concurrent systems. In: A. Lingas, R. Karlsson, S. Carlsson (eds.), Automata, Languages and Programming, ICALP 93, Lect. Notes Comp. Sci. 700 (1993) 532-543.

Using Reflection to Explain and Enhance Type Theory*

Robert L. Constable
Cornell University
Department of Computer Science
Ithaca, NY

January 20, 1994

Abstract

The five lectures at Marktoberdorf on which these notes are based were about the architecture of *problem solving environments* which use theorem provers. Experience with these systems over the past two decades has shown that the prover must be *extensible*, yet it must be kept *safe*. We examine a way to safely add new decision procedures to the Nuprl prover. It relies on a reflection mechanism and is applicable to any tactic-oriented prover with sufficient reflection. The lectures explain reflection in the setting of constructive type theory, the core logic of Nuprl.

Keywords: decision procedures, HOL, Löb's theorem, Nuprl, problem-solving environments, reflection, tactics, tautologies, theorem provers, type theory

*Work supported in part by NSF grant CCR-9244739 and ONR contract N00014-92-J-1764.

1 Introduction

1.1 Computation in Mathematics

The last two decades appear to be part of a new Golden Age of mathematics. We have seen the Four Color Problem fall, the Mordell Conjecture as well, and now this year Wiles has announced a proof of Fermat's Last Theorem, the most famous problem of them all. Soon the P=NP problem may be the most well-known open problem in all of mathematics, and like some among the old Hilbert problems, it deals with computation in a fundamental way.

As we reflect on these connections to computation, we might wonder whether we are in fact entering into a New Age of mathematics in which the computer plays a fundamental and striking role. In the solution of the Four Color Problem, the computer was an essential *tool* as it has been in numerical mathematics since mid century. For a variety of reasons, computers are becoming indispensable in many branches of mathematics.

These lectures will be about the theory and application of a kind of computer system, one that surely is a tool, but one whose designers and intimate users expect to become more, to become a *partner* in some sense. This is because these systems, which we call *proof development systems* or *problem solving environments* combine many tools into a single environment that provides coordinated and integrated assistance with a variety of tasks: formula and text entry and editing, numerical computation, symbolic computation, theorem and definition storage and retrieval from a data base, hypertext reading and browsing facilities for the data base, and proof checking and proof development assistance. There will also soon be communication facilities for monitoring activity at other development centers with their own systems and at repository centers for results in various branches of mathematics.

The combination of these tools, especially the proof checkers and theorem provers accessing the large data bases will create the feeling of a system that is an assistant in the enterprise of communicating mathematics, and in some cases even in the exciting part, discovering new results.

1.2 Technical Issues

The technical problem we solve in these lectures is related to questions that arise in the design and deployment of *problem solving environments* (PSE).

Since these systems interact with users at the limits of what they understand, there is a need to extend the problem solving mechanisms over time; new knowledge is added (say definitions and theorems) as well as new techniques (say new symbolic algorithms for simplifying algebraic expressions). We call these *open systems.*

In the case of theorem provers, but to some extent in all PSE's, extensions of the system require great care lest their reliability suffer. The earliest provers did not allow any user extension of their *methods.* Users could add only new knowledge. The Edinburgh LCF system pioneered the idea of tactics which allow new methods as well.

The LCF tactic mechanism enabled users to securely add new proof methods. The capabilities of the prover could be expanded by users and was not the exclusive province of the designers ("power to the people" is a slogan of this community). Many other provers have now adopted this approach [13, 15, 24, 41].

Experience from the 1970s and earlier showed also that decision procedures are very effective in building proofs. The Stanford Pascal Verifier [30], PL/CV [11], and EHDM [44] among others used such procedures. In particular the PL/CV-arith [11], congruence closure [32], sup-inf [46] are from this period and all are actively used today. New decision procedures are always being studied, monotone closure [36], real-closed fields [33], etc.

Here are typical uses of **arith**.

$$x < x^2 \ \& \ x \neq 0 \Rightarrow 2 \leq x \vee x < 0 \text{ by } \mathbf{arith}$$

$$\text{or } (x - y \leq y \ \& \ y \leq x + 1 \ \& \ x - 1 \leq z \ \& \ z \leq x + 1 \ \& \ y - 1 \leq z \ \& \ z \leq y + 1) \Rightarrow$$
$$(x = y) \vee (y = z) \vee (z = x) \text{ by } \mathbf{arith}$$

Nuprl was the first tactic-oriented prover to incorporate decision procedures. Some people have worried that this will compromise the integrity of the system, because even though the **arith** and congruence closure algorithms were informally proved correct, adding these mechanisms is *dangerous.* They substantially increase the complexity of the prover and raise the burden of proof that the system is correct. This danger makes us very reluctant to add more procedures in this *raw* way,

Our experience has thus created a *design dilemma:* **How to securely add new decision procedures?** One approach is taken by HOL. Decision procedures are used freely when the system executes in *unsafe mode*, then those steps previously solved by the procedure are later reduced to primitive rules when the whole proof is done in *safe mode.* This is a secure method. But it does not represent what people do when they rely on such procedures, and it is costly in terms of machine resources. For Nuprl this approach would be especially costly since that system saves proof trees for user inspection and for control of the proof process.

Another approach would be to verify the decision procedure inside the system. The easiest way is to verify a purely "mathematical" algorithm and then recode it. But his violates the maxim that one should "execute exactly what was verified". One could attempt to verify the algorithm exactly as coded in the implementation language. This approach is being followed by E. Gunter for HOL90 where the implementation language is SML-NJ. This is a very large task which resembles the reflection work in Nuprl [3, 4, 6] on which this paper is based, but one must rely on the stability of a large programming language and must define a *substantially* more complex system.

1.3 Summary of a solution

A critical step in solving the design problem is to observe that for type theories with an identifiable computation system, as with *constructive type theories*, it is possible to write tactics and decision procedures directly in the logic itself. This is true for Nuprl whose computation system is a functional programming language in the spirit of ML [25], which is the language of choice for writing tactics. In this case the decision procedure is both proven correct and directly executable.

A critical part of writing tactics inside a type theory is to have a model of the logic inside it. Again this can be done in rich type theories and has been done for Nuprl.

Even with these two pieces in place, we still do not have a solution. At this stage one could write and verify decision procedures in the logic, and one may apply them to same logic. For example, in Nuprl we can define the syntax and proof rules of the Intuitionistic Propositional Calculus (IPC) abstractly. We can prove the decidability of provability of IPC in Nuprl and instantiate the result for the class of propositional Nuprl formulas, which form an instance of IPC. But *still* these decision procedures are beyond our reach; we need a way to *link* this result to the Nuprl logic itself.

The final step of the solution I am describing here [5] relies on the Nuprl *reflection rule*, a unique feature of the system [3, 4, 6, 31]. This rule says that to prove a goal G under hypothesis H, it suffices to show that the reflected sequent, $\ulcorner H \vdash G \urcorner$, is provable. That is,

$$H \vdash G \ \text{ if } \exists p : \mathit{Proof}.\ p \text{ proves } \ulcorner H \vdash G \urcorner.$$

Now we can apply the verified decision procedure to $H \vdash G$ as we show in section 5. This rule reqnires restrictions to avoid contradictions from Löb's theorem (see section 4.7).

1.4 Plan of the Paper

We will begin with an informal introduction to Type Theory. This is the theory underlying the logic we use to formalize decision procedures. We then look at a formalization of this theory, then at its reflection and finally at applications.

2 An Introduction to Naive Type Theory

The informal language of mathematics uses types and sets, but when mathematicians want to be rigorous about a concept, they tend to rely on the 100 year old tradition of reducing it to concepts in pure set theory. In these lectures, we will find the rigorous concepts in type theory. For pedagogical reasons, we will often mention the set theory version of a concept as well.

The idea of a type is built up inductively (as is the Zermelo hierarchical concept of set). We start with *primitive types* and *type constructors.*

2.1 Primitive Types

void is a type with no elements
N is the type of natural numbers; $0, 1, 2, \ldots$

There will be other primitive types introduced later. Notice, in set theory we usually have only two primitive sets, φ, the empty set, and some infinite set (usually ω).

2.2 Compound Types

We build new types using *type constructors.* These tell us how to construct various kinds of objects. (In set theory, there is only one kind, *sets*).

The type constructors we choose are motivated both by mathematical and computational considerations. So we will see a tight relationship to the notion of *type* in programming languages. The notes by C.A.R. Hoare, *Notes on Data Structuring* [27], make the point well.

2.2.1 Cartesian products

If A and B are types, then so is their product, written $A \times B$. There will be many *formation rules* of this form, so we adopt a simple convention for stating them. We write

$$\frac{A \text{ is a } Type,\; B \text{ is a } Type}{A \times B \text{ is a } Type.}$$

The elements of a product are pairs, $< a, b >$. Specifically if a belongs to A and b belongs to B, then $< a, b >$ belongs to $A \times B$. We abbreviate this by writing

$$\frac{a \in A,\; b \in B}{< a, b > \in A \times B.}$$

In programming languages these types are generalized to n-ary products, say $A_1 \times A_2 \times \ldots \times A_n$ and called *structures* or *records.*

We say that $< a, b >=< c, d >$ in $A \times B$ iff $a = c$ in A and $b = d$ in B.

In set theory, equality is uniform, but in type theory we define equality with each constructor.

There is essentially only one way to decompose pairs. We say things like, "take the first element of the pair P," symbolically we might say $first(P)$ or $1of(P)$. We can also "take the second element of P," $second(P)$ or $2of(P)$.

2.2.2 Function Space

We use the words "function space" as well as "function type" for historical reasons. If A and B are types, then $A \to B$ is the type of *computable functions* from A to B. These are given by rules which are defined on each a in A and which produce a *unique value.* We summarize by

$$\frac{A \text{ is a } Type, \; B \text{ is a } Type}{A \to B \text{ is a } Type}$$

The notation we use informally comes from mathematics texts, e.g. Bourbaki's *Algebra.* We write expressions like $x \mapsto b$ or $x \overset{f}{\mapsto} b$, the later gives a name to the function. For example, $x \mapsto x^2$ is the squaring function on numbers.

If b computes to an element of B when x has value a in A for each a, then we say

$$(x \mapsto b) \in A \to B.$$

If f, g are functions, we define their equality as

$$\boxed{f = g \quad \text{iff } f(x) = g(x) \quad \text{for all} \quad x \text{ in } A.}$$

If f is a function from A to B and $a \in A$, we write $f(a)$ for the value of the function.

2.2.3 Disjoint Unions (also called Discriminated Unions)

Forming the union of two sets, say $x \cup y$, is a basic operation in set theory. It is basic in type theory as well, *but* for computational purposes, we want to discriminate based on which type an element is in. To accomplish this we put tags on the elements to keep them disjoint. Here we use tag_l and tag_r as the tags, but anything will do.

$$\frac{A \text{ is a } Type,\ B \text{ is a } Type}{A + B \text{ is a } Type}$$

The membership rules are

$$\frac{a \in A}{tag_l(a) \in A + B} \qquad \frac{b \in B}{tag_r(b) \in A + B}$$

We say that $tag_l(a) = tag_l(a')$ iff $a = a'$ and likewise for $tag_r(b)$.

We can now use a case statement to detect the tags and use expressions like

$$\begin{array}{lll} \text{if } x = tag_l(z) & \text{then} \ldots & \text{some expression in } z \ldots \\ \text{if } x = tag_r(z) & \text{then} \ldots & \text{some expression in } z \ldots \end{array}$$

in defining other objects. The test for $tag_l(z)$ or $tag_r(z)$ is computable.

2.2.4 Inductive Types

Defining types in terms of themselves is quite common in programming, often pointers are used in the definition (in Pascal and Ada) but in languages like ML, direct recursive definitions are possible. For example, a list of numbers, L can be introduced by a definition like

$$\textit{define type} \qquad L = N + (N \times L).$$

In due course, we will give conditions telling when such definitions are sensible, but for now let us understand how elements of such a type are created. Basically a type of this kind will be sensible just when we understand exactly what elements belong to the type.

Clearly elements like $tag_l(0), tag_l(1), \ldots$, are elements. Given them it is clear that

$$tag_r(< 0, tag_l(0) >), tag_r(< 1, tag_l(0) >)$$

and generally

$$tag_r(< n, tag_l(m) >)$$

are elements. But given these, we can also build

$$tag_r(< k, tag_r(< n, tag_l(m) >) >).$$

In general, we can build elements in any of the types

$$N + N \times Y$$
$$N + N \times (N + N \times Y)$$
$$N + N \times (N + N \times (N + N \times Y))$$

where we do not use elements of the type Y which is just used as a place holder.

The key question about this process is whether we want to allow *anything else* in the type L. Our decision is **no**, we want only elements obtained by this finite process.

In set theory, we can understand similar inductive definitions, say a *set* L such that $L = N \cup (N \times L)$, as least fixed points of monotonic operators $F : Set \rightarrow Set$. In general, given such an operator F, we say that the set inductively defined by F is the least *F–closed* set, call it $I(F)$. We define it as

$$I(F) = \cap\{Y | F(Y) \subseteq Y\}.$$

We use set theory as a *guide* to justify recursive type definitions. Let Type be the collection of (small) types. It is a type itself, but does *not* include Type, i.e. $Type \notin Type$. We take Type as a new primitive (large) type.

For the sake of defining monotonicity, we introduce a very simple *subtyping relation*, $S \subseteq T$. This holds just when the elements of S are also elements of T, and the equality on S and T is the same. For example,

$$\text{if} \quad S \subseteq T \quad \text{then} \quad N + N \times S \subseteq N + N \times T.$$

Definition Given $F : Type \rightarrow Type$ such that if $T_1 \subseteq T_2$ then $F(T_1) \subseteq F(T_2)$, then write $\mu X.F(X)$ as the *type inductively defined by F*.

To construct elements of $\mu X.F(X)$, we basically just *unwind* the definition. That is,

$$\text{if} \quad t \in F(\mu X.F(X)) \quad \text{then} \quad t \in \mu X.F(X).$$

We say that $t_1 = t_2$ in $\mu X.F(X)$ iff $t_1 = t_2$ in $F(\mu X.F(X))$.

The power of recursive types comes from the fact that we can define total computable functions over them very elegantly, and we can prove properties of elements recursively. (These are the same idea as we will show later.) Recursive definitions are given by this term,

$$\mu\text{–}ind(a; f, z.b)$$

called a ***recursor*** or ***recursive-form.*** It obeys the computation rule

$\mu\text{–}ind(a; f, z.b)$ evaluates in one step to $b[a/z, (y \mapsto ind(y; f, z.b))/f]$.

(Note in this rule we use the notation $b[s/x, t/y]$ to mean that we substitute s for x and t for y in b.)

Typing

The way we figure out a type for this form is given by a simple rule. We say that

$$\mu\text{–}ind(a; f, z.b) \qquad \text{is in type } B$$

provided that $a \in \mu X.F(X)$, and if when Y is a Type, and $Y \subseteq \mu X.F(X)$, and z belongs to $F(Y)$, and f maps Y to B, then b is of type B.

Induction

The principle of inductive reasoning over $\mu X.F(X)$ is just this.

μ–induction

Let $R = \mu X.F(X)$ and assume that Y is a subtype of R and that for all x in Y, $P(x)$ is true. (This is the induction hypothesis.) Then if we can show $\forall z : F(Y).P(z)$, we can conclude

$$\forall x : R.P(x).$$

With this principle and the form μ–*ind*, we can write programs over recursive types and prove properties of them. The approach presented here is quite abstract, so it applies to a large variety of specific programming languages. It also stands on its own as a mathematical theory of types.

2.2.5 Subset Types

Among the objects of mathematics we consider are *propositions*. For example, $0 =_N 0$ and $0 < 1$ are true propositions about the primitive type N. We also consider *propositional forms* such as $x =_N y$ or $x < y$. These are sometimes called *predicates*.

The type of all (small) propositions is Prop. *Propositional functions* on a type A are elements of the type $A \rightarrow Prop$.

We also need types that can be restricted by predicates such as $\{x = N \mid x = 0 \text{ or } x = 1\}$. This type behaves like the *Booleans*.

The general formation rule is this. If A is a type and $B : A \rightarrow Prop$, then $\{x : A \mid B(x)\}$ is a *subtype of* A.

The elements of $\{x : A \mid B(x)\}$ are those a in A for which $B(a)$ is true.

Sometimes we state the formation in terms of predicates, so if P is a proposition for any x in A, then $\{x : A \mid P\}$ is a subtype of A.

We clearly have $\{x : A \mid P\} \subseteq A$.

2.3 Dependent Types

2.3.1 Dependent Products

Suppose you are writing a business application and you wish to construct a type representing the date:

$$\begin{aligned} Month &= \{1, \ldots, 12\} \\ Day &= \{1, \ldots, 31\} \\ \\ Date &= Month \times Day \end{aligned}$$

We would need a way to check for valid dates. Currently, $< 2, 31 >$ is a perfectly legal member of *Date*, although it is not a valid date. One thing we can do is to define

$$\begin{aligned} Day(1) &= \{1,\ldots,31\} \\ Day(2) &= \{1,\ldots,29\} \\ &\vdots \\ Day(12) &= \{1,\ldots,31\} \end{aligned}$$

and we will now write our data type as

$$Date = m : Month \times Day(m).$$

We mean by this that the second element of the pair belongs to the type indexed by the first element. Now, $< 2, 20 >$ is a legal date since $20 \in Day(2)$, and $< 2, 31 >$ is illegal because $31 \notin Day(2)$.

Many programming languages implement this or a similar concept in a limited way. An example is Pascal's *variant records.* While Pascal requires the indexing element to be of scalar type, we will allow it to be of any type.

We can see that what we are doing is making a more general type. It is very similar to $A \times B$. Let us call this type *prod(A; x.B)*. We can display this as $x : A \times B$. The typing rules are:

$$\frac{E \vdash a \in A \quad E \vdash b \in B[a/x]}{E \vdash < a, b > \in prod(A; x.B)}$$

$$\frac{E \vdash p : prod(A; x.B) \quad E, u : A, v : B[u/x] \vdash t \in T}{E \vdash spread(p; u, v.t) \in T}$$

Note that we haven't added any elements. We've just added some new typing rules.

2.3.2 Dependent Function Spaces

If we allow B to be a family in the type $A \to B$, we get a new type, denoted by $fun(A; x.B)$, or $x : A \to B$, which generalizes the type $A \to B$. The rules are:

$$\frac{\mathsf{E}, y : A \vdash b[y/x] \in B[y/x]}{\mathsf{E} \vdash (x \mapsto b) \in fun(A; x.B)} \; new\ y$$

$$\frac{\mathsf{E} \vdash f \in fun(A; x.B) \quad \mathsf{E} \vdash a \in A}{\mathsf{E} \vdash f(a) \in B[a/x]}$$

Example Back to our example Dates. We see that m : Month $\to$ Day$[m]$ is just $fun(\mathsf{Month}; m.\mathsf{Day})$, where Day is a family of twelve types. And $\lambda(x.maxday[x])$ is a term in it.

3 Formalizing Type Theory

Nuprl presents a particular formalization of type theory. There are several others [17, 22, 23, 24, 26, 42]. At the core are ideas from deBruijn [17] and Martin-Löf [34] extended with the notions of *inductive types* and *direct computation* which give Nuprl its unique architecture.

The formalization comes in two parts. First we lay down the structure of *terms* and then *evaluation*.

terms
evaluation

This is the functional programming core of the theory, on top of it, we define the notion of type and proof. These are the basis of logic and mathematics.

type
proof

3.1 Terms

The informal language we used in Naive Type Theory does not reveal any pattern to the notation. We see forms like $x \mapsto b$, $f(a)$, $A \to B$, $\forall x\!:\!A.B$, U_i consisting of infix operators, parenthesized arguments, prefix operators, subscripts, etc. We would like a regular pattern to the terms that is a basis for analyzing and manipulating them. A Lisp-like syntax, $(f\ a_1 \ldots a_n)$ might be adequate, but does not provide the flexibility we want for dealing with binding structure; Lisp only allows $(lambda\ (x_1 \ldots x_n)\, b)$ for binding. So the term structure in Nuprl has the form

$$op(\bar{v}_1.t_1;\ldots;\bar{v}_n.t_n)$$

where op is an operator name, $\bar{v}_i$ is a list of binding variables governing the subterm t_i which is then *scope*. Every occurrence of a variable from $\bar{v}_i$ in t_i is called *bound*. The t_i are subterms.

For display purposes, if $\bar{v}_i$ is empty, we do not show it, and if $n = 0$ we drop the parentheses. Here is a list of all the Nuprl terms we need

term	display form
function($A; x.B$)	$x : A \rightarrow B[x]$
function($A; B$)	$A \rightarrow B$
lambda($x.b$)	$\lambda x.b$
apply($f; a$)	$f(a)$
product($a; x.B$)	$x : A \times B[x]$
product($A; B$)	$A \times B$
pair($a; b$)	$< a, b >$
spread($p; u, v.t$)	$p.u$
	$p.v$
rec($t.F$)	let *rec* $t = F(t)$
rec_ind($h; f, z.t$)	$Y(t)$
void	φ

Syntactic variety comes from the display forms, written in the right column. These are like LaTeXprint macros, but they are also used for input as well.

Syntactic Theory

To understand how to compute with terms, we first need a complete account of certain aspects of the syntax. The main idea we need is the notion of substitution of a term s for occurrences of a variable x in a term t, written

$$t[s/x].$$

We need a very precise definition of this concept. To get it we require the definition of α-equality and renaming of bound variables. We want an account that is mathematically rigorous, yet transparently clear to users. To illustrate how this is accomplished, I refer to the working material by W. Aitken *A Formal Introduction to the Lambda Calculus* [2]. This theory is easy to express in Naive Type Theory. Here are some highlights.

Definition The λterms are the following inductively defined class.

λterms

Variables $x, y, z, \ldots$
If f, a are λterms and v is a variable then

$ap(f; a)$

$\lambda(v.a)$

are λterms

Induction Principle

The induction principle corresponding to this class is given by

$$\frac{P(v) \text{ for } v \text{ a variable} \qquad \begin{array}{c}\text{for all } f, a, b \text{ λterms and } x \text{ a variable} \\ P(f) \& P(a) \Rightarrow P(ap(f; a)) \text{ and } P(b) \Rightarrow P(\lambda(x.b))\end{array}}{\forall t : \lambda\text{term}.\ P(t)}$$

Inductive Definitions

Objects can be defined inductively over the class of λterms using the same induction principle. Here is the definition of the set of free variables, $FV(t)$, of a term t.

$$\begin{array}{lll} FV(t) = \textbf{case} & t \text{ is a variable } v & \textbf{then } \{v\} \\ & t \text{ is an application } ap(p; q) & \textbf{then } FV(p) \cup FV(q) \\ & t \text{ is an abstraction } \lambda(v.b) & \textbf{then } FV(b) - \{v\}. \end{array}$$

Based on the syntactic theory, we can define exactly the kind of substitution we want. The interesting feature is that it must rename bound variables to avoid capture and rename them in a minimal way, i. e. rename as few variables as possible. For example, in

$$\lambda(x.\lambda(y.ap(z; x)))$$

if we substitute x for z, then we must rename the bound occurrences of x, but we do not want to rename them to y, because this will cause y to be renamed in the scope of x.

Aitken gives a precise definition of all this. He uses the concept $t[s/x]t'$ defined as:

Definition We say that t' is a substitution of s for x in t, written $t[s/x]t'$, iff

$x[s/x]s$

$v[s/x]v \quad \text{if } v \neq x$

$ap(p;q)[s/x]ap(p';q') \quad \text{if } p[s/x]p' \text{ and } q[s/x]q'$

$\lambda(x.r)[s/x]\ \lambda(x'.r') \quad \text{if } \lambda(x.r) =_\alpha \lambda(x'.r')$

$\lambda(v.r)[s/x]\ \lambda(v'.r') \quad \text{if } x \neq v,\ \ v' \notin FV(s) \ \text{ or } \ x \notin FV(r) \ \&$
$\qquad \exists r''.\ \lambda(v.r) =_\alpha \lambda(v'.r'') \text{ and } r''[s/x]r'.$

Theorem: $\forall t, s$:LTerm. $\forall x$:Var. $\exists t'$:LTerm

$$t[s/x]t' \ \ \& \ \ t' \text{ is a minimal renaming}$$

From a (constructive) proof of this $\forall\exists$ formula, we can extract a function to compute t' from t and s.

3.2 Evaluation

Inductive Definition

The terms of Nuprl denote data, programs and types; but programs and types can be data too. So we must say how to evaluate every term. There are many ways to approach this mathematically. We can give a denotational semantics, or a reduction (rewrite) semantics, or a structural operational semantics or something else. We choose to base the account on inductively defined relations and partial functions in the style of [13, 35, 39, 43], sometimes called "natural semantics". We use a lazy evaluator [1, 13, 35].

We define first a relation "evaluates to," t *evals_to* t'. Then we define *val* as a map

$$val : \{x : term \mid \exists t : term \ \ x \ evals\_to \ t\} \rightarrow term.$$

We give some cases of the evaluation relation to illustrate the method.

- $\dfrac{f \ evals\_to\ \lambda(\mathsf{x.b}) \quad b[a/x]\ evals\_to\ c}{\mathsf{ap(f;a)}\ evals\_to\ c}$
- $\dfrac{}{\lambda(\mathsf{x.b})\ evals\_to\ \lambda(\mathsf{x.b})}$
- $\dfrac{}{\mathsf{inl(a)}\ evals\_to\ \mathsf{inl(a)}}$
- $\dfrac{d\ evals\_to\ \mathsf{inl(a)} \quad t_1[a/u]\ evals\_to\ c}{\mathsf{decide(d;\ u.}t_1\mathsf{;\ v.}t_2\mathsf{)}\ evals\_to\ c}$
- $\dfrac{d\ evals\_to\ \mathsf{inr(b)} \quad t_2[b/v]\ evals\_to\ c}{\mathsf{decide(d;\ u.}t_1\mathsf{;\ v.}t_2\mathsf{)}\ evals\_to\ c}$

Properties of Evaluation

We can prove rather easily these properties of evaluation.

Theorem:

1. t *evals_to* t' & t *evals_to* t'' implies $t' = t''$.
   (deterministic)

2. t *evals_to* t' implies t' is a value.
   (value producing)

3. If t is a value then t *evals_to* t.
   (idempotent)

Evaluation fragments

In order to allow for the addition of new operators in a systematic way, we index the evaluation process by the operator name. So the actual evaluation relation has the form $t\ \textit{evals\_to}_{op}(f)\, t'$ where f is a parameter. We just gave these fragments for some of the operators. The evaluation relation is then defined as

Definition: t *evals_to* t' iff $t\ \textit{evals\_to}_{opname(t)}\ (\textit{evals\_to})\ t'$

3.3 Types and membership

typehood

Essentially types are collections of terms. This is how they look from "outside the theory." Inside the theory we see types as methods of construction; to specify a type is to say how to build a certain category of mathematical object. In order to say anything interesting about these objects, we need also to say when we regard two of them as equal. We follow Martin-Löf's method for doing this.

Definition: To specify a type

- Designate a canonical name for it (type is a value)
- Say how canonical elements are formed (a method of construction is specified)
- Say when two elements are equal

Definition: Membership

If A is a type, then $a \in A$ means that a *evals_to* a' and a' canonical of type A. Since A is a type, we know what its canonical members are.

Examples of type construction

A new type

Let **1** be a type, the unit type

Let $\bullet$ be a value in **1**, equal to itself, $\bullet =_1 \bullet$ or write $\bullet = \bullet$ in **1**.

product(1; 1) is a type pair(•; •) is an element
function(1; 1) is a type λ(x.x) λ(x.•) are elements.

Also, λ(x.x)=λ(x.•) in function(1; 1)

Membership

The notion of type membership, written $a \in A$ for A a type, is defined by the mutually recursive inductive definition of typehood given above. It is easy to see that membership is not decidable.

Let Φ_i be the *ith* Turing machine in some acceptable indexing,

Define

$$H_i(x) := \text{if } \Phi_i(i) \text{ halts in } x \text{ steps then void else } \mathbf{1} \text{ fi}$$

$$\lambda(x.\bullet) \in x : N \to H_i(x) \quad \text{iff} \quad \Phi_i(i) \text{ diverges}$$

3.4 Derivations of membership judgments

We need some method of systematically showing that $a \in A$. Since the definition is inductive, we can resort to some standard presentation method for membership in inductively defined types as discussed in section 2 on Naive Type Theory. The result is a calculus or logic for making what Martin-Löf calls *membership judgments.* The form of this judgment is

$$\vdash a \in A$$

which is read: we know that a belongs to type A.

The definition of

$$\vdash \lambda(x.b) \in A \to B$$

suggest that we also want *hypothetical judgments*, such as

$$x : A \vdash b \in B.$$

In general, a *sequent*

$$x_1 : A_1, x_2 : A_2, \ldots, x_n : A_n \vdash g \in G$$

means: if A_1 is a type, x_1 an element, A_2 a type for x_1 in $A_1, \ldots$, then G is a type with element g.

Rules

We present the inductive definition in terms of rules for sequents. We organize them *top-down*, e. g.

$$\begin{array}{l} \vdash \lambda(x.b) \in A \to B \quad \text{by} \ \_\_\_\_\_\_ \\ \quad x : A \vdash b \in B \end{array}$$

Nuprl supports a problem-solving style that is *top-down* or *goal-driven* (like Prolog and λProlog [38]). In this style, the goals is

$$\vdash G$$

thought of as

$$\vdash ? \in G$$

starting with G we try to synthesize g.

Rules in this style look like

$$\begin{array}{l} \vdash A \to B \ \text{ by } \ D0\,[\mathit{new}\ x] \\ x : A \vdash B \end{array}$$

The rule name Di means decompose hypothesis i or if $i = 0$ decompose the conclusion.

Sample Proof

$\vdash A \rightarrow \mathbf{1}$ by $D0$ [*new* y]
$y : A \vdash \mathbf{1}$ by $\bullet$
$\vdash \bullet \in \mathbf{1}$

Extract $\lambda(y.\bullet)$ as element.

Proof Trees

In Nuprl, derivations are organized into proof trees. Here is a diagram.

$$\vdash G \;\; by\ r_0$$

$$H_1 \vdash G_1 \;\; by\ r_1 \qquad\qquad\qquad H_2 \vdash G_2 \;\; by\ r_2$$

$$H_{11} \vdash G_{11} \qquad H_{12} \vdash G_{12} \qquad\qquad H_{21} \vdash G_{21} \;\; by\ r_{21}$$

We can define these trees inductively in Naive Type Theory as follows:

sequent = (term list) × term
justification = {rule names} + ...
proof = sequent × justification × proof list

3.5 Completeness

The terms, evaluation fragments, and rules for type membership are intended to formalize Naive Type Theory. The display notation and the informal semantics we have given suggest the connections between formal and informal accounts. I want to just mention some of the highlights by referring to the presentation in section 2.

Primitive Types

Nuprl has void for the empty type and takes the integers, int, as primitive. But N can be defined inductively or as a subset of int, e. g.

$$N = \{z : \mathsf{int} \mid 0 \leq z\}$$

Compound Types

The display forms indicate the correspondence between the type constructors and Nuprl terms, for example the Cartesian product, $A \times B$, is product(A; B). The tags for discriminated unions are formalized as inr(b), inl(a). The informal function notation, $x \mapsto b$ is formalized as λ(x.b) of course.

The inductive type $\mu x.F(x)$ is formalized as rec(x.F(x)), and the recursor, $\mu-ind(a; f, z.b)$ is rec_ind(a; f, z.b).

Universes

The formal type theory must make some distinctions not seen in the naive version. In order for types to be values, they need to belong to a type. We might try adding a single new type, say *Type*, for this purpose. But then it must belong to a type. We might try attaining closure by postulating.

$$Type \in Type.$$

But this causes many troubles. It allows us to inhabit every type [21] and permits non-terminating functions in the function spaces [28]. So following the traditions of predicative type theory [45, 34, 19, 20], we introduce a hierarchy of types. Following [35] we call them *universes*. They are denoted $\mathsf{U}_1, \mathsf{U}_2, \ldots$, and we have that

$$\mathsf{U}_i \in \mathsf{U}_{i+1} \quad \text{and}$$
$$\text{if } A \in \mathsf{U}_i \text{ then } A \in \mathsf{U}_j \text{ for } i < j.$$

Logic is introduced into type theory by taking *propositions-as-types*. This is well explained elsewhere, e. g. [12, 13, 17, 23, 34, 35]. According to this principle,

$$\mathsf{Prop}_i = \mathsf{U}_i.$$

4 Reflecting Formal Type Theory

4.1 The Basic Ideas

We used the Naive Type Theory of section 2 to present the Nuprl type theory, e. g. *term* is an inductive type. Nuprl is an approximation to the naive theory. A natural question to ask is whether we could use Nuprl to "define itself." We know from Gödel's incompleteness theorem and Tarski's truth theorem that we cannot define Nuprl completely inside itself. But how close can we come?

It is easy to see that the basic syntax and the evaluation semantics can be completely defined. We already know similar results from experience defining Lisp evaluation in Lisp.

It is also easy to imagine that we can define the notion of a proof internally and the concept of a tactic. This again is clear from the ability to define these concepts nicely in ML as actually done in LCF, HOL, and Nuprl.

What is new and exciting here is this:

- We can define *terms*, *evaluations*, and *proof* naturally by just transcribing the informal definition. The formal definition clarifies the logic.

- The internal definition allows us to write theorem-proving tactics as Nuprl programs and prove properties of them.
- The internal account of proof enables us to define a new proof rule, the *reflection rule*, which is the basis for solving the design problem posed by decision procedures (see the introduction).

In this section we will examine the first two points and treat the last one in the next section. Most of this is conceptually clear, but the details are critical. I refer you to other writings for most of those. There are however a few subtle points to notice.

In reflecting the class *term* we will define an internal inductive type, Term. We have shown elsewhere that Term represents *term*. One tricky point here is to avoid the following infinite regress. For each operator name, like *int* or *spread*, we add a new operator name that denotes it, say **int** and **spread**. (We actually use opid{int:opname} and opid{spread:opname}.) Now we have new operators, **int** and **spread**, and we need names for them; we might try **int1** and **spread1** or something. But then we would need **int2** and **spread2**, etc. How to stop the regress? First notice this; all reflected operators have the form opid{o:opname}, and this is the generic form of operators. So they get reflected like any other operator, essentially op{i:family} is represented as *a pair*, <reflected operator, reflection of the list of pairs of index and index family name>.

Another subtle point arises with reflecting evaluation. We do this by adding new operators Evals_To and Val. But we need to define evaluation fragments for them which are self-referential. This is explained in Aitken's thesis.

One of the most subtle points comes up in the definition of *proof*. The reflected concept uses PreProof and Proof. We need to specify how tactics build proofs, and this must be part of the definition of *proof*. But a tactic is a function that builds elements of Proof. So the definition of the informal concept of *proof* requires that we anticipate the definition of Proof and refer to it. This requires a delicate fixed point construction that appears in [6].

Dealing with these points correctly and formally takes a lot of care and relies on many details. But using the mechanism is quite natural. It is much like replacing ML or SML in a *tactic-oriented* prover with Nuprl's programming language.

4.2 Terms

Recall that terms have the form

$$op(\bar{v}_1.t_1; \ldots; \bar{v}_n.t_n)$$

where *op* is an operator. Operators actually have structure because some of them come in *families*. For example, universes (displayed as U_i) are nullary operators indexed by natural numbers. Generally an operator has the form

$$o\{i : f\}$$

where *o* is an *operator name*, and $i : f$ is a *tagged parameter* consisting of *index family*, f. For universes, the operator is

$$\mathsf{univ}\{\mathsf{i} : \mathsf{nat}\}.$$

To reflect operators, we need to reflect operator names and index families. The *index families* include **opname, ifname, var, nat**. There are others. In order to specify the type representing these families, we need the new type constructor IFam(x); if x represents an index family name, then IFam(x) is a type representing the corresponding family. For example, IFam(ifid(nat)) is the type $\{z : \mathsf{int} \mid 0 \leq z\}$.

Notice, we need an operator to represent the index family name; this operator is ifid ("index family id"). It produces an operator name inside Nuprl corresponding to the various index families. In the Nuprl literature, we use these definitions

$$IFName = \mathsf{IFam(ifid(ifname))}$$
$$OpName = \mathsf{IFam(ifid(opname))}.$$

The primitive type *OpName* is used to represent the operator names. Its elements look like *opid{p}* for some tagged parameter *p*. For example, the representation of the opname univ is opid{univ:opname}.

To illustrate the level of precision that we achieve with this machinery, let us look at the official definition of variables. The terms called variables are

$$\mathsf{var}\{\mathsf{v} : \mathsf{var}\}$$

These are formed with the var operator name indexed by elements from a mathematical class, an index family (also named var). At this level, we could choose any discrete mathematical class for var, say character strings. To be precise, we say that var is the class of finite sequences of *variable characters*. We require a subset of the ASCII characters as specified in the Nuprl manual.

Now to represent the variables *inside* Nuprl, we choose one of the Nuprl types that will represent this index family, and axiomatize it to be equal to IFam(ifid(var)). We have chosen to represent the variable characters by the type N_{256}, natural numbers from 0 to 256. Variables are lists of these. The definition and axioms are

$$VarChar = N_{256}$$
$$\mathsf{IFam(ifid(var))} = VarChar\ list.$$

Let us take a closer look at **OpName**. This is a new type. What are its elements? Intuitively the elements are the terms that represent the actual operator names. Here are some of those names.

operator	representation
var	opid{var}
univ	opid{univ}
natnum	opid{natnum}
void	opid{void}
lambda	opid{lambda}

We use the convention that opid{x} stands for opid{x:opname}. This can be done for any fixed number of operator names because we can specify in the definition which index family is associated with which parameter. So in practice, we can eliminate the tags. They are needed to deal with the open-ended nature of the class OpName. This type must be open-ended because the mathematical class of actual operator names is open-ended.

The operator opid{x} produces canonical values. These are terms that evaluate to themselves.

Here is the official definition of Term.

Definition

Parameter = i:IFName × IFam(i)
OP = OpName × Parameter list
VarName = VarChar list
Term = rec(t.OP ×(VarName list × t) list)

This tells us everything we need to know about terms. From the inductive type definition we get a recursor defining objects inductively over Term and we get an induction principle.

The small theory of the lambda calculus can be seen as a special case of a theory of Term if we just take these definitions and display forms.

Display $<<$ opid{var}, $[a_1, \ldots, a_n]$ $>$, nil $>$ *as* $implode([a_1, \ldots, a_n])$
Define $\lambda(v.t)$ *as* $<<$ opid{lambda}, nil $>$, $[< [v], t >]$ $>$
$ap(f; a)$ *as* $<<$ opid{apply}, nil $>$, $[<$ nil, $f >, <$ nil, $a >]$ $>$

where $[a_1, \ldots, a_n]$ is a notation for lists and implode takes a list of characters to an atom.

4.3 Sequents

Once we have represented terms, it is quite an easy matter to represent a sequent $H \vdash G$. The goal G is a term and H is a list of terms. Actually, since some of the hypotheses are treated specially, we need to tag them with a boolean value, and hypotheses are labelled. So the actual definitions are these.

Definition

$$\text{Hypothesis} = \mathsf{VarName} \times (\mathsf{Term} \times \mathsf{Boolean})$$
$$\text{Sequent} = \mathsf{Hypothesis\ \ list} \times \mathsf{Term}$$

4.4 Primitive Proofs

Essentially, a proof is a finite tree whose notes are sequents and justifications connected in certain ways. The definitions can be given in the form

$$\mathsf{PreProof} = \mathsf{rec(p.\ Sequent} \times \mathsf{(Justification} \times \mathsf{p\ list))}$$
$$\mathsf{Proof} = \{\mathsf{p{:}PreProof} \mid \mathrm{IsAProof(p)}\}$$

The justifications explain why the goal sequent follows from the subgoals. We allow incomplete proofs; they are recognizable because the justification is a special element and no subgoals are generated.

Another kind of justification is a primitive rule name. These are the simple justifications, and if we were not interested in automating the development of proofs, this would be all that is necessary to completely define the concept of a proof. The only condition to be checked by the predicate *IsAProof* would be that the right subgoals were generated by the rules. We call such proofs *primitive*.

4.5 Tactics and Automation

Nuprl uses *tactics* to automate parts of the proof development process. A tactic is a function that builds part of a proof; it is given a sequent, and it produces a proof whose goal is the sequent. The resulting proof might be incomplete (usually is), and it might not be possible to complete it if the approach is a dead end.

The general idea of *tactic-oriented theorem proving* was pioneered by Edinburgh LCF [25]. Nuprl has extended this method by introducing the concept of a *tactic-tree* and by introducing the concept of a transformation tactic. We will not discuss the later idea.

The idea of a tactic-tree is that a justification for a proof step can be a tactic. The subproof built is not displayed, but the frontier of the subproof is listed as the new subgoals to be proved. This makes the tactic look like a new proof rule. So, by writing tactics, users extend the logic, but in a totally *safe* way. The tactics actually build primitive proofs, so there is no increase in the axiomatic basis of the theory. When a proof is complete, there is a primitive proof built in the background.

Here is an example of the internal proof produced by a tactic called `ChainThruHyps`.

1. $A \Rightarrow B_1$ 2. $B_1 \Rightarrow B_2$ 3. $B_2 \Rightarrow B_3$

$\vdash A \Rightarrow B_3$ by `ChainThruHyps`

4.A
$\vdash B_3$ by $D1$
$\vdash A$ by Hyp

5.B_1
$\vdash B_3$ by $D2$
$\vdash B_1$ by Hyp

6.B_2
$\vdash B_3$ by $D3$
$\vdash B_2$ by Hyp

7.B_3
$\vdash B_3$ by Hyp

The rule used in chaining is

$$\boxed{\begin{array}{l} H, f : A \Rightarrow B, H' \vdash G \;\; by\ D\ f \\ \quad H, f : A \Rightarrow B, H' \vdash A \\ \quad H, f : A \Rightarrow B, H', B \vdash G \end{array}}$$

The more conventional form is

$$\frac{H \vdash A \qquad\qquad H, A \Rightarrow B, B \vdash G}{H, A \Rightarrow B \vdash G}$$

4.6 Tactic-tree Proofs

Tactic-tree proofs are just proofs which allow tactic applications as justifications. Since they can be reduced to primitive proofs by executing the tactic, one might think that it is not necessary to give a reflected account of them. But there are reasons for reflecting.

The first reason is that we want to provide a theoretical basis for the idea that if we prove that the tactic works, then we can treat it as a new proof rule. This will both raise the level of abstraction of proofs and save the resources involved with executing the tactic call. Doing this requires that we have a formal definition of a tactic-tree proof since it is exactly objects in this class which will serve as the more abstract proofs.

A second reason to formalize the concept is that we wish to provide a rigorous account of what Nuprl actually does. This account will be a *formal explanation* of the system. This second reason influences the way we formalize the concept.

A very interesting problem arises as soon as we try to formalize tactic trees because we are now entering the realm of self-reference. A tactic is a Nuprl function, so it must operate on some internal data type, something we will eventually call **Proof**. But we use the tactic in explaining the top level notion of proof, the concept we have been denoting as *proof*. This concept does not even make sense until we have said how Proof represents *proof*. This involves some level of reflection already. *

If we are at a node with sequent s_0 whose subgoals are generated by t as justification, this means that $t(s_0)$ is Proof and the frontier of $t(s_0)$, say $s_1 \ldots, s_n$ represent the sequents which are subgoals of s_0. This would suggest that justifications are

$$1 + \text{primitive\_rules} + \text{Proof}$$

In actual practice, we do not write a function application as a justification. We care about the exact computational behavior of the algorithm, so we really write a term t which denotes a function (Sequent $\rightarrow$ Proof). Since we want to analyze the term, it is more useful, if somewhat more complex, to use an element from Term.†

We spell this out precisely using these abbreviations. Let the *premises* of a proof be the unproven sequents at the leaves, the frontier of the proof tree. Given this frontier of sequents, let $\bar{p}_{1\ldots n}$ denote the sequence of them. Given a proof p, let $root(p)$ be the root of the tree, which is the main goal or the theorem being proved. If j is a proof rule, then let $instance(j, s, roots(\bar{p}_{1\ldots n}))$ be a predicate that checks that the roots of the subtrees $\bar{p}_{1\ldots n}$ are the correct subgoals generated from s by j. Finally we use the relation j *reps* x to mean that the term j *represents the pre–proof* x.

When we write these concepts in bold letters,

say $\mathbf{premises}(x)$, $\mathbf{root}(p)$, $\mathbf{instance}(j, s, \mathbf{roots}(x))$,

they apply to the internal notion of proof, Proof.

Now we have these definitions.

Definition

pre-proof = sequent × justification × pre-proof list

justification = $1 + \text{rule} + Term$

proof = $\{p : pre\!-\!proof \mid is\_a\_proof(p)\}$

*Let us assume that the type **Proof** exists and it represents *proof*. Then a justification could be a function $t \in$ **Sequent $\rightarrow$ Proof.**

†In a system like HOL based on Standard ML, say SML/HOL, these issues don't arise. In this system, the tactics are not part of the logic.

$is\_a\_proof(< s, j, \bar{p}_1 \ldots n >) =$

case j of

premise $\rightarrow \quad n = 0$

rule $\rightarrow \quad instance(j, s, roots(\bar{p}_{1\ldots n}))$

tactic $\rightarrow \quad \exists x : pre-proof.$

$root(x) = s \ \&$

$premises(x) = roots(\bar{p}_{1\ldots n})$

$\& \ j \ reps \ x \ \& \ is\_a\_proof(x)$

where $is\_a\_proof(p_i)$.

Tactic-tree Diagram

$< s, j, [P_1, \ldots, P_n] >$

$j = t$ a Term, tactic text

$\exists x$:pre-proof.______

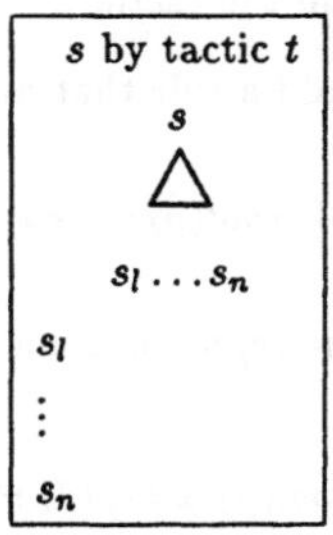

Tactic-trees Internally

Definition

PreProof == rec(PP. Sequent × Justification × PP list)

Justification == 1 + Rule + Term

Proof == $\{p$: PreProof | IsAProof(p)$\}$ where

$$
\begin{array}{l}
\text{IsAProof}(< S, J, [P_1, \ldots, P_n] >) = \\
\quad \text{case } J \text{ of} \\
\qquad \textbf{premise} \rightarrow \; n = 0 \\
\qquad \textbf{rule} \rightarrow \; \textbf{\textit{instance}}(J, S, \textbf{\textit{roots}}(P_1, \ldots, P_n)) \\
\qquad \textbf{tactic} \rightarrow \; \exists x : \mathsf{PreProof}. \\
\qquad\qquad \textbf{\textit{root}}(x) = S \;\; \& \\
\qquad\qquad \textbf{\textit{premises}}(x) = \textbf{\textit{roots}}(P_1, \ldots, P_n) \\
\qquad\qquad \& \; J \; \textit{Reps} \; x \;\; \& \;\; \textit{IsAProof}(x) \\
\quad \text{where } \textit{IsAProof}(p_i).
\end{array}
$$

Note the internal Reps map.

4.7 Reflection Rule

The material in the section is from the article *The Semantics of Reflected Proof* [6]. We want to take full advantage of the fact that we can reason about tactics. In particular, if we demonstrate that a tactic t generates a proof of some sequent s, we would like to be able to cite this demonstration rather than run the tactic.

Our general plan for doing this is to add a rule that essentially asserts that if

$$\exists p\text{: } \mathsf{Proof}.\ root(p) = rep(H \vdash G)$$

then $H \vdash G$ is true, where $rep(s)$ is the representation of a sequent s. The rule might be

$$
\begin{array}{ll}
\textit{Refl} & H \vdash G \text{ by Reflection} \\
& \quad \vdash \exists p : \mathsf{Proof}.\ root(p) = rep(H \vdash G).
\end{array}
$$

But the rule is not sound. From Löb's theorem on proof predicates [7], we suspect that this is not valid. Aitken discusses the situation in his thesis [3]. The idea is this. Given

$$\boxed{\exists p : \mathsf{Proof}.\ root(p) = rep(H \vdash G) \mid R}$$

to conclude $H \vdash G$ we need a new rule, say $Refl$, but this is not part of $proof$. If we add it, we get a new notion of $proof_0$, say $proof_1$. If we allow reflection over these, we get $proof_2$, etc. Can we close this off? Almost! We can close it syntactically, but what does it mean? Notice that the extractor which is needed to build a canonical proof from $\boxed{\text{R}}$ depends on this meaning.

Reflected Pre-Proofs

To get *syntactic closure* under reflection, we want a rule like *Refl* above. Where Proof reps proof with Proof in the rule itself. To accomplish this, we first parameterize proof to *proof*(Q) where Q is the term used in the rule. This forces us to consider Proof(Q).

Syntactic Closure of Proof Definition

Define *proof*(Q) to allow the rule

$$\boxed{\begin{array}{l} H \vdash G \text{ by Refl}, i, t \\ \qquad \vdash \exists p : \ Q \ .root(p) = rep(H \vdash G) \\ \qquad\qquad\quad \uparrow \end{array}}$$

In the [6] we show how to take a *syntactic fixed point* through Q.

Here is another approach.

Add to the library a definition

$$\begin{array}{ll} \text{PF} == & \text{Term representing} \\ & \text{proof(PF)} \end{array}$$

This is straightforward from PreProof since PF is just a term.

The actual rule uses an indexing mechanism to keep track of how deeply the use of reflection is nested. The function *level*(p) gives the maximum of the level numbers occurring in reflection rules of p. Using this we incorporate the rule in the definition of *proof* as follows.

$$proof = \{p : pre-proof \mid is\_a\_proof(p)\}$$

$$\begin{array}{l} \text{is\_a\_proof}(< s, j, [p_1, \ldots, p_n] >) \\ \quad \text{case } j \text{ of} \\ \qquad \textbf{premise} \rightarrow \quad n = 0 \\ \qquad \textbf{rule} \rightarrow \quad \ldots \\ \qquad \textbf{tactic} \rightarrow \quad \ldots \\ \qquad \textbf{refl} \rightarrow \quad p_l = \vdash \exists p : PF.\ root(p) = rep(s) \quad (where\ j = < i, t >) \\ \qquad\qquad\qquad \&\ level(p) < i \\ \qquad\qquad\qquad \&\ t\ reps\ roots(\bar{p}_{2 \ldots n})\ \&\ is\_a\_proof(p) \\ \quad \text{where } is\_a\_proof(p_i) \text{ for } i = 2, \ldots, n. \end{array}$$

4.8 Validity Theorem

Definition A proof is valid if the goal is true when the premises are.

Theorem: Proofs are valid.

Pf: By induction on level, n, and for each n by induction on the subproof relation.

Let $\Delta = (s, r, \Delta_1, \ldots, \Delta_k)$ be a proof whose premises are true. Let $\bar{\Delta} = (\Delta_1, \ldots, \Delta_k)$.

Consider the cases for rule r.

1. $r = \bullet$, s is a premise, hence true.

2. r is a primitive rule, then s is true by validity of the primitive rules.

3. r is a tactic with term t. Then t represents a subproof Δ' which is valid by induction. The premises of Δ' are roots of $\bar{\Delta}$, they are true, so the goal of Δ' is, which is s.

4. r is Refl i, t. So the goal of Δ_1 is true by induction. This means that there is a proof Δ' *of level less than* i; thus Δ' is valid. The premises of Δ' are the roots of $\Delta_2, \ldots, \Delta_k$, hence true, so s is true as well. QED.

5 Applications

We will demonstrate an application of reflection to the task of adding sound decision procedures to tactic-oriented provers. For the sake of illustration, we take a very simple and familiar example — the decision problem for classical tautologies. A more realistic example would be decision procedures commonly used such as *arith* or *sup−inf*, but those are too complex to treat in these lectures, and the tautology procedure is used in some provers, see [18].

5.1 Sample Decision Procedure

The tableau decision procedure for classical propositional formulas is especially well understood (and still current, see the account in [40]). The idea is that given a formula, we try to systematically falsify it by building up a truth assignment to the subformulas, eventually including the atomic ones (the variables). If we succeed, then the formula is not a tautology as the falsifying assignment attests. If we fail in this systematic attempt, then we can show that the resulting formula is valid and has a propositional proof, that is, one using only rules of the propositional calculus. To illustrate, consider the following formula.

Decision Procedure

Show that

$$((A \Rightarrow B_1) \mathbin{\&} (B_1 \Rightarrow B_2) \mathbin{\&} (B_2 \Rightarrow B_3)) \Rightarrow (A \Rightarrow B_3) \quad \text{is a tautology.}$$

Try to assign it the truth-value **F**

$T(A \Rightarrow B_1 \,\&\, B_1 \Rightarrow B_2 \,\&\, B_2 \Rightarrow B_3) \quad F(A \Rightarrow B_3)$

Decomposing $(A \Rightarrow B_3)$ in the same way we get: TA, FB_3

Now force the assignment to the other atoms in the natural way:

$T(A \Rightarrow B_1)$

$T(B_1 \Rightarrow B_2)$

$T(B_2 \Rightarrow B_3)$, TA, FB_3

Now consider these observations.

FA is not possible, so TB_1

FB_1 is not possible, so TB_2

FB_2 is not possible, so TB_3 which contradicts FB_3

5.2 Formalizing the Decision Procedure

There are many ways to present the tableau decision procedure formally. In Nuprl we have looked at two of them [9, 14]. Caldwell's account [9] is based on the approach in [14].

The methods referred to above can also be extended to the full set of connectives and to a notion of propositional proof derived from the internal definition of Proof. The details of how to do this are in [5].

Define PropSeq as a subtype of Term and PropProof as a subtype of Proof. Informally these are the definitions.

Definition: PropSeq = $\{s$: Sequent | hypotheses of s are variable declarations, the goal of s is a proposition built from $\&, \vee, \Rightarrow$, *False* using the declared variables.$\}$

PropProof = $\{p$: *Proof* | the goal of p is PropSeq and all rules are primitive and are rules of the classical propositional calculus.$\}$

As an abbreviation let *P or not* mean $P \vee \neg P$, then based on the proof method in [5, 14] we can constructively prove:

Theorem (*dp*):

$$\forall s : \text{PropSeq}.\ (\exists \text{p} : \text{PropProof}.\text{root(p)} = \text{s}) \text{ or not}$$

Given a sequent $H \vdash G$ which is propositional, we have that the computable function extracted from *dp*, called here *extract*(*dp*), decides whether or not there is a propositional proof.

5.3 Applying the Formal Decision Procedure

Let $d = extract(dp)\,(rep(H \vdash G))$. The value of d is either inl(p) for some proof p or inr(f) for some function f. In the case of inl(p), we have $p \in \exists p : \mathsf{PropProof}.root(p) = rep(H \vdash G)$. Notice that P =decide(d; u.u; v.v). Let $\hat{d}$ be the term decide(p; u.u; v.v).

We can now prove $H \vdash G$ by reflection if it is true.

$$\begin{array}{l} H \vdash G \text{ by Refl 1, nil} \\ \quad \vdash \exists p : \mathsf{Proof}.root(p) = rep(H \vdash G) \text{ by } D\ \hat{d} \end{array}$$

We can write a tactic

Decide(dp)

which checks d for inl(p), writes $\hat{d}$ and uses it as the witness, or fails.

$$H \vdash G \text{ by Decide(dp).}$$

This simple example illustrates the essential ideas behind the method of using relfection to justify decision procedures and for type checking in programming languages [10]. I think that this will be an important technique for modern provers that employ decision procedures. There are many other uses of reflection, and I recommend 'looking at articles by Boyer and Moore [8], Weyrauch [50], Howe [29], as well as [3, 6, 16, 37, 47, 48].

Acknowledgements

I want to thank Kate Ricks for preparing this manuscript in LaTeX and for cheerfully tolerating such complexity in her first such project. I appreciate Stuart Allen's comments on an earlier draft.

References

[1] S. Abramsky. The lazy λ-calculus. In D. A. Turner, editor, *Logical Foundations of Functional Programming*, pages 65–116. Addison-Wesley, 1990.

[2] W. Aitken. A formal introduction to the lambda calculus. Computer Science Dept., Cornell University, Ithaca, NY, 1993.

[3] W. Aitken. *Metaprogramming in Nuprl Using Reflection.* PhD thesis, Computer Science Dept., Cornell University, Ithaca, NY, 1994.

[4] W. Aitken and R. L. Constable. Reflecting on Nuprl Lessons 1-4. Technical report, Cornell University, Computer Science Dept., Ithaca, NY, 1992. To appear.

[5] W. Aitken, R. L. Constable, and J. Underwood. Using reflected decision procedures. Technical report, Computer Science Dept., Cornell University, Ithaca, NY, 1993.

[6] S. F. Allen, R. L. Constable, D. J. Howe, and W. Aitken. The Semantics of Reflected Proof. In *Proc. of Fifth Symp. on Logic in Comp. Sci.*, pages 95–197. IEEE, June 1990.

[7] G. S. Boolos and R. C. Jeffrey. *Computability and Logic, Second Edition.* Cambridge University Press, 1980.

[8] R. S. Boyer and J. S. Moore. Metafunctions: Proving them correct and using them efficiently as new proof procedures. In *The Correctness Problem in Computer Science*, pages 103–84. Academic Press, New York, 1981.

[9] J. Caldwell. A constructive proof of propositional completeness in Nuprl. implementation notes, June 1993.

[10] L. Cardelli and P. Longo. A semantic basis for Quest. In *Journal of Functional Programming*, pages 1:417–458, 1991.

[11] T. Chan. An algorithm for checking PL/CV arithmetic inferences. In G. Goos and J. Hartmanis, editors, *An Introduction to the PL/CV Programming Logic,* Lecture Notes in Computer Science, Vol. 135, pages 227–264. Springer-Verlag, 1982.

[12] R. L. Constable. Assigning meaning to proofs: a semantic basis for problem solving environments. *Constructive Methods of Computing Science,* NATO ASI Series, Vol. F55, pages 63–91, 1989.

[13] R. L. Constable et al. *Implementing Mathematics with the Nuprl Development System.* Prentice-Hall, NJ, 1986.

[14] R. L. Constable and D. J. Howe. Implementing metamathematics as an approach to automatic theorem proving. In R. Banerji, editor, *Formal Techniques in Artificial Intelligence: A Source Book*, pages 45–76. Elsevier Science Publishers (North-Holland), 1990.

[15] T. Coquand and G. Huet. The Calculus of Constructions. *Information and Computation*, 76:95–120, 1988.

[16] M. Davis and J. T. Schwartz. Metamathematical extensibility for theorem verifiers and proof checkers. *Comp. Math. with Applications*, 5:217–230, 1979.

[17] N. deBruijn. A survey of the project Automath. In *To H.B. Curry: Essays in Combinatory Logic, Lambda Calculus, and Formalism*, pages 589–606. Academic Press, 1980.

[18] R. Dyckhoff. Contraction-free sequent calculi for intuitionistic logic. In *The Journal of Symbolic Logic*, pages Vol.57, Number 3, September 1992.

[19] S. Feferman. Polymorphic typed lambda-calculi in a type free axiomatic framework. *Contemporary Mathematics*, 106:101–135, 1990.

[20] S. Feferman. Logics for termination and correctness of functional programs. Technical report, Stanford University, Dept. of Mathematics, 1993.

[21] J.-Y. Girard. The system F of variable types: Fifteen years later. *Theoretical Computer Science*, 45:159–192, 1986.

[22] J.-Y. Girard. *Proof Theory and Logical Complexity, Volume 1.* Bibliopolis, Napoli, 1987.

[23] J. Y. Girard, P. Taylor, and Y. Lafont. *Proofs and Types.* Cambridge Tracts in Computer Science, Vol. 7. Cambridge University Press, 1989.

[24] M. Gordon and T. Melham. *Introduction to HOL.* University Press, Cambridge, 1993.

[25] M. Gordon, R. Milner, and C. Wadsworth. *Edinburgh LCF: a mechanized logic of computation,* Lecture Notes in Computer Science, Vol. 78. Springer-Verlag, NY, 1979.

[26] S. Hayashi and H. Nakano. *PX: A Computational Logic.* MIT Press, Cambridge, MA, 1988.

[27] C. Hoare. Notes on data structuring. In *Structured Programming.* Academic Press, New York, 1972.

[28] D. J. Howe. The computational behaviour of Girard's paradox. In *Proc. of Second Symp. on Logic in Comp. Sci.*, pages 205–214. IEEE, June 1987.

[29] D. J. Howe. Computational metatheory in Nuprl. In E. Lusk and R. Overbeek, editors, *9th International Conference on Automated Deduction,* Lecture Notes in Computer Science, Vol. 310, pages 238–257. Springer-Verlag, New York, 1988.

[30] S. Igarashi, R. London, and D. Luckham. Automatic program verification I: a logical basis and its implementation. *Acta Informatica*, 4:145–82, 1975.

[31] T. B. Knoblock and R. L. Constable. Formalized metareasoning in type theory. In *Proc. of the First Symp. on Logic in Comp. Sci.*, pages 237–248. IEEE, 1986.

[32] D. Kozen. *Complexity of Finitely Presented Algebras.* PhD thesis, Computer Science Department, Cornell University, Ithaca, New York, 1977.

[33] D. Kozen, M. Ben-Or, and J. Reif. The complexity of elementary algebra and geometry. In *Proc. 16th ACM Symp. Theory of Comput.*, pages 457–464, 1984. Invited special issue *J. Comput. Syst. Sci.* 32(2):251-264, 1985.

[34] P. Martin-Löf. An intuitionistic theory of types: predicative part. In *Logic Colloquium '73*, pages 73–118. North-Holland, Amsterdam, 1973.

[35] P. Martin-Löf. Constructive mathematics and computer programming. In *Sixth International Congress for Logic, Methodology, and Philosophy of Science*, pages 153–75. North-Holland, Amsterdam, 1982.

[36] D. A. McAllester. *ONTIC: A Knowledge Representation System for Mathematics.* MIT Press, Cambrige, Mass., 1989.

[37] J. McCarthy. A basis for a mathematical theory of computation. *Comput. Program. and Formal Syst.*, pages 33–70, 1963. Amsterdam:North-Holland.

[38] D. Miller and G. Nadathur. A logic programming approach to manipulating formulas and programs. In *IEEE Symp. on Logic Programming.* ACM, 1987.

[39] R. Milner, M. Tofte, and R. Harper. *The Definition of Standard ML.* The MIT Press, 1991.

[40] A. Nerode and R. Shore. *Logic for Applications.* Springer-Verlag, New York, 1994.

[41] S. Owre, J. Rushby, and N. Shankar. PVS: An integrated approach to specification and verification. Preprint, January, 1992.

[42] F. Pfenning. Elf: a language for logic definition and verified metaprogramming. In *Proc. of Fourth Symp. on Logic in Comp. Sci.*, pages 313–321, 1989.

[43] G. Plotkin. A structural approach to operational semantics. Technical Report DAIMI-FN-19, Aarhus University, Aarhus University, Computer Science Dept., Denmark, 1981.

[44] J. Rushby, F. von Henke, and S. Owre. An introduction to formal specification and verification using EHDM. Technical Report CSL-91-2, Computer Science Laboratory, SRI International, February, 1991.

[45] B. Russell. Mathematical logic as based on a theory of types. *Am. J. Math.*, 30:222–62, 1908.

[46] R. Shostak. A practical decision procedure for arithmetic with function symbols. *J. Assoc. Comput. Mach.*, 26:351–360, 1979.

[47] B. Smith. Reflection and semantics in lisp. *Principles of Programming Languages*, pages 23–35, 1984.

[48] C. Smorynski. *Self-Reference and Modal Logic*. Springer-Verlag, NY, 1985.

[49] A. Troelstra. *Metamathematical Investigation of Intuitionistic Mathematics,* Lecture Notes in Mathematics, Vol. 344. Springer-Verlag, 1973.

[50] R. Weyrauch. Prolegomena to a theory of formal reasoning. *Artificial Intelligence*, 13:133–170, 1980.

On Geometry of Interaction

Jean-Yves Girard

Laboratoire de Mathématiques Discrètes
UPR 9016 – **CNRS**
163, Avenue de Luminy, Case 930
F-13288 Marseille Cedex 09
`girard@lmd.univ-mrs.fr`

Abstract. The paper expounds geometry of interaction, for the first time in the full case, i.e. for all connectives of linear logic, including additives and constants. The interpretation is done within a C^*-algebra which is induced by the rule of resolution of logic programming, and therefore the execution formula can be presented as a simple logic programming loop. Part of the data is public (shared channels) but part of it can be viewed as private dialect (defined up to isomorphism) that cannot be shared during interaction, thus illustrating the theme of communication without understanding. One can prove a nilpotency (i.e. termination) theorem for this semantics, and also its soundness w.r.t. a slight modification of familiar sequent calculus in the case of exponential-free conclusions.

1 Introduction

1.1 Towards a monist duality

Geometry of interaction is a new form of semantics. In order to understand what is achieved, one has to discuss the more traditional forms of semantics.

1.1.1 Classical model theory

The oldest view about logic is that of an external observer : there is a preexisting reality (mathematical, let us say) that we try to understand (e.g. by proving theorems). This form of *dualism* is backed by the so-called completeness theorem of Gödel (1930), which says that a formula is provable iff it is true in all models (i.e. in all realisations). There is strong heterogeneity in the duality world/observer (or model/proof) proposed by model-theory, since the latter is extremely finite whereas the former is infinite. Hilbert's attempt at reducing the gap between the two actors failed because of the renowned incompleteness theorems, also due to Gödel (1931), whose basic meaning is that infinity cannot be eliminated.

A paradoxical situation arose because Gentzen proved in 1934 a *cut-elimination* theorem for a formulation of logic of his own, *sequent calculus*, yielding the lineaments of an elimination of infinity in the style proposed by Hilbert. The paradox was mainly ideological, since the applications that Gentzen gave of his method to Peano arithmetic (in 1936 and later), make use of very strong forms of infinity... and therefore achieved no elimination of infinity, but for true believers.
The cut-elimination procedure introduced by Gentzen was a pure syntactical rewriting technique for proofs, enabling one to eliminate infinitary notions from finitary theorems, but whose termination could not be proved without even stronger infinitary techniques... Such a hybrid animal had a difficult life, and in particular could not find his status within the narrow duality models/proofs.

1.1.2 The semantics of proofs

Of course classical model-theory does not refuse the observer as a minor part of the reality (the same is true for classical physics): it makes the assumption of an objective reality where notions like *true, false*... make sense. A formula A is true or false in the world, and if I prove A it is also true that I prove A. Brouwer, by introducing *intuitionism*, radically changed the classical paradigm, by excluding the external reality and focusing on the interaction of proofs (the first consequence was taxonomical: the creation of *classical* logic, the new name for ordinary simple-minded logic).
Instead of explaining a proof Π of $A \Rightarrow B$ as *the justification that whenever A is true then B is also true*, intuitionism takes the viewpoint of the function which, given as input a proof Σ of A, yields a proof $\Pi(\Sigma)$ of B as output; the basic example is the identity function, which maps a proof of A to itself, and which can be seen as a proof of the basic tautology $A \Rightarrow A$. This functional viewpoint yields the so-called *semantics of proofs*.
This change of viewpoint should not be too easily styled subjectivistic, even if this was the ideology of Brouwer. This must be seen as a critique of simple-minded realism, analogous to relativity theory, which considers time as the quantified result (and no longer the cause) of interaction. A pure subjectivistic reading is so-called *realizability* which interprets the functions of semantics of proofs as purely syntactical operations, taking the code (e.g. the Gödel number) of a proof of A to the code of a proof of B. But this interpretation is slightly regressive, since the functions involved in the semantics of proofs have a very high degree of *naturality*, which is conspicuous in the following extraordinary fact: whether we take Gentzen's sequent calculus or neighbouring paradigms (lambda-calculi, natural deduction), equipped with the rewriting used for eliminating infinity (cut-elimination or β-conversion or normalization), the functions which are induced by proofs are exactly the canonical morphisms of a *cartesian closed category*, a beautiful kind of category (even if one had to wait until 1969 to find the first truly interesting

example, *Scott domains*).
It must be observed that intuitionism does not formally contradict the duality models/proofs. For instance there is still a notion of model with a completeness theorem for intuitionistic logic. But the problem is not that one cannot import classical model-theory in the intuitionistic world: it is that it becomes inoperant. Take for instance the notion of *consistency*: in classical logic, in conformity with Hilbert's view, the consistency of A produces a model of A, i.e. an object; but intuitionistic consistency only produces the mockery of an object, a so-called *Kripke model*. One should therefore (but for very limited technical purposes) avoid systematic import/export of concepts, which too often look like a horseshoe on a tire, not to speak of a gasoline diet for a horse.
The central point of intuitionism is indeed *constructivity* (which can take the more ideological dress of *constructivism*, that we shall not discuss). Like classical logic (and like any reliable bank), one should be able *on request*, to exhibit something; no longer a model, but some explicit information[1]. Typically, a proof of a disjunction $A \vee B$, could on request be replaced with a proof of A or with a proof of B. From this viewpoint, classical logic which allows the principle $A \vee \neg A$ without being able to tell which one holds, is "inconsistent" (in the common interpretation, as in the famous saying by King François 1er: *women are inconsistent*)[2].

Technically speaking, this exhibition is made possible by the fact that the cut-free proofs (or the normal terms in λ-calculus) that are the outputs of cut-elimination are proved by very restricted means: typically, in the case of intuitionistic disjunction, the only cut-free way to get $A \vee B$ is to get it from A or to get it from B.
But this explanation would be incomplete, if one were not stressing its "on request" aspect: in real life, a proof of $A \vee B$ is *never* a proof of A or a proof of B (only a masochist would state $A \vee B$ when he knows A). In other terms, a proof of $A \vee B$ is a proof of A or B, but we don't know which one; only on request, we can be more explicit, and tell you which side holds. But this requires a painful work (typically cut-elimination) to replace everywhere the implicit by the explicit.

Here we have to be very precise, *mathematics cannot be explicit*: as soon as one departs from trivial elementary facts like $2+2=4$, mathematics (and all forms of reasoning) become abstract, i.e. *implicit*. This implicit character is conspicuous by the use of *lemmas*, which are combined by transitivity, by means of the rule of *Modus Ponens*: from A and $A \Rightarrow B$ deduce B, rewritten under the form of the *CUT*-rule of Gentzen (we adopt here the formalism of

[1] In classical logic the model is constructed from the absence of a proof, whereas constructivity tries to extract information from the presence of a proof

[2] Souvent femme varie/Bien fol est qui s'y fie

linear logic, see below):

$$\frac{\vdash \Gamma, A \quad \vdash A^{\perp}, \Delta}{\vdash \Gamma, \Delta} CUT$$

Without the CUT-rule, it is practically impossible to make any deduction (typically, if I know that $27 \cdot 37 = 999$, I cannot use the lemma "commutativity of multiplication" to infer $37 \cdot 27 = 999$). Hence the truly paradoxical result of Gentzen is that this essential rule can be eliminated (for pure logic, i.e. predicate calculi), i.e. that mathematics can (within some limits) be explicated... not by a human being, but surely by a machine.
These observations are the starting point for applications to computer science: the choice L/R between the two premises of a disjunction can be used to represent a boolean datum, a proof of an implication can represent a computable function, the CUT-rule can take care of application of a function to an argument and cut-elimination, suitably implemented, can handle evaluation. The existence of a categorical model basically asserts the robustness of this approach (e.g. independence of any implementation). This is the basis for ***functional programming***, more precisely ***typed*** λ***-calculus*** an experimental language popular among theoreticians.

1.1.3 The monist duality

The intuitionistic world replaces the relation *model of* $\neg A$*/proof of* A of classical logic with the relation *proof of* $A \Rightarrow B$*/proof of* A. This is not a duality: to get a duality, one at least needs an involutive negation, and such a thing does not exist in the intuitionistic world. Linear logic ([G86]) is based on a refined analysis of the categorical semantics of intuitionistic proofs (replacement of Scott domains by ***coherent spaces***), and individualises new logical connectives. The basic point is to remove the rule of contraction of Gentzen (which amounts to saying that from A one can deduce $A \wedge A$) and also the rule of weakening (which allows fake hypotheses), in which case the basic symmetries missing in intuitionistic logic are restored; these symmetries are expressed by the involutive connective of ***linear negation*** $A^{\perp}$. It is now possible to think of ***monist*** duality, ***proof of*** $A^{\perp}$***/proof of*** A based on the CUT-rule, with the essential difficulty that A and $A^{\perp}$ cannot be simultaneously provable (because in the CUT-rule, Γ and Δ never happen to both empty). In the classical case, the duality is of a very strange kind: a proof of A and a model of $\neg A$ cannot simultaneously exist, in particular classical duality cannot account for the difference between two proofs of A, since there is model of $\neg A$ on which to compare them. If we want duality to account for the distinction between proofs of the same formula, then the two partners should not be exclusive one of another... The construction of a satisfactory monist duality is therefore delicate (we have to replace ***proof of*** A with something slightly more liberal) and has been attacked by various methods, from categorical semantics to game semantics, without yet any definite answer. *Geometry of interaction* is one of the major approaches to this question.

1.2 Linear Logic

Coherent semantics ([G86]) is built in analogy to linear algebra. The basic constructions of linear algebra can be mimicked by coherent semantics, and yield the basic linear connectives:

- There is an involutive duality (anaologous to vector space duality), which induces *linear negation* $A^\perp$;

- There are operations analogous to the tensor product of vector spaces, and which yield the so-called *multiplicative* connectives $\otimes, ⅋, \multimap$; ⅋, which is the disjunction *par*, is the dual of $\otimes$, which is the conjunction *times* (the tensors here are strongly non self-dual); the *linear implication* $\multimap$ can be defined by $A \multimap B := A^\perp ⅋ B$. To the multiplicative universe should be attached the dual constants 1 and $\perp$.

- There are operations analogous to the direct sum of vector spaces, and which yield the so-called *additive* connectives, $\&, \oplus$; &, which is the conjunction *with* is the dual of $\oplus$, which is the disjunction *plus* ; here too the sums are strongly non-involutive. To the additive universe should be attached the dual constants $\top$ and 0.

- There are operations analogous to symmetric tensor algebras, and which yield the so-called *exponential* connectives !, ?. ! (*of course*) is the dual of ? (*why not*).

The main categorical properties of these connectives are expressed by a certain number of canonical isomorphisms:

- Involutivity of negation (which allow in practice to ignore double negation symbols);

- Commutativity of $\otimes$, ⅋, &, $\oplus$;

- Associativity of $\otimes$, ⅋, &, $\oplus$;

- Distributivity of $\otimes$ over $\oplus$ (and dually of ⅋ over &) ;

- Exponentiation isomorphisms $!(A\&B) \simeq !A \otimes !B$ (and dually $?(A \oplus B) \simeq ?A \oplus ?B$) ;

- Neutrality of the constants 1 w.r.t. $\otimes$, $\perp$ w.r.t. ⅋, $\top$ w.r.t. &, 0 w.r.t. $\oplus$, together with $!\top \simeq 1$ and $?0 \simeq \perp$.

1.3 Linear sequent calculus

Linear logic is organised into a sequent calculus, (in which we can in particular derive the canonical proofs of the isomorphisms just mentioned). Its standard syntax is one-sided, with defined negation (and implication). Sequents are of the form $\vdash A_1, \ldots, A_n$, where $A_1, \ldots, A_n$ is a sequence of formulas. The rules are organised in three groups:

1.3.1 Identity/negation

$$\frac{}{\vdash A, A^\perp}ID \qquad \frac{\vdash \Gamma, A \quad \vdash A^\perp, \Delta}{\vdash \Gamma, \Delta}CUT$$

This group asserts that A is A, the only absolute evidence of logic; this fact is expressed by two rules. In the one-sided calculus, the identity can be seen as the definition of negation.

1.3.2 Structure

$$\frac{\vdash \Gamma, A, B, \Delta}{\vdash \Gamma, B, A, \Delta}X$$

a rule that can be avoided if one considers sequents as multisets instead of sequences. The two other traditional structural rules

$$\frac{\vdash \Gamma}{\vdash A, \Gamma}w? \qquad \frac{\vdash A, A, \Gamma}{\vdash A, \Gamma}c?$$

also called *weakening* and *contraction* are not allowed in linear logic. These rules are essential in classical and (when written in two-sided style) in intuitionistic logics.

1.3.3 Logic

$$\frac{\vdash \Gamma, A \quad \vdash \Delta, B}{\vdash \Gamma, \Delta, A \otimes B}\otimes \qquad \frac{\vdash A, B, \Gamma}{\vdash A \wp B, \Gamma}\wp \qquad \frac{}{\vdash 1}1 \qquad \frac{\vdash \Gamma}{\vdash \bot, \Gamma}\bot$$

The rule for $\otimes$ concatenates (adds up) the contexts; if the context is a price to pay (by destruction) to get a formula, a natural price for $A \otimes B$ is the sum of price for A and of a price for B.

$$\frac{\vdash \Gamma, A \quad \vdash \Gamma, B}{\vdash \Gamma, A \& B}\& \qquad \frac{\vdash A, \Gamma}{\vdash A \oplus B, \Gamma}\oplus_1 \qquad \frac{\vdash B, \Gamma}{\vdash A \oplus B, \Gamma}\oplus_2 \qquad \frac{}{\vdash \top, \Gamma}\top$$

The rule for & assumes that the two contexts, i.e. prices, are the same ; in other terms a possible price for $A\&B$ is a price for which one can get any of A and B, which means that if both of them are simulataneously available (as expected from a conjunction) for this price, only one of them (up to our choice) will eventually be bought.

$$\frac{\vdash ?\Gamma, A}{\vdash ?\Gamma, !A}\,! \quad \frac{\vdash A, \Gamma}{\vdash ?A, \Gamma}\,d? \quad \frac{\vdash \Gamma}{\vdash ?A, \Gamma}\,w? \quad \frac{\vdash ?A, ?A, \Gamma}{\vdash ?A, \Gamma}\,c?$$

These rules are called ***promotion, dereliction, weakening, contraction.*** The interpretation in terms of prices (or resources) is no longer very convincing for exponentials (especially because of the contraction rule): this is because the exponential group is a "classical" group, which enables one to instillate traditional classical or intuitionistic features inside a calculus which otherwise (in spite of the novelty of its connectives) would not be expressive enough. Indeed exponentials allow weakening and contraction, but no longer as structural (i.e. universal) rules, but as controlled logical rules: indeed the role of exponentials is precisely to control the use of these two rules.

The main connective of linear logic is linear implication (which does not appear in the official right-handed syntax) and which is extremely different from the familiar intuitionistic implication: from A and $A \multimap B$ one can still derive B... but we have lost A in the process. This strong difference with preexistant logical systems is due to the disparition of the contraction rule which enabled one to produce two copies of A from one. This fact is responsible for the resource-sensitive character of linear logic. This is also expressed as a form of *causality*: $A \multimap B$ is the action "from A get B" which involves, like in physics, a reaction: the destruction of the cause. Now observe that among the rules for exponentials, reappears the contraction rule, but now limited to the holders of the front symbol ?. This is why the interpretation of intuitionistic implication as $!A \multimap B$ is possible. $!A$ has the meaning of A ***ad libitum***, i.e. allows us from a single use of A to its unlimited reuse. One can also say, in analogy with quantum mechanics, that linear logic is about microscopic facts and that the exponentials ensure the transition with the macroscopic world... but such tantalizing analogies should be taken with care ; in particular quantum non-determinism has not yet find a definite analogue in terms of linear logic.

The idea of causality is that of a reciprocal annihilation of A and $A^{\perp}$ by the CUT -rule ; this goes very well with our idea of a monist (or homogeneous) duality. (Notice the change w.r.t. intuitionistic logic, where there inputs and outputs ; linear logic says that these roles are exchanged by linear negation, and what is called input or output depends on the user). But how does this annihilation work ?

1.4 Proof-nets

Although the basic formalisms for explication are basically equivalent, it makes in practice a lot of difference to work with natural deduction (or typed λ-calculus) instead of sequent calculus. If sequent calculus remains the best possible formulation of logic, its cut-elimination procedure spends too much time on bureaucratic details, typically permutation of rules. Natural deduction has the immense advantage (in the absence of disjunction and existence) of being insensitive to the order of rules, and therefore is much

more efficient than sequent calculus. The idea was therefore to build a kind of natural deduction for linear logic, the main difficulty being that the tree form of natural deduction (so successful in the absence of disjunction and existence) could no longer be exploited in the presence of an involutive negation, for which the distinction input/output or hypothesis/conclusion no longer makes sense.

Proof-nets are this "linear natural deduction", and their discovery has been a long process: first limited to multiplicatives (without the constants), they have been extended to quantifiers, and more recently to additives [G94]. The main difficulty came because of the graph-like structure of such unfamiliar proofs: given a graph $\mathcal{G}$ which pretends to be a proof-net (a so-called *proof-structure*), can we decide whether this claim is grounded ? This is the problem of correctness criterions: the solutions found are of the form $\mathcal{G}$ *successfully passes a certain set of tests.* Soon after the introduction of multiplicative proof-nets in [G86], the paper [G86A] introduced a major dualist idea, namely that the tests to be passed by a proof of A are like *virtual* proofs of its negation $A^{\perp}$; "virtual" means that these tests (which in general do not represent proofs at all) are "dense" in a set containing all proofs of $A^{\perp}$ if such objects were to exist.

Proofs and tests are therefore homogeneous in nature: both can be seen as finite "wirings" of the atoms of formulas, and cut-elimination basically amounted to connect the wires and to "follow the current". Mathematically such wires are permutations and cut-elimination can be written as a pure operation between permutations, the only role of logic being to guarantee that a certain expression makes sense, i.e. converges. It "only" remained to extend this paradigm to full logic...

1.5 Geometry of interaction

The philosophy is the following: elimination of infinity (or better: explication) is not the implementation of a clever, but artificial algorithm. On the contrary, there is an intrinsic dynamics of interaction (expressing the physical reciprocal annihilation of two antagonist actors), and logic is here only as a comment on this "physical" phenomenon. The comment is about:

- Specifications of the antagonistic actors in the interaction (in analogy with the shape of plugs in electricity which are not responsible for the passage of the current, but which limits in principle the plugging of an *acceptor of 22OV* to a *giver of 22OV*) ; the specification is therefore a formula.

- The building of a step-by-step justification of the specification ; this justification is what we usually see as a syntactical proof.

The problem was to find the right kind of mathematical objects. The multiplicative case was handled by means of finite permutations. A permutation of n can be represented by an isometry of the finite-dimensional Hilbert space C^n. The need to represent the contraction rule for the connective ? (and whose dynamics is duplication) leads one to replace finite dimension by "continuous" dimension. In other terms this means that the general case will be handled by means of certain operators (technically: partial isometries) of the standard separable Hilbert space $\mathcal{H}$. Indeed we are interpreting proofs by objects of a C^*-algebra (acting on the Hilbert space, but consisting only of very specific operators). This C^*-algebraic aspect is not technically essential, but it was the constant leading intuition for the whole program, and without a backbone made of solid mathematics (presumably not yet used in a significant way) nothing would have existed.
Cut-free proofs are interpreted by such operators, the meaning being that of a static linear input/output machinery. Typically the linear logic axiom $\vdash A, A^\perp$ is represented by an extension cord, whose abstract definition is to have two extremities complementarily labelled or shaped (here $A, A^\perp$) and whose physical action is to transfer the input in A as an output to $A^\perp$ and vice versa: such an operator can be written as a 2×2 anti-diagonal matrix. In general all logical rules are interpreted by $*$-isos, and in case of binary rules, summations. Typically, in [G88] the interpretation was based on the $*$-isos induced by an isometry of $\mathcal{H} \oplus \mathcal{H}$ into $\mathcal{H}$ and $\mathcal{H} \otimes \mathcal{H}$ onto $\mathcal{H}$.
To achieve a dynamical effect, we must give inputs to our operators ; this is achieved in presence of CUT: the general paradigm of interpretation is that of a pair (U, σ) (with $\sigma = 0$ in the cut-free case). The partial isometry σ is hermitian (i.e. it is a partial symmetry) and expresses a feedback. This is our way of saying that CUT is a physical plugging. Take the example of a cut between $\vdash \Gamma, A$ and $\vdash A^\perp, \Delta$: the two operators V and W enjoying these specifications are indeed square matrices (respectively labelled by the indices Γ, A and $A^\perp, \Delta$). We can join the sets of indices into $\Gamma, A, A^\perp, \Delta$, and form U (which corresponds to a "disjoint sum" of V and W). Now we can introduce the partial symmetry σ which exchanges the two opposite labels and is zero everywhere else. From the pair (U, σ) it is possible to express the feedback, namely that the inputs coming through A and $A^\perp$ are equal to the outputs coming out of the same channels, but flipped by σ. This amounts to writing a linear equation (whose parameters are the inputs labelled by the remaining free plugs Γ, Δ). For instance assume that U maps a direct sum $\mathcal{H} \oplus \mathcal{H}$ to itself, and that σ is a feedback on the second coordinate (i.e. σ^2 is the projection on the second coordinate) ; then the problem is, given $x \in \mathcal{H}$,

to find $y, a \in \mathcal{H}$ such that the equation:

$$U(x \oplus \sigma(a)) = y \oplus a$$

holds ; a sufficient condition for a solution is the invertibility of $1 - \sigma U$, in which case the *execution formula*

$$RES(U, \sigma) := (1 - \sigma^2)U(1 - \sigma U)^{-1}(1 - \sigma^2)$$

expresses the solution, i.e. the input/output dependency of the remaining plugs (i.e. y as a function of x).
A stronger existence condition is the *nilpotency* of σU, i.e. that some power $(\sigma U)^n$ is zero, a condition which is experimentally true for all pairs (U, σ) arising from logical proofs[3]. In this case the central part of the execution formula can be written as a finite sum:

$$U(1 - \sigma U)^{-1} = U + U\sigma U + U\sigma U\sigma U + \ldots$$

the length of the sum being equal to the *order of nilpotency* n of σU. Observe that n can be extremely big: the basic way to get a high order of nilpotency n is to make n cuts with identity axioms, which is not very surprising ! But the logical rules, interpreted by $*$-isos will alter the pattern, since if $*$-isos cannot affect invariants like orders of nilpotency etc., they surely can dramatically affect the apparent size (i.e. the description) of objects: think for instance that the sum $U_1 + \ldots + U_{100}$ of 100 isometric copies of U can easily be recovered from a single operator of the form $U \otimes 1$ which has a more compact definition (this is the basic idea behind the interpretation of !). In fact this change of size is so dramatic that the order of nilpotency can hardly be predicted from the pair (U, σ), and that the nilpotency of all pairs $\Pi^\bullet \sigma$ coming from proofs in standard logical systems cannot be proved within usual mathematics: this nilpotency is related to the termination of cut-elimination, and therefore implies the consistency of various systems (in consequence, by incompleteness it cannot be proved without a heavy use of infinitary notions).

The *execution formula* does not quite correspond to syntactical cut-elimination, but it is not too far ; in particular for proofs of sufficiently simple formulas, the correspondance is exact and that's enough. The interest of this approach for computer science was later confirmed by its applications to optimal reduction in λ-calculus due to Gonthier ([GAL92]).

1.6 The case of additives

The original paper [G88] only dealt with multiplicatives, exponentials and quantifiers of any order, which was enough to modelize extant typed λ-calculi.

---

[3] For non-logical systems like *pure λ-calculus* there is still a notion of *weak nilpotency*, see [G88A]

However some essential elements were missing, typically the treatment of additive connectives. The situation remained the same for several years, until a satisfactory extension of ***proof-nets*** to the additive case was found [G94]. The main novelty consists in assigning boolean ***weights*** to the links of a proof-nets. Since geometry of interaction uses a C^*-algebraic formulation and boolean algebras are basically commutative C^*-algebras, there is no major obstacle to this extension ; but the paradigm is (slightly) altered:

- In [G88] a cut-free proof Π of a n-ary sequent was interpreted by a $n \times n$ matrix with entries in a C^*-algebra Λ^*. This operator could be seen as acting on a n-ary direct sum $\mathcal{H} \oplus \ldots \oplus \mathcal{H}$. Now, our operators will still act on a n-ary direct sum, but of n spaces $\mathcal{H} \otimes \mathcal{H}'$ where $\mathcal{H}'$ is a Hilbert space, seen as a space of ***auxiliary messages.*** This notion of auxiliary message space is the main novelty.

- Of course $\mathcal{H} \otimes \mathcal{H}'$ is isomorphic to $\mathcal{H}$, but this is not very helpful, since $\mathcal{H}$ and $\mathcal{H}'$ are handled in quite different ways: typically everything dealing with $\mathcal{H}'$ is up to isomorphism... what we express by introducing an equivalence relation: being ***variants.*** All constructions are compatible with that equivalence, but one, namely the promotion-rule for ! which strongly resists to the proof-net spirit (in [G86], the promotion rule is accomodated with a "box", i.e. treated as a global entity ; this globality is perhaps inherent to this rule and technically expressed by a failure of the variance principle). When we interpret a conjunctive rule, $\otimes$ or &, a common dialect has to be produced, but due to the possibility of inner automorphisms of the dialect part, there is no common dialect at all. The only possibilities are to make generic operations, tensorization of dialects or summation. On the other hand $\mathcal{H}$ represents (shared) ***channels*** of communication. The paradigm of interaction is therefore ***communication without understanding through shared channels.***

- As in [G88] the ***execution*** formula does not quite correspond to syntactical normalisation ; in other terms certain operations (typically erasings) are not performed. But the soundness of the execution formula w.r.t. cut-elimination still holds under a reasonable restriction on the proven formula. Notice that the syntax has to be adapted to prove this result (introduction of the ***inflated calculus***) and that everything looks simpler if the exponentials are slightly modified (without any change of the global expressive power).

1.7 Closing the system

The interpretation given here, although perhaps not definitive, omits no logical connective. But there is still an essential lingering problem: the problem of "closing the system". This means being able to build a duality between proofs of A and proofs of $A^{\perp}$. We already noticed that the notion of proof

has to be liberalized to something wider (like the tests for $A^{\perp}$ in the case of multiplicative proof-nets) but homogeneous to proofs. This is for instance what we do in our definition of ***weak types***, see definition 13. But the problem is that, among the elements of the weak type $A^{\bullet}$ interpreting A there is no way to distinguish those objects which are interpretations of proofs. Surely the notion of ***orthogonality*** introduced in definition 12 and central in the notion of a weak type, is too lax. What is missing is that $U \perp V$ is completely symmetrical in the partners, and that there is no way to tell the wheats (real proofs) from the tares (tests). This could be fixed by defining (but how ?) an output $< U, V >$ of the interaction, sufficiently antisymmetric so that a distinguished value -say 1- for $< U, V >$ would exclude the same value for $< V, U >$. It would of course remain to prove completeness, namely that if $< U, V >= 1$ for all $V \in A^{\perp\bullet}$, then U is the interpretation of a proof of A, a widely open program...
This is where geometry of interaction merges with ***game semantics***: the weak orthogonality between U, V means that U, V can be seen as strategies for the two players inside the same game, and our output $< U, V >$ is the result of the game, won by U when the result is 1. Therefore the problem of closing the system is closely related to the search for a complete game semantics for linear logic... (the connection between linear logic and game semantics was initiated by Blass, see [B92]).

1.8 Notes on the style

Geometry of interaction is most naturally handled by interpreting logical rules by $*$-isomorphisms ; this yields surely more elegant proofs, but it obscures the concrete interpretation. So we prefer to follow a down to earth description of the interpretation. An unexpected feature will help us: the C^*-algebras used can in fact be interpreted in terms of ***logic programming***, since the basic operators are very elementary PROLOG programs, and composition is resolution ! This ultimate explanation of the dynamics of logic in terms of (some theoretical) logic programming cannot be without consequence: we could for instance try to use linear logic is a typing (i.e. specification) discipline for such logic programs... this promising connection is therefore a strong reason to remain concrete.

Although our main inspiration was our (yet unwritten) work on additive proof-nets, proof-nets will not at all appear below: first this would make too many simultaneous novelties, second certain details (inflated proof-nets, simultaneous treatment of additives and quantifiers) have not yet been fixed. Finally, we shall give many examples, especially in the section on exponentials ; it must not be inferred from the elementary character of these examples that the global construction is trivial: like in λ-calculus, all atomic steps are elementary, but the combination of few such steps easily becomes... explosive.

2 The algebra of resolution

2.1 Resolution

By a *term language* T, we mean the set of all terms t that can be obtained from variables $x_1, \ldots, x_n$ by means of a finite stock of function letters ; we assume that T contains at least one constant (0-ary function letter), so that the set of ground (i.e. closed) terms is non-empty. By a *language* L, we mean the set of all atoms $pt_1 \ldots t_n$, where p is a n-ary predicate letter varying through a finite set of such symbols, given with their arity. An important case is when the n predicates of L are of the same arity m, in which case we shall use the notation $T^m \cdot n$ for L.

Remark 1

The basic need of geometry of interaction is to have two constants g and d and a binary function $\odot$. The predicate letters correspond to the indices of matrices, i.e. a $n \times n$ matrix makes use of $p_1, \ldots, p_n$ (in the sequel we shall rather use formulas as indices : $p_A, p_B, \ldots$, which supposes that they are pairwise distinct, a standard bureaucratic problem, solved by replacing formulas by occurrences etc.). These predicates will be binary, although in the presence of a binary function, the arity of the predicates hardly matters.

Definition 1 (Unification)
Let L be a language ; we say that two expressions (terms or atoms) e and e' are unifiable *when there is a substitution (called a* unifier *of e and e') θ of terms for the variables occuring in e and e' such that $e\theta = e'\theta$.*

Remark 2
In such a case, there is a *most general unifier* (mgu), i.e. a unifier θ of e and e' such that any other unifier θ' can can be written as the composition $\theta\theta''$ of θ with a substitution θ''. Most general unifiers are unique up to renaming of variables.

Definition 2 (Clauses)
Let L be a language ; by a (rudimentary) clause *in L, we mean a sequent $P \mapsto Q$, where P and Q are atoms of L* with the same variables.

Remark 3
Clauses are considered up to the renaming of their variables : we do not distinguish between $p(x) \mapsto q(x)$ and $p(y) \mapsto q(y)$ (but we distinguish between $p(x \odot y) \mapsto q(x \odot y)$ and $p(y \odot x) \mapsto q(x \odot y)$; in other words the variables of clauses are bound, and other notations like $\forall x_1 \ldots x_n P \Rightarrow Q$ could be considered.

Definition 3 (Resolution)
If $P \mapsto Q$ and $P' \mapsto Q'$ are clauses, then we can assume w.l.o.g. that they have no variable in common and

- *either* Q *and* P' *are not unifiable, in which case we say that* resolution *of the clauses* $P \mapsto Q$ *and* $P' \mapsto Q'$ *(in this order)* fails
- *or* Q *and* P' *have a mgu* θ, *in which case resolution succeeds, with the* resolvant $P\theta \mapsto Q'\theta$; *in that case we introduce the notation*

$$(P \mapsto Q) \cdot (P' \mapsto Q') := P\theta \mapsto Q'\theta.$$

Remark 4
Let us introduce the formal clause 0, and extend resolution by setting $(P \mapsto Q) \cdot (P' \mapsto Q') := 0$ when unification fails ; then it turns out that resolution is associative. All the clauses $P \mapsto P$ are idempotent. Finally the operation defined on clauses by $(P \mapsto Q)^* := Q \mapsto P$ is an anti-involution : $(R \cdot R')^* = R'^* \cdot R^*$.

Remark 5
In case L has a single (let us say : binary) predicate symbol p, the clause $I := p(x,y) \mapsto p(x,y)$ is neutral w.r.t. resolution : if x, y are not free in r, s, t, u then $p(x,y)$ unifies with $p(r,s)$ by means of the mgu $\theta(x) = r, \theta(y) = s$, hence $I(p(r,s) \mapsto p(t,u)) = p(x,y)\theta \mapsto p(t,u)\theta = p(r,s)\ p(t,u), \ldots$

2.2 The algebra $\Lambda^*(L)$

Definition 4 (The algebra)
Let $\lambda^*(L)$ *be the set of all (finite) formal linear combinations*

$$\sum \alpha_i \cdot (P_i \mapsto Q_i),$$

with the scalar α_i *in* C ; $\lambda^*(L)$ *is obviously equipped with*

- *A structure of complex vector space.*
- *A structure of complex algebra, the multiplication being extended by bilinearity from resolution :*

$$(\sum \alpha_i \cdot (P_i \mapsto Q_i))(\sum \beta_j \cdot (R_j \mapsto S_j)) := \sum \alpha_i\beta_j \cdot (P_i \mapsto Q_i))(R_j \mapsto S_j)$$

- *An identity : for instance the identity of* $T^2 \cdot n$ *is* $\sum p_i(x,y) \mapsto p_i(x,y)$ *(*x, y*, distinct variables) : easy consequence of remark 5.*
- *An (anti-)involution defined by*

$$(\sum \alpha_i \cdot (P_i \mapsto Q_i))^* := \sum \overline{\alpha_i} \cdot (Q_i \mapsto P_i).$$

In other terms $\lambda^*(L)$ bears all the features of a C^*-algebra, but for the norm features. Although we have no deep use for this[4], it is of interest to make a C^*-algebra out of $\lambda^*(L)$. Observe that in a C^*-algebra clauses must be *partial isometries* (since RR^* is an idempotent), and therefore $\|\sum \alpha_i \cdot (P_i \mapsto Q_i)\|$ is bounded by $\sum |\alpha_i|$ for any C^*-seminorm. We shall therefore be able to equip $\lambda^*(L)$ with its greatest C^*-seminorm, provided we can show that there is at least one such norm. For this its obviously suffices to show that $\lambda^*(L)$ operates on a Hilbert space.

Proposition 1
$\lambda^*(L)$ acts on the Hilbert space $\ell^2(G)$, where G is the set of ground propositions of L in such a way that the sum of $\lambda^*(L)$ is interpreted by sum of operators, the scalar multiplication of $\lambda^*(L)$ by scalar multiplication of operators, the product $\lambda^*(L)$ by composition of operators, and the involution * of $\lambda^*(L)$ by adjunction of operators.

PROOF: let G be the set of ground propositions in L ; to any clause $P \mapsto Q$ in L we can associate a bijection $|P \mapsto Q|$ from a subset of G to a subset of G : let g_0 be a fixed object in G, we set $|P \mapsto Q|(g) := g'$ if $(P \mapsto Q)(g \mapsto g_0) = g' \mapsto g_0$, $|P \mapsto Q|(g)$ undefined otherwise. It is immediate that

$$|P \mapsto Q| \circ |P' \mapsto Q'| = |(P \mapsto Q) \cdot (P' \mapsto Q')|$$

an equation between partial functions that persists when resolution fails, if we interpret 0 as the fully undefined function. The partial functions $|P \mapsto Q|$ induce operators (partial isometries) of the Hilbert space $\ell^2(G)$:

$$|P \mapsto Q|(\sum \alpha_i . g_i) := \sum \alpha_i \cdot |P \mapsto Q|(g_i).$$

This immediately extends to all elements of $\lambda^*(L)$ which are therefore ascribed operators on $\ell^2(G)$, with the obvious properties that we have stated. □

Definition 5 (The completion)
$\Lambda^*(L)$ is defined to be the C^*-algebra obtained by completing $\lambda^*(L)$ w.r.t. its greatest C^*-seminorm.

The following are useful hints as to the relation between functions, predicates one one hand and algebraic manipulations on the other hand :

Proposition 2
$\lambda^*(T^m \cdot n)$ is (isomorphic to) the algebra of $n \times n$-matrices with entries in $\lambda^*(T^m)$.

---

[4] **Danos & Regnier [DR93] have proved a result on geometry of interaction which actually makes use of the Hilbert space ℓ^2, hence one should not consider too quickly that the full C^*-algebraic viewpoint is irrelevant to geometry of interaction.**

PROOF: let us assume for instance $m = 2$; the basic idea is that the matrix M_{ij} whose entries are all zero but the one of index ij, which is 1 (i.e. the clause $p(x,y) \mapsto p(x,y)$), is represented by the clause $p_i(x,y) \mapsto p_j(x,y)$ of $\lambda^*(T^2 \cdot n)$. □

Proposition 3
$\lambda^*(T^m)$ contains an isomorphic copy of the m-ary tensor power of $\lambda^*(T)$.

PROOF: typically if $m = 2$, p is unary and q is binary, then we can define an isomorphism ϕ from $\lambda^*(T) \otimes \lambda^*(T)$ into $\lambda^*(T^2)$, by

$$\phi((p(t) \mapsto p(u)) \otimes (p(t') \mapsto p(u'))) := q(t,t') \mapsto q(u,u')$$

This isomorphism is of course not surjective. □

2.3 The execution formula

Definition 6 (Loops)
Let us fix the language L. We adapt the main definitions and notions coming from [G88].

- *A* message *is any finite sum $\sum(P_i \mapsto P_i)$, with the P_i and P_j not unifiable for $i \neq j$. A message is therefore a projection of $\Lambda^*(L)$, and messages commute with each other. We abbreviate the message $P_i \mapsto P_i$ in P_i.*
- *A* wiring *is any finite sum $\sum(P_i \mapsto Q_i)$, with the P_i and P_j not unifiable for $i \neq j$ and the Q_i and Q_j not unifiable for $i \neq j$; each of the clauses $(P_i \mapsto Q_i)$ is called a* wire. *A finite message is therefore a partial isometry of $\Lambda^*(L)$. (In [G88], wirings were called "observables".) If w and w' are wirings then ww' is a wiring.*
- *If m is a message and w is a wiring, there is a (non-unique) message m', such that $m'w = wm$: take $m' = wmw^*$ (PROOF: $wmw^*w = ww^*wm = wm$. □) ; in other terms wirings* propagate *messages.*
- *A* loop *(U, σ) is a pair of wirings such that σ is hermitian, i.e. $\sigma = \sigma^*$ (in particular σ^2 is a projection and $\sigma^3 = \sigma$.*
- *A loop* converges *when σU is nilpotent, i.e. when $(\sigma U)^n = 0$ for some n. The* execution

  of the loop is the element

$$EX(U,\sigma) := U(1-\sigma U)^{-1} := U\sum_{k=0}^{n-1}(\sigma U)^k$$

  and the result *of the execution is defined as*

$$RES(U,\sigma) := (1-\sigma^2)U(1-\sigma U)^{-1}(1-\sigma^2).$$

The output is a wiring, whereas the execution is not a wiring. Observe that (with $V := EX(U, \sigma)$) $U = V(1 + \sigma V)^{-1}$.

Remark 6
Wirings correspond to very specific PROLOG programs. (Here we do not refer to the actual language PROLOG, but rather to the idea of resolution, independently of any implementation). Each *wire* $P_i \mapsto Q_i$ in $\sum(P_i \mapsto Q_i)$ can be seen as a clause in a program : the program consists in the clauses $P_i \mapsto Q_i$. Such programs are very peculiar :

- The *tail* (i.e. the body) of the clause consists of a single literal,
- The same variables occur in the head and in the tail,
- The heads are pairwise non-unifiable,
- The tails are pairwise non-unifiable.

But one has to be (slightly) subtler to relate execution in our sense to the execution of a logic program. In order to interpret a loop, we systematically duplicate all predicate symbols $p, q, r, \ldots$ into pairs (p^-, p^+) (p^- for inputs, p^+ for outputs) ; therefore an atom P receives two interpretations P^- and P^+. The pair (U, σ) yields the following program, consisting in the clauses

- Clauses $P^- \mapsto Q^+$ for each wire $P \mapsto Q$ in U,
- Clauses $P^+ \mapsto Q^-$ for each wire $P \mapsto Q$ in σ.

The execution of (U, σ) consists in finding all clauses $P^\epsilon \mapsto Q^{\epsilon'}$ which are consequences of this program by means of resolution. The result of the execution consists in retaining only those clauses $P^- \mapsto Q^+$ that cannot be unified (by prefixing and/or suffixing) with a clause coming from σ.

Exercise 1
By suitable modifications of the language L, express the execution formula as a logic programming problem such that

- *L has only three unary predicates e, c, s*
- *(U, σ) is interpreted by only four kinds of clauses*
  - *clauses* $e(t) \mapsto c(u)$,
  - *clauses* $e(t) \mapsto s(u)$,
  - *clauses* $c(t) \mapsto c(u)$,
  - *clauses* $c(t) \mapsto s(u)$,

in such a way that the result of the execution corresponds to the consequences of the program (by means of resolution) of the form $e(t) \mapsto c(u)$.

Remark 7
The execution formula can also be seen as the solution of a linear equation on the Hilbert space on which $\lambda^*(L)$ operates :

- U produces, given an input $h \in \mathcal{H}$ an output $U(h) \in \mathcal{H}$;
- σ feedbacks certain outputs h' of U to inputs $\sigma(h') \in \mathcal{H}$;
- $1 - \sigma^2$ is a projection corresponding to the subspace $\mathcal{H}'$ on which we want to observe the external behaviour of the loop ;
- In other terms, given $h' \in \mathcal{H}'$ we are looking for $h \in \mathcal{H}$ such that $(1-\sigma^2)(h) = h'$ and $(\sigma U)(h) = h - h'$; we eventually keep the output value $h'' := (1-\sigma^2)(U(h))$.

As a linear equation this can be written :

- $(\sigma U)(h) = h - h'$, equivalently $h' = (1 - \sigma U)(h)$;
- h' is well defined as soon as $1 - \sigma u$ is invertible, which is the case when σU is nilpotent ; in that case $h = (1-\sigma U)^{-1}(h')$, which can also be written $h = (1-\sigma U)^{-1}(1-\sigma^2)(h')$;
- Therefore
$$h'' = (1-\sigma^2)U(1-\sigma U)^{-1}(1-\sigma^2)(h'),$$
i.e. $h'' = RES(U, \sigma)(h')$.

3 The interpretation of the multiplicative–additive fragment

3.1 Laminated wirings

Strictly speaking, the interpretation can be made within a fixed algebra $\Lambda^*(L)$, for instance with one constant, one binary function and one unary predicate... but this not very user-friendly. In practice we shall need two constants g and d and one function letter $\odot$, together with a variable number of predicates to reflect the structure of *sequents*, which have a variable number of formulas. Also these predicates will be chosen *binary*, for reasons that will be explained below. In the sequel T will be the term language $\{g, d, \odot\}$, and $T^2 \cdot n$ will be the language built from the terms of T by means of n binary predicates, $p_1, \ldots, p_n$. We shall use the following notational convention for terms : $t_1 \ldots t_k$ is short for $t_1 \odot (\ldots \odot (t_{k-1} \odot t_k) \ldots)$, so that tuv is the same as $t(uv)$.

Definition 7 (Variants)
Let U and V be wirings in $\Lambda^*(T^2 \cdot n)$; U and V are said to be variants *if there exist* wirings W *and* W' *of* $\Lambda^*(T^2 \cdot n)$ *such that :*

- *$U = W^*VW$ and $V = W'^*UW'$*
- *W (resp. W') is a sum of wirings of the form $\sum_{i=1}^{n} p_i(x,t) \mapsto p_i(x,u)$ with x not free in t,u ; this means that W and W', as operators, are of the form $Id \otimes Z$.*

The notion of being variants is clearly an equivalence relation, noted $\sim$.

Proposition 4
If $U \sim V$ and σ is a sum of wires of the form $p(t,x) \mapsto q(u,x)$ (x not occuring in t,u), then :

- *$U\sigma$ is nilpotent iff $V\sigma$ is nilpotent ;*
- *In this case, $EX(U,\sigma) \sim EX(V,\sigma)$ and $RES(U,\sigma) \sim RES(V,\sigma)$.*

PROOF: immediate. □

Definition 8 (Changing dialects, see remark 8 below)
Let U be a wiring in $\Lambda^*(T^2 \cdot n)$;

- *$\otimes_1(U)$ is defined as follows : every wire $p(t,u) \mapsto q(t',u)$ of U is replaced with the wire $p(t,uy) \mapsto q(t',uy)$, where y is a fresh variable.*
- *$\otimes_2(U)$ is defined as follows : every wire $p(t,u) \mapsto q(t',u)$ of U is replaced with the wire $p(t,xu) \mapsto q(t',xu)$, where x is a fresh variable.*
- *$\&_1(U)$ is defined as follows : every wire $p(t,u) \mapsto q(t',u)$ of U is replaced with the wire $p(t,ug) \mapsto q(t',ug)$.*
- *$\&_2(U)$ is defined as follows : every wire $p(t,u) \mapsto q(t',u)$ of U is replaced with the wire $p(t,ud \mapsto q(t',ud)$.*

Proposition 5

- $\otimes_1(U) \sim \otimes_2(U)$
- $\&_1(U) \sim \&_2(U)$
- $\otimes_1(U) \sim \otimes_1(\otimes_1(U))$

PROOF: let us verify the third fact : let $W := \sum_{i=1}^{n} p_i(x,(yz)z) \mapsto p_i(x,yz)$ and $W' := \sum_{i=1}^{n} p_i(x,yz_1z_2) \mapsto p_i(x,(yz_1)z_2)$;
then $\otimes_1(U) = W^* \otimes_1(\otimes_1(U))W$ and $\otimes_1(\otimes_1(U)) = W'^* \otimes_1(U)W'$. □

Definition 9 (Laminated wirings)

A wiring U in $\Lambda^*(T^2 \cdot n)$ is said to be laminated *if it is a sum of wires the form $p(t,u) \mapsto q(t',u)$ and if $U \sim \otimes_1(U)$.*

Remark 8

A proof Π of a n-ary sequent using m cuts will be interpreted by a pair $(\Pi^{\bullet}, \sigma)$ of laminated wirings in $\Lambda^*(T^2 \cdot (2m+n))$. We can see $\Pi^{\bullet}$ as an operator on the Hilbert space $(\mathcal{H}\otimes\mathcal{H})\oplus\ldots\oplus(\mathcal{H}\otimes\mathcal{H})$, a n-ary direct sum of spaces $\mathcal{H}\otimes\mathcal{H}$, which can also be seen as $(\mathcal{H}\oplus\ldots\oplus\mathcal{H})\otimes\mathcal{H}$. Now, in $(\mathcal{H}\oplus\ldots\oplus\mathcal{H})\otimes\mathcal{H}$ the two components are treated in a very different spirit :

- The first component $(\mathcal{H}\oplus\ldots\oplus\mathcal{H})$ is seen as a space of *shared* messages, the *channels* ; typically the decomposition of $(\mathcal{H}\oplus\ldots\oplus\mathcal{H})$ into its summands is part of this public knowledge.
- The second component $\mathcal{H}$ is seen as a space of *private* messages : a private *dialect* useful only for $\Pi^{\bullet}$ but that cannot be communicated to the environment ; the dialect is a typical additive creation, coming from the need of avoiding overlap in the treatment of the &-rule.
- This distinction between these two components is a consequence of the *lamination* of $\Pi^{\bullet}$: by definition 9 every wire of $\Pi^{\bullet}$ is of the form $p(t,u) \mapsto q(t',u)$; if we make $\Pi^{\bullet}$ operate on $(\mathcal{H}\oplus\ldots\oplus\mathcal{H})\otimes\mathcal{H}$, this means that $\Pi^{\bullet}(1\otimes m) = (1\otimes m)\Pi^{\bullet}$, i.e. that $\Pi^{\bullet}$ belongs to the *commutant* of the set of messages of the form $1\otimes m$.

In fact the privacy of the dialect is expressed by the fact that all constructions of geometry of interaction indeed deal with equivalence classes modulo $\sim$. In the case of binary rules, this fact limits the possible ways of merging dialects. If we want to combine U and V, we shall try to replace them by *variants* U' and V' whose respective dialects are in certain relation. We basically can use the constructions of definition 8, and these two possibilities lead to the interpretations of the $\otimes$- and &-rules, that we now sketch : let us assume that U and V are laminated wirings interpreting cut-free proofs of $\vdash \Gamma, A$ and $\vdash \Delta, B$; then

- The first operation is to create a common dialect ; this is done by :
  - in the case of a $\otimes$ -rule, tensorization of the respective dialects : let $U' := \otimes_1(U)$, $V' := \otimes_2(V)$.
  - in the case of a & -rule, summation of the respective dialects : $U' := \&_1(U)$, $V' := \&_2(V)$.
- The next operation is to glue together A and B ; this is done by merging (in $U'+V'$) the predicates p_A and p_B respectively associated to A and B into a new predicate p_C (where C is $A\otimes B$ or $A\&B$, depending on the rule), as follows :
  - every occurrence $p_A(t,u)$ of p_A in U' is replaced with an occurrence of $p_C(tg,u)$,

- every occurrence $p_B(t, u)$ of p_B in V' is replaced with an occurrence of $p_C(td, u)$.

the result W of this operation is the desired interpretation.

To end with our informal description of the interpretation, let us consider the case of a CUT-rule : let us assume that U and V interpret cut-free proofs of $\vdash \Gamma, A$ and $\vdash \Delta, A^{\perp}$; then

- We define $U' := \otimes_1(U)$, $V' := \otimes_2(V)$ as in the case of a $\otimes$-rule, and we sum them ;
- We introduce σ as the sum $(p_A(x, y) \mapsto p_{A^{\perp}}(x, y)) + (p_{A^{\perp}}(x, y) \mapsto p_A(x, y))$. The pair $(U' + V', \sigma)$ is the desired interpretation.

Remark 9
The rules for $\otimes$ and & share the last step, which consists in merging two predicates p_A and p_B by means of renaming of the channels ; the choice of the constants g and d to do so is arbitrary, but once it has been set, one cannot touch it. The main difference between a $\otimes$-rule and a &-rule is the way in which the respective dialects are merged

- In the case of the $\otimes$, this is splendid ignorance (the dialects now commute with each other) : this is achieved by replacing u by uy and v by xv ;
- Whereas in the case of the &, the dialects are just disjoint (i.e. incompatible). This is achieved by replacing u by ug and v by vd.

Observe that we could as well choose isomorphic solutions for the first step, for instance :

- Exchange (in the case of $\otimes$) $\otimes_1$ with $\otimes_2$.
- Replace (in the case of &) $\&_1$ with $\&_2$.

This flexibility contrasts with the (relative) rigidity of the choice involved when merging predicates. This is because channels are public whereas dialects are private.

3.2 Interpretation of proofs in the multiplicative-additive case

Definition 10 (Pattern)
The general pattern is as follows :

- *We are given a proof* Π *of a sequent* $\vdash \Gamma$*, containing cuts on formulas* Δ *; in* Δ *formulas are coupled in pairs* $(B, B^{\perp})$ *; in fact we use the formulas of* Δ *as labels for the cuts performed rather than actual formulas. The expression* "Let Π be a proof of $\vdash [\Delta], \Gamma$" *is short for* "Let Π be a proof of $\vdash \Gamma$, containing cuts on the formulas Δ".

- *We introduce the algebra* $\Lambda^*(\Delta, \Gamma)$ *:*

  - *the function letters are the constants* d *and* g *and the binary* $\odot$ *;*
  - *the predicate letters are all binary :* p_A *for all formula* A *in* Δ *and* Γ*. We assume that these formulas are pairwise distinct (if not, they can be distinguished by adding indices).*
  - *an important particular case is* $\Delta = \emptyset$*, in which case we use the notation* $\Lambda^*(\Gamma)$ *; if* Γ *were empty as well (a case that will never occur) then we would need to define* $\Lambda^*(\emptyset)$*, that can conveniently be taken as the field* C *of scalars.*

- *We define* $\sigma_{\Delta;\Gamma}$ *in the algebra* $\Lambda^*(\Delta, \Gamma)$ *by*

$$\sigma_{\Delta;\Gamma} := \sum_{B \in \Delta} p_B(x, y) \mapsto p_{B^{\perp}}(x, y)$$

  x, y *are distinct variables, the sum is taken over all cut formulas ; observe that* $\sigma = \sigma^*$*, and that* σ^2 *is a projection. Therefore* $1 - \sigma^2$ *is a projection too, and*

$$\forall u \in \Lambda^*(\Delta, \Gamma), \quad (1 - \sigma^2)u(1 - \sigma^2) \in \Lambda^*(\Gamma)$$

  in other terms prefixing and suffixing with $1 - \sigma^2$ *acts like a restriction, see remark 7.*

- *We interpret* Π *by a pair* $(\Pi^{\bullet}, \sigma_{\Delta;\Gamma})$ *in the algebra* $\Lambda^*(\Delta, \Gamma)$ *: the definition of* $\Pi^{\bullet}$ *is by induction.*

Definition 11 **($\Pi^{\bullet}$)**
Let Π *be a proof of* $\vdash [\Delta], \Gamma$ *; we associate to* Π *its interpretation* $(\Pi^{\bullet}, \sigma_{\Delta;\Gamma})$*, where* $\Pi^{\bullet}$ *is defined by induction :*

Case 11.1 *If* Π *consists in the axiom* $\vdash A, A^{\perp}$*, then* $\Pi^{\bullet} := \sigma_{A,A^{\perp}}$*; i.e.*

$$\Pi^{\bullet} := (p_A(x, y) \mapsto p_{A^{\perp}}(x, y)) + (p_{A^{\perp}}(x, y) \mapsto p_A(x, y))$$

Case 11.2 *If* Π *is obtained from* Π_1 *and* Π_2 *by means of a CUT-rule with cut-formulas* $A, A^{\perp}$*, then*

- *let us replace in* $\Pi_1^{\bullet}$ *all atoms* $p_C(t, u)$ *by* $p_C(t, uy)$ *(*y *fresh variable) ; the result is called* U_1 *;*

- *let us replace in* $\Pi_2^\bullet$ *all atoms* $p_C(t,v)$ *by* $p_C(t,xv)$ *(x fresh variable) ; the result is called* U_2 *; then*

$$\Pi^\bullet := U_1 + U_2.$$

Of course the essential feature of this step is the modification of the component σ *: if* Π_1 *and* Π_2 *respectively prove* $\vdash [\Delta_1], \Gamma_1, A$ *and* $\vdash [\Delta_2], \Gamma_2, A^\perp$*, then* Π *is a proof of* $\vdash [\Delta_1, \Delta_2, A, A^\perp], \Gamma_1, \Gamma_2$.

Case 11.3 *If* Π *is obtained from* Π_1 *by means of an exchange rule, then*

$$\Pi^\bullet := \Pi_1^\bullet.$$

Case 11.4 *If* Π *is obtained from* Π_1 *and* Π_2 *by means of a* $\otimes$*-rule introducing* $A \otimes B$*, then*

- *let us replace in* $\Pi_1^\bullet$ *all atoms* $p_C(t,u)$ *by* $p_C(t,uy)$ *(y fresh variable) when C is distinct from A, and all atoms* $p_A(t,u)$ *by* $p_{A\otimes B}(tg,uy)$ *; the result is called* U_1 *;*

- *let us replace in* $\Pi_2^\bullet$ *all atoms* $p_C(t,v)$ *by* $p_C(t,xv)$ *(x fresh variable) when C is distinct from B, and all atoms* $p_B(t,v)$ *by* $p_{A\otimes B}(td,xv)$ *; the result is called* U_2 *; then*

$$\Pi^\bullet := U_1 + U_2.$$

Case 11.5 *If* Π *is obtained from* Π_1 *by means of a* $\wp$*-rule, then let us replace in* $\Pi_1^\bullet$ *all atoms* $p_A(t,u)$ *by* $p_{A\wp B}(tg,u)$ *and all atoms* $p_B(t,v)$ *by* $p_{A\wp B}(td,v)$ *; the result is by definition* $\Pi^\bullet$.

Case 11.6 *If* Π *is obtained from* Π_1 *and* Π_2 *by means of a* &*-rule introducing* $A\&B$*, then*

- *let us replace in* $\Pi_1^\bullet$ *all atoms* $p_C(t,u)$ *by* $p_C(t,ug)$ *when C is distinct from A, and all atoms* $p_A(t,u)$ *by* $p_{A\&B}(tg,ug)$ *; the result is called* U_1 *;*

- *let us replace in* $\Pi_2^\bullet$ *all atoms* $p_C(t,v)$ *by* $p_C(t,vd)$ *when C is distinct from B, and all atoms* $p_B(t,v)$ *by* $p_{A\&B}(td,vd)$ *; the result is called* U_2 *; then*

$$\Pi^\bullet := U_1 + U_2.$$

Case 11.7
- *If* Π *is obtained from* Π_1 *by means of a* $\oplus_1$*-rule, then let us replace in* $\Pi_1^\bullet$ *all atoms* $p_A(t,u)$ *by* $p_{A\oplus B}(tg,u)$ *; the result is by definition* $\Pi^\bullet$.

- *If Π is obtained from Π_1 by means of a $\oplus_2$-rule, then let us replace in $\Pi_1^\bullet$ all atoms $p_B(t,u)$ by $p_{A\oplus B}(td,u)$; the result is by definition $\Pi^\bullet$.*

Case 11.8 *If Π is obtained from Π' by means of $\forall$-rule (first or second order), then $\Pi^\bullet := \Pi'^\bullet$.*

Case 11.9 *If Π is obtained from Π' by means of $\exists$-rule (first or second order), then $\Pi^\bullet := \Pi'^\bullet$.*

Remark 10
Definition 11 uses several times rather arbitrary constructions to merge dialects (the cases 11.4 and 11.6, but also the constant g in the case 11.1) . In fact in all these cases, we might have chosen variants, without any difference. This is because of proposition 4 will allows one to replace $\Pi^\bullet$ with a variant.

Example 1
Cut-elimination replaces the cut

$$\dfrac{\vdash \Gamma, A \quad \dfrac{}{\vdash B^\perp, B}\,Id}{\vdash \Gamma, B}\,CUT$$

(with A and B different occurences of the same formula) by the original proof Π of $\vdash \Gamma, A$. The original proof is interpreted as

$$(\otimes_1(\Pi^\bullet) + \otimes_2(\sigma_{B^\perp,B\ ;\Gamma,A}), \sigma_{A,B^\perp\ ;\Gamma,B}).$$

Nilpotency is easily shown $((U\sigma)^2 = 0)$, and the result of execution is $(\otimes_1(\Pi^\bullet)$ in which p_A has been replaced with p_B, i.e. it is a variant of the interpretation of the modified proof. This shows, in this basic, but essential example, that execution corresponds to cut-elimination, up to variance.

3.3 The nilpotency theorem

Our first goal is to prove that $EX(\Pi^\bullet, \sigma)$ $(= \Pi^\bullet(\sigma\Pi^\bullet)^{-}1$, see definition 6) makes sense, i.e., $\Pi^\bullet\sigma$ is *nilpotent* . This will be done by an adaptation of the results from [G88].

Theorem 1 (Transitivity of cut)
In the algebra $\Lambda^*(\Delta, \Delta', \Gamma)$ let us define σ, τ as respectively $\sigma_{\Delta;\Delta',\Gamma}$ and $\sigma_{\Delta';\Delta,\Gamma}$ (so that $\sigma_{\Delta,\Delta';\Gamma} = \sigma + \tau$) ; let U be a wiring

- *If $U\sigma$ is nilpotent, $RES(U,\sigma)$ $(= (1-\sigma^2)EX(U,\sigma)(1-\sigma^2))$ is a wiring.*
- *$(\sigma+\tau)U$ is nilpotent iff σU and $\tau \cdot RES(U,\sigma)$ are nilpotent ;*
- *In that case $RES(U, \sigma+\tau) = RES(RES(U,\sigma),\tau)$.*

PROOF: see [G88], lemmas 4 and 5, where this result was called *associativity of cut.* □

Theorem 2 (Nilpotency)
$\sigma\Pi^\bullet$ is nilpotent.

PROOF: the proof is adapted from [G88]. (We neglect all features related to quantifiers, which are quite the same.) We first define :

Definition 12 (Weak orthogonality)
Let U and V be laminated wirings in $\Lambda^*(T^2)$ (an algebra with a single binary predicate p; U and V are said to be weakly orthogonal *if $\otimes_1(U) \cdot \otimes_2(V)$ is nilpotent. In that case we introduce the notation $U \perp V$.*

The definition makes sense because of the following :

Proposition 6

- *If $U \perp V$, then $V \perp U$.*
- *If $U \sim U'$ and $V \sim V'$ and $U \perp V$, then $U' \perp V'$.*

Definition 13 (Weak types)

Given a subset X of $\Lambda^*(T^2)$, we define $X^\perp := \{U \; ; \forall V \in X, \; U \perp V\}$. A weak type *is any set X of laminated wirings such that $X = X^{\perp\perp}$. The connectives of linear logic can be interpreted as operations on weak types :*

Case 13.1 *If X and Y are weak types, then we define the weak type $X \otimes Y$ as $Z^{\perp\perp}$, where Z consists of all $U' + V'$, where U' and V' are constructed from $U \in X$ and $V \in Y$ as follows :*

- *Replace in U all atoms $p(t,u)$ by $p(tg,uy)$.*
- *Replace in U all atoms $p(t,v)$ by $p(td,xv)$.*

Case 13.2 *If X and Y are weak types, then we define the weak type $X \wp Y$ as $(X^\perp \otimes Y^\perp)^\perp$.*

Case 13.3 *If X and Y are weak types, then we define the weak type $X \oplus Y$ as $Z^{\perp\perp}$, where Z consists of : all U' and all V', where U' and V' are constructed from $U \in X$ and $V \in Y$ as follows :*

- *Replace in U all atoms $p(t,u)$ by $p(tg,u)$.*
- *Replace in U all atoms $p(t,v)$ by $p(td,v)$.*

Case 13.4 *If X and Y are weak types, then we define the weak type $X\&Y$ as $(X^\perp \oplus Y^\perp)^\perp$.*

Remark 11
The interpretation by weak types is very degenerated ; in particular, it easy to check that & and ⊕ are not separated by this rough interpretation.

Definition 14 (Weak types associated to formulas and sequents)
Let us associate an arbitrary weak type $\alpha^\bullet$ to each atom α of a language in linear logic ; then each formula A built from such atoms by means of linear negation, and the binary connectives $\otimes, \wp, \&, \oplus$ immediately gets interpreted by a weak type $A^\bullet$. (This also extends to quantifiers of any order, provided one works on weak type parameters as in definition 3 of [G88].) The definition of $A^\bullet$ immediately induces a definition of $\Gamma^\bullet$, when $\vdash \Gamma$ is a sequent :
assume that $\Gamma = A_1, \ldots, A_n$, and let U be a laminated wiring in $\Lambda^*(\Gamma)$;

- *Given laminated wirings $V_1, \ldots, V_n$ in $A_1^{\perp\bullet}, \ldots, A_n^{\perp\bullet}$ respectively, we can rename the predicates so that $V_i \in \Lambda^*(A_i)$, and introduce $V_1' + \ldots + V_n'$ as the result of applying $n-1$ cases 11.4 to the V_i ; the object depends in which order the rules are performed, but different choices yield* variants.
- *Then $U \in \Gamma^\bullet$ iff for all $V_1, \ldots, V_n$ in $A_1^{\perp\bullet}, \ldots, A_n^{\perp\bullet}$ respectively, $U \perp V_1' + \ldots + V_n'$.*

of course the definition has two particular cases of interest :

- *The particular case $n = 1$ yields $(\vdash A)^\bullet = A^\bullet$;*
- *The particular case $n = 0$ yields $\vdash {}^\bullet = \mathcal{C}$, see definition 10.*

Lemma 2.1 *If U, V are variants and $U \in A^\bullet$ (resp. $U \in \Gamma^\bullet$), then $V \in A^\bullet$ (resp. $V \in \Gamma^\bullet$).*

PROOF: immediate □

Lemma 2.2 *$U \in (\Gamma, A)^\bullet$ iff for all $V \in A^{\perp\bullet}$ the "cut" $(U_1 + V_1, \sigma_{A,A^\perp;\Gamma})$ (defined as in case 11.2) is such that : $(U_1 + V_1)\sigma_{A,A^\perp;\Gamma}$ is nilpotent and the result $RES(U_1 + V_1, \sigma_{A,A^\perp;\Gamma})$ is in $\Gamma^\bullet$.*

PROOF: easy, see lemma 7 of [G88]. □

The nilpotency theorem follows from the more precise :

Theorem 3
If Π is a proof of $\vdash [\Delta], \Gamma$, then $\Pi^\bullet\sigma_{\Delta;\Gamma}$ is nilpotent and $RES(\Pi^\bullet, \sigma_{\Delta;\Gamma}) \in \Gamma^\bullet$.

PROOF: this theorem is proved by induction on Π ; there is a general pattern that can be followed for all cases. We do this in detail for the first case (the case of a &-rule, which is the truly new case), but we shall only treat simplified versions of the other cases. The pattern basically reduces the general case to the cut-free case (i.e. when the premise(s) is (are) cut-free), and then the cut-free case to a context-free case. To do this in very rigourous way, it would be natural to enlarge the set of rules with an additional one : for each $V \in\vdash \Gamma^\bullet$ add the axiom Γ whose proof Π_V is interpreted as $\Pi_V^\bullet := V$ (in the spirit of model-theory, where new constants for the elements of the model are introduced).

Case 3.1 • We first treat a very limited case : the last rule of Π is a &-rule between proofs Π_1 of $\vdash A_1$ and Π_2 of $\vdash A_2$, (so that $\Delta = \emptyset$ and $\Gamma = A_1 \& A_2$). We have to show that for all U' obtained from $U \in A_1^{\perp\bullet}$ (resp. all V' obtained from $V \in A_2^{\perp\bullet}$) by means of definition 13.3, then $\Pi^\bullet \perp U'$ (resp. $\Pi^\bullet \perp V'$). But this immediately reduces to $\Pi_1^\bullet \perp U$ (resp. $\Pi_2^\bullet \perp V$).

- Then we treat the case where the premises are cut-free : the last rule of Π is a &-rule between proofs Π_1 of $\vdash \Phi, A_1$ and Π_2 of $\vdash \Phi, A_2$, (so that $\Delta = \emptyset$ and $\Gamma = \Phi, A_1 \& A_2$). We can argue by induction on the cardinality of Φ :

  – the case $\Phi = \emptyset$ has already been treated ;

  – if $\Phi = B, \Psi$, then by lemma 2.2, $\Pi^\bullet \in (\Phi, A_1 \& A_2)^\bullet$ (resp. $\Pi_1^\bullet \in (\Phi, A_1)^\bullet$, $\Pi_2^\bullet \in (\Phi, A_2)^\bullet$) iff for all $V \in B^{\perp\bullet}$ $\Pi^\bullet \sigma_{B,B^\perp;\Psi,A_1\&A_2}$ is nilpotent and $RES(\Pi^\bullet, \sigma_{B,B^\perp;\Psi,A_1\&A_2}) \in (\Psi, A_1 \& A_2)^\bullet$ (resp. $\Pi_1^\bullet \sigma_{B,B^\perp;\Psi,A_1}$ is nilpotent and $RES(\Pi_1^\bullet, \sigma_{B,B^\perp;\Psi,A_1}) \in (\Psi, A_1)^\bullet$, $\Pi_2^\bullet \sigma_{B,B^\perp;\Psi,A_2}$ is nilpotent and

  $$RES(\Pi_2^\bullet, \sigma_{B,B^\perp;\Psi,A_2}) \in (\Psi, A_2)^\bullet).$$

  The result follows from the induction hypothesis on the size of Φ and the fact that

  * $\Pi^\bullet \sigma_{B,B^\perp;\Psi,A_1\&A_2}$ is nilpotent iff $\Pi_1^\bullet \sigma_{B,B^\perp;\Psi,A_1}$ and $\Pi_2^\bullet \sigma_{B,B^\perp;\Psi,A_2}$ are nilpotent ;
  * if we apply definition 11.6 to $RES(\Pi_1^\bullet, \sigma_{B,B^\perp;\Psi,A_1})$ and $RES(\Pi_2^\bullet, \sigma_{B,B^\perp;\Psi,A_2})$, then we get a variant of $RES(\Pi^\bullet, \sigma_{B,B^\perp;\Psi,A_1\&A_2})$.

- It remains to treat the general case : the last rule of Π is a &-rule between proofs Π_1 of $\vdash [\Delta_1], \Phi, A_1$ and Π_2 of $\vdash [\Delta_2], \Phi, A_2$, (so that $\Delta = \Delta_1, \Delta_2$ and $\Gamma = \Phi, A_1 \& A_2$). ¿From the induction hypothesis, $\Pi_i^\bullet \sigma_{\Delta_i;\Phi,A_i}$ is nilpotent and $RES(\Pi_i^\bullet, \sigma_{\Delta_i;\Phi,A_i}) \in (\Phi, A_i)^\bullet$ for $i = 1, 2$. Let $u_i := RES(\Pi_i^\bullet, \sigma_{\Delta_i;\Phi,A_i})$; then it is easily seen that $\Pi^\bullet \sigma_{\Delta;\Phi,A\&B}$ is nilpotent and that by definition 11.6 applied to u_1 and u_2, $RES(\Pi^\bullet, \sigma_{\Delta;\Phi,A\&B}) \in (\Phi, A\&B)^\bullet$.

Case 3.2 If Π is the axiom $\vdash A, A^\perp$; for reasons of labelling, we prefer to rename the first A as B; then

$$\Pi^\bullet := (p_B(x,y) \mapsto p_{A^\perp}(x,y)) + (p_{A^\perp}(x,y) \mapsto p_B(x,y))$$

select $U = \sum p_A(t_i, v_i) \mapsto p_A(u_i, v_i)$ in $A^\bullet$; then these two wirings are respectively modified into

$$\otimes_1(\Pi^\bullet) := (p_B(x, yz) \mapsto p_{A^\perp}(x, yz)) + (p_{A^\perp}(x, yz) \mapsto p_B(x, yz))$$

and $\otimes_2(U) = \sum p_A(t_i, y'v_i) \mapsto p_A(u_i, y'v_i)$. The nilpotency of $(\otimes_1(\Pi^\bullet) + \otimes_2(U))\sigma_{A,A^\perp;B}$ is rather immediate, and the result is easily seen to be $\sum p_B(t_i, yv_i) \mapsto p_B(u_i, yv_i)$, a variant of U ; the result follows from lemmas 2.2 and 2.1.

Case 3.3 The last rule of Π is a CUT-rule between proofs Π_1 of $\vdash A$ and Π_2 of $\vdash A^\perp$. (This extremely simplified case can only occur because we have extended the set of all proofs by adding a lot of new axioms ; also remark that the reduction of the general case to this case makes a heavy use of transitivity of cut). Due to the peculiar definition of $\vdash^\bullet$, it suffices to show that $\Pi^\bullet\sigma_{A,A^\perp;}$ is nilpotent ; but this immediately reduces to the orthogonality of $\Pi_1^\bullet$ and $\Pi_2^\bullet$, which is precisely the induction hypothesis.

Case 3.4 The last rule of Π is a $\oplus_1$-rule applied to a proof Π_1 of $\vdash A$. This simplified case follows immediately from definition 13.3. The case of a $\oplus_2$-rule is similar.

Case 3.5 The last rule of Π is a $\otimes$-rule between proofs Π_1 of $\vdash A$ and Π_2 of $\vdash B$. This simplified case follows from definition 13.1.

Case 3.6 The last rule of Π is a $\wp$-rule applied to a proof Π_1 of $\vdash A, B$. By definition 13.2 we must show that for all $U \in A^{\perp\bullet}$ and $V \in B^{\perp\bullet}$, then (with U' and V' as in definition 13.3) $\Pi^\bullet \perp U' + V'$. But this is easily reduced to $\Pi_1^\bullet \perp \otimes_1(U) + \otimes_2(V)$, i.e. the induction hypothesis.

Case 3.7 The remaining cases are left to the reader. □

Remark 12
Although the result proved (nilpotency) is rather weak, it is "proof-theoretically strong" : for instance, in the presence of exponentials (next subsection) and second order quantification, it cannot be proved inside second-order arithmetic : this is because the orders of nilpotency are related to the length of normalisation steps (see [DR93] for instance).

3.4 Inflated sequent calculus

Our next goal is to prove that the result $RES(\Pi^\bullet, \sigma)$ of the execution is $\Pi_1^\bullet$, where Π_1 is obtained from Π by cut-elimination. Unfortunately

- We can only expect a *variant* of $\Pi_1^\bullet$;

- Worse, the result $RES(\Pi^\bullet, \sigma)$ contains "unerased" information, what we called a *beard* in [G88]. This second feature makes the precise statement of a theorem quite delicate.

Since of the two partners -geometry of interaction and cut-elimination-, the former (although very recent) is the most natural, we shall therefore (slightly) modify sequent calculus : the result is called *inflated* sequent calculus. The reason for this terminology is that the calculus contains unessential rules corresponding to the erasings that geometry of interaction, in its lazy spirit, does not perform. We introduce a new symbol, $\flat$, that we call *flat* ; this symbol is treated differently from the other formulas of linear logic : $\flat$ cannot be combined ($\flat \wp \flat$, $\flat^\perp$ are not allowed) ; this leaves only the possibility of using $\flat$ inside sequents, like in $\vdash \flat$, $\vdash \flat, A$, $\vdash \flat, A, \flat$ etc. The specific rules of $\flat$ are the following :

$$\frac{}{\vdash \flat, \Gamma}\flat \qquad \frac{\vdash \flat, \flat, \Gamma}{\vdash \flat, \Gamma}c\flat \qquad \frac{\vdash \flat, \Gamma}{\vdash \flat, \flat, \Gamma}i\flat \qquad \frac{\vdash \flat, \Gamma \quad \vdash \Gamma}{\vdash \Gamma}s\flat$$

- The axiom scheme $\vdash \flat, \Gamma$ for any Γ.
- The contraction rule : from $\vdash \flat, \flat, \Gamma$ deduce $\vdash \flat, \Gamma$.
- The inflation rule : from $\vdash \flat, \Gamma$ deduce $\vdash \flat, \flat, \Gamma$.
- The summation rule : from $\vdash \flat, \Gamma$ and $\vdash \Gamma$ deduce $\vdash \Gamma$.

These rules are basically minor structural rules.

Remark 13

- $\flat$ can be faithfully interpreted as the constant $\top$ of linear logic : the axiom is the usual axiom for $\top$ and the rules are derivable... but this is not the point.

- Contraction and inflation mean that we do not count the multiplicity of $\flat$. This means that in reality, we are working with two kinds of sequents : usual ones (without $\flat$) and *flat* ones (with at least one $\flat$). Of course we get a much simpler presentation with only one kind of sequents, and this explains our choice.

- Among the derivable rules for $\flat$: another form of summation : from $\vdash \flat, \Gamma$ and $\vdash \flat, \Gamma$ deduce $\vdash \flat, \Gamma$.

Definition 15 (Geometry of interaction of $\flat$)
A proof Π of $\vdash [\Delta], \Gamma$ is interpreted as a laminated wire in $\Lambda^*(\Delta, \Gamma - \flat)$, where $\Gamma - \flat$ is obtained from Γ by the removal of all $\flat$ (observe that since $\flat$ cannot be negated, it does not occur in Δ).

- *An axiom* $\vdash \flat, \Gamma$ *is interpreted by* $0 \in \Lambda^*(\Gamma - \flat)$.
- *Contraction and inflation are interpreted identically.*
- *If* Π *follows from* Π_1 *and* Π_2 *by means of a summation : then* $\Pi^\bullet := \&_1(\Pi_1^\bullet) + \&_2(\Pi_2^\bullet)$.

Remark 14
The basic idea behind the symbol "$\flat$" is that a proof of a flat sequent $\vdash \flat, \Gamma$ comes from a "real" proof in which some essential part has been erased. Flat proofs interact with real ones by means of summation : summation typically occurs in the case of cut-elimination on an additive formula. For instance assume that we perform a cut between a proof of $\vdash A \oplus B$ (obtained from a proof Π of A by a $\oplus_1$-rule) and a proof of of a sequent $\vdash A^\perp \& B^\perp, \Gamma$ coming from proofs Π_1 of $\vdash A^\perp, \Gamma$ and Π_2 of $\vdash B^\perp, \Gamma$. The traditional way of eliminating the cut is to replace it with the proof Π' obtained by a cut between Π and Π_1, but this does not quite correspond to geometry of interaction. This is why in the *inflated* calculus, our cut is replaced by the *summation* of Π' and the proof of $\vdash \flat, \Gamma$ obtained by a cut between the axiom $\vdash \flat, B$ and Π_2.

It is not surprising that the inflated calculus enjoys cut-elimination... In fact we are rather interested in the details of the cut-elimination procedure that we now sketch :

Definition 16 (Cut-elimination procedure)

A cut

$$\dfrac{\dfrac{\cdots}{\vdash \Gamma, A}R \qquad \dfrac{\cdots}{\vdash \Delta, A^\perp}S}{\vdash \Gamma, \Delta}CUT$$

is replaced as follows :

- *If the rule R is not an introduction of A or an axiom, then R is performed after the cut, for instance :*
  - *If R is a summation rule, with premises* $\vdash \Gamma, A$ *and* $\vdash \Gamma, A, \flat$, *then our cut is replaced with :*

$$\dfrac{\dfrac{\vdash \Gamma, A \qquad \vdash \Delta, A^\perp}{\vdash \Gamma, \Delta}CUT \qquad \dfrac{\vdash \Gamma, A, \flat \qquad \vdash \Delta, A^\perp}{\vdash \Gamma, \Delta, \flat}CUT}{\vdash \Gamma, \Delta}s\flat$$

- *If the rule S is not an introduction of* $A^\perp$ *or an axiom, then S is performed after the cut, as above.*

- *If both R and S are introductions of the cut-formulas or axioms, then there are several cases, including :*

  - *when R is the identity axiom $\vdash A, A^\perp$, the proof is replaced with*
    $$\frac{\qquad}{\vdash \Delta, A^\perp}\,S \quad ;$$
  - *when R is the axiom $\vdash \Phi, A, \flat$, the solution depends on S, typically :*
    * *if S is the axiom $\vdash \Psi, A^\perp, \flat$, then we replace the proof with the axiom $\vdash \Phi, \Psi, \flat, \flat$;*
    * *if S is a $\otimes$-rule, with premises $\vdash \Psi, B^\perp$ and $\vdash \Xi, C^\perp$, then our proof is replaced with two cuts between the axiom $\vdash \Phi, B, C, \flat$ and the premises of the $\otimes$-rule ;*
    * *if S is a $\wp$-rule, with premise $\vdash \Delta, B^\perp, C^\perp$, then our proof is replaced with two cuts between the axioms $\vdash \Phi, B, \flat$ and $\vdash C, \flat$ and the premise of the $\wp$-rule ; this yields a proof of $\vdash \Phi, \Delta, \flat, \flat$ that we contract into $\vdash \Phi, \Delta, \flat$;*
    * *if S is a &-rule, with premises $\vdash \Delta, B^\perp$ and $\vdash \Delta, C^\perp$, then we cut these premises respectively with the axioms $\vdash \Phi, B^\perp, \flat$ and $\vdash \Phi, C^\perp, \flat$, yielding two proofs of $\vdash \Phi, \Delta, \flat$ to which we can apply summation, as seen in remark 13 ;*
    * *if S is a $\oplus_1$-rule, with premise $\vdash \Delta, B^\perp$, then our proof is replaced with a a cut between this premise and the axiom $\vdash \Phi, B, \flat$...*
  - *when R and S are both logical rules, then we get several cases, including :*
    * *if R is a $\wp$-rule, with premise $\vdash \Gamma, B, C$ and S is a $\otimes$-rule, with premises $\vdash \Psi, B^\perp$ and $\vdash \Xi, C^\perp$, then we replace our proof with*
    $$\dfrac{\dfrac{\vdash \Gamma, B, C \qquad \vdash \Psi, B^\perp}{\vdash \Gamma, \Psi, C}\,CUT \qquad\qquad \vdash \Xi, C^\perp}{\vdash \Gamma, \Psi, \Xi}\,CUT$$
    * *if R is a $\oplus_1$-rule, with premise $\vdash \Gamma, B$ and S is a &-rule, with premises $\vdash \Delta, B^\perp$ and $\vdash \Delta, C^\perp$, then we replace our proof with*
    $$\dfrac{\dfrac{\vdash \Gamma, B \qquad \vdash \Delta, B^\perp}{\vdash \Gamma, \Delta}\,CUT \qquad \dfrac{\dfrac{\qquad}{\vdash \Gamma, C, \flat}\,\flat \qquad \vdash \Delta, C^\perp}{\vdash \Gamma, \Delta, \flat}\,CUT}{\vdash \Gamma, \Delta}\,s\flat$$

Proposition 7
Cut-elimination holds for the inflated calculus, using the procedure sketched in definition 16.

PROOF: boring and straightforward ; however notice that the treatment of a cut on an additive formula makes a significant use of flat sequents. □

Theorem 4 (Soundness)
If a cut-free proof Π *of* $\vdash \Gamma$ *is obtained from a proof* Σ *of* $\vdash [\Delta], \Gamma$ *by means of the transformations sketched in definition 16, then* $\Pi^\bullet$ *is a variant of* $RES(\Sigma^\bullet, \sigma_{\Delta;\Gamma})$.

Lemma 4.1 *If a proof* Π *of* $\vdash [\Delta'], \Gamma$ *is obtained from a proof* Σ *of* $\vdash [\Delta], \Gamma$ *by one step of definition 16, then*

$$RES(\Pi^\bullet, \sigma_{\Delta';\Gamma}) \sim RES(\Sigma^\bullet, \sigma_{\Delta;\Gamma}).$$

PROOF: we treat only a few distinguished cases :

- Assume that Σ is obtained from Σ_1, Σ_2 respectively proving $\vdash [\Delta], \Phi, A$, and $\vdash A^\perp, A$, (an identity axiom that we note $\vdash A^\perp, B$) via a cut on A. Then $\Sigma^\bullet$ is of the form $\otimes_1(\Sigma_1^\bullet) + \otimes_2(\Sigma_2^\bullet)$. An application of transitivity of cut yields

  $$RES(\Sigma^\bullet, \sigma_{\Delta,A,A^\perp;\Phi,B}) = RES(RES(\Sigma^\bullet, \sigma_{\Delta;A,A^\perp,\Phi,B}), \sigma_{A,A^\perp;\Delta,\Phi,B}).$$

  The result is therefore a consequence of the fact that $RES(\otimes_1(U) + \otimes_2(\Sigma_2^\bullet), \sigma_{A,A^\perp;\Delta,\Phi,B})$ is a variant of U (with $U := RES(\Sigma_1, \sigma_{\Delta;\Phi,A})$).

- Assume that Σ is obtained from $\Sigma_1, \Sigma_2, \Sigma_3$, respectively proving $\vdash [\Delta_1], \Phi, B, C, \vdash [\Delta_2], \Psi, B^\perp$ and $\vdash [\Delta_3], \Xi, C^\perp$, by a $\wp$-rule and a $\otimes$-rule followed by a cut, then the construction of $\Sigma^\bullet$ involves two steps :

  - the formation of the sum $U := \otimes_1(\Sigma_1^\bullet) + \otimes_2(\otimes_1(\Sigma_2^\bullet) + \otimes_2(\Sigma_3^\bullet))$; this sum is a variant of $\Pi^\bullet = \otimes_1(\otimes_1(\Sigma_1^\bullet) + \otimes_2(\Sigma_2^\bullet)) + \otimes_2(\Sigma_3^\bullet)$
  - the merge in U of p_B and p_C into p_A and of $p_{B^\perp}$ and $p_{C^\perp}$ into $p_{A^\perp}$.

  If we define $\Lambda := \Delta_1, \Delta_2, \Delta_3$, then it is immediate that

  $$RES(\Sigma^\bullet, \sigma_{\Lambda,A,A^\perp;\Gamma}) = RES(U, \sigma_{\Lambda,B,B^\perp,C,C^\perp;\Gamma}).$$

  But $\Lambda, B, B^\perp, C, C^\perp = \Delta'$ so $RES(U, \sigma_{\Lambda,B,B^\perp,C,C^\perp;\Gamma})$ is a variant of $RES(\Pi^\bullet, \sigma_{\Delta';\Gamma})$.

- Assume that Σ is obtained from $\Sigma_1, \Sigma_2, \Sigma_3$, respectively proving $\vdash [\Delta_1], \Phi, B, C, \vdash [\Delta_2], \Psi, B^\perp$ and $\vdash [\Delta_3], \Psi, C^\perp$, by a $\oplus_1$-rule and a &-rule followed by a cut, then the construction of $\Sigma^\bullet$ involves two steps :

- the formation of the sum $U := \otimes_1(\Sigma_1^\bullet) + \otimes_2(\&_1(\Sigma_2^\bullet) + \&_2(\Sigma_3^\bullet))$; this sum is a variant of $\Pi^\bullet = \&_1(\otimes_1(\Sigma_1^\bullet) + \otimes_2(\Sigma_2^\bullet)) + \&_2(\otimes_2(\Sigma_3^\bullet))$
- the merge in U of p_B and p_C into p_A and of $p_{B^\perp}$ and $p_{C^\perp}$ into $p_{A^\perp}$.

If we define $\Lambda := \Delta_1, \Delta_2, \Delta_3$, then it is immediate that

$$RES(\Sigma^\bullet, \sigma_{\Lambda,A,A^\perp;\Gamma}) = RES(U, \sigma_{\Lambda,B,B^\perp,C,C^\perp;\Gamma})$$

. But $\Lambda, B, B^\perp, C, C^\perp = \Delta'$ so $RES(U, \sigma_{\Lambda,B,B^\perp,C,C^\perp;\Gamma})$ is a variant of $RES(\Pi^\bullet, \sigma_{\Delta';\Gamma})$.

- All the petty cuts between an axiom for $\flat$ and a logical rule (or an axiom for $\flat$) are easily treated : this is because the axiom for flat is interpreted by 0.

- The endless list of commutations of the *CUT*-rule is easily handled : such steps introduce variants. □

PROOF: the proof of theorem 4 is immediate from the lemma. □

Whether or not we have totally achieved our task, surely execution corresponds exactly to normalisation, but in a slightly exotic sequent calculus. This calculus contains more cut-free proofs than usual, and therefore its use might be problematic. But the question is solved by the following proposition :

Proposition 8

In the inflated calculus, the booleans remain standard : there is a sequent Γ such that the set $\Gamma^\bullet$ of all $\Pi^\bullet$ where Π varies through cut-free proofs of Γ has exactly two elements.

PROOF: we have to give a precise meaning to the proposition.

- We can decide to represent booleans by a sequent $\vdash \alpha^\perp, \alpha \oplus \alpha$, where α is a given atomic formula.

- This formula has only two cut-free proofs in the usual sequent calculus (Identity axiom and a $\oplus_1$-rule, identity axiom and a $\oplus_1$-rule) ; these two proofs can be taken as the two booleans. Geometry of interaction obviously interprets them differently.

- In the inflated calculus, the summation rule offers more possibilities of cut-free proofs ; however it is easily proved that the flat summands must have interpretation 0. □

Therefore the execution formula is consistent with usual syntactic manipulations : this means that for any boolean question that we can ask of the output of some cut-elimination, both methods will yield the same answer. See [G88] for a discussion.

3.5 The neutral elements

The interpretation is easily extended to the additive neutral elements 0 and $\top$. The only rule -the axiom $\vdash \Gamma, \top$- is interpreted by the null wiring 0, and all results still hold.

But the case of the multiplicative units is much more delicate ; the only "natural" choice for interpreting the rule of introduction of $\perp$ is the trivial one : if a proof Π of $\vdash \Gamma, \perp$ comes from a proof $\vdash \Pi_1$ of Γ by the $\perp$-rule, let $\Pi^\bullet := \Pi_1^\bullet$. This choice has the following consequence : given two proofs Π_1 and Π_2 of $\vdash \Gamma$ (this sequent is written below as $\vdash \Gamma_i$ to distinguish the two proofs), the proofs

$$\dfrac{\dfrac{\vdash \Gamma_1}{\vdash \Gamma, \perp}\perp \quad \dfrac{\vdash \Gamma_2}{\vdash \Gamma, \perp}\perp}{\vdash \Gamma, \perp \& \perp}\& \qquad \text{and} \qquad \dfrac{\dfrac{\vdash \Gamma_2}{\vdash \Gamma, \perp}\perp \quad \dfrac{\vdash \Gamma_1}{\vdash \Gamma, \perp}\perp}{\vdash \Gamma, \perp \& \perp}\&$$

are interpreted by variants, i.e. are not distinguished. In fact proposition 8 would become false. In terms of inflated calculus, a new problematic case arises, namely a cut

$$\dfrac{\dfrac{}{\vdash 1, \flat}\flat \quad \dfrac{\vdash \Gamma}{\vdash \perp, \Gamma}\perp}{\vdash \Gamma, \flat}CUT$$

for which the only natural cut-elimination would be to add a rule of weakening on $\flat$ (from $\vdash \Gamma$, deduce $\vdash \Gamma, \flat$) which contradicts our idea that a flat proof is a proof in which something is actually missing : this rule would acknowledge ordinary proofs as flat ones, thus destroying all our construction. In other terms, the "natural" geometry of interaction for $\perp$ leads to accept the rule : from $\vdash \Gamma$ and $\vdash \Gamma$ deduce $\vdash \Gamma$, which is a strong form of summation. With such a rule, booleans would no longer be standard (any formal sum of booleans would be accepted).

Instead, we modifiy the $\perp$-rule into : from $\vdash \Gamma, A$ deduce $\vdash \Gamma, A, \perp$, in which one of the formulas in the premise is distinguished. Strictly speaking, there is no need to distinguish A, if we stick to the viewpoint of sequents as sequences (instead of multisets). With this modified rule, the problem of a cut

$$\dfrac{\dfrac{}{\vdash 1, \flat}\flat \quad \dfrac{\vdash \Gamma, A}{\vdash \perp, \Gamma, A}\perp}{\vdash \Gamma, A, \flat}CUT$$

disappears ; for instance the cut can be replaced with a cut between $\vdash \Gamma, A$ and the flat axiom $\vdash A^\perp, A, \flat$ ...

Definition 17
The rules for the multiplicative units are interpreted as follows :

- *If* Π *is the axiom* $\vdash 1$, *then* $\Pi^\bullet$ *is the message* $p_1(x, y)$ *;*
- *If the proof* Π *of* $\vdash [\Delta], \Gamma, A, \bot$ *is obtained from a proof* Π_1 *of* $\vdash [\Delta], \Gamma, A$ *by a* $\bot$*-rule, then* $\Pi^\bullet$ *is defined in terms of* $\Pi_1^\bullet$ *:*
  - *in* $\Pi_1^\bullet$ *replace any wire* $p_B(t, u) \mapsto p_A(t', u)$ *by a wire* $p_B(t, u) \mapsto p_\bot(t', u)$ *;*
  - *add to the result of this replacement the wire* $p_\bot(x, y) \mapsto p_A(x, y)$.

Remark 15

- One should prove the analogue of theorem 4 which can be done without difficulty. In case of extreme laziness, observe that a (slightly more complicated) geometry of interaction of multiplicative neutrals can easily be given by means of the second-order translation of $\bot$ as $\exists\alpha(\alpha \otimes \alpha^\bot)$, for which the previous section yields a geometry of interaction. Our modified $\bot$-rule can be translated in second-order linear logic by

$$\dfrac{\dfrac{\vdash \Gamma, A \quad \dfrac{}{\vdash A^\bot, A}\,ID}{\vdash \Gamma, A \otimes A^\bot, A}\otimes}{\vdash \Gamma, \exists\alpha(\alpha \otimes \alpha^\bot), A}\exists$$

- In terms of proof-nets, the "natural choice" for $\bot$ corresponds to proof-nets where the weakened formula ($\bot$ or $?A$) is physically disconnected. There is no reasonable correctness criterion for such nets :
  - In the case of a multiplicative formula A using only multiplicative units as atoms, such a choice would lead to identify all proofs of A ;
  - Hence the correctness problem for such proof-nets contains the decision problem for the constant-only multiplicative fragment, which is known to be **NP**-complete, see [LW92].
  - But the general shape of the known criterions is **coNP**, hence the existence of a criterion of the same shape is very unlikely...

  Our solution corresponds to a version of proof-nets in which weakened formulas are attached to a formula of the net.

- The geometry of interaction of $\bot$ (which is the geometry of the weakening rule) shows for the first time a non-contrived non-commutativity :

the formula to the left of $\perp$ in the conclusion of the rule actually matters. One could imagine another version in which $\perp$ is physically linked to its rightmost neighbour. This non-commutative reading is controversial anyway : we could also decide to say that the rule involves two formulas, $\perp$ and A.

- The multiplicative neutrals display another originality w.r.t. geometry of interaction : in the absence of $\perp$ and 1, Π^* is always a sum of two adjoint wirings $W + W^*$. The interpretation of the axiom for 1 is a single wire, whereas the $\perp$-rule introduces non self-adjoint wirings...

4 The case of exponentials

4.1 Strict exponentials

We describe here a modification of the exponential rules ; the modification is close to some extant experimental systems, see e.g. [A93, MM93]. The system has a weak form of promotion, which is corrected by an additional rule for ?.

$$\frac{\vdash \Gamma, A}{\vdash ?\Gamma, !A}\,! \qquad \frac{\vdash A, \Gamma}{\vdash ?A, \Gamma}\,d? \qquad \frac{\vdash ?A, ?A, \Gamma}{\vdash ?A, \Gamma}\,c? \qquad \frac{\vdash ??A, \Gamma}{\vdash ?A, \Gamma}\,??$$

These rules are respectively called *(weak) promotion, dereliction, contraction, digging.* They only ensure a variant of the subformula property ($??A$ subformula of $?A$) which may look strange at first sight, but is not more artificial than the familiar definition which allows $A[t/x]$ to be a subformula of $\forall x A$. This modification of the notion of subformula yields infinitely many subformulas for a propositional linear formula, in accordance with the undecidability of propositional linear logic, [LMSS90]. The rule of promotion is understood in the obvious way : if Γ is $A_1, \ldots, A_n$, then $?\Gamma$ is $?A_1, \ldots, ?A_n$.

Now observe that the rule of weakening :

$$\frac{\vdash \Gamma}{\vdash ?A, \Gamma}\,w?$$

is absent. Of course the rule is not derivable, and therefore, the *strict* exponential defined by the rules above do not recover the full expressive power of linear logic. But this rather inessential, since there is an operation allowing weakening, namely the formation of $1\&A$. Usual exponentials can be recovered by means of $!(1\&A)$, $?(\perp \oplus A)$. This kind of splitting in two steps of exponentials has been studied by Jacobs in [J93]. We are therefore interpreting a slight modification of linear logic, corresponding to extant experimental systems. If we adopt here the modifications of [A93, MM93] and [J93], this is because geometry of interaction is particularly simple in this setting. But also, since geometry of interaction is in many senses the most natural semantics, it might be seen as backing the syntactical variants proposed in these works.

Exercise 2 *Prove the cut-elimination theorem for this variant of the exponential rules. Of course one has to define an* ad hoc *cut-elimination procedure.*

4.2 The pattern

The general pattern is as follows :

we want to allow reuse ; the absolutely weakest form of reuse is expressed by the principle $\vdash ?A^{\perp}, A \otimes A$. The interpretation $\Pi^{\bullet}$ of the (natural) proof of this principle must have the following property : if U interprets a proof of A and $!U$ is the result of the promotion of U, then a cut between $!U$ and $\Pi^{\bullet}$ yields after normalisation a variant of V, where V is obtained by means of definition 11.4 from U and U in the respective roles of $\Pi_1^{\bullet}$ and $\Pi_2^{\bullet}$. In the formation of V the essential step is the formation of $\otimes_1(U) + \otimes_2(U)$, which is the sum of two variants. The execution formula is able to extract two variants of U from the sole $!U$. Building variants basically involves non-laminated wirings, and therefore the execution formula is unable to achieve the task. In particular the solution $!U := U$ is inadequate. Of course, the problem is easily solved if we could drop lamination..., but this would destroy some essential features of our construction. But we can also mimick non-laminated wires by laminated ones : we are eventually lead to shift the dialectal components ux of $\otimes_1(U)$ from the private part to the public part.
The context-free promotion $!U$ is interpreted as follows : in U replace all atoms $p_A(t,u)$ with atoms $p_{!A}(txu, z)$; This construction is the result of two steps :

- We first form $\otimes_1(U)$, by replacing in U all atoms $p_B(t,u)$ with atoms $p_B(t, xu)$.
- Then we replace in $\otimes_1(U)$ all atoms $p_B(t, xu)$ with $p_{!A}(txu, z)$. The final z is present only because our predicates need a second argument : in reality there is no dialect in this case.

4.3 Interpretation of the exponential rules

In what follows $t_1 \ldots \widehat{t_i} \ldots t_n$ is short for $t_1 \ldots t_{i-1} t_{i+1} \ldots t_n$.

Definition 18 (def. 11 cont^d)

Case 18.1 *Assume that the proof* Π *of* $\vdash [\Delta], ?\Gamma, !A$ *is obtained from a proof* Π_1 *of* $\vdash [\Delta], \Gamma, A$ *by a promotion rule ; we define* $\Pi^{\bullet}$ *in terms of* $\Pi_1^{\bullet}$. *Let us assume that* Γ *is* $C_1, \ldots, C_n$ *(in this order). Replace in* $\Pi_1^{\bullet}$ *:*

- *all atoms* $p_A(t,u)$ *with atoms* $p_{!A}(txuy_1 \dots y_n, z)$;
- *all atoms* $p_B(t,u)$ *with atoms* $p_B(txuy_1 \dots y_n, z)$ *when* B *is in* Δ ;
- *all atoms* $p_{C_i}(t,u)$ *with atoms* $p_{?C_i}(t(uy_1 \dots \widehat{y_i} \dots y_n x)y_i, z)$.

($x, y_1, \dots, y_n, z$ *are fresh variables). as usual* x, x', x'', z *are fresh variables. The result is by definition* $\Pi^\bullet$.

Case 18.2 *Assume that the proof* Π *of* $\vdash [\Delta], ?A, \Gamma$ *is obtained from a proof* Π_1 *of* $\vdash [\Delta], A, \Gamma$ *by a dereliction rule ; we define* $\Pi^\bullet$ *by replacing in* $\Pi_1^\bullet$.

- *All atoms* $p_A(t,u)$ *with atoms* $p_{?A}(tgz, uz)$;
- *All atoms* $p_B(t,u)$ *(when* $B \neq A$*) with atoms* $p_B(t, uz)$.

Case 18.3 *Assume that the proof* Π *of* $\vdash [\Delta], ?A, \Gamma$ *is obtained from a proof* Π_1 *of* $\vdash [\Delta], ?A_1, ?A_2, \Gamma$ *by a contraction rule (we note* $?A_1, ?A_2$ *two distinct occurences of* A*) ; we define* $\Pi^\bullet$ *in terms of* $\Pi_1^\bullet$.

- *First put the atoms* $p_{?A_i}(t,u)$ *"in the form* $p(tt't'', u)$*" ; this means that we form*

$$W := \sum_{i=1}^{2} p_{?A_i}(xx'x'', y) + \sum_{C \in \Gamma} p_C(x,y)$$

  and replace $\Pi_1^\bullet$ *with* $U := W\Pi_1^\bullet W$.

- *Then we replace in* U :
  - *All atoms* $p_{?A_1}(tt't'', u)$ *with atoms* $p_{?A}(t(gt')t'', u)$;
  - *All atoms* $p_{?A_2}(tt't'', u)$ *with atoms* $p_{?A}(t(dt')t'', u)$.

the result is by definition $\Pi^\bullet$.

Case 18.4 *Assume that the proof* Π *of* $\vdash [\Delta], ?A, \Gamma$ *is obtained from a proof* Π_1 *of* $\vdash [\Delta], ??A, \Gamma$ *by a digging rule ; we define* $\Pi^\bullet$ *in terms of* $\Pi_1^\bullet$.

- *First put the atoms* $p_{??A}(t,u)$ *"in the form* $p((tt't'')uu', v)$*" ; this means that we form*

$$W := p_{??A}((xx'x'')yy', z) + \sum_{C \in \Gamma} p_C(x,y)$$

  and replace $\Pi_1^\bullet$ *with* $U := W\Pi_1^\bullet W$.

- *Then replace in* U *all atoms* $p_{??A}((tt't'')uu', v)$ *with* $p_{?A}(t(t'uu')t'', v)$.

Remark 16

The rule of promotion is the first violation of the principle that all our constructions are compatible with $\sim$. This is perhaps the ultimate meaning of the "!-box".

4.4 Basic examples

We shall treat certain basic cuts

$$\dfrac{\dfrac{\cdots}{\vdash \Gamma, !A}\,R \qquad \dfrac{\cdots}{\vdash ?A^{\perp}, \Delta}\,S}{\vdash \Gamma, \Delta}\,CUT$$

on an exponential $!A$, in certain cases :

- Γ is empty and R is the promotion rule applied to a cut-free proof Π of $\vdash A$.
- S is a $!$, $d?$, $c?$ or $??$-rule, applied to a cut-free proof Σ of $\vdash \Delta'$ and with main formula $?A^{\perp}$ in case $S \neq !$.

Example 2
Cut-elimination replaces the cut

$$\dfrac{\dfrac{\vdash A}{\vdash !A}\,! \qquad \dfrac{\vdash A^{\perp}, B}{\vdash ?A^{\perp}, !B}\,!}{\vdash !B}\,CUT$$

with

$$\dfrac{\dfrac{\vdash A \quad \vdash A^{\perp}, B}{\vdash B}\,CUT}{\vdash !B}\,!$$

The interpretation of the original proof is $(U + V, \sigma_{!A,?A^{\perp};!B})$, where :

- U is obtained by replacing in $\Pi^{\bullet}$ all atoms $p_A(t, u)$ with $p_{!A}(txu, zz')$;
- V is obtained by replacing in $\Sigma^{\bullet}$:
  - all atoms $p_{A^{\perp}}(t, u)$ with atoms $p_{?A^{\perp}}(t(ux)y, zz')$
  - all atoms $p_B(t, u)$ with atoms $p_{!B}(txuy, zz')$.

The interpretation of the modified proof is $(W + Y, \sigma_{A,A^{\perp};!B})$, where :

- W is obtained by replacing in $\Pi^{\bullet}$ all atoms $p_A(t, u)$ with $p_A(txyu, z)$;
- Y is obtained by replacing in $\Sigma^{\bullet}$ all atoms $p_{A^{\perp}}(t, u)$ with $p_{A^{\perp}}(txuy', z)$ and all atoms $p_B(t, u)$ with $p_{!B}(txuy', z)$.

Observe that :

- $RES(U + V, \sigma_{!A,?A^\perp;!B}) = RES(U_1 + V, \sigma_{!A,?A^\perp;!B})$, where U_1 is obtained from U by replacing all atoms $p_{!A}(txu, zz')$ with atoms $p_{!A}(t(x'x'')u, zz')$;
- $RES(U_1 + V, \sigma_{!A,?A^\perp;!B}) = RES(Z, \sigma_{A,A^\perp;!B})$, where Z is obtained from $U_1 + V$ by replacing :
  - all atoms $p_{!A}(t(x'x'')u, zz')$ with atoms $p_A(tx''x'u, zz')$
  - all atoms $p_{?A^\perp}(t(ux)y, zz')$ with atoms $p_{?A^\perp}(tuxy, zz')$.
- $Z = \otimes_1(W + Y)$.

therefore $RES(U + V, \sigma_{!A,?A^\perp;!B}) = \otimes_1(RES(W + Y, \sigma_{A,A^\perp;!B}))$. This shows the soundness of this particular cut-elimination step.

Exercise 3
Extend example 2 to the more general case of n cuts between :

- *Cut-free proofs of the sequents* $\vdash !A_1, \ldots, \vdash !A_n$
- *A cut free proof of the sequent* $\vdash !A_1, \ldots, !A_n$ *ending with a promotion rule.*

Example 3
Cut-elimination replaces the cut

$$\dfrac{\dfrac{\vdash A}{\vdash !A}\,! \qquad \dfrac{\vdash A^\perp, \Phi}{\vdash ?A^\perp, \Phi}\,d?}{\vdash \Phi}\,CUT$$

with

$$\dfrac{\vdash A \quad \vdash A^\perp, \Phi}{\vdash \Phi}\,CUT$$

The interpretation of the original proof is $(U + V, \sigma_{!A,?A^\perp;\Phi})$, where U is as in example 2 ; and V is obtained by replacing in $\Sigma^\bullet$

- All atoms $p_A^\perp(t, u)$ with atoms $p_{?A^\perp}(tgz, (uz)z')$;
- All atoms $p_B(t, u)$ (when $B \neq A^\perp$) with atoms $p_B(t, (uz)z')$.

whereas the modified proof is interpreted by $(\otimes_1(\Pi^\bullet) + \otimes_2(\Sigma^\bullet), \sigma_{A,A^\perp;\Phi})$. Now observe that :

- $RES(U + V, \sigma_{!A,?A^\perp;\Phi}) = RES(U + V, \sigma_{!A,?A^\perp;\Phi})$
- $RES(U + V, \sigma_{!A,?A^\perp;\Phi}) = RES(U_1 + V, \sigma_{!A,?A^\perp;\Phi})$, where U_1 is obtained from U by replacing all atoms $p_{!A}(txu, zz')$ with $p_{!A}(tgu, (z_1u)z')$;

- $RES(U_1+V, \sigma_{!A,?A^\perp;\Phi}) = RES(W, \sigma_{A,A^\perp;\Phi})$, where W is obtained from U_1+V by replacing all atoms $p_{!A}(tgu, (z_1u)z')$ with atoms $p_A(t, (z_1u)z')$ and all atoms $p_{?A^\perp}(tgz, (uz)z')$ with atoms $p_{A^\perp}(t, (uz)z')$;
- $W = \otimes_2(\otimes_1(\Pi^\bullet) + \otimes_2(\Sigma^\bullet))$.

therefore $RES(U + V, \sigma_{!A,?A^\perp;\Phi}) = \otimes_2(RES(\otimes_1(\Pi^\bullet) + \otimes_2(\Sigma^\bullet), \sigma_{A,A^\perp;\Phi}))$. This shows the soundness of this particular cut-elimination step.

Example 4
Cut-elimination replaces the cut

$$\dfrac{\dfrac{\vdash A}{\vdash !A}! \qquad \dfrac{\vdash ?A_1^\perp, ?A_2^\perp, \Phi}{\vdash ?A^\perp, \Phi}c?}{\vdash \Phi}CUT$$

with

$$\dfrac{\dfrac{\vdash A}{\vdash !A_2}! \qquad \dfrac{\dfrac{\vdash A}{\vdash !A_1}! \qquad \vdash ?A_1^\perp, ?A_2^\perp, \Phi}{\vdash ?A_2^\perp, \Phi}CUT}{\vdash \Phi}CUT$$

The interpretation of the original proof is $(U + V, \sigma_{!A,?A^\perp;\Phi})$, where U is as in example 2 ; and V is obtained from $\Sigma^\bullet$ as follows :

- First define $V' := \otimes_2(\Sigma^\bullet)$;
- First put the atoms $p_{?A_i}(t, u)$ of V' "in the form $p(tt't'', u)$" ; this means that we form as in definition 18.3

$$W := p_{?A_1}(xx'x'', y) + p_{?A_2}(xx'x'', y) + \sum_{C\in\Phi} p_C(x, y)$$

and replace V' with $V'' := WV'W$;

- Then we replace in V'' :
  - All atoms $p_{?A_1}(tt't'', u)$ with atoms $p_{?A}(t(gt')t'', u)$;
  - All atoms $p_{?A_2}(tt't'', u)$ with atoms $p_{?A}(t(dt')t'', u)$.

  the result is called V.

The interpretation of the modified proof is $(X_1 + X_2 + Y, \sigma_{!A_1,?A_1^\perp,!A_2,?A_2^\perp;\Phi})$ where :

- $X_1 = U_1$
- $X_2 = \otimes_2(U_2)$
- $Y = \otimes_2(\otimes_2(\Sigma^\bullet))$

where U_i is obtained from U by replacing the predicate $p_{!A}$ by $p_{!A_i}$. Now observe that U (which is of the form $\otimes_1(U')$ is also of the form $\otimes_2(\otimes_2(U''))$. $X_1 + X_2 + Y$ is therefore of the form $\otimes_2(\otimes_2(Z))$, which is a variant of $\otimes_2(Z)$ by proposition 5. Therefore by proposition 4 we can take as interpretation of our modified proof the pair $(X_1' + X_2' + Y', \sigma_{!A_1,?A_1^\perp,!A_2,?A_2^\perp;\Phi})$, where :

- $X_1 = \otimes_2(X_1')$
- $X_2' = U_2$
- $Y = \otimes_2(\Sigma^\bullet)$

But it is immediate to see that $RES(X_1' + X_2' + Y', \sigma_{!A_1,?A_1^\perp,!A_2,?A_2^\perp;\Phi}) = RES(X_1'' + X_2' + Y', \sigma_{!A_1,?A_1^\perp,!A_2,?A_2^\perp;\Phi})$ where X_1'' is obtained from X_1' by replacing all atoms $p_{A_1}(t,z)$ with atoms $p_{A_1}(t,zz')$; but observe that $X_1'' = \otimes_2(X_1')$, hence $X_1'' = X_1 = U_1$. Now observe that all the wires in X_1 and X_2' are of the form $p_{A_i}(tt't'',u)$, hence $RES(X_1 + X_2' + Y', \sigma_{!A_1,?A_1^\perp,!A_2,?A_2^\perp;\Phi}) = RES(X_1 + X_2' + Y'', \sigma_{!A_1,?A_1^\perp,!A_2,?A_2^\perp;\Phi})$, where Y'' is obtained by putting the atoms $p_{?A_i}(t,u)$ of Y' "in the form $p(tt't'',u)$" ; but then $Y'' = V''$. We are left with $RES(U_1 + U_2 + V'', \sigma_{!A_1,?A_1^\perp,!A_2,?A_2^\perp;\Phi})$. Obviously this expression is unchanged if we merge the $p_{?A_i}$ into $p_{?A}$ and the $p_{!A_i}$ into $p_{!A}$. This merge changes V'' into V, $U_1 + U_2+$ into ZUZ and $\sigma_{!A_1,?A_1^\perp,!A_2,?A_2^\perp;\Phi}$ into $Z\sigma_{!A,?A^\perp;\Phi}Z$, where Z is the projection

$$Z := p_{?A}(x(gx')x", y) + p_{?A}(x(dx')x", y) + \sum_{C \in \Gamma} p_C(x,y)$$

Observe that $V = ZVZ$, hence $RES(ZUZ + V, Z\sigma_{!A,?A^\perp;\Phi}Z) = Z \cdot RES(U + V, \sigma_{!A,?A^\perp;\Phi}) \cdot Z = RES(U + V, \sigma_{!A,?A^\perp;\Phi})$ and thus we have eventually proved soundness.

Example 5
Cut-elimination replaces the cut

$$\dfrac{\dfrac{\vdash A}{\vdash !A}\,! \qquad \dfrac{\vdash ??A^\perp, \Phi}{\vdash ?A^\perp, \Phi}\,??}{\vdash \Phi}\,CUT$$

with

$$\dfrac{\dfrac{\dfrac{\vdash A}{\vdash !A}!}{\vdash !!A}! \quad \vdash ??A^{\perp}, \Phi}{\vdash \Phi} CUT$$

The interpretation of the original proof is $(U + V, \sigma_{!A,?A^{\perp};\Phi})$, where U is as in example 2 ; and V is obtained from $\Sigma^{\bullet}$ as follows :

- First put the atoms $p_{??A}(t, u)$ "in the form $p((tt't'')uu', v)$" ; this means that we form

$$W := p_{??A}((xx'x'')yy', z) + \sum_{C \in \Phi} p_C(x, y)$$

  as in definition 18.4 and replace $\Pi_1^{\bullet}$ with $V' := W\Sigma^{\bullet}W$.

- Then replace in V' all atoms $p_{??A}((tt't'')uu', v)$ with $p_{?A}(t(t'uu')t'', vy)$;

the result is by definition V. The modified proof is interpreted as $(X + Y, \sigma_{!!A,??A^{\perp};\Phi})$, with :

- X is obtained from $\Pi^{\bullet}$ by replacing all atoms $p_A(t, u)$ with atoms $p_{!!A}((txu)yz, z')$;
- $Y = \otimes_2(\Sigma^{\bullet})$

First observe that $RES(X+Y, \sigma_{!!A,??A^{\perp};\Phi}) = RES(X' + \otimes_2(V'), \sigma_{!A,?A^{\perp};\Phi})$, where X' is obtained from $\Pi^{\bullet}$ by replacing all atoms $p_A(t, u)$ with atoms $p_{!A}(t(xyz)u, z')$. This is because $X' = ZXZ^*$ and $\otimes_2(V') = ZYZ^*$ with

$$\begin{aligned} Z &= p_{!!A}((xx'x'')yy', z) \\ &\mapsto p_{!A}(x(x'yy')x'', z) + p_{??A^{\perp}}((xx'x'')yy', z) \\ &\mapsto p_{??A^{\perp}}(x(x'yy')x'', z) + \sum_{C \in \Phi} p_C(x, y) \end{aligned}$$

Now observe that

$$RES(X' + \otimes_2(V'), \sigma_{!A,?A^{\perp};\Phi}) = RES(X' + WVW, \sigma_{!A,?A^{\perp};\Phi})$$

but

$$\begin{aligned} RES(X' + WVW, \sigma_{!A,?A^{\perp};\Phi}) &= RES(WX'W + V, \sigma_{!A,?A^{\perp};\Phi}) \\ &= RES(U + V, \sigma_{!A,?A^{\perp};\Phi}) \end{aligned}$$

This proves soundness in this last case.

4.5 Nilpotency for exponentials

It remains to extend the nilpotency theorem to the full case ; this offers no difficulty of principle (basically our previous treatment of exponentials in [G88] suitably modified in the spirit of section 3).

For instance the nilpotency theorem basically needs an extension of definition 13 :

Definition 19 (def. 13 cont^d)

Case 19.1 *If X is a weak type, then we define the weak type $!X$ as $Z^{\perp\perp}$, where Z consists of all $!U$ obtained from some $U \in X$ by means of definition 18.1 (in the simplified case $\Gamma = \Delta = \emptyset$).*

Case 19.2 *If X is a weak type, then we define the weak type $?X$ as $(!X^{\perp})^{\perp}$.*

This definition is the key to an unproblematic extension of the nilpotency theorem to the full case (left without hypocrisy to the reader).

Theorem 5 (Nilpotency)
$\sigma\Pi^{\bullet}$ is nilpotent.

4.6 Soundness for exponentials

Soundness is more delicate ; in fact we can only prove soundness under a strong restriction on Γ.

Theorem 6 (Limited soundness)

If a cut-free proof Π of $\vdash \Gamma$ is obtained from a proof Σ of $\vdash [\Delta]\Gamma$ by means of the transformations sketched in definition 16, and if Γ contains no exponential and no second order existential quantifier, then $\Pi^{\bullet}$ is a variant of $RES(\Sigma^{\bullet}, \sigma_{\Delta;\Gamma})$.

PROOF: the proof is an imitation of theorem 1 ii) of [G88]. We give some hints :

- The restriction on Γ makes it possible to consider a limited form of cut-elimination, where an exponential cut is eliminated only when the premise containing $!A$, let us say $\Phi, !A$, comes from a !-rule, with Φ empty. One can show that cut-elimination works with this restricted algorithm.

- The examples 2 (including its n-ary generalisation of exercise 3), 3, 4, 5 are the basic steps of the verification of the soundness of this limited cut-elimination algorithm.

- Among the specific technicalities of this extension, we need to develop the inflated calculus in presence of exponentials. For instance the promotion rule in presence of $\flat$ is understood as follows : $?\Gamma = ?(\Gamma - \flat), \flat$. One also needs to define the cut-elimination between a flat axiom and an exponential rule. This is straightforward :
  - A cut between the flat axiom $\vdash \Gamma, !A, \flat$ and a $d?$-rule is replaced with a cut between the flat axiom $\vdash \Gamma, A, \flat$ and the premise of the $d?$-rule ;
  - A cut between the flat axiom $\vdash \Gamma, !A, \flat$ and a $c?$-rule is replaced with two cuts between the flat axioms $\vdash \Gamma, !A, \flat$ and $\vdash !A, \flat$ and the premise of the $c?$-rule ;
  - A cut between the flat axiom $\vdash \Gamma, !A, \flat$ and a $??$-rule is replaced with a cut between the flat axiom $\vdash \Gamma, !!A, \flat$ and the premise of the $??$-rule ;
  - A cut between a promotion $!A$ and the flat axiom $\vdash \Gamma, ?A^{\perp}, \flat$ is replaced with the flat axiom $\vdash \Gamma, \flat$.
- One must also prove the soundness w.r.t. the full cut-elimination procedure. The only way to do so would be to work with ***proof-nets*** in some variant of [G94] (with boxes for !), then to prove a Church-Rosser property, and the fact that the interpretation of a cut-free proof of Γ depends (up to variance) only on the associated net. This seems to be unproblematic, but we quailed in the place of the burden... □

Remark 17

- In [G88] we were able to prove a little more, namely ! was allowed in Γ. The problem here is that reduction above a promotion rule would need a more general notion of variant. If we come back to definition 7, we see that the essential point is the possible choices for W and W'. We can perfectly define, ***for each type*** A a set of wirings (inducing an equivalence $\sim_A$, coarser than $\sim$). For instance for $!A$ we could consider all wirings $\sum p_{!A}(xyt, z) \mapsto p_{!A}(xyu, z)$ to enlarge the notion of variant in that case. No doubt that with this extension, we can allow ! in Γ.
- What about a complete soundness ? Surely, one needs a liberalised notion of variant, as above, typically to take care of rather arbitrary choices made in the case of the ?-rules, but is this enough ? Presumably not, and we honestly don't know whether the inflated calculus (perhaps improved with additional principles) can cope with the situation. In fact we have conflicting intuitions :
  - Dr Jekill thinks that the syntax can be adapted to cope with the geometrical interpretation ;

  - Mr Hyde thinks that there is something basically infinite in exponentials and that for this reason, there is an irreducible global configuration, the !-box.

- Anyway, we do not want to achieve soundness by replacing the rather natural notion of being variants (with the possibility of modifying the possible W, W') with a notion of observational equivalence : it should be rather immediate to check that two objects are variants.

References

[A93] Asperti, A.: Linear Logic, Comonads and Optimal Reductions, to appear in Fundamenta Informaticae, Special Issue devoted to Categories in Computer Science.

[B92] Blass, A.: A game semantics for linear logic, Annals Pure Appl. Logic 56, pp. 183–220, 1992.

[DR93] Danos, V. & Regnier, L.: GOI 2.0. manuscript, 1993.

[G86] Girard, J.-Y.: Linear Logic, Theoretical Computer Science 50.1, pp. 1–102, 1987.

[G86A] Girard, J.-Y.: Multiplicatives, Rendiconti del Seminario Matematico dell'Università e Policlinico di Torino, special issue on Logic and Computer Science, pp. 11–33, 1987.

[G87] Girard, J.-Y.: Towards a Geometry of Interaction, Categories in Computer Science and Logic, Contemporary Mathematics 92, AMS 1989, pp. 69–108.

[G88] Girard, J.-Y.: Geometry of Interaction I : interpretation of system F, Proceedings Logic Colloquium '88, eds. Ferro & al., pp. 221–260, North Holland 1989.

[G88A] Girard, J.-Y.: Geometry of Interaction II : deadlock-free algorithms, Proceedings of COLOG-88, eds Martin-Löf & Mints, pp. 76–93, LNCS 417, Springer-Verlag 1990.

[G94] Girard, J.-Y.: Proof-nets for additives, manuscript, 1994.

[GAL92] Gonthier, G. & Abadi, M. & Levy, J.-J.: The Geometry of Optimal Lambda Reduction, Proc. 19-th Annual ACM Symposium on Principles of Programming Languages, Albuquerque, New Mexico, ACM Press, New York, 1992.

[J93] Jacobs, B.: Semantics of weakening and contraction, Annals of Pure and Applied Logic 69, 1994, pp. 73–106.

[LMSS90] Lincoln, P. & Mitchell, J. & Scedrov, A. & Shankar N.: Decision Problems in Propositional Linear Logic, Annals of Pure and Applied Logic 56, 1992, pp. 239–311.

[LW92] Lincoln, P. & Winkler, T.: Constant-only Multiplicative Linear Logic is NP-complete, Theoretical Computer Science 135, 1994, pp. 155–159.

[MM93] Martini, S. & Masini A.: On the fine structure of the exponential rules, to appear in Advances in Linear Logic, Cambridge University Press 1995.

Behavioural Specifications[1]

Rolf Hennicker, Martin Wirsing

Institut für Informatik
Ludwig-Maximilians-Universität München
Leopoldstr. 11B
D-80802 München, Germany

Abstract. In these lectures an observational approach to the semantics of algebraic specification is presented. Observability is defined in a logical way: an algebra is a behavioural model of a specification SP if it satisfies the axioms w.r.t. an observational interpretation of the equality relation.
The advantages of this notion are proof-theoretic ones: the proof system for observational first-order formulae needs just one additional (infinitary) proof schema, the so-called context-induction. The completeness of this proof system for behavioural properties of a specification is shown. Examples for context-induction proofs are given and it is shown how by the theorem of Bidoit & Hennicker context-inductive properties can be proven without explicit use of context-induction.
An ASL-like kernel language for modular behavioural specification is introduced and the correctness of refinements of specifications is studied. Sound and complete proof systems for properties of structured specifications and the validity of behavioural refinements are given. Moreover, behavioural refinement is transitive; the refinement of a subspecification SP0 induces a correct refinement of the full specification SP if SP is behaviourally complete w.r.t. SP0.

1 Introduction

Observability plays an important role in programming. For example, the notion of correct change of data structures can be based on this concept; other applications are the notion of equivalence between concurrent processes and the abstraction from single step transitions to input-output operational semantics.

One of the main reasons for the interest in observability in the area of software specification is a problem with the axiomatization of many standard data structures. In contrast with the idea that requirements should describe those properties which have to be satisfied by implementations many classical refinements of, for instance, set-like data structures do not preserve all requirements axioms; a careful analysis shows that only the

[1] This work has been partially sponsored by the DFG project SPECTRUM.

observable consequences of the axioms are preserved. Therefore by using an observational approach to data refinement one is able to overcome this drawback. Moreover, using an observational approach several abstractions from a given problem can be studied depending on the notion of observation.

Since the beginning of the eighties observational approaches have found continuous interest in the area of algebraic specifications by e.g. Goguen, Meseguer, Sannella, Wirsing, Reichel, Schoett, Hennicker, Orejas, and many others (see e.g. [Sannella, Wirsing 83], [Reichel 85], [Sannella, Tarlecki 87], [Schoett 87], [Orejas, Nivela 90], [Bernot, Bidoit 91], [Hennicker 91]). One can distinguish two main approaches to the semantics of observability. In the so-called *observational semantics* one considers a notion of observational equivalence between *algebras*, in the *behavioural* semantics one defines the semantics of a specification based on the notion of behavioural equality between *terms*. Comparing both approaches one can see that any behavioural model is observationally equivalent to a fully abstract standard model of a given specification SP and therefore behavioural semantics is stronger than observational semantics where any observational model is observationally equivalent to some standard model of SP (which in general is not fully abstract).

In this paper we present the behavioural approach because of its proof-theoretic elegance: The system for proving behavioural properties of a specification needs just an additional (infinitary) proof schema, the so-called context induction. In fact, the behaviourally valid properties are exactly the (standard first-order) properties that hold in all fully abstract models of the specification. Two techniques (originally introduced by Bidoit and Hennicker) are given for replacing the infinitary proof rule by finitary methods: finitary context induction which is based on the syntactic structure of the observable contexts and an even stronger schema which allows one to axiomatize the fully abstract equality such that proofs of behavioural properties can be performed without context induction.

In the second part of the paper an ASL-like kernel language [Wirsing 86] for structured specifications is studied which includes specification operators for forming the class of behavioural models and - as an auxiliary operator - the class of observable quotients of a specification. With the help of this operator observational abstraction can be defined as a derived operator. For proving first-order properties of structured specifications we associate to each specification expression particular (so-called non-logical) axioms and rules that extend the classical natural deduction style proof systems for first-order logic.

It is shown that for each (closed) specification expression the associated proof system is sound and complete.

Moreover, the refinement relation between specifications is studied. A sound and complete proof system is given which extends the proof system of [Wirsing 91] by two additional rules for behaviour and observational quotient. The proof rule for the refinement of a class of behavioural models uses the operator for observational quotient and therefore gives another reason for introducing this operator. Finally, a typical way of constructing behavioural implementations is considered.

The paper is organized as follows:
In section 2 basic definitions for algebraic specifications are given. In section 3 observational and behavioural semantics are presented. As main new result it is shown that the class of behavioural models of a specification SP is characterized as the class of all algebras which are observationally equivalent to a fully abstract model of SP.
In section 4 behavioural proofs are studied. It is shown that the behavioural theory of SP can be axiomatized using the axioms of SP together with an axiom schema for infinitary context induction. Soundness and completeness results for behavioural logic are given.
In section 5 a finitary proof-technique for proving behavioural properties in practice is derived, the so-called principle of context induction [Hennicker 91]. Examples for context induction proofs are given. The proofs suggest that context induction is rather difficult to perform by hand and that the use of a system such as ISAR is recommended as support [Bauer, Hennicker 93]. Another remedy is provided by the theorem of Bidoit and Hennicker which asserts that under reasonable assumptions context inductive properties can be proven without explicit use of context induction [Bidoit, Hennicker 92].
In section 6 structured behavioural specifications are considered. In particular, a sound and complete proof system for structured specifications is given extending the system of [Wirsing 91].
In section 7 behavioural implementations are studied, a sound and complete proof system for behavioural implementations is provided and vertical composition theorems are presented.

2 Algebraic Preliminaries

In this section the basic notions of algebraic specifications and the underlying assumptions and notations which will be used here are biefly summarized (for more details see e.g. [Ehrig, Mahr 85]).

A (many sorted) *signature* Σ is a pair (S, F) where S is a set of *sorts* and F is a set of *function symbols*. To every function symbol $f \in F$ a functionality $s_1, \dots, s_n \to s$ with $s_1, \dots, s_n \in S$ is associated. If $n = 0$ then f is called *constant* of sort s. A *signature morphism* σ: $S \to S'$ between two signatures $\Sigma = (S, F)$ and $\Sigma' = (S', F')$ is a pair $(\sigma_{sorts}, \sigma_{functs})$ of mappings σ_{sorts}: $S \to S'$, σ_{functs}: $F \to F'$ such that for all $f \in F$ with functionality $s_1 \times \dots \times s_n \to s$, $\sigma_{functs}(f)$ has functionality $\sigma_{sorts}(s_1) \times \dots \times \sigma_{sorts}(s_n) \to s_{sorts}(s)$. If $S \subseteq S'$ and $F \subseteq F'$ then Σ is called *subsignature* of Σ' (denoted by $\Sigma \subseteq \Sigma'$).

A *total* Σ*-algebra* $A = ((A_s)_{s\in S}, (f^A)_{f\in F})$ over a signature $\Sigma = (S, F)$ consists of a family of non empty carrier sets $(A_s)_{s\in S}$ and a family of (total) functions $(f^A)_{f\in F}$ such that f^A: $A_{s_1} \times \dots \times A_{s_n} \to A_s$ if f has functionality $s_1, \dots, s_n \to s$ (if $n = 0$ then f^A denotes a constant object of A_s). The class of all Σ-algebras is denoted by *Alg(Σ)*. Sometimes we will also consider signatures which have in addition to S and F a set P of predicate symbols where any predicate $p \in P$ has an arity $s_1 \times \dots \times s_n$ with $s_1, \dots, s_n \in S$. Then a Σ-*structure* is a Σ-algebra A where any predicate $p \in P$ with arity $s_1 \times \dots \times s_n$ is interpreted by a relation $p^A \subseteq A_{s_1} \times \dots \times A_{s_n}$. The class of all Σ-structures is denoted by *Struct(Σ)*.

The *restriction* of a Σ'-algebra A w.r.t. a signature morphism σ: $\Sigma \to \Sigma'$ is denoted by $A|_\sigma$ i.e. $(A|_\sigma)_s =_{def} A_{\sigma(s)}$ for all $s \in S$ and $f^{A|_\sigma} =_{def} \sigma(f)^A$. In particular, the restriction of A to a subsignature $\Sigma \subseteq \Sigma'$ is denoted by $A|_\Sigma$. Analogously, if C is a class of Σ'-algebras then the *restriction of C w.r.t. a signature morphism* σ: $\Sigma \to \Sigma'$ is the class $C|_\sigma = \{A|_\sigma \mid A \in C\}$ and the *restriction of C to a subsignature* $\Sigma \subseteq \Sigma'$ is the class $C|_\Sigma = \{A|_\Sigma \mid A \in C\}$.

The *term algebra* $T(\Sigma, X)$ over an S-sorted family $X = (X_s)_{s\in S}$ of sets X_s of variables of sort s has as carriers the sets $T(\Sigma, X)_s$ of *terms* of sort s. For any term $t \in T(\Sigma, X)$, Var(t) denotes the set of variables occurring in t.

If A is a total Σ-algebra then a *valuation* α: $X \to A$ is a family of mappings $(\alpha_s: X_s \to A_s)_{s\in S}$. The *interpretation* associated with α is the family $I_\alpha = (I_{\alpha s})_{s\in S}$ of functions $I_{\alpha s}$: $T(\Sigma, X)_s \to A_s$, inductively defined by

(1) $I_{\alpha s}(x) =_{def} \alpha_s(x)$ if $x \in X_s$,

(2) $I_{\alpha s}(f(t_1, \dots, t_n)) =_{def} f^A(I_{\alpha s_1}(t_1), \dots, I_{\alpha s_n}(t_n))$ if f has functionality $s_1, \dots, s_n \to s$.

3 Observational and Behavioural Semantics

3.1 Syntax and Semantics of Behavioural Specifications

For the behavioural specification of properties of data structures and software systems we will use arbitrary first-order formulas.

Definition 3.1 Let $\Sigma = (S, F)$ be a signature and $X = (X_s)_{s \in S}$ an S-sorted family of sets X_s of variables of sort s. The set WFF(Σ, X) of *(well-formed) Σ-formulas* is inductively defined by:

(0) If $t, r \in T(\Sigma, X)_s$ are terms of sort s then $t = r$ is a Σ-formula (called *equation*).

(1) If φ is a Σ-formula then $\neg\varphi$ is a Σ-formula.

(2) If φ_1 and φ_2 are Σ-formulas then $\varphi_1 \wedge \varphi_2$ is a Σ-formula.

(3) If φ is a Σ-formula then $\forall x{:}s.\varphi$ is a Σ-formula.

The set of free variables of a Σ-formula is defined as usual. A Σ-formula without free variables is called Σ-sentence. ♦

Notation In the sequel we will use the following abbreviations for Σ-formulas:

- If V is a set of variables $x_1, \ldots, x_n$ of sorts $s_1, \ldots, s_n$ then $\forall V.\varphi =_{def} \forall x_1{:}s_1 \ldots .\forall x_n{:}s_n.\varphi$
- Formulas with existential quantifier $\exists$ or with the connectives $\vee$, $\Rightarrow$, etc. are defined as usual. ♦

Σ-formulas are used as axioms of behavioural specifications. In addition to standard first-order specifications behavioural specifications contain a distinguished set Obs of observable sorts which determine the observable objects of data structures (see also [Reichel 85], [Orejas, Nivela 90] and others).

Definition 3.2 A *behavioural specification* SP = (Σ, Obs, E) consists of a signature $\Sigma = (S, F)$, a distinguished subset $Obs \subseteq S$ of *observable sorts* and a set E of Σ-sentences, called *axioms* of SP. ♦

Example 3.3

1. The following specification SETNAT describes properties of finite sets of natural numbers. Since the sort "bool" for the truth values is declared as an observable sort, sets can be observed via the membership test "iselem".

spec SETNAT =
  sorts {bool, nat, set}
  obs-sorts {bool, nat}
  opns {true: → bool, false: → bool, or: bool, bool → bool,
    zero: → nat, succ: nat → nat, eq: nat, nat → bool,
    empty: → set, add: nat, set → set, iselem: nat, set → bool}
  axioms {∀ b:bool, x, y:nat, s:set.
    or(true, b) = true ∧ or(false, b) = b ∧
    eq(zero, zero) = true ∧ eq(zero, succ(x)) = false ∧
    eq(succ(x), succ(y)) = eq(x, y) ∧eq(x, y) = eq(y, x)∧
    iselem(x, empty) = false ∧ iselem(x, add(y, s)) = or(eq(x,y), iselem(x, s)) ∧
    add(x, add(x, s)) = add(x, s) ∧ add(x, add(y, s)) = add(y, add(x, s))}.
endspec

2. The specification STATE describes environments (also called states) of a set of identifiers with values in an arbitrary set of data where the sorts "id" and "data" are specified as observable sorts. The constant "init" denotes the initial state, the operation "update" assigns a value to an identifier and the operation "lookup" delivers the current value of an identifier. Since the sort "state" is not observable states can only be observed via the "lookup" operation.

spec STATE =
  sorts {id, data, state}
  obs-sorts {id, data}
  opns {d0: → data,
    init: → state,
    update: id, data, state → state,
    lookup: id, state → data}
  axioms { ∀ x, y : id, d, e : data, s : state .
    lookup(x, init) = d0 ∧
    (x = y ⇒ lookup(x, update(y, d, s)) = d) ∧
    (x ≠ y ⇒ lookup(x, update(y, d, s)) = lookup(x, s)) ∧
    (x = y ⇒ update(x, d, update(y, e, s)) = update(x, d, s)) ∧
    (x ≠ y ⇒ update(x, d, update(y, e, s)) = update(y, e, update(x, d, s))) }
endspec

For the definition of the semantics of a behavioural specification SP we will discuss two different approaches: the observational and the behavioural semantics. The observational semantics will be defined by constructing the closure of the standard model class of SP under an observational equivalence relation on algebras. On the other hand, the behavioural semantics of SP will be directly defined in terms of a behavioural satisfaction relation which relaxes the standard satisfaction relation such that equations are not interpreted as identities but as behavioural equivalences of the objects of algebras (cf. also [Reichel 85], [Orejas, Nivela 90], [Bernot, Bidoit 91] for similar notions of behavioural satisfaction).

Let us first consider the definition of the observational semantics. It is based on the standard interpretation of the satisfaction relation $\models$ used in first-order predicate calculus:

Definition 3.4 Let SP = (Σ, Obs, E) be a behavioural specification. The *model class* Mod(SP) of SP consists of all Σ-algebras A which satisfy the axioms of SP, i.e.

$$\mathrm{Mod}(SP) =_{def} \{A \in \mathrm{Alg}(\Sigma) \mid A \models \phi \text{ for all } \phi \in E\}.$$

The algebras $A \in \mathrm{Mod}(SP)$ are called *standard models* of SP.

We say that SP is *consistent* if E is consistent in first-order logic, i.e. not all Σ-formulas can be deduced from E by the axioms and rules of a sound and complete system for first-order logic. ♦

Remark

(1) Because of the completeness of first-order logic, consistency is equivalent with satisfiability: SP is consistent iff $\mathrm{Mod}(SP) \neq \emptyset$.

(2) The model class notation is compatible with the usual notation in first-order logic and algebraic specification theory. We write equivalently,
$\mathrm{Mod}(\Sigma, Obs, E) = \mathrm{Mod}(\Sigma, E) = \mathrm{Mod}_\Sigma(E) = \{A \in \mathrm{Alg}(\Sigma) \mid A \models \varphi \text{ for all } \varphi \in E\}$.

For instance the algebra $P_{fin}(N)$ of finite subsets of N is a standard model of the specification SETNAT.

From the observational point of view we are not only interested in the standard models of a specification but also in all algebras which have the same observable behaviour as a standard model because such algebras represent observationally correct implementations. Algebras with the same observable behaviour are called *observationally equivalent* [Sannella, Wirsing 83] :

Definition 3.5 Let Σ = (S, F) be a signature, $Obs \subseteq S$ be a set of observable sorts and let A, B be two Σ-algebras. A and B are called *observationally equivalent* w.r.t. Obs (denoted by $A \sim_{Obs} B$) if there exists a family $Y = (Y_s)_{s \in S}$ of sets Y_s of variables of sort s and two surjective valuations $\alpha 1: Y \to A$ and $\beta 1: Y \to B$ such that for all terms $t, r \in T(\Sigma, Y)_s$ of observable sort $s \in Obs$ the following holds:

$$I_{\alpha 1}(t) = I_{\alpha 1}(r) \quad \text{if and only if} \quad I_{\beta 1}(t) = I_{\beta 1}(r). \ \blacklozenge$$

Fact 3.6 $\sim_{Obs}$ is an equivalence relation on the class Alg(Σ) of all Σ-algebras. ♦

Example 3.7 Let N^* be the usual algebra of finite sequences of natural numbers (over the signature of SETNAT). N^* is observationally equivalent to the standard model $P_{fin}(N)$ of finite subsets of N. But note that N^* is not a standard model of SETNAT because sequences of natural numbers do not satisfy the characteristic set equations specified in SETNAT. ♦

This example shows that model classes of behavioural specifications are in general not closed under observational equivalence. Hence we define the observational semantics of a behavioural specification SP as the closure of the model class of SP under observational equivalence:

Definition 3.8

1. Let Σ = (S, F) be a signature and Obs ⊆ S be a set of observable sorts. For any class C ⊆ Alg(Σ) of Σ-algebras, Obs(C) denotes the closure of C under observational equivalence w.r.t. Obs, i.e.

   $$\mathrm{Obs}(C) =_{def} \{B \in \mathrm{Alg}(\Sigma) \mid B \sim_{Obs} A \text{ for some } A \in C\}.$$

2. Let SP = (Σ, Obs, E) be a behavioural specification. Then Obs(SP) $=_{def}$ Obs(Mod(SP)) is the class of *observational models* of SP. ♦

We will now define the behavioural semantics of SP. The basic idea behind behavioural semantics is that the equations occurring in the first-order axioms of a specification are not interpreted as identities but as behavioural equivalences of objects where two objects are called behaviourally equivalent if they cannot be distinguished by "observable experiments". Formally, observable experiments are represented by observable contexts:

Definition 3.9 Let Σ = (S, F) be a signature, Obs ⊆ S be a set of observable sorts, $X = (X_s)_{s \in S}$ be an S-sorted family of countably infinite sets X_s of variables of sorts s and let $Z = \{z_s \mid s \in S\}$ be an S-sorted set of variables such that $z_s \notin X_s$ for all s ∈ S.

1. A term c ∈ T(Σ, X ∪ Z) is called *context over* Σ (or *Σ-context*), if c contains (besides variables in X) exactly one variable $z_s \in Z$ and z_s occurs only once in c. To indicate the variable z_s occurring in c we also write $c[z_s]$ instead of c.

Similarly, if $x_1,...,x_n$ of sorts $s_1, ..., s_n$ are the other free variables (in X) in c, we write often $c[z_s; x_1, ..., x_n]$ instead of c.

2. The application of a context $c[z_s]$ to a term $t \in T(\Sigma, X)$ of sort s is defined by the substitution of z_s by t. Instead of $c[t/z_s]$ we also write briefly c[t].
3. A Σ-context c is called *observable Σ-context* if the (result) sort of c belongs to the set Obs of observable sorts. The set of observable Σ-contexts is denoted by $C(\Sigma)^{Obs}$.
4. If $c[z_s]$ is a Σ-context and α, α_c: $X \to A$ are valuations, then $\alpha+\alpha_c$: $X \to A$ is a valuation defined by $(\alpha+\alpha_c)(x) =_{def} \alpha_c(x)$ if $x \in Var(c)$, $(\alpha+\alpha_c)(x) =_{def} \alpha(x)$ if $x \notin Var(c)$.

   If $c[z_s; x_1,..., x_n]$ is a Σ-context with free variables $x_1,..., x_n$ of sorts $s_1,...,s_n$, and u_s, v_s are new variables of sort s (such that $z_s, u_s, v_s, x_1,...,x_n$ are pairwise different), then

   (*) $\forall x_1: s_1 ... \forall x_n: s_n.\ c[u_s; x_1,...,x_n] = c[v_s, x_1,...,x_n]$

   is the *context equation associated to c*; if $V = Var(c)\backslash\{z_s\}$ we write shortly $\forall V.c[u_s]= c[v_s]$ instead of (*). ♦

For instance, if we consider the signature of the specification STATE with observable sorts "id" and "data" then the following are observable contexts: lookup(x, z_{state}), lookup(x, update(y, d, z_{state})), etc.

Definition 3.10 Let $\Sigma = (S, F)$ be a signature, $Obs \subseteq S$ be a set of observable sorts and let A be a Σ-algebra.

For any $s \in S$, two elements $a, b \in A_s$ are called *behaviourally equivalent* (denoted by $a \approx_{Obs} b$) if for all observable Σ–contexts $c[z_s]$ with $Var(c) = V \cup \{z_s\}$ and for all valuations α with $\alpha(u_s) = a$ and $\alpha(v_s) = b$ the associated context equation holds in A, i.e.

$$A, \alpha \models \forall V.c[u_s] = c[v_s],$$

where u_s, v_s are variables of sort s not occurring in c. ♦

Equivalently, $a,b \in A_s$ are behaviourally equivalent, if for all observable Σ–contexts $c[z_s]$, for all valuations α with $\alpha(u_s) = a$ and $\alpha(v_s) = b$ and for all valuations $\alpha_c : X \to A$,

$$A, \alpha+\alpha_c \models c[u_s] = c[v_s]$$

We are now prepared to define the behavioural satisfaction relation for arbitrary first-order formulas. As already mentioned above the difference to the standard case concerns only the interpretation of equations. For non-atomic formulas the behavioural satisfaction relation is defined by the usual inductive definition:

Definition 3.11 Let Σ, Obs, X be as in the above definition, A be a Σ-algebra, α: $X \rightarrow A$ be a valuation, let $t, r \in T(\Sigma, X)_s$ be terms of sort s and $\varphi, \varphi_1, \varphi_2$ Σ-formulas. The *behavioural satisfaction relation* (denoted by $\models_{beh}$) is inductively defined as follows:

(0) $A, \alpha \models_{beh} t = r$ if $I_\alpha(t) \approx_{Obs} I_\alpha(r)$

(1) $A, \alpha \models_{beh} \neg\varphi$ if $A, \alpha \not\models_{beh} \varphi$.

(2) $A, \alpha \models_{beh} \varphi_1 \wedge \varphi_2$ if $A, \alpha \models_{beh} \varphi_1$ and $A, \alpha \models_{beh} \varphi_2$.

(3) $A, \alpha \models_{beh} \forall x{:}s.\varphi$ if for all valuations α_0: $X \rightarrow A$ with $\alpha_0(z) = \alpha(z)$ if z is different from x, the following holds: $A, \alpha_0 \models_{beh} \varphi$.

(4) $A \models_{beh} \varphi$ if $A, \alpha \models_{beh} \varphi$ for all valuations α: $X \rightarrow A$. ♦

The following fact gives an equivalent definition to condition (0) from above which is an immediate consequence of the definition of behavioural equivalence of objects:

Fact 3.12 Let Σ, Obs, X, A, α, t, r be as in the above definition.
The following conditions are equivalent:

(1) $A, \alpha \models_{beh} t = r$;

(2) for all observable Σ-contexts $c[z_s]$ with variables $Var(c) = \{z_s\} \cup V$ and V disjoint from the variables of t and r, $A, \alpha \models \forall V.\ c[t] = c[r]$;

(3) for all observable Σ-contexts $c[z_s]$ with variables $Var(c) = \{z_s\} \cup V$ and V disjoint from the variables of t and r and for all valuations α_c: $V \rightarrow A$,
$$A, \alpha+\alpha_c \models c[t] = c[r];$$

(4) there exists a surjective valuation $\alpha_1: X \rightarrow A$ such that for all observable contexts $c[z_s]$ with $Var(c) = \{z_s\} \cup V$ and $V \subseteq X$ disjoint from the variables of t and r the following holds: $A, \alpha+\alpha_1 \models c[t] = c[r]$. ♦

The behavioural semantics of a specification SP is given by the behaviour class of SP:

Definition 3.13 Let SP = (Σ, Obs, E) be a behavioural specification. The *behaviour class* Beh(SP) of SP consists of all Σ-algebras which behaviourally satisfy the axioms of SP, i.e.

$$Beh(SP) =_{def} \{A \in Alg(\Sigma) \mid A \models_{beh} \phi \text{ for all } \phi \in E\}.$$

The algebras $A \in$ Beh(SP) are called *behavioural models* of SP. ♦

For instance, both algebras N^* and $P_{fin}(N)$ are behavioural models of SETNAT.

If SP is a behavioural specification where all axioms of SP are conditional equations with observable premises then the observational and the behavioural semantics of SP coincide, i.e. Obs(SP) = Beh(SP) (cf. [Reichel 85]). However the following example shows that for specifications with arbitrary first-order axioms there may exist standard models (which by definition are observational models as well) but which are no behavioural models:

Example 3.14 Consider the following specification:

spec DEMO =
  sorts { s, s' }
  obs-sorts { s }
  opns { a, b: $\to$ s, c, d : $\to$ s', f: s' $\to$ s }
  axioms { f(c) = f(d), c = d $\Rightarrow$ a = b}
endspec

Then a Σ-algebra A in which all constants are interpreted as different objects and where f takes the same value for c and d is a standard model of the specification but not a behavioural model : since the only observable context for c and d is $f(z_{s'})$, the equation c = d is behaviourally satisfied by A. On the other hand, since a and b are of observable sort, a $\neq$ b is behaviourally satisfied by A, i.e. A is not a behavioural model of the specification. ♦

3.2 Characterization of Behavioural Semantics

The aim of this section is to show that the behaviour class of a specification SP coincides with the closure of the class of fully abstract (standard) models of SP under observational equivalence. This result has various consequences: First, it implies that the behavioural semantics of a specification is a subclass of the observational semantics. Second, this means that standard models which are no behavioural models of a specification (cf. Example 3.14) are not fully abstract. A further consequence is that a specification SP is behaviourally consistent (i.e. has a behavioural model) if and only if SP has a (standard) fully abstract model. Thereby, following the definition of [Milner 77], a Σ-algebra A is called fully abstract if all elements of A which cannot be distinguished by observable experiments (i.e. are behaviourally equivalent) are identical:

Definition 3.15 Let $\Sigma = (S, F)$ be a signature, Obs $\subseteq$ S be a set of observable sorts. A Σ-algebra A is called *fully abstract* w.r.t. Obs if for all a, b $\in$ A the following holds:

$$a = b \text{ if and only if } a \approx_{Obs} b. \ ♦$$

For instance, the algebra $P_{fin}(N)$ of finite subsets of N is fully abstract. The next lemma shows that one can construct for any algebra A an observationally equivalent fully abstract algebra:

Lemma 3.16 Let $\Sigma = (S, F)$ be a signature, Obs $\subseteq$ S be a set of observable sorts and let A be a Σ-algebra.

1. $\approx_{Obs}$ is a congruence relation on A.
2. $A/\approx_{Obs}$ is fully abstract w.r.t. Obs (where $A/\approx_{Obs}$ denotes the quotient of A w.r.t. $\approx_{Obs}$).
3. A $\sim_{Obs}$ $A/\approx_{Obs}$ are observationally equivalent w.r.t. Obs. ♦

The following lemma provides the key for showing that observationally equivalent algebras satisfy behaviourally the same formulas (cf. Corollary 3.19, 1):

Lemma 3.17 Let A $\sim_{Obs}$ B be observationally equivalent Σ-algebras w.r.t. a set Obs of observable sorts and w.r.t. $\alpha 1$: Y $\rightarrow$ A and $\beta 1$: Y $\rightarrow$ B. Then for all Σ-formulas φ the following holds:

$$A, \alpha \models_{beh} \varphi \text{ if and only if } B, \beta \models_{beh} \varphi$$

for all valuations α: X $\rightarrow$ A, β: X $\rightarrow$ B which satisfy the following compatibility property (*):

(*) For any $x \in X$ there exists a term $t_x \in T(\Sigma, Y)$ with $\alpha(x) = I_{\alpha 1}(t_x)$ and $\beta(x) = I_{\beta 1}(t_x)$.
♦

The next lemma implies that for fully abstract algebras there is no difference between the standard and the behavioural satisfaction relation (cf. Corollary 3.19, 2):

Lemma 3.18 Let A be a fully abstract Σ-algebra w.r.t. a set Obs of observable sorts. Then for all Σ-formulas φ the following holds:

$$A, \alpha \models_{beh} \varphi \text{ if and only if } A, \alpha \models \varphi$$

for all valuations α: X $\rightarrow$ A. ♦

The following corollary is a consequence of Lemma 3.17 and Lemma 3.18. The third part of the corollary is essential for the proof of Theorem 3.21.

Corollary 3.19 Let $A \sim_{Obs} B$ be observationally equivalent Σ-algebras w.r.t. a set Obs of observable sorts. Then for all Σ-formulas φ the following holds:

1. $A \models_{beh} \varphi$ if and only if $B \models_{beh} \varphi$.
2. If B is fully abstract w.r.t. Obs then
   $B \models_{beh} \varphi$ if and only if $B \models \varphi$.
3. If B is fully abstract w.r.t. Obs then
   $A \models_{beh} \varphi$ if and only if $B \models \varphi$. ♦

Definition 3.20 Let SP = (Σ, Obs, E) be a behavioural specification. Then FA(SP) denotes the class of all standard models of SP which are fully abstract w.r.t. Obs. ♦

We can now state the central theorem of this section which provides the desired characterization of behavioural semantics in terms of fully abstract standard models and observational equivalence of algebras:

Theorem 3.21
Let SP = (Σ, Obs, E) be a behavioural specification and A a Σ-algebra.
Then the following assertions are equivalent:
(1) $A \in$ Beh(SP),
(2) $A \in$ Obs(FA(SP)),
(3) $A/\!\approx_{Obs} \in$ FA(SP),
(4) $A/\!\approx_{Obs} \in$ Mod(SP). ♦

4 Behavioural Proofs

In this section we will study proof techniques which allow to derive behavioural properties of specifications. In practice, behavioural proofs are not simple because one has usually to reason about infinitely many observable contexts. As a main result we will show that under certain conditions behavioural proofs can be reduced to standard proofs. Behavioural properties of a specification are expressed by behavioural theorems which are those formulas that are behaviourally satisfied by all behavioural models of a specification. Analogously to the standard case the set of behavioural theorems over a specification SP is called the behavioural theory of SP.

Definition 4.1 Let $C \subseteq Alg(\Sigma)$ be a class of Σ-algebras. Then the (standard) *theory* Th(C) of C is the set of all Σ-formulas which are satisfied (in the standard sense) by all algebras of C, i.e.

$$Th(C) =_{def} \{\varphi \mid A \models \varphi \text{ for all } A \in C\}.$$

Analogously the *behavioural theory* BehTh(C) of C is the set of all Σ-formulas which are behaviourally satisfied by all algebras of C, i.e.

$$BehTh(C) =_{def} \{\varphi \mid A \models_{beh} \varphi \text{ for all } A \in C\}.$$

In particular, for any behavioural specification SP, Th(Mod(SP)) is called the (standard) *theory* of SP and BehTh(Beh(SP)) is called the *behavioural theory* of SP. ♦

The following theorem provides a first reduction of behavioural theories to standard theories. It says that the behavioural theory of a specification SP coincides with the standard theory of the class of fully abstract models of SP.

Theorem 4.2
Let SP = (Σ, Obs, E) be a behavioural specification. Then:

$$BehTh(Beh(SP)) = Th(FA(SP)). \ ♦$$

In the next step we are interested in a proof system for the class FA(SP). For that purpose we need a proof rule for full abstractness. Since this property is characterized by an infinite number of contexts we cannot characterize it by a (finitary) axiom but we need an infinitary induction rule which allows one to prove behavioural theorems by induction on the structure of contexts. In a further step we will provide conditions under which the infinitely many of premises can be replaced by a finite number. Let us first consider "infinitary context induction" :

Definition 4.3 Let SP = (Σ, Obs, E) be a behavioural specification and let Π be a sound and complete proof system for the standard theory of SP. Then the proof system Π_{beh} consists of the proof system Π together with the following rule of *infinitary context induction* for any two variables $u, v \in X_s$, $s \in S$, and any Σ-formula φ:

$$[iCI_{\Sigma, Obs}] \quad \frac{\forall V.\ c[u] = c[v] \text{ for all } c \in C(\Sigma)^{Obs} \text{ with } u, v \notin Var(c),\ V = Var(c)\backslash\{z_s\}}{u = v}$$

In the following we write SP $\vdash_{beh}$ t = r if t = r is deducible using the axioms and rules of Π_{beh}. We call Π_{beh} behavioural logic. ♦

Theorem 4.4

Let SP be a behavioural specification. Then Π_{beh} is sound for FA(SP) and Beh(SP). ♦

To prove the completeness of the deduction system we use the "omitting types theorem". First we need some definitions.

Definition 4.5.

Let T and Γ be sets of Σ-formulas.

(1) A Σ-algebra A *realizes* Γ,
if there exists a valuation $\alpha: X \to A$ such that $A, \alpha \models \gamma$ for all $\gamma \in \Gamma$.

(2) A Σ-formula φ is *consistent with T* if there exists a Σ-algebra $A \in Mod_\Sigma(T)$, a valuation $\alpha: X \to A$ such that $A, \alpha \models \varphi$; i.e. if A realizes $\{\varphi\}$.

(3) T *locally omits* Γ, if for each Σ-formula φ consistent with T there exists $\gamma \in \Gamma$ such that $\varphi \wedge \neg\gamma$ is consistent with T.

(4) A Σ-algebra A *omits* Γ if A does not realize Γ, i.e. if for all valuations $\alpha: X \to A$ there exists $\gamma \in \Gamma$ such that $A, \alpha \models \neg\gamma$. ♦

Then the omitting types theorem reads as follows (cf. e.g [Chang, Keisler 90], page 84).

Theorem 4.6 ("Extended omitting types theorem")

Let Σ be a countable signature, T a consistent set of Σ-sentences and for each $i \in \mathbf{N}$, Γ_i a set of Σ-formulas with only finitely many free variables.

If T locally omits each Γ_i then T has a countable Σ-model which omits each Γ_i. ♦

For the analysis of the fully abstract models of a specification we introduce the following closure property and the characteristic set Γ_{iCI} of Σ-formulas:

Definition 4.7

Let SP = (Σ, Obs, E) a behavioural specification and T a set of Σ'-sentences with $\Sigma' \subseteq \Sigma$.

(1) T is called *closed w.r.t. $[iCI_{\Sigma, Obs}]$* if for all $s \in$ Obs the following holds:
if $T \models \forall V.\ c[u] = c[v]$ for all $c[z_s] \in C(\Sigma)^{Obs}$ with $u, v \notin Var(c)$ and $V = Var(c) \backslash \{z_s\}$ then $T \models u = v$.

(2) $\Gamma_{iCI\ \Sigma, Obs}$ is the following family of sets of Σ-formulas: For each $s \in S \backslash Obs$,
$\{u \neq v\} \cup$
$\{ \forall V.\ c[u] = c[v] \mid c[z] \in C(\Sigma)^{Obs}$ with $u, v \notin Var(c)$, $V = Var(c) \backslash \{z_s\} \}$ ♦

In the following we often omit the index Σ, Obs if it can be deduced from the context.

Γ_{iCI} consists of sets of Σ-formulas which have only two free variables u and v. Therefore we can apply the omitting types theorem.

Lemma 4.8
Let $\Sigma \subseteq \Sigma'$ and let T be a set of Σ'-sentences. If T is closed w.r.t. $[iCI_{\Sigma,Obs}]$, then T locally omits each set in $\Gamma_{iCI_{\Sigma,Obs}}$. ♦

Corollary 4.9
Let SP = (Σ, Obs, E) be a consistent behavioural specification. If E is closed w.r.t. $[iCI_{\Sigma,Obs}]$ then SP has a fully abstract Σ-model. ♦

Theorem 4.10 ("Completeness of behavioural logic")
A behavioural spec SP = (Σ, Obs, E) is consistent in behavioural logic if and only if SP has a fully abstract model. ♦

Corollary 4.11
Let SP = (Σ, Obs, E) a behavioural specification. Then for all Σ-formulas φ the following holds: $SP \vdash_{beh} \varphi$ if, and only if $Beh(SP) \models_{beh} \varphi$. ♦

5 Context Induction: A Behavioural Proof Technique

In order to apply the rule of infinitary context induction in practical examples one has to prove a formula for infinitely many observable contexts. For this purpose an appropriate proof technique is provided by the *principle of context induction.* Since in many cases context induction proofs need an iteration where also contexts of non observable sorts are considered, the principle of context induction is formulated for an arbitrary distinguished subset $S_d \subseteq S$ of the sorts of a signature Σ = (S, F). The set of Σ-contexts of sort $s \in S_d$ is denoted by $C(\Sigma)^{S_d}$.

Definition 5.1 Let Σ = (S, F) be a signature and $S_d \subseteq S$.

1. A *$C(\Sigma)^{S_d}$-context property* P is a property that ranges over all contexts in $C(\Sigma)^{S_d}$; we say that P is true if it holds for all $c \in C(\Sigma)^{S_d}$, i.e. for all $c \in C(\Sigma)^{S_d}$, P(c) holds.

2. A $C(\Sigma)^{S_d}$-context property P is called *inductive* if the following conditions are satisfied:

(0) $P(z_s)$ is true for all trivial contexts $z_s \in C(\Sigma)^{S_d}$ (which consist just of the variable z_s of some sort $s \in S_d$).

(1) For all contexts of the form $f(t_1, ..., c, ..., t_n)$ with a function symbol $f \in F$, $f: s_1 \times ... \times s_n \rightarrow s$, $s \in S_d$, terms $t_1, ..., t_n \in T(\Sigma, X)$ and a context $c \in T(\Sigma, X \cup Z)$ the following implication is true: If $P(c')$ is true for all subcontexts $c' \leq c$, $c' \in C(\Sigma)^{S_d}$ then $P(f(t_1, ..., c, ..., t_n))$ is true. ♦

Proposition 5.2 A $C(\Sigma)^{S_d}$-context property P is true if and only if it is inductive. ♦

The following proposition says that we can apply the principle of context induction for proving behavioural theorems:

Proposition 5.3 [Hennicker 91] Let SP = (Σ, Obs, E) be a behavioural specification and $t, r \in T(\Sigma, X)_s$. Then:

$t = r \in BehTh(Beh(SP))$ if and only if the following $C(\Sigma)^{Obs}$-context property P is inductive:

For all $c[z_s] \in C(\Sigma)^{Obs}$, $P(c[z_s])$ where

$P(c[z_s])$ is true if and only if the following holds:
if s is the sort of t and r, then $SP \vdash_{beh} c[t] = c[r]$. ♦

Example 5.4 The following specification CONTAINER describes containers with arbitrary elements. The operation "isin" allows to test whether an element is in a container. Since "bool" is declared as an observable sort containers can be observed via the operation "isin". (Note that in the axioms $x \neq y$ stands for $\neg(x = y)$.)

```
spec CONTAINER =
  sorts {bool, elem, container}
  obs-sorts   {bool, elem}
  opns  {true: → bool, false: → bool,
         empty: → container,
         put: elem, container → container,
         isin: elem, container → bool}
  axioms   {∀ b:bool, ct:container. ∀x, y:elem.
            isin(x, empty) = false ∧
            (x =y ⇒ isin(x, put(y, ct)) = true) ∧
            (x ≠ y ⇒ isin(x, put(y, ct)) = isin(x, ct))}
endspec
```

We will now prove that the order of putting elements into a container is irrelevant, i.e. that the following equation is a behavioural theorem of CONTAINER:

$$put(x, put(y, ct)) = put(y, put(x, ct)) \in BehTh(Beh(CONTAINER)).$$

By Proposition 5.3 it is enough to show that the following property P is an inductive context property for the set of observable contexts:

For all $c[z_s] \in C(\Sigma)^{Obs}$, $P(c[z_s])$ = true if and only if the following holds:

if s = container, then

CONTAINER $\vdash_{beh}$ c[put(x, put(y, ct))] = c[put(y, put(x, ct))].

We have to prove that the conditions (0) and (1) of Definition 5.1 (2) are satisfied w.r.t. P:

(P0) The trivial observable contexts are z_{bool} and z_{elem}. $P(z_{bool})$ and $P(z_{elem})$ are satisfied since bool ≠ container and elem ≠ container.

(P1) One has to consider all contexts of the form "isin(t, $c[z_s]$)" where t is an arbitrary term of sort "elem" and $c[z_s]$ is a context of sort "container". (By induction hypothesis it is not necessary to consider contexts of the observable sort elem at the first argument position of isin.)

For the proof of (P1) it is enough to show (by a nesting of context induction) that the following context property Q is inductive for the set of contexts of sort "container":

For all $c[z_s] \in C(\Sigma)^{\{container\}}$, $Q(c[z_s])$ = true if and only if the following holds:

if s = container, then

CONTAINER $\vdash_{beh}$ isin(x', c[put(x, put(y, ct))]) = isin(x', c[put(y, put(x, ct))]).

We have to prove that the conditions (0) and (1) of Definition 5.1 (2) are satisfied w.r.t. Q:

(Q0) For the proof of $Q(z_{container})$ we have to show:

CONTAINER $\vdash_{beh}$ isin(x', put(x, put(y, ct))) = isin(x', put(y, put(x, ct))).

Proof:

It is enough to prove the following two implications from which the required equation can be easily derived:

CONTAINER $\vdash_{beh}$ x' = x ⇒ isin(x', put(x, put(y, ct))) = isin(x', put(y, put(x, ct)))

and

CONTAINER $\vdash_{beh}$ x' ≠ x ⇒ isin(x', put(x, put(y, ct))) = isin(x', put(y, put(x, ct)))

As an example we will prove the first implication. (The second one can be derived similarly.) Since
CONTAINER $\vdash_{beh}$ x' = x ⇒ isin(x', put(x, put(y, ct))) = true
it is enough to prove
CONTAINER $\vdash_{beh}$ x' = x ∧ x' = y ⇒ isin(x', put(y, put(x, ct))) = true
and
CONTAINER $\vdash_{beh}$ x' = x ∧ x' ≠ y ⇒ isin(x', put(y, put(x, ct))) = true.
The first implication is an immediate consequence of
CONTAINER $\vdash_{beh}$ x' = y ⇒ isin(x', put(y, put(x, ct))) = true.
The second implication can be derived as follows: Since
CONTAINER $\vdash_{beh}$ x' ≠ y ⇒ isin(x', put(y, put(x, ct))) = isin(x', put(x, ct))
and
CONTAINER $\vdash_{beh}$ x' = x ⇒ isin(x', put(x, ct)) = true
we obtain
CONTAINER $\vdash_{beh}$ x' = x ∧ x' ≠ y ⇒ isin(x', put(y, put(x, ct))) = true.

(Q1) It is enough to consider all contexts of the form "put(t, c[z_s])" where t is an arbitrary term of sort "elem" and c[z_s] is a context of sort "container". (By the overall induction hypothesis for P it is not necessary to consider contexts of the observable sort elem at the first argument position of put.)
Then we have to show:
CONTAINER $\vdash_{beh}$
    isin(x', put(y', c[put(x, put(y, ct))])) = isin(x', put(y', c[put(y, put(x, ct))]))
whereby it can be assumed (by induction hypothesis) that
CONTAINER $\vdash_{beh}$ isin(x', c[put(x, put(y, ct))]) = isin(x', c[put(y, put(x, ct))]).

Proof:
For the proof of the desired equation it is enough to prove the following two implications:
CONTAINER $\vdash_{beh}$ x' = y' ⇒
    isin(x', put(y', c[put(x, put(y, ct))])) = isin(x', put(y', c[put(y, put(x, ct))]))
and
CONTAINER $\vdash_{beh}$ x' ≠ y' ⇒
    isin(x', put(y', c[put(y, put(x, ct))])) = isin(x', put(y', c[put(y, put(x, ct))]))

The first case is immediately done because under the premise x' = y' both terms of the conclusion reduce to "true".
In the second case we have
CONTAINER $\vdash_{beh}$ x' ≠ y' ⇒

isin(x', put(y', c[put(x, put(y, ct))])) = isin(x', c[put(x, put(y, ct))])

and
CONTAINER $\vdash_{beh}$ x' ≠ y' ⇒

isin(x', put(y', c[put(x, put(y, ct))])) = isin(x', c[put(y, put(x, ct))]).

Since, by induction hypothesis,
CONTAINER $\vdash_{beh}$ isin(x', c[put(x, put(y, ct))]) = isin(x', c[put(y, put(x, ct))]) we can immediately derive the second implication from above. ♦

For proving behavioural theorems over a specification we will now consider the approach of [Bidoit, Hennicker 94] where it is shown that under certain conditions the infinite set of premises in [$iCI_{\Sigma, Obs}$] (which characterize the full abstractness) can be replaced by a finite set of premises.
In the remainder of this section we assume for simplicity that SP has exactly one non-observable sort s, i.e. the sorts S of SP are divided into Obs ∪ {s} (see [Bidoit, Hennicker 94] for the general case with arbitrary many non-observable sorts) . Then the set F of operation symbols of SP can be split into three parts:

- The set F_0 of operations $0 : s_1, ..., s_n \to s_d$ with $s_d \in$ Obs and at least one s_i equal to s. These operations "observe" the sort of interest.

- The set F_I of operations $I : s_1, ..., s_m \to s$ with at least one s_i equal to s. These operations "produce" values of interest.

- The set of all other operations. This set contains the operations involving only observable sorts in their domains and codomains and those operations producing values of interest which are either constant or involve only observable sorts in their domain.

The following definition is a simplified (but stronger) version of the so-called "observability kernel" introduced in [Bidoit, Hennicker 94]:

Definition 5.5 The observability kernel OK[SP] of SP consists of the following formulas for all $0 \in F_0$ and $I \in F_I$:

$$\forall Z_L, Z_R : s.$$
$$([\forall\{x_1, ..., x_n\}.\ 0(x_1, ..., Z_L, ..., x_n) = 0(x_1, ..., Z_R, ..., x_n)] \Rightarrow$$
$$[\forall\{x_1, ..., x_n, y_1, ..., y_k\}.\ 0(x_1, ..., I(y_1, ..., Z_L ..., y_k), ..., x_n) =$$
$$0(x_1, ..., I(y_1, ..., Z_R, ..., y_k), ..., x_n)]) \quad \blacklozenge$$

The observability kernel provides a sufficient condition under which the infinite characterization of full abstractness can be expressed by a finitary formula. More precisely, if a specification SP satisfies the observability kernel OK[SP] then the equality x = y in the fully abstract models of SP can be specified by the following finitary formula (for the elements of the non-observable sort s):

$$[\mathrm{FA}(s)] \quad \forall x, y : s.\ ([\bigwedge_{0 \in F_0} \forall\{x_1, ..., x_n\}.\ 0(x_1, ..., x, ..., x_n) = 0(x_1, ..., y, ..., x_n)] \Rightarrow x = y)$$

The class of the fully abstract models of SP can then be axiomatized by the axioms of SP together with the axiom FA(s). Since the behavioural theory of a specification coincides with the standard theory of the class of the fully abstract models of SP (cf. Theorem 4.2) we can then prove behavioural theorems by reasoning in standard first-order logic:

Theorem 5.6 [Bidoit, Hennicker 94]
Let SP = (Σ, Obs, E), SP-part = (Σ, Obs, E ∪ FA(s)) and let φ be an arbitrary Σ-formula. If SP $\models$ OK[SP] then φ is a behavioural theorem of SP (i.e. φ ∈ BehTh(Beh(SP))) if, and only if φ is a standard theorem of SP-part (i.e. φ ∈ Th(Mod(SP-part))).
(The additional axiom of SP-part can be defined as a *partitioned by deduction rule* of the Larch Prover (cf. [Garland, Guttag 89]). This was the reason for choosing the name SP-part). ♦

Example 5.7 Consider the specification CONTAINER of Example 5.4, where the sort "container" is the only non observable sort. According to our separation of operation symbols from above we obtain F_0 = {isin} and F_I = {put}. Hence the observability kernel of CONTAINER consists of the following formula

$$\forall Z_L, Z_R : \text{container}.$$
$$([\forall x : \text{elem}.\ \text{isin}(x, Z_L) = \text{isin}(x, Z_R)] \Rightarrow$$
$$[\forall x, y : \text{elem}.\ \text{isin}(x, \text{put}(y, Z_L)) = \text{isin}(x, \text{put}(y, Z_R))])$$

and the specification CONTAINER-part has the following additional axiom:

(*) ∀ ct1, ct2 : container. ([∀ x : elem. isin(x, ct1) = isin(x, ct2)] ⇒ ct1 = ct2).

Proof of the observability kernel

For proving the observability kernel it is enough to show

CONTAINER $\vdash$ $isin(x, Z_L) = isin(x, Z_R) \Rightarrow isin(x, put(y, Z_L)) = isin(x, put(y, Z_R))$

For this it suffices to prove the following two implications:

CONTAINER $\vdash$ $(x = y \wedge isin(x, Z_L) = isin(x, Z_R)) \Rightarrow$
$isin(x, put(y, Z_L)) = isin(x, put(y, Z_R))$

and

CONTAINER $\vdash$ $(x \neq y \wedge isin(x, Z_L) = isin(x, Z_R)) \Rightarrow$
$isin(x, put(y, Z_L)) = isin(x, put(y, Z_R))$

The first implication can be proved as follows:

Since CONTAINER $\vdash x = y \Rightarrow isin(x, put(y, Z_L)) = true$

and CONTAINER $\vdash x = y \Rightarrow isin(x, put(y, Z_R)) = true$

we have

CONTAINER $\vdash x = y \Rightarrow isin(x, put(y, Z_L)) = isin(x, put(y, Z_R))$.

Hence

CONTAINER $\vdash$ $(x = y \wedge isin(x, Z_L) = isin(x, Z_R)) \Rightarrow$
$isin(x, put(y, Z_L)) = isin(x, put(y, Z_R))$.

The second implication can be proved as follows:

Since CONTAINER $\vdash x \neq y \Rightarrow isin(x, put(y, Z_L)) = isin(x, Z_L)$

and CONTAINER $\vdash x \neq y \Rightarrow isin(x, put(y, Z_R)) = isin(x, Z_R)$

we have

CONTAINER $\vdash$ $(x \neq y \wedge isin(x, Z_L) = isin(x, Z_R)) \Rightarrow$
$isin(x, put(y, Z_L)) = isin(x, put(y, Z_R))$.

Proof of behavioural theorems

In order to prove behavioural theorems for containers it is now enough, by Theorem 5.6, to prove (standard) theorems over CONTAINER-part. As an example we can now prove very efficiently that the order of putting elements into a container is irrelevant, i.e. that

$$put(x, put(y, ct)) = put(y, put(x, ct))$$

is a behavioural theorem of CONTAINER.

For the prove of this equation we first can derive (in the same way as it is done in Example 5.4 for proving the base (Q0) of the context induction proof for property Q):

CONTAINER-part $\vdash \forall x' : elem.\ isin(x', put(x, put(y, ct))) = isin(x', put(y, put(x, ct)))$.

Then we can apply the axiom (*) of CONTAINER-part and we obtain:

CONTAINER-part $\vdash put(x, put(y, ct)) = put(y, put(x, ct))$. ♦

6 Structured Behavioural Specifications

In this section we will introduce a "behaviour" operator for structured specifications. The definition of the behaviour operator is motivated by the results of the last section where we have seen that the behaviour class of a (flat) specification SP is the closure of the class of fully abstract models of SP under observational equivalence. Hence we define for any structured specification SP and set Obs of observable sorts a specification "beh SP on Obs" with semantics given by the observational closure of the fully abstract models of SP.

Besides the behaviour operator we assume several other specification building operators, like renaming of specifications, introduction of reachability constraints and the quotient of a specification w.r.t. an observational congruence. Most of these operators were already introduced in ASL (cf. [Wirsing 86]) or similarly in [Sannella, Tarlecki 87]. In the following definition "Spec" denotes the type of all specification expressions, "Formulas" denots the type of all sets of first-order Σ-formulas, "Sig" denotes the type of all signatures, "Opns" ("Sorts" resp.) the type of all sets of operation symbols (sorts resp.) and "Renaming" denotes the type of all *bijective* signature morphisms.

Syntax of structured specifications:

(0) (. , .) : Spec, Formulas → Spec	("basic")
(1) **rename** . **by** . : Spec, Renaming → Spec	("renaming")
(2) . | . : Spec, Sign → Spec	("export")
(3) . + . : Spec, Spec → Spec	("combine")
(4) **reach** . **on** . : Spec, Opns→ Spec	("reach")
(5) **beh** . **on** . : Spec, Sorts → Spec	("behaviour")
(6) . /≈. : Spec, Sorts → Spec	("observability quotient")

A (well-formed) closed specification expression without free variables is called "specification". In the following we consider only closed specification expressions, i.e. specifications. All results will remain true for arbitrary expressions interpreted in an appropriate environment. The semantics of a structured specification SP is given by its signature, denoted by *sig(SP)*, and by its class of models, denoted by *Mod(SP)*:

Semantics:

$sig((\Sigma, E)) = \Sigma$,
$sig($**rename** SP **by** $\sigma) = \Sigma'$, if $\sigma: sig(SP) \to \Sigma'$,
$sig(SP|_\Sigma) = \Sigma$,
$sig(SP + SP') = sig(SP) \cup sig(SP')$,
$sig($**reach** SP on $Cons) = sig(SP)$
$sig($**beh** SP **on** $Obs) = sig(SP)$,
$sig(SP/{\approx_{Obs}}) = sig(SP)$,

$Mod((\Sigma, E)) = \{A \in Alg(\Sigma) \mid A \models e \text{ for all } e \in E\}$,
$Mod($**rename** SP **by** $\sigma) = \{A \in Alg(\Sigma') \mid A|_\sigma \in Mod(SP)\}$,
$Mod(SP|_\Sigma) =$
$\{A \in Alg(\Sigma) \mid$ there exists $B \in Mod(SP)$ such that A is isomorphic to $B|_\Sigma\}$,
$Mod(SP + SP') = \{A \in Alg(sig(SP+SP')) \mid A|_{sig(SP)} \in Mod(SP),\ A|_{sig(SP')} \in Mod(SP')\}$,
$Mod($**reach** SP **on** $Cons) =$
$\{A \in Mod(SP) \mid$ for all $a \in A_s$ there exists $t \in (T_{Cons})_s$ and $\alpha: X' \to A$ with $a = I_\alpha(t)\}$
where $T_{Cons} = T((S, Cons), X')$ with $X'_s = \emptyset$ if s is the result sort of some constructor $f \in Cons$, $X'_s = X_s$ otherwise,
$Mod($**beh** SP **on** $Obs) = \{A \in Alg(sig(SP)) \mid A \sim_{Obs} B$ for some $B \in FA(SP)\}$,
$Mod(SP/{\approx_{Obs}}) =$
$\{A \in Alg(sig(SP)) \mid$ there exists $B \in Mod(SP)$ such that A is isomorphic to $B/{\approx_{Obs}}\}$.

As a derived operator we define the *enrichment* of a specification SP by a set S of sorts, a set F of operation symbols and a set E of axioms by:

enrich SP **by sorts** S **opns** F **axioms** $E =_{def} SP + ((sorts(SP) \cup S, opns(SP) \cup F), E)$.

The class of observational models of SP can also be defined as derived operator:

obs SP **on** Obs $=_{def}$ **beh** $(SP/{\approx_{Obs}})$ **on** Obs

This is justified by the following theorem:

Theorem 6.1 (Characterization of observational equivalence)
Let $SP = (\Sigma, E)$, $Obs \subseteq S$, and let A be a Σ–algebra.
Then the following assertions are equivalent:
(1) $A \sim_{Obs} B$ for some $B \in Mod(SP)$,
(2) $A \in Mod($**beh**$(SP/{\approx_{Obs}})$ **on** $Obs)$,
(3) $A/{\approx_{Obs}} \in Mod(SP/{\approx_{Obs}})$. ♦

For proving theorems which are valid in all models of a structured specification SP we extend the first-order proof system with particular axioms and rules of SP. Technically we need to consider the following aspects:

(1) Due to the export operator a specification SP may contain "hidden symbols". These are function symbols or sorts occurring in the signature of some subspecification SP_0 of SP but not in sig(SP). W.l.o.g. we always assume that the hidden symbols of two different subspecifications of a specification expression are disjoint, i.e. if $SP = C[SP_0, SP_1]$ then hidden $(SP_0) \cap$ hidden $(SP_1) = \emptyset$ (where hidden(SP_i) denotes the set of hidden symbols of SP_i). This is possible since hidden symbols can be renamed without changing the model class of a specification:
if r: sig(SP) $\rightarrow \Sigma'$ is a bijective renaming with r(x) = x for all $x \in \Sigma$ then $Mod(SP|_\Sigma) = Mod((\textbf{rename } SP \textbf{ by } r)|_\Sigma)$.

(2) For the definition of behaviour and observable quotient we have used two different congruence relations, the equality $=^A$ of an algebra and the behavioural equality $\approx_{Obs}{}^A$. Therefore we distinguish in the proof system several equality symbols, usually by indexing them with a specification expression, e.g. $=_{FA(SP1)}$. However, in the proof system for SP there is exactly one non-indexed equality symbol "=" that represents the equality in the models of SP.

(3) In order to treat adequately the different congruences (induced by the equality symbols) we consider the equality and the indexed equalities as sorted binary relation symbols which are part of the signature. Thus the interpretation of a signature is now a structure not an algebra. Moreover, in the proof system we note explicitely the defining congruence axioms of each (indexed) equality symbol.

First, we define the hidden symbols of a specification and the congruence axioms:

Definition 6.2

(1) Let SP, SP' be structured specifications, Σ, Σ' be signatures, E a set of Σ-formulas, σ: sig(SP)$\rightarrow\Sigma'$ a renaming, Cons a set of constructor symbols, Obs a set of sorts. Then

1.1) hidden(Σ , E) = $\emptyset$,

1.2) hidden (**rename** SP **by** σ) = hidden(SP) if σ(sig(SP)) $\cap$ hidden(SP) = $\emptyset$,

1.3) hidden($SP|_\Sigma$) = hidden(SP) $\cup$ (sig(SP)$\setminus\Sigma$) $\cup$ $\{=_s \mid s \in sorts(SP) \setminus sorts(\Sigma)\}$,

1.4) hidden(SP+SP') = hidden (SP) $\cup$ hidden(SP') if hidden(SP) $\cap$ hidden(SP') = $\emptyset$,

1.5) hidden(**reach** SP **on** Cons) = hidden(SP),

1.6) hidden(**beh** SP **on** Obs) = hidden(SP) $\cup$ $\{=_{FA(SP),s} \mid s \in$ sorts(SP)$\}$,

1.7) hidden(SP/$\approx_{Obs}$) = hidden(SP) $\cup$ $\{=_{SP,s} \mid s \in$ sorts(SP)$\}$.

(2) For any signature $\Sigma = (S,F)$ and any S-sorted family $\{=_s \mid s \in S\}$ of binary relation symbols the congruence axioms Cong(Σ, =) are defined as follows:

(Reflexivity) $\forall x{:}s.\ x =_s x$ for each $s \in S$,

(Symmetry) $\forall x, y{:}s.\ x =_s y \Rightarrow y =_s x$ for each $s \in S$,

(Transitivity) $\forall x, y, z{:}s.\ x =_s y \wedge y =_s z \Rightarrow x =_s z$ for each $s \in S$,

(Compatibility) $\forall x_1, y_1{:}s_1, \ldots, x_n, y_n{:}s_n.$

$x_1 =_{s1} y_1 \wedge \ldots \wedge x_n =_{sn} y_n \Rightarrow f(x_1,\ldots,x_n) =_s f(y_1,\ldots,y_n)$

for each $f{:}\ s_1 \times \ldots \times s_n \rightarrow s \in F$.

In the following we usually omit the index and write "=" instead of "$=_s$". ♦

Now we can define the proof system Π_{SP} for each structured specification:

Definition 6.3 (Proof system Π_{SP})

Let Π be a proof system for first-order logic (without equality).

(1) $\Pi_{(\Sigma,E)} =_{def} \Pi \cup$ Cong(Σ, =) $\cup$ E,

(2) $\Pi_{\textbf{rename}\ SP\ \textbf{by}\ \sigma} =_{def} \Pi_{SP}$ [$\sigma(x)/x$ for all $x \in$ sig(SP)]

if σ(sig(SP)) $\cap$ hidden (SP) = $\emptyset$,

(3) $\Pi_{SP|\Sigma} =_{def} \Pi_{SP}$,

(4) $\Pi_{SP+SP'} = \Pi_{SP} \cup \Pi_{SP'}$

if hidden(SP) $\cap$ hidden(SP') = $\emptyset$,

(5) $\Pi_{\textbf{reach}\ SP\ \textbf{on}\ Cons} = \Pi_{SP} \cup$ "infinitary induction"

$$\left\{ [iI_{sig(SP),Cons,s}]\ \frac{\varphi[t/x] \text{ for all } t \in (T_{Cons})_s}{\forall x{:}s.\ \varphi} \ \middle|\ s \in \text{sorts(sig(SP))} \right\}$$

(6) $\Pi_{\textbf{beh}\ SP\ \textbf{on}\ Obs} =_{def} \Pi_{SP}[=_{FA(SP)}/=] \cup$ Cong(sig(SP), =) $\cup$

$\{[iCI_{sig(SP),Obs}\ [=_{FA(SP)}/=]\}$ "infinitary context induction" $\cup$

$\{[=_{Obs}]\ \forall x,y{:}s.\ x =_{SP} y \iff x = y \mid s \in Obs\}$

"observable equality and equality are the same on observable sorts".

(7) $\Pi_{SP/\approx Obs} =_{def} \Pi_{SP}[=_{SP}/=] \cup$ Cong(sig(SP), =) $\cup$

$\{[iCI_{sig(SP),Obs}]\}$ "infinitary context induction for the quotient" $\cup$

$\{[=_{Obs}]\ \forall x,y{:}s.\ x =_{SP} y \iff x = y \mid s \in Obs\}$

"observable equality and equality are the same on observable sorts".

For any Σ-formula φ with $sig(SP) \subseteq \Sigma$ we write $SP \vdash \varphi$ if φ can be deduced by the axioms and rules of Π_{SP}. ♦

Remark

An equivalent proof system for the quotient operator is induced by the following axiom schema and rule:

[$\approx_{Obs}$-compatibility] $\{\forall x,y{:}s. (x = y \Rightarrow \forall V.c[x] =_{SP} c[y] \mid c \in C(\Sigma)^{Obs},\ s \in sorts(SP)\}$

$$[\approx_{Obs}\text{-}iCI_{sig(SP),Obs}] \qquad \frac{\forall V.c[x] =_{SP} c[y] \text{ for all } c \in C(\Sigma)^{Obs}}{x = y}$$

where $V = Var(c)\setminus\{z_s\}$, $x, y \notin V$ for each $c \in C(\Sigma)^{Obs}$. ♦

Corollary 6.4

Let SP be a specification, $Obs \subseteq sorts(sig(SP))$. The following proof systems are equivalent:

(1) $\Pi_{SP/\approx Obs}$

(2) $\Pi_{SP[=_{SP}/=]} \cup [\approx_{Obs}\text{-compatibility}] \cup [\approx_{Obs}\text{-}iCI_{sig(SP),Obs}]$ ♦

The "loose models" LMod(SP) of a specification SP are those $sig(SP) \cup \{=\}$-structures A that can be made into models of SP by building the quotient $A/{=^A}$ w.r.t. the sig(SP)-congruence $=^A$:

Definition 6.5

Let SP be a specification with signature Σ.

(1) $\Sigma_= =_{def} \Sigma \cup \{=_s \mid s \in S\}$

where $=_s$ is a binary relation symbol with arguments in s x s.

(2) $LMod(SP) =_{def}$

$\{A \in Struct(sig(SP)_=) \mid A \models Cong(sig(SP),=) \text{ and } (A|_{sig(SP)})/{=^A} \in Mod(SP)\}$

(3) Let A be a Σ-structure and $\sim = (\sim_s)_{s \in S}$ with $\sim_s \subseteq A_s \times A_s$ a family of binary relations on A. Then $A{+}{\sim}$ is a $\Sigma_=$-structure with $(A{+}{\sim})|_\Sigma =_{def} A$ and for each $s \in S$, $a, b \in A_s$: $a =_s^{A+\sim} b$ if, and only if $a \sim_s b$. ♦

The definition of LMod(SP) implies directly the following relation between Mod(SP) and LMod(SP):

Lemma 6.6

Let SP, SP_1, SP_2 be specifications, $\Sigma \subseteq sig(SP)$ a signature, E a set of Σ-sentences, Cons $\subseteq$ opns(SP) a set of constructor symbols, Obs a set of sorts.

(1) For any sig(SP)-formula φ and any $sig(SP)_=$-structure $A \in LMod(SP)$, $A \models \varphi$ if, and only if $A|_{sig(SP)}/=^A \models \varphi$.

(2) $LMod((\Sigma, E)) = \{A \in Struct(\Sigma_=) \mid A \models E \text{ and } A \models Cong(\Sigma, =)\}$,

$LMod(\textbf{rename } SP \textbf{ by } \sigma) = \{A \in Struct(\sigma(sig(SP)_=)) \mid A|_{\sigma_=} \in LMod(SP)\}$,

where $\sigma_=: sig(SP)_= \to \sigma(sig(SP))_=$ is the canonical extension of σ by equality: $\sigma_=(x) =_{def} \sigma(x)$, if $x \in sig(SP)$, and $\sigma_=(=_s) =_{def} =_{\sigma(s)}$ for each $s \in sorts(SP)$.

$LMod(SP|_\Sigma) = \{A|_{\Sigma_=} \mid A \in LMod(SP)\}$,

$LMod(SP_1+SP_2) =$

$\{A \in Struct(sig(SP_1+SP_2)_=) \mid A|_{sig(SP_i)_=} \in LMod(SP_i) \text{ for } i=1, 2\}$,

$LMod(\textbf{reach } SP \textbf{ on } Cons) =$

$\{A \in LMod(SP) \mid$ for all $s \in sorts(SP)$, and for all $a \in A_s$, there exists $t \in (T_{Cons})_s$ and $\alpha: X' \to A$ with $\alpha(x) = a$ such that $A, \alpha \models x = t\}$,

$LMod(\textbf{beh } SP \textbf{ on } Obs) =$

$\{A \in Struct(sig(SP)_=) \mid =^A$ is a sig(SP)-congruence, $=^A \subseteq \approx_{Obs}{}^A$ and $(A|_{sig(SP)})+\approx_{Obs}{}^A \in LMod(SP)\}$,

$LMod(SP/\approx_{Obs}) =$

$\{A \in Struct(sig(SP)_=) \mid (=^A) = \approx_{Obs}{}^A$ and there exists $B \in LMod(SP)$ such that $B|_{sig(SP)} = A|_{sig(SP)}$ and $(=^B) \subseteq (=^A)\}$.

(3) Let $\Sigma = sig(SP) \cup hidden(SP)$.

For any $sig(SP)_=$-structure $A \in LMod(SP)$ there exists a Σ-structure A' such that

(i) A' is an expansion of A, i.e. $A'|_{sig(SP)_=} = A$

(ii) A' satisfies all axioms and rules of Π_{SP}. ♦

Definition 6.7

Let Σ be a signature, SP a specification with $sig(SP) \subseteq \Sigma$ and $hidden(SP) \cap \Sigma = \emptyset$, T a set of Σ-sentences, A a Σ-structure, φ a Σ-formula.

(1) A is an *SP-model* if $A|_{sig(SP) \cup \{=\}} \in LMod(SP)$.

(2) A is an *SP-model of T* if A is an SP-model such that $A \models T$.

(3) $T \models \varphi$ if for all Σ-structures A with $A \models T$ we have $A \models \varphi$.

(4) T is Π-*consistent* if not all Σ-formulas are deducible from T by the axioms and rules of Π.

(5) Let [r] be a proof rule of the form

$$\frac{\varphi_i \text{ for all } i \in I}{\varphi}$$

Then T is *[r] -closed* if for all formulas φ_i ($i \in I$), φ which match [r] the following holds: if $T \models \varphi_i$ for all $i \in I$ then $T \models \varphi$.

(6) T is Π_{SP} *-closed,* if T satisfies all axioms of Π_{SP} and is [r]-closed for all rules [r] of Π_{SP}.

(7) Π_{SP} is *sound (w.r.t. LMod(SP))* if for any Σ-formula φ the following holds: if φ is derivable by the axioms and rules of Π_{SP} then φ holds in all SP-models.

(8) Π_{SP} is *complete (w.r.t. LMod(SP))* if any Π_{SP} -consistent set of Σ-sentences T has an SP-model. ♦

Definition 6.8 ($\Gamma_{\Sigma,Cons}$)

Let $\Sigma = (S, F)$ be a signature, $Cons \subseteq F$.

$\Gamma_{\Sigma,Cons,s} =_{def} \{\exists Var(t).\ x \neq t \mid t \in (T_{Cons})_s \text{ with } x \notin Var(t)\}$ for each $s \in S$,
$\Gamma_{\Sigma,Cons} = (\Gamma_{\Sigma,Cons,s})_{s \in S}$. ♦

Lemma 6.9

Let T be a set of Σ'-sentences, $Cons \subseteq opns(\Sigma)$, $\Sigma \subseteq \Sigma'$.
If T is $[iI_{\Sigma,Cons,s}]$-closed for each $s \in sorts(\Sigma)$, then T locally omits all $\Gamma_{\Sigma,Cons,s}$. ♦

Lemma 6.10

Let Σ' be a signature with $(sig(SP)_= \cup hidden(SP)) \subseteq \Sigma'$.

(1) If T is a consistent set of Σ'-sentences that locally omits all Γ_r for any infinitary rule [r] of Π_{SP} and that satisfies all axioms of Π_{SP} then there exists a Σ'-structure that satisfies T, omits all Γ_r for any infinitary rule [r] of SP and that satisfies all axioms of Π_{SP}.

(2) Let A be a Σ'-structure with the following properties:
(i) A omits Γ_r for all infinitary rules [r] of Π_{SP},
(ii) A satisfies all axioms of Π_{SP}.
Then $A|_{sig(SP) \cup \{=\}} \in LMod(SP)$. ♦

Theorem 6.11 (Completeness of Π_{SP})
Let Σ be a signature, SP a specification with $sig(SP)_= \subseteq \Sigma$ and $\Sigma \cap hidden(SP) = \emptyset$.
A set T of Σ-sentences is Π_{SP}-consistent if, and only if T has an SP-model. ♦

Corollary 6.12
Let SP be a specification and φ a sig(SP)-formula. Then:
φ is deducible in Π_{SP} if, and only if $Mod(SP) \models \varphi$. ♦

7 Behavioural Implementations

The power of the specification building operators of the last section allows us to define a simple implementation relation for structured specifications by model class inclusion:

Definition 7.1
Let SP and SP1 be (structured) specifications. SP1 is called *implementation* of SP (denoted by SP ~~> SP1) if

- sig(SP1) = sig(SP) and
- $Mod(SP) \supseteq Mod(SP1)$ ♦

Fact 7.2
All specification operators are monotonic w.r.t. ~~>.
In particular, if SP ~~> SP1 then **beh** SP **on** Obs ~~> **beh** SP1 **on** Obs. ♦

Although the implementation notion is defined for arbitrary structured specifications, we are particularly interested in implementation relations where the specification to be implemented is obtained by behavioural abstraction. We say that SP1 is a *behavioural implementation* of SP with respect to a set Obs of observable sorts, if **beh** SP **on** Obs ~~> SP1. This definition is motivated by the fact that in many cases an implementation SP1 does not exactly satisfy all the properties of an abstract specification SP but satisfies only the "observable behaviour" of SP. For instance, the usual implementation of sets by lists does not satisfy the characteristic set equations but it satisfies the observable properties of sets if only membership tests $x \in S$ are observable. Another example is the familiar implementation of stacks by arrays with pointers to the top elements. This implementation does not satisfy the stack equation pop(push(x, s)) = s but if one considers only the top elements of stacks to be observable then all observable consequences of this equation (such as top(pop(push(x, s))) = top(s)) are satisfied by the implementation.

7.1 A Proof System for Implementations

In the following we will provide a calculus $\Pi L_{\rightsquigarrow}$ which allows us to prove the correctness of implementations (in particular of behavioural implementations) according to the structure of the abstract specification. In rule (2) SP1+SP is called persistent extension of SP1 if $\mathrm{Mod}(SP1) = \mathrm{Mod}((SP1+SP)|_{sig(SP1)})$. Rule (4) uses an infinitary disjunction which requires that (the values of) all models of SP1 are reachable by the given constructors.

Proof system for implementations:

(0) $$\frac{SP1 \vdash E}{(\Sigma, E) \text{ \~{}\~{}> } SP1} \quad \text{if } sig(SP1) = \Sigma$$

(1) $$\frac{SP \text{ \~{}\~{}> } \textbf{rename } SP1 \textbf{ by } \sigma^{-1}}{\textbf{rename } SP \textbf{ by } \sigma \text{ \~{}\~{}> } SP1} \quad \text{if } \sigma : sig(SP) \rightarrow sig(SP1) \text{ is bijective}$$

(2) $$\frac{SP \text{ \~{}\~{}> } SP1 + SP}{SP|_\Sigma \text{ \~{}\~{}> } SP1} \quad \text{if } sig(SP1) = \Sigma \text{ and } SP1{+}SP \text{ is a persistent extension of } SP1.$$

(3) $$\frac{SP \text{ \~{}\~{}> } SP1|_{sig(SP)},\; SP' \text{ \~{}\~{}> } SP1|_{sig(SP')}}{SP + SP' \text{ \~{}\~{}> } SP1} \quad \text{if } sig(SP1) = sig(SP{+}SP')$$

(4) $$\frac{SP \text{ \~{}\~{}> } SP1,\; SP1 \models \forall x{:}s.\ \bigvee_{t \in T_{Cons}} x = t \quad \text{for all } s \in sorts(sig(SP1))}{\textbf{reach } SP \textbf{ on } Cons \text{ \~{}\~{}> } SP1}$$

(5) $$\frac{SP \text{ \~{}\~{}> } SP1/\approx_{Obs}}{\textbf{beh } SP \textbf{ on } Obs \text{ \~{}\~{}> } SP1}$$

(6) $$\frac{SP \text{ \~{}\~{}> } SP2,\; SP1 = SP2/\approx_{Obs} \text{ are semantically equivalent}}{SP/\approx_{Obs} \text{ \~{}\~{}> } SP1}$$

Note that the above proof system contains two technical difficulties which are not addressed in this paper:

1.) Applying rule (2) for "export" requires to prove the persistency of SP1 + SP w.r.t. SP1. One has to show that each model of SP1 can be extended by interpretations of the symbols in $sig(SP)\backslash\Sigma$ in such a way that the extension is a model of SP. In general, this has to be done by a semantical construction. However, if the axioms for the hidden function symbols are in a constructive form such as explicit or implicit definition then it is well-known that persistency can be checked by syntactic conditions (under mild assumptions).

2.) Applying rule (4) requires the proof of an infinite disjunction. In practice, this reduces to prove "sufficient completeness" of SP1 w.r.t. Cons (see e.g. [Wirsing 90]).

For proving the completeness of the above proof system for implementations the following lemma is used:

Lemma 7.3
Let SP, SP1 be specifications such that $SP/\approx_{Obs}$ ~~> SP1.
Then ((**beh** SP1 **on** Obs)+SP)$/\approx_{Obs}$ and SP1 are semantically equivalent. ♦

Theorem 7.4
The proof system $\Pi_{\text{~~>}}$ is sound and complete if the proof system Π_{SP} used in rule (0) is sound and complete for each specification SP. In this case an implementation relation SP ~~> SP1 holds if, and only if it can be derived by the rules of $\Pi_{\text{~~>}}$. ♦

Example 7.5
The specification STATE1 to be implemented is the following structured specification of states:

spec STATE1 = **reach** (ID+STATE) **on** {init, update}

where ID is a basic specification of identifiers and STATE is the specification from Example 3.3.2.

We give a behavioural implementation of STATE1 by a specification HISTORY which implements states by sequences of pairs consisting of an identifier and its associated value. In contrast to the abstract specification of states each sequence stores not only the current value of an identifier x but also all previous values of x. (Such implementations are particularly useful if one wants to retrieve old states of a system or, more concretely, if one wants to test and to analyze the state transitions performed by the execution of an imperative program.)
The specification HISTORY is built on top of the specification ID and of a specification SEQDATA of finite sequences over the underlying data values. The function "history"

computes for a given identifier x and some state s the history of all environments of x, i.e. the sequence of all previous values of x.

```
spec HISTORY = reach enrich (ID + SEQDATA) by
      sorts  {state}
      opns   {init : → state,
              <. , .> : id, data : → state,
              .&. : state, state → state,
              update : id, data, state → state,
              lookup : id, state → data,
              history : id, state → seqdata }
      axioms {∀x, y : id, d : data, s, t, u : state.
              s & init = s  ∧ init & s = s ∧
              (s & t) & u = s & (t & u) ∧
              update(x, d, s) = < x , d > & s  ∧
              lookup(x, init) = d0 ∧
              (x = y ⇒ lookup(x,  < y,  d > & s) = d) ∧
              (x ≠ y  ⇒ lookup(x,  < y,  d > & s) = lookup(x, s)) ∧
              history(x, init) = append(d0, emptyseq) ∧
              (x = y  ⇒ history (x, < y,  d  > & s) = append(d, history(x, s))) ∧
              (x ≠ y  ⇒ history (x, < y,  d  > & s)  =  history(x, s)) }
   on {init, <. , .>, .&.}
endspec
```

The specification HISTORY is a behavioural implementation of STATE1 if we consider only the sorts "id" and "data" as observable sorts of STATE1, i.e.

beh STATE1 **on** {id, data} ~~> $\mathrm{HISTORY}|_{\mathrm{sig(STATE1)}}$.

Informally this follows from the fact that using the operations of STATE1 states can only be observed via the lookup operation and hence one cannot access old values of an identifier. (Note that in contrast to the abstract operations of STATE1 the history function of the implementation is explicitly defined for accessing old values.) The following scheme shows how a structured "bottom up" implementation proof can be performed by the rules of our calculus $\Pi L_{\text{-->}}$. Thereby we use the following abbreviations:

Obs stands for {id, data}, Σ for sig(STATE1), Cons for {init, update}. Moreover, STATE-axioms denotes the axioms of the specification STATE (cf. Example 3.3.2) and ID-axioms denotes the axioms of ID.

$$\frac{\dfrac{\dfrac{(\text{HISTORY}|_\Sigma)/\!\approx_{Obs}|_{sig(ID)} \vdash \text{ID-axioms},\ (\text{HISTORY}|_\Sigma)/\!\approx_{Obs} \vdash \text{STATE-axioms}}{\text{ID} \leadsto (\text{HISTORY}|_\Sigma)/\!\approx_{Obs}|_{sig(ID)},\ \text{STATE} \leadsto (\text{HISTORY}|_\Sigma)/\!\approx_{Obs}}}{\text{ID+STATE} \leadsto (\text{HISTORY}|_\Sigma)/\!\approx_{Obs},\ (\text{HISTORY}|_\Sigma)/\!\approx_{Obs} \models \forall x{:}s \bigvee_{t \in T_{Cons}} x = t\ (*)}}{\dfrac{\text{STATE1} \leadsto (\text{HISTORY}|_\Sigma)/\!\approx_{Obs}}{\textbf{beh}\ \text{STATE1}\ \textbf{on}\ \text{Obs} \leadsto \text{HISTORY}|_\Sigma}}$$

In the last step one has to prove (*) and $(\text{HISTORY}|_\Sigma)/\!\approx_{Obs}|_{sig(ID)} \vdash$ ID-axioms and $(\text{HISTORY}|_\Sigma)/\!\approx_{Obs} \vdash$ STATE-axioms.
The proof of the ID-axioms is simple because ID is a subspecification of HISTORY. The proof of the STATE-axioms is non-trivial because the two axioms

$(x = y \Rightarrow \text{update}(x, d, \text{update}(y, e, s)) = \text{update}(x, d, s)) \wedge$
$(x \neq y \Rightarrow \text{update}(x, d, \text{update}(y, e, s)) = \text{update}(y, e, \text{update}(x, d, s)))$

are not valid in the implementation. However, one can show by context induction that for all observable contexts c over the signature of state

$(x = y \Rightarrow c[\text{update}(x, d, \text{update}(y, e, s))] = c[\text{update}(x, d, s)]) \wedge$
$(x \neq y \Rightarrow c[\text{update}(x, d, \text{update}(y, e, s))] = c[\text{update}(y, e, \text{update}(x, d, s))])$

holds in HISTORY and therefore, using (3) and (7) of the proof system for structured specifications (cf. Definition 6.3), also in $(\text{HISTORY}|_\Sigma)/\!\approx_{obs}$. Then one can derive the desired result using infinitary context induction for the quotient (cf. (7) in Definition 6.3). ♦

7.2 Vertical Composition of Behavioural Implementations

An important technique for the construction of implementations is the principle of stepwise refinement where an implementation is constructed in several steps by a series of design decisions such as choice of data representation, choice of algorithms etc. In our framework we are particularly interested in a stepwise development of behavioural implementations. Hence the situation to be considered is the following one:
Given an abstract specification SP for which we want to construct an implementation, we first define the observable behaviour of SP (which has to be satisfied by an implementation) by applying the behaviour operator to SP. Then, we choose a specification SP1 as a basis of the first implementation step and we construct a first behavioural implementation on top of SP1 (by applying some structuring operators to

SP1). In a next step SP1 is behaviourally implemented by a more concrete specification, say SP2, in the same way until an implementation is obtained which can be translated into an executable program.

In the following we consider two typical implementation constructions and we show the compatibility of these constructions with behavioural implementations. Let us first assume that the underlying specification SP1 (used for the implementation) comprises already all sorts and operations of the abstract specification SP, i.e. the implementation is constructed by restricting SP1 to the signature of SP. Then we obtain the following "vertical" composition theorem:

Theorem 7.6
If **beh** SP **on** Obs ~~> $SP1|_{sig(SP)}$ and **beh** SP1 **on** Obs ~~> SP2 then
beh SP **on** Obs ~~> $SP2|_{sig(SP)}$. ♦

Let us now consider the more general case where (some of) the sorts and operations of SP are defined on top of SP1 by an appropriate enrichment Δ. Since often some constructors are defined for the representations of abstract values we then apply the reach operator to the enriched specification and finally we restrict to the abstract signature. In order to formulate the next vertical composition theorem we need the notion of "observational sufficient completeness":

Definition 7.7
Let Δ = **sorts** S **opns** F **axioms** E be an enrichment of some specification with signature $\Sigma = (S_0, F_0)$, let Cons $\subseteq F \cup F_0$ be a set of constructors and let Obs $\subseteq S_0$ be a set of observable sorts. Moreover, let $X = (X_s)_{s \in S_0 \cup S}$ be a family of sets X_s of variables of sort s with $X_s = \emptyset$ if s is the result sort of some constructor f ∈ Cons, X_s is countably infinite otherwise and let $X_0 = (X_s)_{s \in S_0}$.

Then Δ is called *observationally sufficiently complete* w.r.t. Cons and Obs if the following holds:

For all $t \in T((S \cup S_0, F \cup F_0), X)_s$ with s ∈ Obs there exists $p \in T(\Sigma, X_0)_s$ such that **reach** $((S \cup S_0, F \cup F_0), E)$ **on** Cons $\models t = p$. ♦

Theorem 7.8
If **beh** SP **on** Obs ~~> (**reach** (**enrich** SP1 **by** Δ) **on** Cons)$|_{sig(SP)}$ and **beh** SP1 **on** Obs ~~> SP2 and if Δ is observationally sufficiently complete w.r.t. Cons and Obs and Δ is a persistent enrichment of SP1 w.r.t. Cons, i.e. Mod(SP1) = Mod((**reach** (**enrich** SP1 **by** Δ) **on** Cons)$|_{sig(SP1)}$), then

beh SP **on** Obs ~~> (**reach** (**enrich** SP2 **by** Δ) **on** Cons)$|_{sig(SP)}$. ♦

8 Concluding Remarks

In this paper we have presented a behavioural approach to the semantics of algebraic specifications which is based on the notion of behavioural satisfaction with respect to a set of observable sorts. One of the main advantages of our approach is the availability of proof rules, not only for flat specifications but also for structured ones. Also the proof system for implementations of structured specifications of [Wirsing 91] has been extended to the behavioural case.
As a next step, it remains to include behavioural specifications with interfaces in order to obtain a complete approach to modular system design with reuse. In order to to this, composition properties similar to those presented in [Hennicker, Nickl 92] have to be established.

Acknowledgments

Many thanks to Michel Bidoit for stimulating discussions during which many of the ideas presented here were evolved. Thanks also to the members of the COMPASS working group, in particular to Fernando Orejas, Don Sannella and Andrzej Tarlecki, who provided an exciting and (sometimes) critical forum for our ideas. Many thanks by MW to the organizers of the Marktoberdorf Summer School for their patience in awaiting the draft version of this paper and last not least many thanks to Heidrun Walker, Sabine Niedballa and Marianne Diem for providing their assistance in preparing this manuscript.

References

[Astesiano et al. 86] E. Astesiano, G. Reggio, M. Wirsing: Relational specifications and observational semantics. In: I. Gruska, B. Rovan, J. Wiedermann (eds.): Proc. MFCS '86, Mathematical Foundations of Computer Science, Bratislava, Springer Lecture Notes in Computer Science, 233, 209-217, 1986.

[Bauer, Hennicker 93] B. Bauer, R. Hennicker: Proving the correctness of algebraic implementations by the ISAR system. In: A. Miola (ed.): Proc. DISCO '93, International Symposium on Design and Implementation of Symbolic Computation Systems, Springer Lecture Notes in Computer Science, 722, 2-16, 1993.

[Bernot, Bidoit 91] G. Bernot, M. Bidoit: Proving the correctness of algebraically specified software: modularity and observability issues. Laboratoire d'Informatique de l'Ecole Normale Supérieure, Paris, Technical Report, LIENS-91-8, 1991.

[Bidoit, Hennicker 94] M. Bidoit, R. Hennicker: Proving Behavioural Theorems with Standard First-Order Logic. Submitted for publication, 1994.

[Broy, Wirsing 89] M. Broy, M. Wirsing: A modular framework for algebraic specification and implementation. In: J. Diaz, F. Orejas (eds.): Proc. TAPSOFT '89, Springer Lecture Notes in Computer Science, 351, 42-73, 1989.

[Chang, Keisler 90] C. C. Chang, H. J. Keisler: Model Theory. Amsterdam, North-Holland, 3rd edition, 1990.

[Garland, Guttag 89] S. Garland, J. Guttag: An overview of LP, the Larch Prover. In: Proc. of the Third International Conference on Rewriting Techniques and Applications, Springer Lecture Notes in Computer Science, 355, 137-151, 1989.

[Hennicker 91] R. Hennicker: Context induction: a proof principle for behavioural abstractions and algebraic implementations. Formal Aspects of Computing 3 (4), 326-345, 1991.

[Milner 77] R. Milner: Fully abstract models of typed λ-calculi. Theoretical Computer Science 4, 1-22, 1977.

[Nickl, Hennicker 92] F. Nickl, R. Hennicker: A behavioural algebraic framework for modular system design with reuse. Ludwig-Maximilians-Universität München, Technical Report 9206, 1992.

[Orejas, Nivela 90] F. Orejas, M. P. Nivela: Constraints for behavioural specifications. In: H. Ehrig, K. P. Jantke, F. Orejas, H. Reichel (eds.): Recent trends in Data Type Specifications, Proc. 7th Workshop on Specification of Abstract Data Types, Wusterhausen/Dosse, Springer Lecture Notes in Computer Science, 534, 220-245, 1990.

[Reichel 85] H. Reichel: Initial restrictions of behaviour. IFIP Working Conference, The Role of Abstract Models in Information Processing, 1985.

[Sannella, Tarlecki 87] D. T. Sannella, A. Tarlecki: On observational equivalence and algebraic specification. Journal Comp. and Sys. Sciences 34, 150-178, 1987.

[Sannella, Tarlecki 89] D. T. Sannella, A. Tarlecki: Toward formal development of ML programs: foundations and methodology. In: J. Diaz, F. Orejas (eds.): Proc. TAPSOFT '89, Springer Lecture Notes in Computer Science, 352, 375-389, 1989.

[Sannella, Wirsing 83] D. T. Sannella, M. Wirsing. A kernel language for algebraic specifications and implementation. In: M. Karpinski (ed.): Proc. FCT '83, Springer Lecture Notes in Computer Science 158, 413-427, 1983.

[Schoett 87] O. Schoett: Data abstraction and the correctness of modular programming. Ph.D. thesis, CST-42-87, Department of Computer Science, Univ. Edinburgh, 1987.

[Wirsing 86] M. Wirsing: Structured algebraic specifications: a kernel language. Theoretical Computer Science 42, 123-249, 1986.

[Wirsing 90] M. Wirsing: Algebraic specification. In: J. van Leeuwen (ed.): Handbook of Theoretical Computer Science, Amsterdam, North-Holland, 675-788, 1990.

[Wirsing 91] M. Wirsing: Structured specifications: syntax, semantics and proof calculus. In: F. L. Bauer, W. Brauer, H. Schwichtenberg (eds.): Logic and Algebra of Specification, International Summer School Marktoberdorf, 1991, NATO ASI Series F, Vol. 94, Springer, 411-442, 1993.

A Deductive Approach to Logic Programming

Gerhard Jäger

Institut für Informatik und angewandte Mathematik, Universität Bern, Switzerland

Abstract. This article presents a deductive approach to logic programming and relates three- and four-valued models of the completion of general programs to suitable rule based calculi. Special emphasis is put on questions concerning the elimination of cuts and identity axioms.

Keywords. Proof theory, logic programming, three- and four-valued logic, deductive systems, partiality in logic programming, elimination of cuts and identity axioms.

1 Introduction

In studying the foundations of logic programming it is often possible and useful to distinguish between three levels of abstraction:

I. Procedural semantics. It tries to bridge the gap between efficient implementations of logic programming and the general principles behind these implementations. Today most procedural approaches to logic programming are based on some form of SLDNF-resolution which extends the traditional SLD-resolution by the so called *negation as failure* rule[1].

II. Proof theory. It is concerned with the development and analysis of deductive systems for logic programs. A lot of attention is directed to structural properties of proofs and the *proofs as computations* paradigm. In the ideal case there is a close connection between the proof theory and the procedural aspects of logic programming in the sense that query-answering mechanisms can be interpreted as formal proofs and suitable formal proofs (e.g. cut-free and/or identity-free proofs) can be transformed into successful computations with respect to the query-answering mechanism under consideration.

III. Declarative semantics. Semantical considerations about logic programming are often guided by the attempt of constructing suitable minimal models of logic programs. In many case these efforts lead to interesting and

[1] The logical status of this rule is quite delicate and provides a treatment of negative information which is not captured by standard forms of negation like classical or intuitionistic negation.

perspicuous classes of structures, and often the close connections to the model theory of inductive definitions can be exploited in a convincing way. However, there is a high price which has to be paid. In general the logical complexity of minimal models of logic programs is very high and the corresponding semantics is noneffective.

In the following we present rule-based Tait-style[2] calculi $\mathcal{R}_\mathcal{L}(P)$ for logic programs P which are closely related to SLDNF-resolution and the traditional model-theoretic semantics in this area. Special emphasis is put on the role of the identity axioms and cuts in this deductive approach to logic programming. It is pointed out, among other things, that the subsystem of $\mathcal{R}_\mathcal{L}(P)$, which consists of the identity-free derivations, is sound and complete with respect to (i) all three-valued models and (ii) the finite parts of the least three-valued Herbrand model of the completion of P.

We will also see below that there is a natural form of ***partiality in logic programming*** which gives rise to partial completions of logic programs P and to partial calculi $\partial_\mathcal{L}(P)$. Both notions are interesting for us in connection with questions concerning structural properties of proofs in logic programming.

2 Basic notions

In the following we will deal with countable first order languages $\mathcal{L}$ with equality which consist of the following basic symbols:

1. Countably many free variables $(u, v, w, u_1, v_1, w_1, \ldots)$ and countably many bound variables $(x, y, z, x_1, y_1, z_1, \ldots)$;
2. one or more 0-ary function symbols (= constants) and an arbitrary countable number of function symbols of finite arities greater than 0;
3. the symbols $=$ for equality and $\neq$ for inequality;
4. countably many relation symbols $(R, S, T, R_1, S_1, T_1, \ldots)$ of every finite arity greater than 0;
5. the symbol $\sim$ for the formation of complementary relations;
6. the propositional constants $\top$ and $\bot$, the propositional connectives $\vee$ and $\wedge$ and the quantifiers $\exists$ and $\forall$.

As auxiliary symbols we have parentheses and commas. To simplify the notation we do not denote the equality and inequality symbols as relation symbols. Apart from the function and relation symbols the basic vocabulary of all languages, which we will consider, is the same. Each of our first order languages is thus determined by its function and relation symbols.

[2]Tait-style calculi are sometimes also denoted as one-sided sequent calculi.

The *terms* $(a, b, c, a_1, b_1, c_1, \ldots)$ of the language $\mathcal{L}$ are defined as usual. The *literals* $(L, L_1, \ldots)$ of $\mathcal{L}$ are all expressions $R(a_1, \ldots, a_n)$ and $\widetilde{R}(a_1, \ldots, a_n)$ so that R is an n-ary relation symbol of $\mathcal{L}$; the literals $R(a_1, \ldots, a_n)$ are called *positive*, and the literals $\widetilde{R}(a_1, \ldots, a_n)$ are called *negative*. The *atomic formulas* of $\mathcal{L}$ are the literals plus $\top$, $\bot$, $(a = b)$ and $(a \neq b)$. The *formulas* $(A, B, C, A_1, B_1, C_1, \ldots)$ of $\mathcal{L}$ are generated as follows:

1. If A is an atomic formula of $\mathcal{L}$, then A is an $\mathcal{L}$ formula.
2. If A and B are $\mathcal{L}$ formulas, then $(A \vee B)$ and $(A \wedge B)$ are $\mathcal{L}$ formulas.
3. If $A(u)$ is an $\mathcal{L}$ formula, then $\exists x A(x)$ and $\forall x A(x)$ are $\mathcal{L}$ formulas.

The vector notation $\vec{V}$ is used as shorthand for a finite string $V_1, \ldots, V_n$ whose length will be specified by the context. We write $A[\vec{R}, \vec{u}]$ to indicate that all relation symbols of the formula A come from the list $\vec{R}$ and all free variables of A from the list $\vec{u}$; analogously, $a[\vec{u}]$ stands for a term with no variables different from $\vec{u}$. The formula $A(\vec{R}, \vec{u})$ and the term $a(\vec{u})$ may contain other relation symbols and free variables besides $\vec{R}$ and $\vec{u}$.

So far we have no negation for $\mathcal{L}$ formulas. However, it can be easily introduced by means of the complementary relations, the law of double negation and de Morgan's laws. The negation $\neg A$ of an $\mathcal{L}$ formula A is inductively defined as follows:

1. If R is an n-ary relation symbol of $\mathcal{L}$, then we set

$$\neg R(\vec{a}) := \widetilde{R}(\vec{a}) \qquad \text{and} \qquad \neg \widetilde{R}(\vec{a}) := R(\vec{a}).$$

2. For the other formulas we have

$$\begin{array}{rclrcl}
\neg \top & := & \bot, & \neg \bot & := & \top, \\
\neg(a = b) & := & (a \neq b), & \neg(a \neq b) & := & (a = b), \\
\neg(A \vee B) & := & (\neg A \wedge \neg B), & \neg(A \wedge B) & := & (\neg A \vee \neg B), \\
\neg \exists x A(x) & := & \forall x \neg A(x), & \neg \forall x A(x) & := & \exists x \neg A(x).
\end{array}$$

Logical implication $(A \rightarrow B)$ and logical equivalence $(A \leftrightarrow B)$ are defined as $(\neg A \vee B)$ and $((A \rightarrow B) \wedge (B \rightarrow A))$, respectively. In the following we shall omit parentheses whenever the meaning is evident from the context. The complexity of formulas will often be measured in terms of their rank.

Definition 1 *The* rank $rn(A)$ *of an $\mathcal{L}$ formula A is inductively defined as follows:*

1. *If A is an equation or an inequation, then $rn(A) := 0$.*
2. *If A is a propositional constant or a literal, then $rn(A) := 1$.*

3. *If A is a formula $(B \vee C)$ or $(B \wedge C)$ so that $rn(B) = m$ and $rn(C) = n$, then* $rn(A) := \max(m, n) + 1$.

4. *If A is a formula $\exists x B(x)$ or $\forall x B(x)$ so that $rn(B(u)) = n$, then* $rn(A) := n + 1$.

Terms and formulas without free variables are called *closed*. The *equality formulas* $(E, E_1, \ldots)$ of $\mathcal{L}$ are the $\mathcal{L}$ formulas which do not contain relation symbols; the *existential formulas* of $\mathcal{L}$ are the $\mathcal{L}$ formulas which do not contain universal quantifiers. Finally, the *positive formulas* of $\mathcal{L}$ are the $\mathcal{L}$ formulas which do not contain negative literals. Observe, however, that positive formulas may contain inequations.

Following Shepherdson [27], a language $\mathcal{L}$ is called *finite* if its set of function symbols is finite, otherwise it is called *infinite*. Thus finite languages with at least one function symbol of positive arity have an infinite number of closed terms[3].

An $\mathcal{L}$ theory is a (possibly infinite) set of $\mathcal{L}$ formulas. By $\mathsf{K} \vdash A$ we express that the formula A can be deduced from the theory K by the usual axioms and rules of predicate logic. Observe, however, that we do not assume in this case that K contains equality axioms. They will be introduced in a later section and mentioned explicitly if needed.

Now we turn to some of the central notions in our context. *Program clauses* in the language $\mathcal{L}$ are all expressions of the form

$$A :- L_1, \ldots, L_n$$

with $n \geq 0$ so that $L_1, \ldots, L_n$ are (positive or negative) literals and A is a positive literal of $\mathcal{L}$. The atom A is the *head* and the sequence $L_1, \ldots, L_n$ the *body* of this clause. A program clause is a *definite* clause if all literals in its body are positive. A *logic program* in $\mathcal{L}$ is a finite set of program clauses in $\mathcal{L}$, and a *definite program* in $\mathcal{L}$ is a finite set of definite clauses in $\mathcal{L}$.

Suppose that P is an $\mathcal{L}$ program and that there are m clauses in P whose head is of the form $R(\ldots)$ so that the i-th clause is of the form

$$R(a_{i,1}[\vec{v}], \ldots, a_{i,n}[\vec{v}]) :- L_{i,1}[\vec{v}], \ldots, L_{i,k(i)}[\vec{v}]$$

and has $k(i)$ literals in its body. Then the *defining formula* of R with respect to P is defined to be the formula

$$D_R[u_1, \ldots, u_n] \;:=\; \bigvee_{i=1}^{m} \exists \vec{x} \, (\bigwedge_{j=1}^{n} (u_j = a_{i,j}[\vec{x}]) \wedge \bigwedge_{j=1}^{k(i)} L_{i,j}[\vec{x}]).$$

[3]Hence "finite" refers to the number of function symbols and not to the number of closed terms.

The special cases $m = 0$ and $k(i) = 0$ are included by interpreting empty disjunctions as $\bot$ and empty conjunctions as $\top$. The *definition* of R with respect to P is the formula

$$\forall \vec{x}(D_R[\vec{x}] \rightarrow R(\vec{x})),$$

and the *normal form* $norm_{\mathcal{L}}(P)$ of the $\mathcal{L}$ program P is the set of the definitions of the relation symbols R of $\mathcal{L}$ with respect to P. Often we identify P and $norm_{\mathcal{L}}(P)$ and write P instead of $norm_{\mathcal{L}}(P)$.

3 Two-, three- and four-valued structures for $\mathcal{L}$

Classical logic just employs two truth values **t** (true) and **f** (false), and in the classical view every sentence is either true or false. On the other hand, recent research in logic programming indicates that a third truth value **u** (undefined) has its natural place. It is used in connection with logic programs where we know that some statements are true, some statements are false and about some other statements we do not know (yet) whether they are true or false.

Kleene [17] for example suggested to use three-valued logic in order to deal with nonterminating computations. Much later Lassez and Maher [20] and Mycroft [23] applied related ideas to logic programming, and Fitting [8] and Kunen [18, 19] began to develop a theory of logic programming based on three-valued logic.

In this article we go a step further and and work with four truth values. To this end we add to **t**, **f** and **u** a new value **c**, which may be considered as standing for contradiction. This approach is technically not more complicated than the three-valued case but has some advantages in connection with partiality in logic programming (see Section 8). Relevant examples will be studied later, now we confine ourselves to introducing the basic terminology for two-valued, three-valued and four-valued structures.

We follow the presentation of Fitting [9]; similar approaches are also due to Belnap [4] and Ginsberg [10]. The set of truth values is the set $\text{FOUR} := \{0,1\} \times \{0,1\}$. If $\langle \mathsf{x}, \mathsf{y} \rangle \in \text{FOUR}$ is assigned to some statement A, then x represents the degree of evidence for A and y the degree of evidence against A. The usual truth values can be embedded into this framework by setting $\mathbf{t} := \langle 1, 0 \rangle$ and $\mathbf{f} := \langle 0, 1 \rangle$; the third truth value $\mathbf{u}$ is represented by $\langle 0, 0 \rangle$ and $\mathbf{c}$ is $\langle 1, 1 \rangle$. On FOUR two binary relations are defined by

$$\langle \mathsf{x}_1, \mathsf{y}_1 \rangle \preceq \langle \mathsf{x}_2, \mathsf{y}_2 \rangle \quad :\Longleftrightarrow \quad \mathsf{x}_1 \leq \mathsf{x}_2 \quad \text{and} \quad \mathsf{y}_2 \leq \mathsf{y}_1,$$

$$\langle \mathsf{x}_1, \mathsf{y}_1 \rangle \sqsubseteq \langle \mathsf{x}_2, \mathsf{y}_2 \rangle \quad :\Longleftrightarrow \quad \mathsf{x}_1 \leq \mathsf{x}_2 \quad \text{and} \quad \mathsf{y}_1 \leq \mathsf{y}_2$$

where the relation $\leq$ on the right hand side is the usual ordering relation of the natural numbers. The relation $\preceq$ is sometimes denoted as *truth-ordering* and the relation $\sqsubseteq$ as *information-ordering*. Both are partial orderings on FOUR.

On FOUR one defines the following operations which will be used below to interpret the logical connectives:

$$\begin{aligned}
-\langle x, y\rangle &:= \langle y, x\rangle, \\
\langle x_1, y_1\rangle + \langle x_2, y_2\rangle &:= \langle \max(x_1, x_2), \min(y_1, y_2)\rangle, \\
\langle x_1, y_1\rangle \cdot \langle x_2, y_2\rangle &:= \langle \min(x_1, x_2), \max(y_1, y_2)\rangle, \\
\textstyle\sum_{i\in I}\langle x_i, y_i\rangle &:= \langle \sup_{i\in I} x_i, \inf_{i\in I} y_i\rangle, \\
\textstyle\prod_{i\in I}\langle x_i, y_i\rangle &:= \langle \inf_{i\in I} x_i, \sup_{i\in I} y_i\rangle.
\end{aligned}$$

One has to observe that the sets $\{\mathbf{t}, \mathbf{f}\}$ and $\{\mathbf{t}, \mathbf{f}, \mathbf{u}\}$ are closed under these operations. This is not the case for the limit of elements of FOUR which is defined by taking the pointwise suprema:

$$\lim_{i\in I}\langle x_i, y_i\rangle := \langle \sup_{i\in I} x_i, \sup_{i\in I} y_i\rangle.$$

It is clear, however, that all these operations are monotonic on FOUR with respect to the relation $\sqsubseteq$. In addition all operations but the first one are also monotonic on FOUR with respect to $\preceq$.

Definition 2 *1. A* four-valued structure $\mathcal{M}$ *for* $\mathcal{L}$ *consists of a nonempty domain* $|\mathcal{M}|$ *together with assignments* $\mathcal{M}(f)$ *and* $\mathcal{M}(R)$ *to all function symbols* f *and relation symbols* R *of* $\mathcal{L}$ *so that*

(a) $\mathcal{M}(f)$ *is an n-ary function from* $|\mathcal{M}|$ *to* $|\mathcal{M}|$ *if* f *is n-ary,*

(b) $\mathcal{M}(R)$ *is an n-ary function from* $|\mathcal{M}|$ *to* $\{\mathbf{t}, \mathbf{f}, \mathbf{u}, \mathbf{c}\}$ *if* R *is n-ary.*

2. A three-valued structure *for* $\mathcal{L}$ *is a four-valued structure for* $\mathcal{L}$ *so that the functions* $\mathcal{M}(R)$ *do not take the value* $\mathbf{c}$ *for all relation symbols* R *of* $\mathcal{L}$.

3. A two-valued structure *for* $\mathcal{L}$ *is a four-valued structure for* $\mathcal{L}$ *so that the functions* $\mathcal{M}(R)$ *do not take the values* $\mathbf{c}$ *and* $\mathbf{u}$ *for all relation symbols* R *of* $\mathcal{L}$.

If $\mathcal{M}$ is a four-valued $\mathcal{L}$ structure, one introduces the language $\mathcal{L}[\mathcal{M}]$ by adding to $\mathcal{L}$ new constants $\bar{m}$ for all $m \in |\mathcal{M}|$. Yet in order to simplify notation we often write $A[m_1, \ldots, m_n]$ instead of $A[\bar{m}_1, \ldots, \bar{m}_n]$. The value of each closed expression of $\mathcal{L}[\mathcal{M}]$ is now inductively defined as follows.

Definition 3 *Let* $\mathcal{M}$ *be a four-valued* $\mathcal{L}$ *structure. We assign to each closed term* a *and closed formula* A *of* $\mathcal{L}[\mathcal{M}]$ *a value* $\mathcal{M}(a) \in |\mathcal{M}|$ *and a value* $\mathcal{M}(A) \in \{\mathbf{t}, \mathbf{f}, \mathbf{u}, \mathbf{c}\}$.

1. If a *is the term* $\bar{m}$ *for some element* m *of* $|\mathcal{M}|$, *then* $\mathcal{M}(a) := m$.

2. *If a is the term $f(a_1, \ldots, a_n)$ for some n-ary function symbol f and terms $a_1, \ldots, a_n$, then $\mathcal{M}(a) := \mathcal{M}(f)(\mathcal{M}(a_1), \ldots \mathcal{M}(a_n))$.*

3. *If A is the formula $\top$, then $\mathcal{M}(A) :=$ **t**; if A is the formula $(a = b)$, then*
$$\mathcal{M}(A) := \begin{cases} \mathbf{t}, & \text{if } \mathcal{M}(a) = \mathcal{M}(b), \\ \mathbf{f}, & \text{if } \mathcal{M}(a) \neq \mathcal{M}(b). \end{cases}$$

4. *If A is the formula $R(a_1, \ldots, a_n)$ for some n-ary relation symbol R and terms $a_1, \ldots, a_n$, then*
$$\mathcal{M}(A) \ := \ \mathcal{M}(R)(\mathcal{M}(a_1), \ldots, \mathcal{M}(a_n)).$$

5. *If A is the formula $\neg B$ and B the formula $\top$, an equation or a positive literal, then*
$$\mathcal{M}(A) \ := \ -\mathcal{M}(B).$$

6. *If A is a formula $(B \vee C)$, then*
$$\mathcal{M}(A) \ := \mathcal{M}(B) + \mathcal{M}(C).$$

7. *If A is a formula $(B \wedge C)$, then*
$$\mathcal{M}(A) \ := \mathcal{M}(B) \cdot \mathcal{M}(C).$$

8. *If A is a formula $\exists x B(x)$, then*
$$\mathcal{M}(A) \ := \sum_{m \in |\mathcal{M}|} \mathcal{M}(B(m)).$$

9. *If A is a formula $\forall x B(x)$, then*
$$\mathcal{M}(A) \ := \prod_{m \in |\mathcal{M}|} \mathcal{M}(B(m)).$$

Obviously one has $\mathcal{M}(A) \in \{\mathbf{t}, \mathbf{f}, \mathbf{u}\}$ for all three-valued $\mathcal{M}$ and $\mathcal{M}(A) \in \{\mathbf{t}, \mathbf{f}\}$ for all two-valued $\mathcal{M}$. Hence these three-valued and four-valued structures are natural generalizations of the two-valued case. Observe that equality is always handled as the usual two-valued identity. If A is an equality formula of $\mathcal{L}[\mathcal{M}]$, then $\mathcal{M}(A) = \mathbf{t}$ or $\mathcal{M}(A) = \mathbf{f}$, also for three-valued and four-valued structures $\mathcal{M}$.

There are natural notions of extension on the four-valued $\mathcal{L}$ structures which are obtained by lifting the above defined relations on FOUR pointwise to the four-valued $\mathcal{L}$ structures.

Definition 4 *Let $\mathcal{M}$ and $\mathcal{N}$ be four-valued structures for $\mathcal{L}$ which have the same universe and the same interpretations of the function symbols.*

1. *$\mathcal{N}$ is called a* truth-extension *of $\mathcal{M}$ if $\mathcal{M}(R(\vec{m})) \preceq \mathcal{N}(R(\vec{m}))$ for all relation symbols R of $\mathcal{L}$ and $\vec{m} \in |\mathcal{M}|$. In this case we write $\mathcal{M} \preceq \mathcal{N}$.*

2. *$\mathcal{N}$ is said to* extend *$\mathcal{M}$ if we have $\mathcal{M}(R(\vec{m})) \sqsubseteq \mathcal{N}(R(\vec{m}))$ for all relation symbols R of $\mathcal{L}$ and $\vec{m} \in |\mathcal{M}|$. In this case we write $\mathcal{M} \sqsubseteq \mathcal{N}$.*

The relations $\preceq$ and $\sqsubseteq$ are partial orderings on the four-valued structures for $\mathcal{L}$. In addition it is easy to see that the positive $\mathcal{L}$ formulas are monotonic with respect to $\preceq$ and all $\mathcal{L}$ formulas are monotonic with respect to $\sqsubseteq$ in the sense of the following remark.

Remark 5 *Let $\mathcal{M}$ and $\mathcal{N}$ be four-valued $\mathcal{L}$ structures. Then we have for all positive $\mathcal{L}$ formulas A and arbitrary $\mathcal{L}$ formulas B:*

1. $\mathcal{M} \preceq \mathcal{N} \quad \Longrightarrow \quad \mathcal{M}(A) \preceq \mathcal{N}(A)$,
2. $\mathcal{M} \sqsubseteq \mathcal{N} \quad \Longrightarrow \quad \mathcal{M}(B) \sqsubseteq \mathcal{N}(B)$.

As a special case this means that as soon as an $\mathcal{L}[\mathcal{M}]$ formula (a positive $\mathcal{L}[\mathcal{M}]$ formula) has obtained a value **t** or **f** (the value **t**) in a three-valued structure $\mathcal{M}$, it will keep this value in all three-valued extensions (truth-extensions) of $\mathcal{M}$.

Definition 6 *Let $\mathcal{M}$ be a four-valued $\mathcal{L}$ structure and P a logic program in $\mathcal{L}$. Then the P-*envelope *$\mathcal{M}[P]$* of *$\mathcal{M}$ is the structure which is defined as follows:*

1. *$|\mathcal{M}[P]| := |\mathcal{M}|$, and $\mathcal{M}[P](f) := \mathcal{M}(f)$ for all function symbols f of $\mathcal{L}$.*

2. *If R is a relation symbol of $\mathcal{L}$ and $D_R[\vec{u}]$ its defining formula with respect to P, then we set $\mathcal{M}[P](R)(\vec{m}) := \mathcal{M}(D_R[\vec{m}])$ for all $\vec{m} \in |\mathcal{M}|$.*

It follows from this definition that the P-envelope of a two-valued structure is two-valued and that of a three-valued structure three-valued. In general it is not the case that $\mathcal{M}[P]$ is a truth-extension or an extension of $\mathcal{M}$, but the formation of envelopes satisfies the following monotonicity conditions with respect to truth-extensions and extensions.

Remark 7 *Let $\mathcal{M}$ and $\mathcal{N}$ be four-valued $\mathcal{L}$ structures and assume that P is a definite program in $\mathcal{L}$ and Q an arbitrary logic program in $\mathcal{L}$. Then we have:*

1. $\mathcal{M} \preceq \mathcal{N} \quad \Longrightarrow \quad \mathcal{M}[P] \preceq \mathcal{N}[P]$,
2. $\mathcal{M} \sqsubseteq \mathcal{N} \quad \Longrightarrow \quad \mathcal{M}[Q] \sqsubseteq \mathcal{N}[Q]$.

Now we restrict ourselves to two-valued structures and repeat some standard definitions. To this end let $A[\vec{u}]$ be an $\mathcal{L}$ formula, K an $\mathcal{L}$ theory and $\mathcal{M}$ be a two-valued structure for $\mathcal{L}$. Then we have:

1. $A[\vec{u}]$ is *valid* in $\mathcal{M}$ if $\mathcal{M}(A[\vec{m}]) = \mathbf{t}$ for all $\vec{m} \in |\mathcal{M}|$. Then we call $\mathcal{M}$ a *model* of $A[\vec{u}]$ and write $\mathcal{M} \models A[\vec{u}]$.

2. K is *valid* in $\mathcal{M}$ if all elements of K are valid in $\mathcal{M}$. Then we call $\mathcal{M}$ a *model* of K and write $\mathcal{M} \models \mathsf{K}$.

3. $A[\vec{u}]$ is a *logical consequence* of K if $A[\vec{u}]$ is valid in all models of K. Then we write $\mathsf{K} \models A[\vec{u}]$.

Remark 8 *If P is a general logic program in $\mathcal{L}$, then $norm_{\mathcal{L}}(P)$ has a model. Actually, if $\mathcal{M}$ is a two-valued $\mathcal{L}$ structure so that $\mathcal{M}(R)$ is identically* t *for all relation symbols R of $\mathcal{L}$, then $\mathcal{M}$ is a model of $norm_{\mathcal{L}}(P)$.*

The *Herbrand universe* $\mathsf{U}_{\mathcal{L}}$ of $\mathcal{L}$ denotes the collection of all closed terms of $\mathcal{L}$, and by our assumptions on $\mathcal{L}$ we know that $\mathsf{U}_{\mathcal{L}}$ contains at least one element. The *Herbrand base* $\mathsf{B}_{\mathcal{L}}$ of $\mathcal{L}$ is the collection of all closed positive literals of $\mathcal{L}$.

Let k be one of the numbers 2, 3 or 4. The k-valued *Herbrand structures* for $\mathcal{L}$ are the k-valued $\mathcal{L}$ structures so that the domain of these structures is the set $\mathsf{U}_{\mathcal{L}}$ and the function symbols have their obvious interpretations over $\mathsf{U}_{\mathcal{L}}$. Hence every k-valued Herbrand structure $\mathcal{M}$ is characterized by the interpretation of its relation symbols. Two-valued Herbrand structures are often simply denoted as Herbrand structures.

By $\mathsf{U}^2_{\mathcal{L}}$ we denote the two-valued Herbrand structure over the universe $\mathsf{U}_{\mathcal{L}}$ which interprets each relation symbol as identically $\mathbf{f}$. $\mathsf{U}^3_{\mathcal{L}}$ is the three-valued Herbrand structure over $\mathsf{U}_{\mathcal{L}}$ with each relation symbol interpreted as identically $\mathbf{u}$. Thus $\mathsf{U}^2_{\mathcal{L}}$ is the least element of the four-valued Herbrand structures for $\mathcal{L}$ with respect to the $\preceq$ relation, and $\mathsf{U}^3_{\mathcal{L}}$ is the least element with respect to the $\sqsubseteq$ relation.

4 Clark's equational theory

Unification plays a major role in practically all implementations of logic programming environments. In general the most simple form of unification[4] is employed which treats two closed expressions as equal if and only if they are syntactically identical. The corresponding unification theorem goes back to Robinson [24] and states the existence of an algorithm which for any two expressions produces an idempotent most general unifier if they are unifiable and otherwise reports the nonexistence of a unifier.

[4]There are also more subtle versions of unification like *AC*-unification which make use of additional axioms.

Space does not permit to go into details, and only the basic terminology can be repeated. An $\mathcal{L}$ *substitution* θ is a finite set $\{u_1/a_1, \ldots, u_n/a_n\}$ of bindings so that the $\mathcal{L}$ terms a_i are different from the variables u_i for $1 \leq i \leq n$ and u_i is different from u_j for $1 \leq i < j \leq n$. Then the *instance* $\mathcal{Z}\theta$ of an expression $\mathcal{Z}$ is the expression obtained from $\mathcal{Z}$ by simultaneously replacing each occurrence of the variable u_i in $\mathcal{Z}$ by the term a_i $(i = 1, \ldots, n)$. We shall use $\theta, \sigma, \tau, \theta_1, \sigma_1, \tau_1, \ldots$ for substitutions; the empty substitution is denoted by ε.

A *unifier* of two $\mathcal{L}$ expressions $\mathcal{Z}_1$ and $\mathcal{Z}_2$ is an $\mathcal{L}$ substitution θ with the property that $\mathcal{Z}_1\theta$ and $\mathcal{Z}_2\theta$ are identical. This unifier is *most general* if for any other unifier σ of $\mathcal{Z}_1$ and $\mathcal{Z}_2$ there exists a substitution τ so that σ is the composition of θ and τ, i.e. $\sigma = \theta\tau$. For further unexplained notions we refer to Apt [1] and Lloyd [21].

Clark's equational theory $CET_{\mathcal{L}}$ (cf. [6]) may be understood as the axiomatic counterpart of this form of unification. It consists of a series of *equality axioms* asserting that equality is an equivalence relation and that equals may be substituted for equals within any function together with several *freeness axioms* which make sure that no objects are equal unless required by the logic.

The theory $CET_{\mathcal{L}}$ depends on the language $\mathcal{L}$ and consists of the following equality-freeness axioms (1)–(7) for all terms a and n-ary function symbols f and g of $\mathcal{L}$:

(1) $u = u$,

(2) $u = v \rightarrow v = u$,

(3) $(u = v \wedge v = w) \rightarrow u = w$,

(4) $(u_1 = v_1 \wedge \ldots \wedge u_n = v_n) \rightarrow f(u_1, \ldots, u_n) = f(v_1, \ldots, v_n)$,

(5) $f(u_1, \ldots, u_n) = f(v_1, \ldots, v_n) \rightarrow u_i = v_i$ for all $1 \leq i \leq n$,

(6) $f(u_1, \ldots, u_n) \neq g(v_1, \ldots, v_m)$, if f and g are different function symbols of $\mathcal{L}$,

(7) $a \neq u$ if a contains the variable u and is different from u.

$\mathsf{U}^2_{\mathcal{L}}$ is a model of $CET_{\mathcal{L}}$ which is sometimes called the *standard model* of $CET_{\mathcal{L}}$. This is justified by the obvious fact that every model of $CET_{\mathcal{L}}$ contains an isomorphic copy of the standard model. $CET_{\mathcal{L}}$ does not contain equality assertions for relation symbols. Hence, for example, a formula of the form $(a = b \wedge R(a) \rightarrow R(b))$ for some unary relation symbol R is not derivable from $CET_{\mathcal{L}}$.

The relationship between the theory $CET_{\mathcal{L}}$ and the unification problem of first order logic is reflected by the following lemma. Its proof is presented in Clark [6] and Kunen [18] and can be easily verified.

Lemma 9 *Let $a[\vec{u}]$ and $b[\vec{u}]$ be any terms of $\mathcal{L}$. If $a[\vec{u}]$ and $b[\vec{u}]$ are unifiable, then $CET_{\mathcal{L}}$ proves $\exists\vec{x}(a[\vec{x}] = b[\vec{x}])$. If they are not unifiable, then $CET_{\mathcal{L}}$ proves $\forall\vec{x}(a[\vec{x}] \neq b[\vec{x}])$.*

Deeper results about $CET_{\mathcal{L}}$ concern its completeness. A first result states that $CET_{\mathcal{L}}$ is complete if $\mathcal{L}$ is an infinite language. Observe, however, that $CET_{\mathcal{L}}$ is in general not complete for finite languages. Let $\mathcal{L}$, for example, be a language with a constant a and no other function symbols. Then $\mathsf{U}^2_{\mathcal{L}}$ is a model of $\forall x(x = a)$ but $CET_{\mathcal{L}}$ does not prove this equality formula. In order to obtain completeness also for finite languages $\mathcal{L}$ one has to strengthen $CET_{\mathcal{L}}$ by the so called *domain closure axiom* $DCA_{\mathcal{L}}$,

$$DCA_{\mathcal{L}} := \forall x \bigvee_{f\in\mathcal{L}} \exists\vec{y}(x = f(\vec{y})),$$

which says that every element of the universe belongs to the range of some function symbol of $\mathcal{L}$. Then the following theorem follows for example from Shepherdson [27].

Theorem 10 *Let E be a closed equality formula of the language $\mathcal{L}$. Then we have the following equivalence*

$$CET_{\mathcal{L}} \vdash E \iff \mathsf{U}^2_{\mathcal{L}} \models E,$$

provided that $\mathcal{L}$ is infinite. On the other hand, if $\mathcal{L}$ is finite, then one can only show that

$$CET_{\mathcal{L}} + DCA_{\mathcal{L}} \vdash E \iff \mathsf{U}^2_{\mathcal{L}} \models E.$$

This theorem implies for all closed equality formulas E of $\mathcal{L}$ that one has $CET_{\mathcal{L}} \nvdash E$ if and only if $CET_{\mathcal{L}} \vdash \neg E$, provided that $\mathcal{L}$ is infinite. In addition it also implies that $CET_{\mathcal{L}} + DCA_{\mathcal{L}} \nvdash E$ if and only if $CET_{\mathcal{L}} + DCA_{\mathcal{L}} \vdash \neg E$, provided that $\mathcal{L}$ is finite. Hence the decidability of $CET_{\mathcal{L}}$ for infinite $\mathcal{L}$ and $CET_{\mathcal{L}} + DCA_{\mathcal{L}}$ for finite $\mathcal{L}$ follows immediately.

The distinction in Theorem 10 between finite and infinite languages is necessary because of the treatment of universal quantifiers. The following completeness result for closed existential equality formulas does not depend on the language and is proved by straightforward induction on the rank of E.

Theorem 11 *Let E be a closed existential equality formula of the language $\mathcal{L}$. Then we have:*

$$\mathsf{U}^2_{\mathcal{L}} \models E \implies CET_{\mathcal{L}} \vdash E.$$

5 Definite logic programs

In this section we consider definite logic programs and study them from a semantical and syntactical point of view. We expect that the reader has some basic knowledge about elementary inductive definitions and is familiar with SLD-resolution. Good references concerning these topics are for example Hinman [13], Lloyd [21] and Moschovakis [22].

5.1 Least models semantics for definite programs

The model theory of definite programs is straightforward and characterized by the fact that every definite program has a least Herbrand model. Often this distinguished model is called the *meaning* of the program. In the following we deal with two-valued models only; three- and four-valued structures will play an important role later in the discussion of general logic programs.

The following lemma is a key observation for constructing models of logic programs and characterizes models of logic programs P as structures which are truth extensions of their P-envelopes. Its proof is obvious from the definitions in Section 3 and will be omitted.

Lemma 12 *Let P be a logic program in $\mathcal{L}$ and $\mathcal{M}$ a model of $CET_{\mathcal{L}}$. Then we have:*

$$\mathcal{M} \models P \iff \mathcal{M}[P] \preceq \mathcal{M}.$$

For the next definition we need unions of Herbrand structures which are introduced as follows: Let $(\mathcal{M}_i : i \in I)$ be a family of two-valued Herbrand structures for $\mathcal{L}$. Then the *union* $\bigcup_{i \in I} \mathcal{M}_i$ over this family is the two-valued Herbrand structure $\mathcal{N}$ for $\mathcal{L}$ so that

$$\mathcal{N}(R)(\vec{a}) := \sum_{i \in I} \mathcal{M}_i(R)(\vec{a})$$

for all relation symbols R of $\mathcal{L}$ and $\vec{a} \in \mathsf{U}_{\mathcal{L}}$. Hence the positive literal $R(\vec{a})$ is true in $\bigcup_{i \in I} \mathcal{M}_i$ if it is true in some $\mathcal{M}_i$ and false otherwise. The union over an empty index set is the Herbrand structure $\mathsf{U}^2_{\mathcal{L}}$.

The least Herbrand model of a definite logic program can now be easily obtained in terms of the following hierarchy of Herbrand structures. Starting off from $\mathsf{U}^2_{\mathcal{L}}$ the formation of envelopes will be iterated through all ordinals.

Definition 13 *Let P be a definite logic program in $\mathcal{L}$. Then we define by recursion on the ordinals the following (two-valued) Herbrand structures $\mathcal{H}_P^{<\alpha}$ and $\mathcal{H}_P^{\alpha}$ for $\mathcal{L}$:*

$$\mathcal{H}_P^{<\alpha} := \bigcup_{\xi < \alpha} \mathcal{H}_P^{\xi}, \qquad \mathcal{H}_P^{\alpha} := \mathcal{H}_P^{<\alpha}[P], \qquad \mathcal{H}_P := \bigcup_{\alpha \in On} \mathcal{H}_P^{\alpha}.$$

This definition of the Herbrand structures $\mathcal{H}_P^{<\alpha}$ and $\mathcal{H}_P^{\alpha}$ fits perfectly well into the theory of elementary inductive definitions as it is developed in Moschovakis [22]. First one recalls that every logic program P provides for each relation symbol R of $\mathcal{L}$ a defining formula $D_R[\vec{u}]$. If P is definite, then these formulas are positive. Hence we may regard a definite program P as a system of inductive definitions of the relations R over $\mathsf{U}_\mathcal{L}$ in the sense of Moschovakis' *Simultaneous Induction Lemma*.

This implies in particular that such systems of inductive definitions possess least fixed points which form the least Herbrand models of definite logic programs. More precisely, it follows from the general theory or Remark 7 that

$$\alpha \leq \beta \implies \mathcal{H}_P^{<\alpha} \preceq \mathcal{H}_P^{<\beta}$$

for all ordinals α, β and definite programs P. In addition, there is an ordinal α so that $\mathcal{H}_P^{<\alpha} = \mathcal{H}_P^{\alpha} = \mathcal{H}_P$. We can even say more. The definitions with respect to definite programs are positive existential, and therefore the structure $\mathcal{H}_P$ is determined by the union of its finite stages.

Lemma 14 *If P is a definite logic program in $\mathcal{L}$, then we have $\mathcal{H}_P = \mathcal{H}_P^{<\omega}$.*

Lemma 12 and Lemma 14 are applied to show that $\mathcal{H}_P$ is the least Herbrand model of the definite program P. From $\mathcal{H}_P = \mathcal{H}_P^{<\omega}$ we conclude that $\mathcal{H}_P[P] = \mathcal{H}_P^{<\omega}[P] = \mathcal{H}_P^{\omega} \preceq \mathcal{H}_P$ and obtain by Lemma 12 that $\mathcal{H}_P$ is a model of P. If $\mathcal{M}$ is a Herbrand structure so that $\mathcal{M}[P] \preceq \mathcal{M}$, then a straightforward induction on $\alpha \in On$ shows $\mathcal{H}_P^{<\alpha} \preceq \mathcal{M}$, thus $\mathcal{H}_P \preceq \mathcal{M}$. This yields, again by Lemma 12, that all Herbrand models of P are truth-extensions of $\mathcal{H}_P$. Hence we can summarize.

Theorem 15 (Least Herbrand models) *Let P be a definite logic program in $\mathcal{L}$. Then $\mathcal{H}_P$ is the least Herbrand model of P; i.e. $\mathcal{H}_P$ is Herbrand model of P and $\mathcal{H}_P \preceq \mathcal{M}$ for all Herbrand models $\mathcal{M}$ of P. In addition, $\mathcal{H}_P = \mathcal{H}_P^{<\omega}$.*

Apt and van Emden [3] associate a monotonic operator T_P with any definite logic program P which maps subsets of $\mathsf{B}_\mathcal{L}$ to subsets of $\mathsf{B}_\mathcal{L}$. Their sets $\mathsf{T}_P\uparrow\alpha$ contain exactly the closed positive literals which are true in our $\mathcal{H}_P^{\alpha}$, and the least fixed point of T_P corresponds to $\mathcal{H}_P$.

5.2 Rule-based calculi for definite programs

It is easy to assign to every definite logic program P in $\mathcal{L}$ a natural rule-based system $\mathcal{D}_\mathcal{L}(P)$ which is sound and complete with respect to $\mathcal{H}_P$. In $\mathcal{D}_\mathcal{L}(P)$ provability is restricted to positive literals; therefore we shall later extend this system to a calculus $\mathcal{R}_\mathcal{L}(P)$ which deals with arbitrary $\mathcal{L}$ formulas and is suited for the treatment of arbitrary logic programs.

The system $\mathcal{D}_\mathcal{L}(P)$ is a very simple calculus without any logical rules. Instead it has for every definite clause $C :- B_1, \ldots, B_m$ of P and every $\mathcal{L}$ substitution θ a derivation rule

$$\frac{B_1\theta \quad \ldots \quad B_m\theta}{C\theta}$$

with the m premises $B_1\theta, \ldots, B_m\theta$ and the conclusion $C\theta$. Therefore program clauses C :– with empty body play the roles of axioms in the sense that then all $C\theta$ are derivable.

Definition 16 *Let P be a definite logic program and A a positive literal in $\mathcal{L}$. Then $\mathcal{D}_{\mathcal{L}}(P) \vdash^n A$ is defined by induction on $n < \omega$ as follows. Assume that*

1. *C :– $B_1, \ldots, B_m$ is an element of P and θ an $\mathcal{L}$ substitution so that A is of the form $C\theta$;*
2. *for every $i = 1, \ldots, m$ there exists an $n_i < n$ so that $\mathcal{D}_{\mathcal{L}}(P) \vdash^{n_i} B_i\theta$.*

Then we have $\mathcal{D}_{\mathcal{L}}(P) \vdash^n A$.

We write $\mathcal{D}_{\mathcal{L}}(P) \vdash A$ if there exists an $n < \omega$ so that $\mathcal{D}_{\mathcal{L}}(P) \vdash^n A$ and $\mathcal{D}_{\mathcal{L}}(P) \vdash^{<n} A$ if there exists an $m < n$ so that $\mathcal{D}_{\mathcal{L}}(P) \vdash^m A$.

Although derivability by means of SLD-resolution and provability in $\mathcal{D}_{\mathcal{L}}(P)$ are not the same[5], the calculus $\mathcal{D}_{\mathcal{L}}(P)$ captures many central aspects of SLD-resolution. It is left to the reader to study the exact relationship between both approaches.

$\mathcal{D}_{\mathcal{L}}(P)$ is an extremely simple calculus but powerful enough to provide a syntactic counterpart of the Herbrand structures $\mathcal{H}_P^n$. The following theorem describes their exact relationship.

Theorem 17 *Let P be a definite logic program in $\mathcal{L}$ and A a closed positive literal of $\mathcal{L}$. Then we have for all $n < \omega$:*

$$\mathcal{D}_{\mathcal{L}}(P) \vdash^n A \iff \mathcal{H}_P^n \models A.$$

Proof. Trivial induction on n shows that $\mathcal{D}_{\mathcal{L}}(P) \vdash^n A$ implies $\mathcal{H}_P^n \models A$. Also the converse direction is proved by complete induction. We know that A is a formula $R(\vec{a})$ for some relation symbol R and $\vec{a} \in \mathsf{U}_{\mathcal{L}}$. Then we have a certain number of clauses, say p, whose head is of the form $R(\ldots)$. The i-th such clause can be written as

$$R(b_{i,1}[\vec{v}], \ldots, b_{i,m}[\vec{v}]) :- B_{i,1}[\vec{v}], \ldots, B_{i,k(i)}[\vec{v}]\,, \tag{1}$$

and the defining formula $D_R[\vec{u}]$ of R with respect to P is the formula

$$\bigvee_{i=1}^{p} \exists \vec{x} \, (\bigwedge_{j=1}^{m} (u_j = b_{i,j}[\vec{x}]) \land \bigwedge_{j=1}^{k(i)} B_{i,j}[\vec{x}]).$$

[5]One obvious distinction refers to the use of substitutions since SLD-resolution enforces most general unifiers.

Now assume that $\mathcal{H}_P^n \models A$. Then the definition of $\mathcal{H}_P^n$ implies $\mathcal{H}_P^{\leq n} \models D_R[\vec{a}]$, i.e.

$$\mathcal{H}_P^{\leq n} \models a_1 = b_{i,1}[\vec{c}] \wedge \ldots \wedge a_m = b_{i,m}[\vec{c}], \tag{2}$$

$$\mathcal{H}_P^{<n} \models B_{i,1}[\vec{c}] \wedge \ldots \wedge B_{i,k(i)}[\vec{c}] \tag{3}$$

for some $i = 1, \ldots, p$ and suitable terms $\vec{c} \in \mathsf{U}_{\mathcal{L}}$. Hence the induction hypothesis applied to (3) yields $\mathcal{D}_{\mathcal{L}}(P) \vdash^{<n} B_{i,j}[\vec{c}]$ for all $1 \leq j \leq k(i)$. Finally, let θ be the substitution which binds the variables $\vec{v}$ to the closed terms $\vec{c}$. Then the assertion follows in view of (1) and (2) by an application of the rule which corresponds to the i-th clause and the substitution θ. □

Corollary 18 (Soundness and completeness) *We have for all definite logic programs P in $\mathcal{L}$ and all closed positive literals A of $\mathcal{L}$:*

$$\mathcal{D}_{\mathcal{L}}(P) \vdash A \iff \mathcal{H}_P \models A.$$

6 General logic programs

As we have seen in the previous section, there is a perfect correspondence between the model theory and proof theory of definite logic programs. Even better, for a definite program provability coincides with validity in its standard model. This is mainly due to the fact that we can restrict all arguments to positive literals and positive inductive definability by existential definition clauses.

This situation changes dramatically if we study general logic programs. Then we have also to deal with negative literals in the bodies of program clauses and must struggle with several problems which are created thereby. Obviously the theory of positive or monotonic inductive definitions can no longer be applied directly, and also the calculus $\mathcal{D}_{\mathcal{L}}(P)$ has no simple extension which copes with general programs.

6.1 Negative information in logic programs

The main problems in the procedural and declarative approach to general logic programs arise by the presence of negative literals in the bodies of program clauses. They make it necessary to deal with negative literals and reflect the fact that positive information may depend on negative information.

Given a logic program P in $\mathcal{L}$, we have $norm_{\mathcal{L}}(P) \not\vdash L$ for all negative literals L. Therefore it is not possible to deduce negative information from (the normal form of) a logic program by purely logical means. As a consequence, it is not clear how to apply program clauses with negative literals in their body.

There exist many approaches to the treatment of negative information in the context of logic programming, and it is even impossible to list the most relevant publications. We only mention the survey articles by Apt and Bol [2] and Shepherdson [28], which discuss several aspects of negation in logic programming and contain good bibliographies.

One of the most widely accepted declarative semantics for negative information in logic programming is based on the *completion* of logic programs introduced by Clark [6]. Clark surmounts the inability to deduce negative information from a logic program P by passing from P to the following axiomatic extension[6] of P.

Let P be a general logic program in $\mathcal{L}$ and assume that the defining formula of each relation symbol R of $\mathcal{L}$ with respect to P is the formula $D_R[\vec{u}]$. Then we call

$$\forall\vec{x}(D_R[\vec{x}] \leftrightarrow R(\vec{x}))$$

the *completed definition* of R with respect to P. The *completion* $comp_{\mathcal{L}}(P)$ of P is the $\mathcal{L}$ theory which consists of the theory $CET_{\mathcal{L}}$ plus the completed definitions of all relation symbols R of $\mathcal{L}$.

Although not stated explicitly as axioms, the completion $comp_{\mathcal{L}}(P)$ of a logic program P proves the usual equality assertions

$$a_1 = b_1 \wedge \ldots a_n = b_n \wedge R(a_1, \ldots, a_n) \rightarrow R(b_1, \ldots, b_n)$$

for the relation symbols R of $\mathcal{L}$ by means of the axioms of $CET_{\mathcal{L}}$ and the completed definitions of the relation symbols. In general these equality formulas cannot be proved in $CET_{\mathcal{L}} + norm_{\mathcal{L}}(P)$.

Let $\mathcal{L}$ be a language which contains two constants 0 and c and a unary function symbol f. Then we can talk about the natural numbers by representing any $n < \omega$ by the closed term $f^n(0)$ and considering the (definite) program of the following two clauses:

$$N(0) :- \,,$$
$$N(f(u)) :- N(u)\,.$$

The completed definition of the relation symbol N is (provably equivalent to) the formula

$$\forall x[N(x) \leftrightarrow (x = 0 \vee \exists y(x = f(y) \wedge N(y)))].$$

Hence $comp_{\mathcal{L}}(P)$ proves $\neg N(c)$, and we have an example of derivable negative information.

Remark 19 *If P is a definite program in $\mathcal{L}$, then $\mathcal{H}_P$ is not only a model of P but even a model of $comp_{\mathcal{L}}(P)$.*

[6]Various forms of nonmonotonic reasoning by means of axiomatic extensions are discussed for example in Jäger [15].

This follows easily from Lemma 14. Observe, however, that the completion of a general logic program may be inconsistent. To see this, consider the following example.

Example 20 *Let P be the program which consists of the clause $R(0){:}{-}\neg R(0)$ only. Then $comp_{\mathcal{L}}(P)$ proves that $R(0)$ and $\neg R(0)$ are equivalent. Hence $comp_{\mathcal{L}}(P)$ is inconsistent.*

This example also shows that the completions of logic programs in a two-valued context do not provide a meaningful framework for logic programming. Since the completion of the previous program P is inconsistent, all formulas are logical consequences of $comp_{\mathcal{L}}(P)$. On the other hand, no meaningful procedural approach to logic programming would deduce positive literals like $R(0)$ from that P.

6.2 Three-valued and four-valued models of the completions of general logic programs

Problems of this kind can be overcome by turning to three-valued and four-valued semantics and by considering three-valued and four-valued completions of general logic programs.

Definition 21 *Let k be one of the numbers 2, 3 or 4 and assume that $\mathcal{M}$ is a k-valued $\mathcal{L}$ structure, P a general logic program in $\mathcal{L}$ and $A[\vec{u}]$ an $\mathcal{L}$ formula.*

1. *$\mathcal{M}$ is a k-*valued model *of $CET_{\mathcal{L}}$ if $\mathcal{M}(E[\vec{m}]) = \mathbf{t}$ for all $\vec{m} \in |\mathcal{M}|$ and all axioms $E[\vec{u}]$ of $CET_{\mathcal{L}}$. Then we write $\mathcal{M} \models_k CET_{\mathcal{L}}$.*

2. *$\mathcal{M}$ is a k-*valued model *of the completion of P if the following two conditions are satisfied:*

   (a) $\mathcal{M} \models_k CET_{\mathcal{L}}$.

   (b) $\mathcal{M}(D_R[\vec{m}]) \sqsubseteq \mathcal{M}(R(\vec{m}))$ for all $\vec{m} \in |\mathcal{M}|$, all relation symbols R of $\mathcal{L}$ and their defining formulas $D_R[\vec{u}]$ with respect to P.

   *Then we write $\mathcal{M} \models_k comp_{\mathcal{L}}(P)$. $\mathcal{M}$ is called a k-*valued Herbrand model *of the completion of P if it is a k-valued Herbrand structure so that $\mathcal{M} \models_k comp_{\mathcal{L}}(P)$.*

3. *$A[\vec{u}]$ is a k-*valued consequence *of the completion of P if we have $\mathcal{N}(A[\vec{m}]) \in \{\mathbf{t}, \mathbf{c}\}$ for all k-valued models $\mathcal{N}$ of $comp_{\mathcal{L}}(P)$ and all $\vec{m} \in |\mathcal{N}|$. Then we write $comp_{\mathcal{L}}(P) \models_k A[\vec{u}]$.*

This definition implies that $\mathcal{M} \models_2 comp_{\mathcal{L}}(P)$ if and only if $\mathcal{M} \models comp_{\mathcal{L}}(P)$ and $comp_{\mathcal{L}}(P) \models_2 A[\vec{u}]$ if and only if $comp_{\mathcal{L}}(P) \models A[\vec{u}]$. Three-valued and four-valued models of arbitrary theories have not been defined and will not be needed for what follows.

The structure $\mathsf{U}^3_{\mathcal{L}}$ is a three-valued model of the completion of the program P from Example 20, and $\mathsf{U}^3_{\mathcal{L}}(R(0)) = \mathsf{U}^3_{\mathcal{L}}(\neg R(0)) = \mathbf{u}$ in this structure. This implies that neither $R(0)$ nor $\neg R(0)$ are three-valued consequences of $comp_{\mathcal{L}}(P)$.

Now we show that the completions of all logic programs posses three-valued Herbrand models. In the generation of these standard models we try to follow the considerations of Section 5.1 as closely as possible. The additional difficulties caused by the negative literals in the bodys of general programs will be evaded by working in a three-valued rather than a two-valued environment. This strategy, which has also been pursued e.g. in Fitting [8] and Kunen [18], makes it possible to apply the theory of positive inductive definitions also in this more elaborate case.

Again we start from a general lemma about models of the completions of general logic programs. It is a pendant of Lemma 12 and follows from the relevant definitions immediately.

Lemma 22 *Let k be one of the numbers 2, 3 or 4 and assume that $\mathcal{M}$ is a k-valued $\mathcal{L}$ structure and P a general logic program in $\mathcal{L}$. If $\mathcal{M}$ is a k-valued model of $CET_{\mathcal{L}}$, then we have:*

$$\mathcal{M} \models_k comp_{\mathcal{L}}(P) \iff \mathcal{M}[P] \sqsubseteq \mathcal{M}.$$

The three-valued Herbrand models of the completions of general logic programs P are generated by iterating the formation of P-envelopes through all ordinals less than ω_1^{CK}. However, in contrast to the construction in Section 5.1 we do not start with the structure $\mathsf{U}^2_{\mathcal{L}}$ but with the structure $\mathsf{U}^3_{\mathcal{L}}$ and do not work with the union of structures but the following limit operator.

Let $(\mathcal{M}_i : i \in I)$ be a family of four-valued Herbrand structures for $\mathcal{L}$. Then the *limit* $\bigsqcup_{i \in I} \mathcal{M}_i$ of this family is the four-valued Herbrand structure $\mathcal{N}$ for $\mathcal{L}$ so that

$$\mathcal{N}(R)(\vec{a}) := \lim_{i \in I} \mathcal{M}_i(R)(\vec{a})$$

for all relation symbols R of $\mathcal{L}$ and $\vec{a} \in \mathsf{U}_{\mathcal{L}}$. This implies that the degree of evidence for (against) a positive literal $R(\vec{a})$ is 1 in $\bigsqcup_{i \in I} \mathcal{M}_i$, if it is 1 in some $\mathcal{M}_i$, and 0 otherwise. The limit of an empty family of four-valued Herbrand structures is the structure $\mathsf{U}^3_{\mathcal{L}}$.

Definition 23 *Let P be a general logic program in $\mathcal{L}$. Then we define by recursion on the ordinals the following four-valued Herbrand structures $\mathcal{I}_P^{<\alpha}$ and $\mathcal{I}_P^{\alpha}$ for $\mathcal{L}$:*

$$\mathcal{I}_P^{<\alpha} := \bigsqcup_{\xi < \alpha} \mathcal{I}_P^{\xi}, \qquad \mathcal{I}_P^{\alpha} := \mathcal{I}_P^{<\alpha}[P], \qquad \mathcal{I}_P := \bigsqcup_{\alpha \in On} \mathcal{I}_P^{\alpha}.$$

The operator which maps a four-valued structure to its P-envelope is monotonic with respect to the relation $\sqsubseteq$ according to Remark 7. Therefore a trivial induction on the ordinals shows that

$$\alpha \leq \beta \implies \mathcal{I}_P^{\leq\alpha} \sqsubseteq \mathcal{I}_P^{\leq\beta}$$

for all ordinals α, β and general programs P. Hence the usual cardinality arguments yield that $\mathcal{I}_P^{\leq\alpha} = \mathcal{I}_P^{\alpha} = \mathcal{I}_P$ for some $\alpha \in On$. Since $\mathcal{I}_P^{\leq 0}$ is the three-valued structure $\mathsf{U}_{\mathcal{L}}^3$ we further obtain that the structures $\mathcal{I}_P^{\leq\alpha}$ and $\mathcal{I}_P^{\alpha}$ and the structure $\mathcal{I}_P$ are three-valued. The least ordinal α so that $\mathcal{I}_P^{\leq\alpha} = \mathcal{I}_P$ is called the *three-valued closure ordinal of* P.

Now we proceed as in Section 5.1. Since $\mathcal{I}_P = \mathcal{I}_P^{\leq\alpha}$ for some α, we obtain the equation $\mathcal{I}_P[P] = \mathcal{I}_P$, and $\mathcal{I}_P$ is a three-valued model of $comp_{\mathcal{L}}(P)$ in view of Lemma 22. Furthermore, if $\mathcal{M}$ is a four-valued $\mathcal{L}$ structure so that $\mathcal{M}[P] \sqsubseteq \mathcal{M}$, then $\mathcal{I}_P^{\leq\alpha} \sqsubseteq \mathcal{M}$ follows by induction on α. Thus we have the following result.

Theorem 24 (Least three-valued Herbrand models) *Let P be a general logic program in $\mathcal{L}$. Then $\mathcal{I}_P$ is the least three-valued Herbrand model of the completion of P; i.e. $\mathcal{I}_P$ is a three-valued Herbrand model of $comp_{\mathcal{L}}(P)$ and $\mathcal{I}_P \sqsubseteq \mathcal{M}$ for all three-valued and four-valued Herbrand models $\mathcal{M}$ of $comp_{\mathcal{L}}(P)$.*

Remember that we have $\mathcal{H}_P = \mathcal{H}_P^{\leq\omega}$ and $\mathcal{H}_P \models comp_{\mathcal{L}}(P)$ for all definite logic programs P. This means that the least (two-valued) model of the completion of a definite logic program P is determined by the finite stages. It can be read off from the finite stages of $\mathcal{I}_P$ in the sense of the remark below.

Remark 25 *Let P be a definite logic program in $\mathcal{L}$. Then we have for all relation symbols R of $\mathcal{L}$, all $\vec{a} \in \mathsf{U}_{\mathcal{L}}$ and all $n < \omega$:*

1. $\mathcal{H}_P^{\leq n}(R(\vec{a})) = \mathbf{t} \iff \mathcal{I}_P^{\leq n}(R(\vec{a})) = \mathbf{t}$,
2. $\mathcal{H}_P^{\leq\omega}(R(\vec{a})) = \mathbf{t} \iff \mathcal{I}_P^{\leq\omega}(R(\vec{a})) = \mathbf{t}$.

One essential difference between the hierarchies $\mathcal{H}_P^{\leq\alpha}$ and $\mathcal{I}_P^{\leq\alpha}$ refers to their closure ordinals. In view of the previous remark it is clear that the truth of positive literals will always be determined by the finite stages of the hierarchies $\mathcal{I}_P^{\leq\alpha}$. This does not imply, however, that the three-valued closure ordinal is ω since positive literals can switch in value from $\mathbf{u}$ to $\mathbf{f}$ beyond ω. The following example is taken from Fitting [8].

Example 26 *We assume that $\mathcal{L}$ contains two constants* 0 *and* c *and a unary function symbol f but no other function symbols and consider the definite* *program P consisting of the following clauses:*

$$N(0) :- ,$$
$$N(f(u)) :- N(u),$$
$$R(c) :- R(u), N(u),$$
$$R(f(u)) :- R(u).$$

Thus we have $\mathsf{U}_{\mathcal{L}} = \{f^n(0) : n < \omega\} \cup \{f^n(c) : n < \omega\}$, *and it is easily verified that for all* $\alpha \leq \omega$ *and all* $m, n < \omega$ *so that* $m < \alpha \leq n$:

$$\begin{array}{rclrcl} \mathcal{I}_P^{<\alpha}(N(f^m(0)) & = & \mathbf{t}, & \mathcal{I}_P^{<\alpha}(N(f^m(c)) & = & \mathbf{f}, \\ \mathcal{I}_P^{<\alpha}(N(f^n(0)) & = & \mathbf{u}, & \mathcal{I}_P^{<\alpha}(N(f^n(c)) & = & \mathbf{u}, \\ \mathcal{I}_P^{<\alpha}(R(f^m(0)) & = & \mathbf{f}, & \mathcal{I}_P^{<\alpha}(R(f^n(0)) & = & \mathbf{u}, \\ \mathcal{I}_P^{<\alpha}(R(f^m(c)) & = & \mathbf{u}, & \mathcal{I}_P^{<\alpha}(R(f^n(c)) & = & \mathbf{u}. \end{array}$$

As an obvious consequence we obtain for all $\beta \leq \omega$, *all* $k < \omega$ *and all* $m, n < \omega$ *so that* $m < \beta \leq n$ *the following distribution of truth values:*

$$\mathcal{I}_P^{<\omega+\beta}(N(f^k(0)) = \mathbf{t}, \quad \mathcal{I}_P^{<\omega+\beta}(N(f^k(c)) = \mathbf{f},$$

$$\mathcal{I}_P^{<\omega+\beta}(R(f^k(0)) = \mathbf{f}, \quad \mathcal{I}_P^{<\omega+\beta}(R(f^m(c)) = \mathbf{f}, \quad \mathcal{I}_P^{<\omega+\beta}(R(f^n(c)) = \mathbf{u}.$$

In view of these observations we can conclude that $\mathcal{I}_P^{<\gamma} \neq \mathcal{I}_P^{\gamma}$ *for all* $\gamma < \omega+\omega$ *and* $\mathcal{I}_P = \mathcal{I}_P^{<\omega+\omega}$.

By more subtle arguments one can show that there are definite logic programs with significantly larger three-valued closure ordinals. Without going into details we state a theorem which describes the general situation. It follows from standard results about inductive definability which are contained, for example, in Blair [5], Hinman [13] and Moschovakis [22].

Theorem 27 (Three-valued closure ordinals) *1. Every general logic program has a closure ordinal which is less than or equal to the first non-recursive ordinal* ω_1^{CK}.

2. For every ordinal $\alpha \leq \omega_1^{CK}$ *there exists a definite logic program so that its three-valued closure ordinal is* α.

Some simple reflections on the logical complexity of the structures $\mathcal{I}_P$, which is determined by this theorem, make it clear that the least three-valued Herbrand models of a logic program P can only be of limited use for a procedural approach to logic programming. It follows, for example, that in general the collection of all closed formulas true in $\mathcal{I}_P$ is not even first order definable. We agree with Kunen [19] that the "procedural content" of a logic program P is better approached by the finite stages $(\mathcal{I}_P^{<n} : n < \omega)$ of $\mathcal{I}_P$.

This thesis is also supported by the following observations. Taking up an idea of Shepherdson [27] one assigns to each $\mathcal{L}$ formula A and natural number n an equality formula $\mathsf{E}_P^n(A)$, which depends on the given logic program P.

Definition 28 *Let P be a general logic program in $\mathcal{L}$. For every $n < \omega$ and every $\mathcal{L}$ formula A we define the equality formula $\mathsf{E}_P^n(A)$ by induction on n.*

1. *If A is an atomic equality formula, then $\mathsf{E}_P^n(A) := A$.*

2. *If R is a relation symbol of $\mathcal{L}$ and $D_R[\vec{u}]$ the defining formula of R with respect to P, then we set*

$$\begin{array}{rclrcl} \mathsf{E}_P^0(R(\vec{a})) & := & \bot, & \mathsf{E}_P^0(\neg R(\vec{a})) & := & \bot, \\ \mathsf{E}_P^{n+1}(R(\vec{a})) & := & \mathsf{E}_P^n(D_R[\vec{a}]), & \mathsf{E}_P^{n+1}(\neg R(\vec{a})) & := & \mathsf{E}_P^n(\neg D_R[\vec{a}]). \end{array}$$

3. *The propositional connectives and quantifiers are dealt with in the obvious way as usual.*

$$\begin{array}{rcl} \mathsf{E}_P^n(A \vee B) & := & \mathsf{E}_P^n(A) \vee \mathsf{E}_P^n(B), \\ \mathsf{E}_P^n(A \wedge B) & := & \mathsf{E}_P^n(A) \wedge \mathsf{E}_P^n(B), \\ \mathsf{E}_P^n(\exists x A(x)) & := & \exists x \mathsf{E}_P^n(A)(x), \\ \mathsf{E}_P^n(\forall x A(x)) & := & \forall x \mathsf{E}_P^n(A)(x). \end{array}$$

Now the remark below reduces truth of A in the finite stages $\mathcal{I}_P^n$ of the three-valued Herbrand structures $\mathcal{I}_P$ to two-valued validity of the formulas $\mathsf{E}_P^n(A)$. It is easily proved by main induction on n and side induction on the rank of the formulas involved.

Remark 29 *Let P be a general logic program in $\mathcal{L}$. Then we have for all $\mathcal{L}$ formulas A and all $n < \omega$:*

$$\mathcal{I}_P^{\leq n}(A) = \mathbf{t} \iff \mathsf{U}_{\mathcal{L}}^2 \models \mathsf{E}_P^n(A).$$

This remark may be combined with Theorem 10. As a result we have a reduction of truth in the finite stages $\mathcal{I}_P^n$ to purely equational reasoning.

Corollary 30 *Let P be a general logic program in $\mathcal{L}$, A an $\mathcal{L}$ formula and $n < \omega$.*

1. *If $\mathcal{L}$ is infinite, then we have*

$$\mathcal{I}_P^{\leq n}(A) = \mathbf{t} \iff CET_{\mathcal{L}} \vdash \mathsf{E}_P^n(A).$$

2. *If $\mathcal{L}$ is finite, then we have*

$$\mathcal{I}_P^{\leq n}(A) = \mathbf{t} \iff CET_{\mathcal{L}} + DCA_{\mathcal{L}} \vdash \mathsf{E}_P^n(A).$$

7 Deductive systems for general logic programs

Traditionally a logic program is considered as a set of axioms, and its *meaning* is often identified with the set of all consequences with respect to classical, intuitionistic or, more recently, linear logic. Alternatively, it is also possible to replace the *clauses-as-axioms* interpretation of logic programs by a *clauses-as-rules* paradigm (cf. e.g. [12, 16, 25, 29, 30]) so that one is conceptually closer to a procedural understanding of logic programming.

From now on we focus on syntactic and deductive aspects of general logic programs. We begin with introducing a rule-based calculus $\mathcal{R}_{\mathcal{L}}(P)$ for each such program P which will provide the widest framework for the following considerations. The systems $\mathcal{R}_{\mathcal{L}}(P)$ are designed for a proof-theoretic treatment and form a link between logic programming and inductive definability. Later we will also study subsystems of the $\mathcal{R}_{\mathcal{L}}(P)$ which arise naturally in the context of logic programming.

7.1 The rule-based calculi $\mathcal{R}_{\mathcal{L}}(P)$

The following deduction systems are presented in a Tait-style manner. Accordingly, the axioms and derivation rules are formulated for finite sets of $\mathcal{L}$ formulas which have to be interpreted disjunctively. The capital Greek letters $\Gamma, \Delta, \Pi, \Sigma, \ldots$ (possibly with subscripts) denote finite sets of $\mathcal{L}$ formulas, and we write (for example) Γ, Δ, A, B for the union of Γ, Δ and $\{A, B\}$.

If $\mathcal{M}$ is a k-valued $\mathcal{L}$ structure and Γ the finite set $\{A_1[\vec{u}], \ldots, A_n[\vec{u}]\}$ of $\mathcal{L}$ formulas, then $\mathcal{M}(\Gamma)$ is the truth value $\mathcal{M}(\forall\vec{x}(A_1[\vec{x}] \vee \ldots \vee A_n[\vec{x}]))$ of the universal closure of the disjunction of the elements of Γ with respect to $\mathcal{M}$. If P, in addition, is a general logic program in $\mathcal{L}$, then we set

$$comp_{\mathcal{L}}(P) \models_k \Gamma \quad :\Longleftrightarrow \quad comp_{\mathcal{L}}(P) \models_k \forall\vec{x}(A_1[\vec{x}] \vee \ldots \vee A_n[\vec{x}]).$$

The systems $\mathcal{R}_{\mathcal{L}}(P)$ for general logic programs P are extensions of the usual Tait calculus for predicate logic (cf. e.g. Tait [33]) by adding equality and freeness axioms and so called *program rules*, which take care of the program clauses in P. Altogether we have the following five classes of axioms and rules.

I. Logical axioms. For all atomic $\mathcal{L}$ formulas A:

(L1) $\Gamma, \neg A, A,$

(L2) $\Gamma, \top.$

The axioms (L1) are often called *identity axioms*. They will play an important role later.

II. Equality and freeness axioms. For all $\mathcal{L}$ terms $a, b, c, a_1, \ldots, a_n$ and $b_1, \ldots b_n$ and all function symbols f and g of $\mathcal{L}$:

(E1) $\Gamma,\ a = a$,

(E2) $\Gamma,\ a \neq b,\ b = a$,

(E3) $\Gamma,\ a \neq b,\ b \neq c,\ a = c$,

(E4) $\Gamma,\ a_1 \neq b_1,\ \ldots,\ a_n \neq b_n,\ f(a_1, \ldots, a_n) = f(b_1, \ldots, b_n)$,

(E5) $\Gamma,\ f(a_1, \ldots, a_n) \neq f(b_1, \ldots, b_n),\ a_i = b_i$ for all $1 \leq i \leq n$,

(E6) $\Gamma,\ f(\vec{a}) \neq g(\vec{b})$, if f and g are different,

(E7) $\Gamma,\ a \neq b$, if a contains b and is different from b.

III. Logical rules. For all $\mathcal{L}$ formulas A and B, all $\mathcal{L}$ terms a and all free variables u which do not occur in $\Gamma, \forall x A(x)$:

$$\frac{\Gamma,\ A}{\Gamma,\ A \vee B}\ (\vee 1), \qquad \frac{\Gamma,\ B}{\Gamma,\ A \vee B}\ (\vee 2), \qquad \frac{\Gamma,\ A \qquad \Gamma,\ B}{\Gamma,\ A \wedge B}\ (\wedge),$$

$$\frac{\Gamma,\ A(a)}{\Gamma,\ \exists x A(x)}\ (\exists), \qquad \frac{\Gamma,\ A(u)}{\Gamma,\ \forall x A(x)}\ (\forall).$$

IV. Cut rules. For all $\mathcal{L}$ formulas A:

$$\frac{\Gamma,\ A \qquad \Gamma,\ \neg A}{\Gamma}\ (cut).$$

The formulas A and $\neg A$ are called the *cut formulas* of this cut; the rank of a cut is the rank of its cut formulas.

V. Program rules for P. For every relation symbol R of $\mathcal{L}$ and its defining formula $D_R[\vec{u}]$ with respect to the general logic program P and all $\mathcal{L}$ terms $\vec{a}$ we have the following positive and negative program rules:

$$\frac{\Gamma,\ D_R[\vec{a}]}{\Gamma,\ R(\vec{a})}\ (+R), \qquad \frac{\Gamma,\ \neg D_R[\vec{a}]}{\Gamma,\ \neg R(\vec{a})}\ (-R).$$

One should emphasize that the program rules are impredicative in the sense that the rank of the main formula of the premise of such a rule is in general greater than the rank of the main formula of the corresponding conclusion.

Based on these axioms and rules of inference derivability in $\mathcal{R}_\mathcal{L}(P)$ is introduced in the standard way. The notation $\mathcal{R}_\mathcal{L}(P) \vdash^n_r \Gamma$ expresses that Γ is provable in $\mathcal{R}_\mathcal{L}(P)$ by a proof whose length and cut complexity are bounded by n and r, respectively.

Definition 31 *Let P be a general logic program in $\mathcal{L}$ and Γ a finite set of $\mathcal{L}$ formulas. Then we define $\mathcal{R}_{\mathcal{L}}(P) \vdash_r^n \Gamma$ for all $n, r < \omega$ by induction on n.*

1. *If Γ is an axiom of $\mathcal{R}_{\mathcal{L}}(P)$, then we have $\mathcal{R}_{\mathcal{L}}(P) \vdash_r^n \Gamma$ for all $n, r < \omega$.*

2. *If $\mathcal{R}_{\mathcal{L}}(P) \vdash_r^{n_i} \Gamma_i$ and $n_i < n$ for every premise Γ_i of a logical rule, a program rule or a cut of $\mathcal{R}_{\mathcal{L}}(P)$, whose rank is less than r, then we have $\mathcal{R}_{\mathcal{L}}(P) \vdash_r^n \Gamma$ for the conclusion Γ of that rule.*

We write $\mathcal{R}_{\mathcal{L}}(P) \vdash_r \Gamma$ if there exists an $n < \omega$ so that $\mathcal{R}_{\mathcal{L}}(P) \vdash_r^n \Gamma$. Similarly, $\mathcal{R}_{\mathcal{L}}(P) \vdash \Gamma$ is written if there exists an $r < \omega$ so that $\mathcal{R}_{\mathcal{L}}(P) \vdash_r \Gamma$.

Hence $\mathcal{R}_{\mathcal{L}}(P) \vdash_0 \Gamma$ means that there exists a cut-free proof of Γ in $\mathcal{R}_{\mathcal{L}}(P)$. A proof is called *essentially cut-free* if its only cut formulas are equations and inequations. Therefore Γ is essentially cut-free provable in $\mathcal{R}_{\mathcal{L}}(P)$ if and only if $\mathcal{R}_{\mathcal{L}}(P) \vdash_1 \Gamma$.

In Section 9 we will discuss several aspects of cut elimination for the calculi $\mathcal{R}_{\mathcal{L}}(P)$ and some related systems. For the moment we only mention a partial cut elimination theorem for $\mathcal{R}_{\mathcal{L}}(P)$ which is proved by standard techniques, as presented for example in Girard [11], Schütte [26] or Takeuti [34]. Because of the equality and freeness axioms we must not expect that all cuts of rank 0 can be eliminated; the program rules are responsible for difficulties in eliminating cuts with literals as cut formulas.

Theorem 32 (Partial cut elimination) *We have for all general logic programs P in $\mathcal{L}$ and all finite sets Γ of $\mathcal{L}$ formulas:*

$$\mathcal{R}_{\mathcal{L}}(P) \vdash \Gamma \implies \mathcal{R}_{\mathcal{L}}(P) \vdash_2 \Gamma.$$

Each system $\mathcal{R}_{\mathcal{L}}(P)$ is the rule-based version of the completion $comp_{\mathcal{L}}(P)$ of P. This is clear since (i) all axioms of $comp_{\mathcal{L}}(P)$ are derivable in $\mathcal{R}_{\mathcal{L}}(P)$; (ii) all axioms of $\mathcal{R}_{\mathcal{L}}(P)$ follow from $comp_{\mathcal{L}}(P)$ and all rules of $\mathcal{R}_{\mathcal{L}}(P)$ are permissible in $comp_{\mathcal{L}}(P)$. Therefore we have the following theorem.

Theorem 33 (Relative Soundness and completeness) *We have for all general logic programs P and finite sets Γ of $\mathcal{L}$ formulas:*

$$\mathcal{R}_{\mathcal{L}}(P) \vdash \Gamma \iff comp_{\mathcal{L}}(P) \models \Gamma.$$

One has to remember that the theories $comp_{\mathcal{L}}(P)$ are inconsistent for some general logic programs P. Hence the empty set is derivable in the corresponding rule-based systems $\mathcal{R}_{\mathcal{L}}(P)$.

Remark 34 *There are general logic programs P so that the systems $\mathcal{R}_{\mathcal{L}}(P)$ are inconsistent in the sense that the empty set of $\mathcal{L}$ formulas is derivable.*

This can be easily seen by going back to the general logic program P of Example 20. The set of formulas $\{\neg R(0), R(0)\}$ is an identity axiom and therefore provable in $\mathcal{R}_{\mathcal{L}}(P)$. Applications of the positive and negative program rule for P yield $\mathcal{R}_{\mathcal{L}}(P) \vdash R(0)$ and $\mathcal{R}_{\mathcal{L}}(P) \vdash \neg R(0)$. Hence by a cut the empty set is derived.

We end this section by adding an observation concerning $\mathcal{H}_P$ and $\mathcal{R}_{\mathcal{L}}(P)$ for definite programs P. In view of Remark 19 it is obvious that $\mathcal{R}_{\mathcal{L}}(P) \vdash \Gamma$ implies that Γ is valid in $\mathcal{H}_P$. By following the pattern of the proof of Theorem 17 and making use of Theorem 11 one obtains a restricted form of completeness. There are simple examples, however, which show that this theorem is not true for closed formulas which contain universal formulas or negative literals.

Theorem 35 (Positive existential completeness) *We have for all definite logic programs P and closed positive existential formulas A of $\mathcal{L}$:*

$$\mathcal{H}_P \models A \implies \mathcal{R}_{\mathcal{L}}(P) \vdash A.$$

7.2 Identity-free derivations in $\mathcal{R}_{\mathcal{L}}(P)$

Now we turn to the subsystems of the calculi $\mathcal{R}_{\mathcal{L}}(P)$ which deal with identity-free derivations only. They are of great importance in connection with three-valued models of the completions of general logic programs P and are related to various forms of SLDNF-resolution. This approach is first described in Stärk [29], further results are presented for example in Jäger [16] and Stärk [30, 31].

If we consider the system $\mathcal{R}_{\mathcal{L}}(P)$ under the aspect of three-valued logic, we make the following observation: (i) The identity axioms of $\mathcal{R}_{\mathcal{L}}(P)$ are not three-valid; (ii) the equality and freeness axioms of $\mathcal{R}_{\mathcal{L}}(P)$ are valid in all three-valued models of $CET_{\mathcal{L}}$; (iii) if every premise of a propositional rule, quantifier rule or cut rule of $\mathcal{R}_{\mathcal{L}}(P)$ is valid in a three-valued structure $\mathcal{M}$, then the conclusion of that rule is also valid in $\mathcal{M}$. This means that — besides the identity axioms and the program rules — all axioms and rules of $\mathcal{R}_{\mathcal{L}}(P)$ are correct in the three-valued sense. If $\mathcal{M}$ is a three-valued model of the completion of a general program P, then also the program rules are valid in $\mathcal{M}$.

Hence from the point of view of three-valued logic only the identity axioms create some problems. For this reason it is very pleasing that just deleting the identity axioms yields an interesting subsystem of $\mathcal{R}_{\mathcal{L}}(P)$ which appears to be the adequate proof-theoretic framework for the three-valued approach to the completions of logic programs.

Definition 36 *Let P be a general logic program in $\mathcal{L}$ and Γ a finite set of $\mathcal{L}$ formulas. Then we define $\mathcal{R}_{\mathcal{L}}(P) \Vdash^n_r \Gamma$ for all $n, r < \omega$ by induction on n.*

1. *If Γ is an axiom of $\mathcal{R}_{\mathcal{L}}(P)$ but not an identity axiom, then we have $\mathcal{R}_{\mathcal{L}}(P) \Vdash\!\!-^{n}_{r} \Gamma$ for all $n, r < \omega$.*

2. *If $\mathcal{R}_{\mathcal{L}}(P) \Vdash\!\!-^{n_i}_{r} \Gamma_i$ and $n_i < n$ for every premise Γ_i of a logical rule, a program rule or a cut of $\mathcal{R}_{\mathcal{L}}(P)$, whose rank is less than r, then we have $\mathcal{R}_{\mathcal{L}}(P) \Vdash\!\!-^{n}_{r} \Gamma$ for the conclusion Γ of that rule.*

The short hand notations $\mathcal{R}_{\mathcal{L}}(P) \Vdash\!\!-_{r} \Gamma$ and $\mathcal{R}_{\mathcal{L}}(P) \Vdash\!\!- \Gamma$ will be used in the obvious sense. Hence $\mathcal{R}_{\mathcal{L}}(P) \Vdash\!\!- \Gamma$ means that Γ is derivable in $\mathcal{R}_{\mathcal{L}}(P)$ by a proof which does not make use of the identity-axioms.

Remark 37 *The omission of the identity axioms does not affect the equality formulas. We have for all logic programs in $\mathcal{L}$ and all finite sets of equality formulas Γ:*

$$CET_{\mathcal{L}} \vdash \Gamma \implies \mathcal{R}_{\mathcal{L}}(P) \Vdash\!\!- \Gamma.$$

By recalling the previous remarks, one can immediately conclude that the identity-free derivations in $\mathcal{R}_{\mathcal{L}}(P)$ are correct with respect to the three-valued models of $comp_{\mathcal{L}}(P)$. In Stärk [29] also the corresponding completeness result is proved. The approach presented there makes use of the technique of Schütte valuations for which we refer to Girard [11] and Schütte [26].

Theorem 38 (Three-valued soundness and completeness) *Let P be a general logic program in $\mathcal{L}$. Then we have for all finite sets Γ of $\mathcal{L}$ formulas:*

$$\mathcal{R}_{\mathcal{L}}(P) \Vdash\!\!- \Gamma \iff comp_{\mathcal{L}}(P) \models_3 \Gamma.$$

It is possible to tie identity-free derivations in $\mathcal{R}_{\mathcal{L}}(P)$ very closely to finitely iterated program envelopes of three-valued structures. Given a three-valued $\mathcal{L}$ structure $\mathcal{M}$ and a general logic program P in $\mathcal{L}$, we define for all $n < \omega$:

$$\begin{aligned} \mathcal{M}^0_P &:= \mathcal{M}, \\ \mathcal{M}^{n+1}_P &:= \mathcal{M}^n_P[P]. \end{aligned}$$

It is legitimate to consider the formation of envelopes as a semantic counterpart of the program rules. Therefore a formula proved in n steps may be expected to be true in the n-times iterated envelope.

Theorem 39 *Let P be a general logic program in $\mathcal{L}$ and $\mathcal{M}$ a three-valued model of $CET_{\mathcal{L}}$ so that $\mathcal{M} \sqsubseteq \mathcal{M}[P]$. Then we have for all $n, r < \omega$:*

$$\mathcal{R}_{\mathcal{L}}(P) \Vdash\!\!-^{n}_{r} \Gamma \implies \mathcal{M}^n_P(\Gamma) = \mathbf{t}.$$

This assertion is verified by a simple induction on n. Together with the three-valued completeness it implies the following corollary, which was first obtained by Kunen [18]. His proof, however, is purely model-theoretic and makes use of ultra power constructions. In a sense, the above theorem and its corollary are quite remarkable since the structures $\mathcal{M}^n_P$ are in general not models of $comp_{\mathcal{L}}(P)$.

Corollary 40 *Let P be a general logic program in $\mathcal{L}$ and $\mathcal{M}$ a three-valued model of $CET_{\mathcal{L}}$ so that $\mathcal{M} \sqsubseteq \mathcal{M}[P]$. Then $comp_{\mathcal{L}}(P) \models_3 \forall \vec{x} A[\vec{x}]$ implies that there exists an $n < \omega$ so that $\mathcal{M}^n_P \models_3 \forall \vec{x} A[\vec{x}]$.*

In a next step we compare identity-free derivability in $\mathcal{R}_{\mathcal{L}}(P)$ with truth in the finite stages of $\mathcal{I}_P$. For this purpose we make use of some notions and results which have been presented at the end of Section 6.2. A first lemma describes a connection between formulas A and the equality formulas assigned to them according to Definition 28.

Lemma 41 *Let P be a general logic program in $\mathcal{L}$. Then we have for all $\mathcal{L}$ formulas A and all $n < \omega$:*

$$\mathcal{R}_{\mathcal{L}}(P) \Vdash \neg \mathsf{E}^n_P(A), A.$$

The proof of this lemma by main induction on n and side induction on the rank of A is a matter of routine. This relationship between the formulas A and $\mathsf{E}^n_P(A)$ permits the integration of Corollary 30 into the context of $\mathcal{R}_{\mathcal{L}}(P)$ and provides soundness and completeness with respect to the family $(\mathcal{I}^{\leq n}_P : n < \omega)$.

Theorem 42 *Let P be a general logic program in $\mathcal{L}$ and A an $\mathcal{L}$ formula.*

1. *If $\mathcal{L}$ is infinite, then we have*

$$\mathcal{R}_{\mathcal{L}}(P) \Vdash A \iff \sum_{n<\omega} \mathcal{I}^{\leq n}_P(A) = \mathbf{t}.$$

2. *If $\mathcal{L}$ is finite, then we have*

$$\mathcal{R}_{\mathcal{L}}(P) + DCA_{\mathcal{L}} \Vdash A \iff \sum_{n<\omega} \mathcal{I}^{\leq n}_P(A) = \mathbf{t}.$$

Proof. We assume first that $\mathcal{L}$ is infinite. If we have $\mathcal{R}_{\mathcal{L}}(P) \Vdash A$, then Theorem 39 implies $\mathcal{I}^{\leq n}_P(A) = \mathbf{t}$ for some $n < \omega$. On the other hand, if $\mathcal{I}^{\leq n}_P(A) = \mathbf{t}$ for some $n < \omega$, we can apply Corollary 30 and Remark 37 and conclude that $\mathcal{R}_{\mathcal{L}}(P) \Vdash \mathsf{E}^n_P(A)$. The previous lemma and a cut yield the assertion.

In the second case $\mathcal{L}$ is finite. Then we have only to observe that the domain closure axiom $DCA_{\mathcal{L}}$ is valid in all Herbrand structures and can proceed as before. □

Theorem 38 and Theorem 42 provide as a corollary a proof-theoretic approach to an interesting result which exhibits the close connections between truth in the finite stages of $\mathcal{I}_P$ and truth in all models of the completion of P. Its first part is originally due to Kunen and proved in [18].

Corollary 43 *Let P be a general logic program in $\mathcal{L}$ and A an $\mathcal{L}$ formula.*

1. *If $\mathcal{L}$ is infinite, then we have*

$$\sum_{n<\omega} \mathcal{I}_P^{\leq n}(A) = \mathbf{t} \iff comp_{\mathcal{L}}(P) \models_3 A.$$

2. *If $\mathcal{L}$ is finite, then we have*

$$\sum_{n<\omega} \mathcal{I}_P^{\leq n}(A) = \mathbf{t} \iff comp_{\mathcal{L}}(P) \models_3 (DCA_{\mathcal{L}} \rightarrow A).$$

This corollary connects the finite stages of the least three-valued Herbrand model of the completion of a general logic program P with all three-valued models of $comp_{\mathcal{L}}(P)$, not only with all three-valued Herbrand models of $comp_{\mathcal{L}}(P)$. In general it will not be correct if we replace the notion of three-valued consequence by three-valued consequence with respect to Herbrand structures.

The previous results are also interesting in connection with various forms of SLDNF-resolution. It follows from the work of Kunen, Stärk and several others that SLDNF-resolution is sound and complete for large classes of general logic programs P with respect to identity-free derivability in $\mathcal{R}_{\mathcal{L}}(P)$ or (equivalently) validity in the three-valued models of $comp_{\mathcal{L}}(P)$. For more details see [2], [19] or [32].

8 Partiality in logic programming

The omission of the identity axioms has the effect of disconnecting a relation symbol R from its complement $\widetilde{R}$. However, the resulting formalism is extremely weak and sometimes considered unnatural. In addition, on the semantical side we have to deal with additional truth values in order to obtain an adequate model theory for identity-free derivations. Now we want to further the conceptual clarity and present an alternative approach which is based on two-valued logic and works with partial fixed points generated by logic programs.

Let $\mathcal{L}^\sharp$ be the language which results from $\mathcal{L}$ by replacing each relation symbol R of $\mathcal{L}$ by two new independent relation symbols R^+ and R^-, which are of the same arity as R. Then every $\mathcal{L}$ formula A can be translated into $\mathcal{L}^\sharp$ formulas A^+ and A^- as follows:

1. If A is an atomic equality formula, then $A^+ := A$ and $A^- := \neg A$.
2. If A is of the form $R(\vec{a})$ for some n-ary relation symbol of $\mathcal{L}$, then $A^+ := R^+(\vec{a})$ and $A^- := R^-(\vec{a})$.
3. If A is of the form $\widetilde{R}(\vec{a})$ for some n-ary relation symbol of $\mathcal{L}$, then $A^+ := R^-(\vec{a})$ and $A^- := R^+(\vec{a})$.

4. If A is of the form $(B \vee C)$, then $A^+ := (B^+ \vee C^+)$ and $A^- := (B^- \wedge C^-)$.

5. If A is of the form $(B \wedge C)$, then $A^+ := (B^+ \wedge C^+)$ and $A^- := (B^- \vee C^-)$.

6. If A is of the form $\exists x B(x)$, then $A^+ := \exists x B^+(x)$ and $A^- := \forall x B^-(x)$.

7. If A is of the form $\forall x B(x)$, then $A^+ := \forall x B^+(x)$ and $A^- := \exists x B^-(x)$.

If Γ is the set $\{A_1, \ldots, A_n\}$ of $\mathcal{L}$ formulas, then we write Γ^+ for the corresponding set $\{A_1^+, \ldots, A_n^+\}$ of $\mathcal{L}^\sharp$ formulas.

This means that the $\mathcal{L}^\sharp$ formula A^+ is obtained from the $\mathcal{L}$ formula A by changing all positive literals $R(\vec{a})$ into $R^+(\vec{a})$ and all negative literals $\widetilde{R}(\vec{a})$ into $R^-(\vec{a})$; A^- is obtained from A by replacing all positive literals $R(\vec{a})$ in $\neg A$ by $R^+(\vec{a})$ and all negative literals $\widetilde{R}(\vec{a})$ in $\neg A$ by $R^-(\vec{a})$. If there are axioms available which express that the formulas $R^-(\vec{a})$ are the negations of the formulas $R^+(\vec{a})$, then it is obvious that A^+ is logically equivalent to A and A^- to $\neg A$.

Remark 44 *1. If A is an $\mathcal{L}$ formula, then A^+ and A^- are positive $\mathcal{L}^\sharp$ formulas in the sense that they do not contain negative literals.*

2. If B is the negation of the $\mathcal{L}$ formula A, then A^- and B^+ are identical $\mathcal{L}^\sharp$ formulas.

8.1 Four-valued structures for $\mathcal{L}$ versus two-valued structures for $\mathcal{L}^\sharp$

Now we insert some useful semantical considerations and show that the four-valued structures for $\mathcal{L}$ induce two-valued structures for $\mathcal{L}^\sharp$ and vice versa.

Definition 45 *Let $\mathcal{M}$ be a four-valued $\mathcal{L}$ structure. Then $\mathcal{M}^\sharp$ is the two-valued $\mathcal{L}^\sharp$ structure which is defined as follows:*

1. $|\mathcal{M}^\sharp| := |\mathcal{M}|$, and $\mathcal{M}^\sharp(f) := \mathcal{M}(f)$ for all function symbols f of $\mathcal{L}$.

2. For all relation symbols R of $\mathcal{L}$ and $\vec{m} \in |\mathcal{M}|$ we set

$$\mathcal{M}^\sharp(R^+)(\vec{m}) := \begin{cases} \mathbf{t}, & \text{if } \mathcal{M}(R(\vec{m})) \in \{\mathbf{t}, \mathbf{c}\}, \\ \mathbf{f}, & \text{if } \mathcal{M}(R(\vec{m})) \in \{\mathbf{f}, \mathbf{u}\}; \end{cases}$$

$$\mathcal{M}^\sharp(R^-)(\vec{m}) := \begin{cases} \mathbf{t}, & \text{if } \mathcal{M}(R(\vec{m})) \in \{\mathbf{f}, \mathbf{c}\}, \\ \mathbf{f}, & \text{if } \mathcal{M}(R(\vec{m})) \in \{\mathbf{t}, \mathbf{u}\}. \end{cases}$$

Following a similar pattern, each two-valued $\mathcal{L}^\sharp$ structure $\mathcal{M}$ can also be transformed into a four-valued $\mathcal{L}$ structure $\mathcal{M}^\diamond$. Then the relation symbols R^+ and R^- provide the degrees of evidence for and against R.

Definition 46 *Let $\mathcal{M}$ be a two-valued $\mathcal{L}^\sharp$ structure. Then $\mathcal{M}^\diamond$ is the four-valued $\mathcal{L}$ structure which is defined as follows:*

1. $|\mathcal{M}^\diamond| := |\mathcal{M}|$, *and* $\mathcal{M}^\diamond(f) := \mathcal{M}(f)$ *for all function symbols f of $\mathcal{L}$.*
2. *For all relation symbols R of $\mathcal{L}$ and $\vec{m} \in |\mathcal{M}|$ we set*

$$\mathcal{M}^\diamond(R)(\vec{m}) \;:=\; \begin{cases} \mathbf{t}, & \textit{if} \;\; \mathcal{M}(R^+(\vec{m})) = \mathbf{t} \;\; \textit{and} \;\; \mathcal{M}(R^-(\vec{m})) = \mathbf{f}, \\ \mathbf{f}, & \textit{if} \;\; \mathcal{M}(R^+(\vec{m})) = \mathbf{f} \;\; \textit{and} \;\; \mathcal{M}(R^-(\vec{m})) = \mathbf{t}, \\ \mathbf{u}, & \textit{if} \;\; \mathcal{M}(R^+(\vec{m})) = \mathbf{f} \;\; \textit{and} \;\; \mathcal{M}(R^-(\vec{m})) = \mathbf{f}, \\ \mathbf{c}, & \textit{if} \;\; \mathcal{M}(R^+(\vec{m})) = \mathbf{t} \;\; \textit{and} \;\; \mathcal{M}(R^-(\vec{m})) = \mathbf{t}. \end{cases}$$

The previous two constructions are inverse to each other in the strongest possible sense. We have for all four-valued $\mathcal{L}$ structures $\mathcal{M}$ and all two-valued $\mathcal{L}^\sharp$ structures $\mathcal{N}$ that $(\mathcal{M}^\sharp)^\diamond = \mathcal{M}$ and $(\mathcal{N}^\diamond)^\sharp = \mathcal{N}$. Hence it is perfectly legitimate to identify the four-valued structures for the language $\mathcal{L}$ with the two-valued structures for the extension $\mathcal{L}^\sharp$ of $\mathcal{L}$.

Remark 47 *Let $\mathcal{M}$ be a four-valued $\mathcal{L}$ structure. Then $\mathcal{M}$ is three-valued if and only if we have*

$$\mathcal{M}^\sharp \models \widetilde{R^-}(\vec{m}) \vee \widetilde{R^+}(\vec{m})$$

for all relation symbols R of $\mathcal{L}$ and all $\vec{m} \in |\mathcal{M}|$.

In view of this remark it is possible to identify the three-valued $\mathcal{L}$ structures with the two-valued $\mathcal{L}^\sharp$ structures which satisfy the uniqueness condition that all R^+ and R^- are interpreted as disjoint relations.

The following lemma shows how the close connection between four-valued $\mathcal{L}$ structures and two-valued $\mathcal{L}^\sharp$ structures extends to arbitrary $\mathcal{L}$ formulas. It has a standard proof by simultaneous induction on the rank of A.

Lemma 48 *We have for all four-valued $\mathcal{L}$ structures $\mathcal{M}$ and $\mathcal{L}$ formulas A:*

1. $\mathcal{M}(A) = \mathbf{t}$ *if and only if* $\mathcal{M}^\sharp(A^+) = \mathbf{t}$ *and* $\mathcal{M}^\sharp(A^-) = \mathbf{f}$.
2. $\mathcal{M}(A) = \mathbf{f}$ *if and only if* $\mathcal{M}^\sharp(A^+) = \mathbf{f}$ *and* $\mathcal{M}^\sharp(A^-) = \mathbf{t}$.
3. $\mathcal{M}(A) = \mathbf{u}$ *if and only if* $\mathcal{M}^\sharp(A^+) = \mathbf{f}$ *and* $\mathcal{M}^\sharp(A^-) = \mathbf{f}$.
4. $\mathcal{M}(A) = \mathbf{c}$, *if and only if* $\mathcal{M}^\sharp(A^+) = \mathbf{t}$ *and* $\mathcal{M}^\sharp(A^-) = \mathbf{t}$.

Since three-valued and four-valued structures have been interesting for us in connection with models of the completions of general logic programs we now have to find a related two-valued notion based on the extended language $\mathcal{L}^\sharp$. This aim will be achieved by so called partial completions of logic programs.

Definition 49 *Let P be a general logic program in $\mathcal{L}$. Then the* partial completion $comp^{\sharp}_{\mathcal{L}}(P)$ *of P is the $\mathcal{L}^{\sharp}$ theory which consists of the following axioms.*

I. *$CET_{\mathcal{L}}$ plus equality axioms for all relation symbols $R^{\pm}$ of $\mathcal{L}^{\sharp}$:*

$$(a_1 = b_1 \wedge \ldots \wedge a_n = b_n) \rightarrow (R^{\pm}(a_1, \ldots, a_n) \rightarrow R^{\pm}(b_1, \ldots, b_n)).$$

II. *For all relation symbols R of $\mathcal{L}$ and their defining formulas $D_R[\vec{u}]$ with respect to P:*

$$\forall \vec{x}(D^{+}_{R}[\vec{x}] \rightarrow R^{+}(\vec{x})) \quad \textit{and} \quad \forall \vec{x}(D^{-}_{R}[\vec{x}] \rightarrow R^{-}(\vec{x})).$$

If $\mathcal{M}$ is a model of the partial completion of a program P and if we know, in addition, that the relations $\mathcal{M}(R^-)$ are the (set-theoretic) complements of the relations $\mathcal{M}(R^+)$, then the $\mathcal{M}(R^+)$ describe fixed points of the system of inductive definitions which is associated to P. But in general we do not know whether $\mathcal{M}(R^+)$ and $\mathcal{M}(R^-)$ are complementary so that the prefix "partial" is in place. This is similar to the distinction between truth definitions and partial truth definitions as for example in Feferman [7].

Lemma 50 *Let P be a general logic program in $\mathcal{L}$ and $\mathcal{M}$ a four-valued $\mathcal{L}$ structure which is a model of $CET_{\mathcal{L}}$. Then we have:*

$$\mathcal{M}^{\sharp} \models comp^{\sharp}_{\mathcal{L}}(P) \iff \mathcal{M} \models_4 comp_{\mathcal{L}}(P).$$

The proof of this lemma is immediate by Remark 48. We now obtain a theorem which reduces the four-valued model theory of the completion of logic programs to the two-valued case.

Theorem 51 *We have for all general logic programs P in $\mathcal{L}$ and all $\mathcal{L}$ formulas A:*

$$comp^{\sharp}_{\mathcal{L}}(P) \models A^{+} \iff comp_{\mathcal{L}}(P) \models_4 A.$$

8.2 The calculi $\partial_{\mathcal{L}}(P)$

It is now easy to set up deductive systems which correspond to the partial completion of general programs. As in Section 7 we work in extensions of the Tait calculus for predicate logic; the only difference between $\mathcal{R}_{\mathcal{L}}(P)$ and $\partial_{\mathcal{L}}(P)$ is the way in which the programs are incorporated into the calculi.

The systems $\partial_{\mathcal{L}}(P)$ for general logic programs P in $\mathcal{L}$ comprise the logical axioms, the equality and freeness axioms, the logical rules and the cut rules of $\mathcal{R}_{\mathcal{L}}(P)$, but all formulated for $\mathcal{L}^{\sharp}$ instead of $\mathcal{L}$. In addition we have equality axioms for the relation symbols of $\mathcal{L}^{\sharp}$ and partial program rules:

Equality axioms for the relation symbols of $\mathcal{L}^\sharp$. If $a_1, \ldots, a_n$ and $b_1, \ldots, b_n$ are $\mathcal{L}^\sharp$ terms and if Γ is a finite set of $\mathcal{L}^\sharp$ formulas which comprises a set $\{E_i : 1 \leq i \leq m\}$ of equations and inequations so that

$$CET_{\mathcal{L}} \vdash \bigvee_{i=1}^{m} E_i \vee (a_1 = b_1 \wedge \ldots \wedge a_n = b_n),$$

then we have for all n-ary relation symbols $R^{\pm}$ of $\mathcal{L}^\sharp$:

$$\Gamma,\ \widetilde{R^{\pm}}(a_1, \ldots, a_n),\ R^{\pm}(b_1, \ldots, b_n).$$

Partial program rules for P. We have for every relation symbol R of $\mathcal{L}$ and its defining formula $D_R[\vec{u}]$ with respect to the general logic program P and for all $\mathcal{L}$ terms $\vec{a}$:

$$\frac{\Gamma,\ D_R^+[\vec{a}]}{\Gamma,\ R^+(\vec{a})}\ (R^+), \qquad \frac{\Gamma,\ D_R^-[\vec{a}]}{\Gamma,\ R^-(\vec{a})}\ (R^-).$$

There is an important difference between the program rules of $\mathcal{R}_{\mathcal{L}}(P)$ and the partial program rules of $\partial_{\mathcal{L}}(P)$: The formulas $D_R^+[\vec{u}]$ and $D_R^-[\vec{u}]$ in the premises of the partial program rules of $\partial_{\mathcal{L}}(P)$ are positive in all relation symbols whereas the formulas $D_R[\vec{u}]$ and $\neg D_R[\vec{u}]$ in the premises of the program rule of $\mathcal{R}_{\mathcal{L}}(P)$ may contain positive and negative occurrences of R and other relation symbols.

Definition 52 *Let P be a general logic program in $\mathcal{L}$ and Γ a finite set of $\mathcal{L}^\sharp$ formulas. Then we define $\partial_{\mathcal{L}}(P) \vdash_r^n \Gamma$ for all $n, r < \omega$ by induction on n.*

1. *If Γ is an axiom of $\partial_{\mathcal{L}}(P)$, then we have $\partial_{\mathcal{L}}(P) \vdash_r^n \Gamma$ for all $n, r < \omega$.*

2. *If $\partial_{\mathcal{L}}(P) \vdash_r^{n_i} \Gamma_i$ and $n_i < n$ for every premise Γ_i of a logical rule, a partial program rule or a cut of $\partial_{\mathcal{L}}(P)$, whose rank is less than r, then we have $\partial_{\mathcal{L}}(P) \vdash_r^n \Gamma$ for the conclusion Γ of that rule.*

We will make use of the abbreviated forms $\partial_{\mathcal{L}}(P) \vdash_r \Gamma$ and $\partial_{\mathcal{L}}(P) \vdash \Gamma$ in the same sense as before (cf. Section 7).

It is obvious that $\partial_{\mathcal{L}}(P)$ is a Tait-style formalization of the partial completion of P. This means that we have the following theorem which implies, by applying Theorem 51, the equivalence between $\partial_{\mathcal{L}}(P)$ and the four-valued completion of P.

Theorem 53 *Let P be a general logic program in $\mathcal{L}$. Then we have for all finite sets Γ of $\mathcal{L}^\sharp$ formulas:*

$$\partial_{\mathcal{L}}(P) \vdash \Gamma \iff comp_{\mathcal{L}}^\sharp(P) \models \Gamma.$$

Corollary 54 *Let P be a general logic program in $\mathcal{L}$. Then we have for all $\mathcal{L}$ formulas A:*

$$\partial_{\mathcal{L}}(P) \vdash A^{+} \iff comp_{\mathcal{L}}(P) \models_{4} A.$$

Now we turn to two special collections (UNI) and (TOT) of finite sets of $\mathcal{L}^{\sharp}$ formulas. The first of those, (UNI), express that it is not the case that $R^{-}(\vec{a})$ and $R^{+}(\vec{a})$ may be true. It consists of all

$$\Gamma, \widetilde{R^{-}(\vec{a})}, \widetilde{R^{+}(\vec{a})} \qquad \text{(UNI)}$$

so that Γ is a finite set of $\mathcal{L}^{\sharp}$ formulas and R a relation symbol of $\mathcal{L}$. Because of Remark 47 and Corollary 54 the uniqueness axioms are related to three-valued consequences of the completion of programs.

Theorem 55 *Let P be a general logic program in $\mathcal{L}$. Then we have for all $\mathcal{L}$ formulas A:*

$$\partial_{\mathcal{L}}(P) + (\text{UNI}) \vdash A^{+} \iff comp_{\mathcal{L}}(P) \models_{3} A.$$

(TOT) contains the assertions dual to (UNI), namely that all pairs (R^{+}, R^{-}) are total in the sense that at least one of the two formulas $R^{-}(\vec{a})$ and $R^{+}(\vec{a})$ is true:

$$\Gamma, R^{-}(\vec{a}), R^{+}(\vec{a}). \qquad \text{(TOT)}$$

Hence (TOT) and (UNI) together imply that $R^{-}(\vec{a})$ is equivalent to the negation of $R^{+}(\vec{a})$ and vice versa. $\partial_{\mathcal{L}}(P)$ extended by (TOT) and (UNI) therefore proves for every $\mathcal{L}$ formula A that the $\mathcal{L}^{\sharp}$ formulas A and A^{+} and the $\mathcal{L}^{\sharp}$ formulas $\neg A$ and A^{-} are equivalent. A consequence of this observation is the following theorem.

Theorem 56 *Let P be a general logic program in $\mathcal{L}$. Then we have for all finite sets Γ of $\mathcal{L}$ formulas:*

$$\mathcal{R}_{\mathcal{L}}(P) \vdash \Gamma \iff \partial_{\mathcal{L}}(P) + (\text{TOT}) + (\text{UNI}) \vdash \Gamma^{+}.$$

Instead of continuing with $\partial_{\mathcal{L}}(P)$ plus (TOT) and (UNI) we look more closely at the systems $\partial_{\mathcal{L}}(P)$ extended by one of these axiom schemes only. It will turn out that this leads to systems which correspond to the essentially cut-free and identity-free derivations of $\mathcal{R}_{\mathcal{L}}(P)$. A first observation says that (TOT) and (UNI) can be lifted from the atomic formulas to arbitrary formulas. This simple fact is shown by induction on the rank of A.

Remark 57 *We have for all general logic programs P in $\mathcal{L}$ and all $\mathcal{L}$ formulas A:*

1. $\partial_{\mathcal{L}}(P) + (\text{TOT}) \vdash A^{-}, A^{+}$,
2. $\partial_{\mathcal{L}}(P) + (\text{UNI}) \vdash \neg A^{-}, \neg A^{+}$.

9 Elimination of cuts and identity axioms

In this section we study derivations in $\mathcal{R}_{\mathcal{L}}(P)$ which do not make use of cuts or identity axioms and look more closely at the relationship between cut-free and identity-free proofs in $\mathcal{R}_{\mathcal{L}}(P)$ and proofs in $\partial_{\mathcal{L}}(P) + (\text{Tot})$ and $\partial_{\mathcal{L}}(P) + (\text{Uni})$, respectively. Similar questions are also discussed in Hösli and Jäger [14] in a more general context.

First we observe that the system $\partial_{\mathcal{L}}(P)$ only deals with positive information in the following sense:

- Except for the identity axioms and the equality axioms for the relation symbols of $\mathcal{L}^{\sharp}$ we have that the main formulas of the axioms and rules of the system $\partial_{\mathcal{L}}(P)$ are positive.

- If all formulas in the conclusion of a logical rule or a partial program rule of $\partial_{\mathcal{L}}(P)$ are positive, then all formulas in the corresponding premise(s) are positive as well.

Hence negative formulas cannot be introduced in a critical way, and standard techniques imply that all essential cuts can be eliminated.

Theorem 58 (Cut elimination) *We have for all general programs P and finite sets Γ of $\mathcal{L}^{\sharp}$ formulas:*

1. $\partial_{\mathcal{L}}(P) \vdash \Gamma \implies \partial_{\mathcal{L}}(P) \vdash_1 \Gamma.$

2. $\partial_{\mathcal{L}}(P) \Vdash \Gamma \implies \partial_{\mathcal{L}}(P) \Vdash_1 \Gamma.$

9.1 Essentially cut-free derivations and (Tot)

In a next step we turn to the connections between essentially cut-free proofs in $\mathcal{R}_{\mathcal{L}}(P)$ and the axiom-scheme (Tot). In the following lemma the identity axioms of $\mathcal{R}_{\mathcal{L}}(P)$ are taken care of by (Tot), and each direction is proved by straightforward induction.

Lemma 59 *We have for all general programs P and finite sets Γ of $\mathcal{L}$ formulas:*

$$\mathcal{R}_{\mathcal{L}}(P) \vdash_1 \Gamma \iff \partial_{\mathcal{L}}(P) + (\text{Tot}) \Vdash_1 \Gamma^{+}.$$

Since the main formulas of (Tot) are positive we can apply the same argument as for the proof of Theorem 58 in order to show cut-elimination for $\partial_{\mathcal{L}}(P) + (\text{Tot})$.

Theorem 60 (Cut elimination) *We have for all general programs P and finite sets Γ of $\mathcal{L}^{\sharp}$ formulas:*

1. $\partial_{\mathcal{L}}(P) + (\text{Tot}) \vdash \Gamma \implies \partial_{\mathcal{L}}(P) + (\text{Tot}) \vdash_1 \Gamma.$

2. $\partial_{\mathcal{L}}(P) + (\text{TOT}) \Vdash \Gamma \implies \partial_{\mathcal{L}}(P) + (\text{TOT}) \Vdash_1 \Gamma$.

The previous observation about the positivity of the axioms and rules of the theory $\partial_{\mathcal{L}}(P) + (\text{TOT})$ also enables us to eliminate the identity axioms with respect to $\mathcal{L}^\sharp$ from an essentially cut-free proof of a set of positive formulas. The direction from right to left of the following lemma is obvious; the direction from left to right is proved by induction on the length of the derivation.

Lemma 61 *We have for all general programs P and finite sets Γ of positive $\mathcal{L}^\sharp$ formulas:*

$$\partial_{\mathcal{L}}(P) + (\text{TOT}) \vdash_1 \Gamma \iff \partial_{\mathcal{L}}(P) + (\text{TOT}) \Vdash_1 \Gamma.$$

A characterization of the derivations in $\mathcal{R}_{\mathcal{L}}(P)$ which do not make use of essential cuts is now obtained by simply summarizing Lemma 59, Lemma 61 and Theorem 60.

Theorem 62 *Let P be a general program and Γ a finite set of $\mathcal{L}$ formulas. Then $\mathcal{R}_{\mathcal{L}}(P) \vdash_1 \Gamma$ is equivalent to each of the following four assertions:*

$$\partial_{\mathcal{L}}(P) + (\text{TOT}) \vdash \Gamma^+, \qquad \partial_{\mathcal{L}}(P) + (\text{TOT}) \vdash_1 \Gamma^+,$$

$$\partial_{\mathcal{L}}(P) + (\text{TOT}) \Vdash \Gamma^+, \qquad \partial_{\mathcal{L}}(P) + (\text{TOT}) \Vdash_1 \Gamma^+.$$

9.2 Identity-free derivations, (UNI) and positive cuts

From (UNI) we can deduce that not both formulas $\neg A^+$ and $\neg A^-$ can be true. However, in contrast to (TOT), the main formulas of (UNI) are no longer positive. In order to maintain the advantage of working with derivations consisting of positive formulas we replace (UNI) by what we call *positive cuts*.

Positive cuts. For all $\mathcal{L}$ formulas A:

$$\frac{\Gamma, A^+ \qquad \Gamma, A^-}{\Gamma} \ (p\text{-}cut).$$

Please observe that the premises of a positive cut are positive provided that its conclusion is positive. Positive cuts are an alternative formulation for reflecting the fact that A^+ or A^- is not true. The exact relationship between (UNI) and (*p-cut*) is described by the following remark.

Remark 63 *We have for all general programs P:*

1. *$\partial_{\mathcal{L}}(P) + (\text{UNI})$ with or without identity axioms is closed under positive cuts.*

2. *$\partial_{\mathcal{L}}(P) + (p\text{-}cut)$ proves all elements of* (UNI)*; in general the identity axioms are needed in order to derive* (UNI) *from $(p\text{-}cut)$.*

It is now a trivial matter to relate identity-free derivations in $\mathcal{R}_{\mathcal{L}}(P)$ to derivations in $\partial_{\mathcal{L}}(P) + (p\text{-}cut)$: the cuts in $\mathcal{R}_{\mathcal{L}}(P)$ only have to be replaced by positive cuts.

Lemma 64 *We have for all general programs P and finite sets Γ of $\mathcal{L}$ formulas:*

$$\mathcal{R}_{\mathcal{L}}(P) \Vdash \Gamma \iff \partial_{\mathcal{L}}(P) + (p\text{-}cut) \Vdash_1 \Gamma^+.$$

As in the case of Theorem 60 standard methods of proof theory yield cut elimination for $\partial_{\mathcal{L}}(P)+(p\text{-}cut)$. The following theorem states that all essential cuts can be eliminated; of course positive cuts are still permitted.

Theorem 65 (Cut elimination) *We have for all general programs P and finite sets Γ of $\mathcal{L}^\sharp$ formulas:*

1. $\partial_{\mathcal{L}}(P) + (p\text{-}cut) \vdash \Gamma \implies \partial_{\mathcal{L}}(P) + (p\text{-}cut) \vdash_1 \Gamma.$
2. $\partial_{\mathcal{L}}(P) + (p\text{-}cut) \Vdash \Gamma \implies \partial_{\mathcal{L}}(P) + (p\text{-}cut) \Vdash_1 \Gamma.$

We go on with our symmetric treatment of cut-free and identity-free derivations in $\mathcal{R}_{\mathcal{L}}(P)$ and formulate the analogue of Lemma 61 which says that the identity axioms can be eliminated from essentially cut-free proofs in $\partial_{\mathcal{L}}(P) + (p\text{-}cut)$.

Lemma 66 *We have for all general programs P and finite sets Γ of positive $\mathcal{L}^\sharp$ formulas:*

$$\partial_{\mathcal{L}}(P) + (p\text{-}cut) \vdash_1 \Gamma \iff \partial_{\mathcal{L}}(P) + (p\text{-}cut) \Vdash_1 \Gamma.$$

It only remains to sum up Lemma 64, Lemma 66 and Theorem 65. This yields a description of the identity-free derivations in $\mathcal{R}_{\mathcal{L}}(P)$ dual to Theorem 62.

Theorem 67 *Let P be a general program and Γ a finite set of $\mathcal{L}$ formulas. Then $\mathcal{R}_{\mathcal{L}}(P) \Vdash \Gamma$ is equivalent to each of the following four assertions:*

$$\partial_{\mathcal{L}}(P) + (p\text{-}cut) \vdash \Gamma^+, \qquad \partial_{\mathcal{L}}(P) + (p\text{-}cut) \vdash_1 \Gamma^+,$$

$$\partial_{\mathcal{L}}(P) + (p\text{-}cut) \Vdash \Gamma^+, \qquad \partial_{\mathcal{L}}(P) + (p\text{-}cut) \Vdash_1 \Gamma^+.$$

9.3 More about cut-elimination

We end this section with a theorem which states that all identity-free derivations in $\mathcal{R}_\mathcal{L}(P)$ can be transformed in identity-free derivations without essential cuts. This result can then be used to identify the identity-free derivations in $\mathcal{R}_\mathcal{L}(P)$ and $\partial_\mathcal{L}(P)$.

Theorem 68 (Cut elimination) *We have for all general programs P and finite sets Γ of $\mathcal{L}$ formulas:*

$$\mathcal{R}_\mathcal{L}(P) \Vdash \Gamma \implies \mathcal{R}_\mathcal{L}(P) \Vdash_1 \Gamma.$$

PROOF. This proof is based on the following idea: We interpret $\mathcal{R}_\mathcal{L}(P)$ with respect to its identity-free derivations into a ramified system $\mathcal{R}^\star_\mathcal{L}(P)$, eliminate the essential cuts in $\mathcal{R}^\star_\mathcal{L}(P)$ and then go back from $\mathcal{R}^\star_\mathcal{L}(P)$ to $\mathcal{R}_\mathcal{L}(P)$.

Each relation symbol R of $\mathcal{L}$ is replaced in $\mathcal{R}^\star_\mathcal{L}(P)$ by an infinite family $R^0, R^1, \ldots$ of stratified relation symbols. Given a natural number n, we obtain the n-translation of the $\mathcal{L}$ formula A into the formula A^n of $\mathcal{R}^\star_\mathcal{L}(P)$ by replacing all relation symbols R in A by R^n. Then it is easy to see that the definition of rank can be modified so that

$$rn(D^n_R(\vec{a})) < rn(R^{n+1}(\vec{a})) \tag{1}$$

for all relation symbols R, all terms $\vec{a}$ and all natural numbers n; the rank of disjunctions, conjunctions and quantified formulas is defined from its subformulas as before.

The ramified system $\mathcal{R}^\star_\mathcal{L}(P)$ contains the logical axiom (L2), the equality and freeness axioms (E1) — (E7), the logical rules and the cut rules of $\mathcal{R}_\mathcal{L}(P)$, all formulated for the language of $\mathcal{R}^\star_\mathcal{L}(P)$. The identity axioms of $\mathcal{R}_\mathcal{L}(P)$ are omitted and instead of the program rules we have for all natural numbers m and n so that $m < n$ the following *ramified program rules*:

$$\frac{\Gamma,\ D^m_R[\vec{a}]}{\Gamma,\ R^n(\vec{a})} \qquad\qquad \frac{\Gamma,\ \neg D^m_R[\vec{a}]}{\Gamma,\ \neg R^n(\vec{a})}.$$

If Γ is a finite set of $\mathcal{L}$ formulas, then Δ is called n-adequate for Γ if it results from Γ by replacing each occurrence of a relation symbols R in Γ by R^m for some m so that $n \le m$. Then the following is proved by straightforward induction on n:

$$\mathcal{R}_\mathcal{L}(P) \vdash^n \Gamma \text{ and } \Delta \text{ is } n\text{-adequate for } \Gamma \implies \mathcal{R}^\star_\mathcal{L}(P) \vdash \Delta. \tag{2}$$

Because of (1) it is further possible by standard techniques of proof theory to eliminate all cuts of rank greater than 0; i. e.

$$\mathcal{R}^\star_\mathcal{L}(P) \vdash \Gamma \implies \mathcal{R}^\star_\mathcal{L}(P) \vdash_1 \Gamma \tag{3}$$

for all finite sets Γ of formulas of $\mathcal{R}^{\star}_{\mathcal{L}}(P)$. Finally, if Γ is a finite set of formulas of $\mathcal{R}^{\star}_{\mathcal{L}}(P)$, then we write Γ° for the set of $\mathcal{L}$ formulas which results from Γ by replacing all relation symbols R^m by R. Then an easy induction on the length of the derivation shows:

$$\mathcal{R}^{\star}_{\mathcal{L}}(P) \vdash_1 \Gamma \quad \Longrightarrow \quad \mathcal{R}_{\mathcal{L}}(P) \vdash_1 \Gamma^{\circ}. \tag{4}$$

From (2) — (4) we can conclude that all cuts whose rank is greater than 0 can be eliminated from the identity-free derivations in $\mathcal{R}_{\mathcal{L}}(P)$. □

Following exactly the pattern of Lemma 59 and Lemma 64 it becomes clear that

$$\mathcal{R}_{\mathcal{L}}(P) \Vdash_1 \Gamma \iff \partial_{\mathcal{L}}(P) \Vdash_1 \Gamma^{+}$$

for all general programs P and finite sets Γ of $\mathcal{L}$ formulas. In view of Theorem 58 and Theorem 68 we conclude that the identity-free derivations in $\mathcal{R}_{\mathcal{L}}(P)$ and $\partial_{\mathcal{L}}(P)$ can be identified in the following sense.

Theorem 69 *We have for all general programs P and finite sets Γ of $\mathcal{L}$ formulas:*

$$\mathcal{R}_{\mathcal{L}}(P) \Vdash \Gamma \iff \partial_{\mathcal{L}}(P) \Vdash \Gamma^{+}.$$

Altogether we have now a series of equivalent statements which represent some proof-theoretic and model-theoretic aspects of logic programming and SLDNF-resolution. To following final corollary is an immediate consequence of Theorem 38, Theorem 51, Theorem 53, Theorem 55 and Theorem 69.

Corollary 70 *Let P be a general logic program in $\mathcal{L}$ and A an arbitrary $\mathcal{L}$ formula. Then the following assertions are pairwise equivalent:*

$$\mathcal{R}_{\mathcal{L}}(P) \Vdash A, \qquad comp_{\mathcal{L}}(P) \models_3 A, \qquad comp_{\mathcal{L}}(P) \models_4 A,$$

$$\partial_{\mathcal{L}}(P) \vdash A^{+}, \quad \partial_{\mathcal{L}}(P) + (\text{UNI}) \vdash A^{+}, \quad comp^{\natural}_{\mathcal{L}}(P) \models A^{+}.$$

References

[1] K.R. Apt, Logic programming, in: J. van Leeuwen, *Handbook of Theoretical Computer Science, Volume B*, Elsevier, Amsterdam, 1990.

[2] K.R. Apt and R. Bol, Logic programming and negation: a survey, Technical Report, CWI, Amsterdam, 1993.

[3] K.R. Apt and M.H. van Emden, Contributions to the theory of logic programming, *J. ACM 29* (1982).

[4] N.D. Belnap, A useful four-valued logic, in: J.M. Dunn and G. Epstein (eds.), *Modern Uses of Multiple-Valued Logic*, Reidel, 1977.

[5] H.A. Blair, The recursion-theoretic complexity of the semantics of predicate logic as a programming language, *Information and Control 54* (1982).

[6] K.L. Clark, Negation as Failure, in: H. Gallaire and J. Minker (eds.), *Logic and Data Bases*, Plenum Press, New York, 1978.

[7] S. Feferman, Reflecting on incompleteness, *Journal of Symbolic Logic 56* (1991).

[8] M. Fitting, A Kripke-Kleene semantics for logic programs, *Journal of Logic Programming 2* (1985).

[9] M. Fitting, Bilattices and the semantics of logic programming, *Journal of Logic Programming 11* (1991).

[10] M.L. Ginsberg, Multi-valued logics, in: *Proceedings of AAAI-86, Fifth National Conference on Artificial Intelligence*, Morgan Kaufmann, 1986.

[11] J.-Y. Girard, *Proof Theory and Logical Complexity, Volume I*, Bibliopolis, Napoli, 1987.

[12] L. Halnäs and P. Schroeder-Heister, A proof-theoretic approach to logic programming. I. Clauses as rules, *Journal of Logic and computation 1* (1990).

[13] P.G. Hinman, *Recursion-Theoretic Hierarchies*, Springer, Berlin, 1978.

[14] B. Hösli and G. Jäger, About some symmetries of negation, *Journal of Symbolic Logic* (to appear).

[15] G. Jäger, Non-monotonic reasoning by axiomatic extensions, in: J.E. Fenstad, I.T. Frolov and R. Hilpinen (eds.), *Logic, Methodology and Philosophy of Science VIII*, North-Holland, Amsterdam, 1989.

[16] G. Jäger, Some proof-theoretic aspects of logic programming, in: F.L. Bauer, W. Brauer and H. Schwichtenberg (eds.), *Logic and Algebra of Specification*, Springer, Berlin, 1993.

[17] S.C. Kleene, *Introduction to Metamathematics*, Van Nostrand, princeton, 1952.

[18] K. Kunen, Negation in logic programming, *Journal of Logic Programming 4* (1987).

[19] K. Kunen, Signed data dependencies in logic programming, *Journal of Logic Programming 7* (1989).

[20] J.-L. Lassez and M.J. Maher, Optimal fixed points of logic programs, *Theoretical Computer Science 39* (1985).

[21] J.W. Lloyd, *Foundations of Logic Programming*, Springer, Berlin, 1987 (second edition).

[22] Y.N. Moschovakis, *Elementary Induction on Abstract Structures*, North-Holland, Amsterdam, 1974.

[23] A. Mycroft, Logic programs and many-valued logic, in: M. Fontet und K. Mehlhorn (eds.) *Proceedings of Symposium on Theoretical Aspects of Computer Science*, Lecture Notes in Computer Science 166, Springer, Berlin, 1984.

[24] J.A. Robinson, A machine-oriented logic based on the resolution principle, *J. ACM 12* (1965).

[25] P. Schroeder-Heister, Hypothetical reasoning and definitional reflection in logic programming, in: P. Schroeder-Heister (ed.), *Extensions of Logic Programming*, Lecture Notes in Artificial Intelligence 475, Springer, Berlin, 1991.

[26] K. Schütte, *Proof Theory*, Springer, Berlin, 1977.

[27] J.C. Shepherdson, Language and equality theory in logic programming, Technical Report PM-88-08, University of Bristol, 1988.

[28] J.C. Shepherdson, Logics for negation as failure, Technical Report PM-90-02, University of Bristol, 1990.

[29] R.F. Stärk, A complete axiomatization of the three-valued completion of logic programs, *Journal of Logic and Computation 1:6* (1991).

[30] R.F. Stärk, *The Proof Theory of Logic Programs with Negation*, Dissertation, Universität Bern, 1992.

[31] R.F. Stärk, Negation-complete logic programs, in: E. Börger, G. Jäger, H. Kleine Büning, S. Martini and M.M. Richter (eds.), *Computer Science Logic CSL' 92*, Lecture Notes in Computer Science 702, Springer, Berlin, 1993.

[32] R.F. Stärk, Input/output dependencies of normal logic programs, *Journal of Logic and Computation* (to appear).

[33] W.W. Tait, Normal derivability in classical logic, in: J. Barwise (ed.), *The Syntax and Semantics of Infinitary Languages*, Lecture Notes in Mathematics 72, Springer, Berlin, 1968.

[34] G. Takeuti, *Proof Theory*, North-Holland, Amsterdam, 1987 (second edition).

Rewrite Proofs and Computations

Jean-Pierre Jouannaud[1]

Laboratoire de Recherche en Informatique, Bat. 490, Université de Paris Sud, F-91405 Orsay, France, jouannaud@lri.fr

Abstract. Rewriting is a general paradigm for expressing computations in various logics, and we focus here on rewriting techniques in equational logic. When used at the proof level, rewriting provides with a very powerful methodology for proving completeness results, a technique that is illustrated here. We also consider whether important properties of rewrite systems such as confluence and termination can be proved in a modular way. Finally, we stress the links between rewriting and tree automata. Previous surveys include [21; 18; 37; 12; 45; 46]. The present one owes much to [21].

Keywords. completion, confluence, critical pair, ground reducibility, inductive completion, local confluence, modularity, narrowing, order-sorted algebras, rewrite rule, rewriting, term algebra, termination, tree automata.

1 Introduction

The use of equations is traditional in mathematics. Its use in computer science has culminated with the success of algebraic specifications, a method of specifying software by encapsulating data defined abstractly by means of constructors, operations, and equations defining the action of the operations on the constructor expressions. It is also used for the axiomatic description of various calculi used in computer science.

Equations can be used for reasoning, by using Leibnitz's law of replacing equals by equals, a highly non-deterministic way of using equations. In contrast, rewriting uses the equations in one way, hence eliminating one source of non-determinism. This one way use is achieved by rewriting according to a well-founded ordering on terms, hence it is called a reduction. The other sources of non-determinism, choice of an equation, and of a subterm at which to rewrite can be eliminated as well if every expression has a unique irreducible form. In this case, rewriting defines a functional computation.

* This work was parly supported by the "Greco de programmation du CNRS", and the ESPRIT working group COMPASS.

The theory of rewriting which originates from algebra, theorem proving and computability theory centers therefore around the notion of normal form, an expression that cannot be reduced any further. Rewriting is first illustrated in section 2. The necessary algebraic apparatus is described in section 3. Confluence and termination are investigated in sections 4 and 5. How to obtain normal forms is described in section 6. Section 7 then describes all kinds of logical queries that can be solved by rewriting, provided normal forms exist. Section 8 investigates whether the existence of normal forms is a property closed under union of rewrite systems. The recognizability properties of the language of ground terms in normal form, are studies in section 9. Last but not least, we briefly mention several generalizations of rewriting in section 10.

2 Examples

Before we start the journey, we overview the road by means of a few examples taken from various knowledgeable sources. All of them raise questions which can be answered by using the material of this survey.

2.1 Hercules versus Hydra

This is one of Nachum Dershowitz's favorite examples:

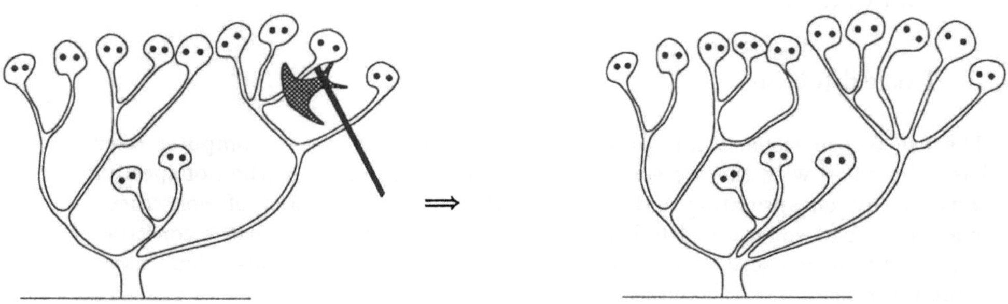

Fig. 1. Hercules versus Hydra

When Hercules chops off a head, finitely many new heads grow out from below: will Hercules eventually defeat Hydra?

This question has been extensively studied by logicians during the century. It is nothing but the uniform termination problem for this game.

2.2 Grecian Urn

This example is again taken from Dershowitz who took it from the greeks.

$$\bullet\bullet \rightarrow \circ \qquad \circ\circ \rightarrow \bullet \qquad \circ\bullet \rightarrow \circ \qquad \bullet\circ \rightarrow \circ$$

The Grecian urn contains 150 black beans, and 75 white beans. Greeks come to the urn, take out two beans and replace them by a single one according to the above rules. Can the colour of the last bean be predicted?

Uniform termination is obvious here. This game is related to an additional property: uniqueness of the resulting bean. The underlying property of the game is called confluence.

2.3 Chamelion Island

This example is due to Claude Marché, who took it from a famous mathematical contest, the "Olympiades".

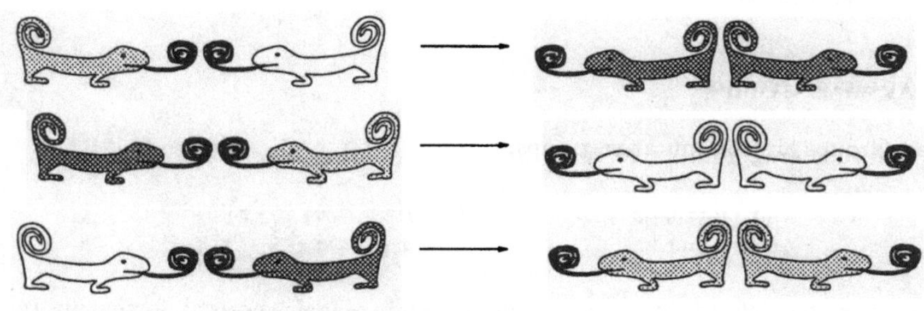

Fig. 2. Chameleons' Island

15 red, 14 blue, and 13 yellow chamelions are randomly wandering on chamelion island, and when two of them meet, they change their colour according to the above rules. Will they reach a stable state, all sharing the same colour?

Uniform termination is no more true for this game: starting from three chamelions of different colours terminates and results into three chamelions of the same colour; in contrast, starting with three pairs of chamelions, each pair of the same colour may result into an infinite sequence of changes, but may also result into a finite sequence ending in six chamelions of the same colour. The question is therefore whether there exist such an end. This is called weak termination.

2.4 Groups

Groups are usually defined by a set of three equations:

$$(x \cdot y) \cdot z = x \cdot (y \cdot z) \qquad x \cdot x^- = 1 \qquad x \cdot 1 = x$$

or, better, by a set of rewrite rules defining normal forms:

$$\begin{array}{rcl@{\qquad}rcl}
(x \cdot y) \cdot z & \rightarrow & x \cdot (y \cdot z) & (x \cdot y)^- & \rightarrow & y^- \cdot x^- \\
x \cdot x^{-1} & \rightarrow & 1 & x^- \cdot x & \rightarrow & 1 \\
x \cdot 1 & \rightarrow & x & 1 \cdot x & \rightarrow & x \\
x^{--} & \rightarrow & x & 1^- & \rightarrow & 1 \\
x \cdot (x^- \cdot y) & \rightarrow & y & x^- \cdot (x \cdot y) & \rightarrow & y
\end{array}$$

This time, we have a set of rules which is both uniformly terminating and also confluent, which allows to compute a unique normal form for any expression, allowing us to compare two expressions by checking their respective normal forms for equality. This procedure is clearly effective, in contrast with replacing equals by equals which is clearly not.

Obtaining the set of rules automatically from the set of equations is called completion. The algorithm is due to Knuth and Bendix as well as the above set of rules for groups [47].

2.5 Abelian Groups

Abelian Groups are commutative groups:

$$\begin{array}{rcl@{\qquad}rcl}
x \cdot y & = & y \cdot x & (x \cdot y) \cdot z & = & x \cdot (y \cdot z) \\
x \cdot x^- & = & 1 & x \cdot 1 & = & x
\end{array}$$

but can also be defined by a set of rules defining again normal forms up to associativity and commutativity. Indeed, these two equations cannot be turned into rules together, since otherwise normal forms could no more be computed in finite time by repeatedly replacing instances of left-hand sides by the corresponding right-hand sides.

$$x \cdot y = y \cdot x \qquad\qquad (x \cdot y) \cdot z = x \cdot (y \cdot z)$$

$$\begin{array}{rcl@{\qquad}rcl@{\qquad}rcl}
x \cdot x^{-1} & \rightarrow & 1 & & & & (x \cdot y)^- & \rightarrow & x^- \cdot y^- \\
x \cdot 1 & \rightarrow & x & 1^- & \rightarrow & 1 & x^{--} & \rightarrow & x
\end{array}$$

This kind of rewriting performs *modulo* associativity and commutativity, two axioms very common in practice. Completion modulo associativity and commutativity is due to Peterson and Stickel [63], and was later studied in more details by Jouannaud and Kirchner [38].

2.6 Commutative Rings

Rings use an additional operation, addition. The ring is commutative when so is its multiplication, as here. We give the rules only.

$$x + y = y + x \qquad (x + y) + z = x + (y + z)$$
$$x \cdot y = y \cdot x \qquad (x \cdot y) \cdot z = x \cdot (y \cdot z)$$

$$x + x^{-1} \rightarrow 0 \qquad x + 0 \rightarrow x$$
$$0^{-} \rightarrow 0 \qquad x^{--} \rightarrow x$$
$$(x + y)^{-} \rightarrow x^{-} + y^{-} \qquad x \cdot (y + z) \rightarrow (x \cdot y) + (x \cdot z)$$
$$x \cdot 1 \rightarrow x \qquad x \cdot y^{-} \rightarrow (x \cdot y)^{-} \qquad x \cdot 0 \rightarrow 0$$

2.7 Boolean Rings

The axioms are given in three parts. The first set expresses the usual connectives in terms of *and* and *exclusive or*. The next four equations state that both connectives are associative and commutative. Then, the remaining rules allow to compute normal forms built up from *and* and *exclusive or*. Note that there is no set of rules defining unique normal forms based on the usual *or* connective.

$$\neg x \rightarrow x \oplus \top \qquad x \vee y \rightarrow (x \wedge y) \oplus x \oplus y \oplus \bot$$
$$x \Rightarrow y \rightarrow (x \wedge y) \oplus x \oplus \top \qquad x \Leftrightarrow y \rightarrow (x \wedge y) \oplus x \oplus y$$

$$x \wedge y = y \wedge x \qquad (x \wedge y) \wedge z = x \wedge (y \wedge z)$$
$$x \oplus y = y \oplus x \qquad (x \oplus y) \oplus z = x \oplus (y \oplus z)$$

$$x \oplus \bot \rightarrow x \qquad x \oplus x \rightarrow \bot$$
$$x \wedge (y \oplus z) \rightarrow (x \wedge y) \oplus (x \wedge z) \qquad (x \oplus y) \wedge z \rightarrow (x \wedge z) \oplus (y \wedge z)$$
$$x \wedge \top \rightarrow x \qquad x \wedge \bot \rightarrow \bot \qquad x \wedge x \rightarrow x$$

The above convergent rewrite system for Boolean Rings was first given by Hsiang, based on previous work by Zegalkin [78], and Stone [71].

2.8 Combinatory Logic

$$I \cdot x \;\rightarrow\; x \qquad (K \cdot x) \cdot y \;\rightarrow\; x \qquad ((S \cdot x) \cdot y) \cdot z \;\rightarrow\; (x \cdot z) \cdot (y \cdot z)$$

From this example, it can then be easily predicted that rules can encode Turing machines. Actually, one single rule suffices to encode a Turing machine [14]. As a consequence, all useful properties of rewrite rules are undecidable, even when restricted to a single rule.

2.9 Programming

The following is a *rewrite program* to sort a list of natural numbers by inserting elements one by one into position, with lists represented in LISP notation and natural numbers in successor notation. *cons* and *nil* are therefore the list constructors, and

s and 0 are the constructors for natural numbers. We will use x and y for arbitrary natural numbers, and l for an arbitrary list of natural numbers.

$$\begin{array}{rr} max(0,x) \to x & min(0,x) \to 0 \\ max(x,0) \to x & min(x,0) \to 0 \\ max(s(x),s(y)) \to s(max(x,y)) & min(s(x),s(y)) \to s(min(x,y)) \end{array}$$

$$\begin{array}{rcl} sort(nil) & \to & nil \\ sort(cons(x,l)) & \to & insert(x, sort(l)) \\ insert(x,nil) & \to & cons(x,nil) \\ insert(x,cons(y,l)) & \to & cons(max(x,y), insert(min(x,y),l)) \end{array}$$

This program sorts numbers in decreasing order. If one wishes to sort them in ascending order, then the definition of *insert* must be rewritten. We therefore write a higher-order definition that will cover both cases by instantiating the two higher-order variables F and G (of type $nat \to nat \to nat$) by appropriate lambda expressions in each function *ascending_*sort and *descending_*sort.

$$\begin{array}{r} ascending\_sort(l) \to new\_sort(l, \lambda xy.min(x,y), \lambda xy.max(x,y) \\ descending\_sort(l) \to new\_sort(l, \lambda xy.max(x,y), \lambda xy.min(x,y) \\ new\_sort(nil, F, G) \to nil \\ new\_sort(cons(x,l), F, G) \to insert(x, new\_sort(l, F, G), F, G)) \\ insert(x, nil, F, G) \to cons(x, nil) \\ insert(x, cons(y,l), F, G) \to cons(Fxy, insert(Gxy, l, F, G)) \end{array}$$

The question is whether confluence and termination proofs for these sets of rules can be performed in a modular way. This is especially important for practical applications, where rules defined in different modules should be proved confluent and terminating in their respective modules. The free use of lambda-expressions for instantiating higher-order variables of course complicates the question. Further complications arise if we assume a type system rich enough to accommodate polymorphic types and if one wants to prove properties of such programs like $new\_sort(l, \lambda xy.x, \lambda xy.y) = l$.

Here, assuming a monomorphic type discipline, we can deduce the termination and confluence properties of the whole set of rules from the termination proof of each set by showing that the interactions between all these sets of rules do not falsify these properties.

3 Equations and Rewrite Rules

We recall here standard notions about terms, substitutions, rewrite relations, equations and rewrite rules.

3.1 Terms

We start with a *signature* $\mathcal{F} = \cup_{n \geq 0} \mathcal{F}_n$ of function symbols. $\mathcal{F}_0 = \mathcal{C}$ is the set of constant symbols. Assuming the sets $\mathcal{F}_n$ are disjoint, each symbol f has the *arity* n if and only if $f \in \mathcal{F}_n$. We also assume a (denumerable) set $\mathcal{X}$ of variable symbols.

The set of (first-order) *terms* $\mathcal{T}(\mathcal{F},\mathcal{X})$ over $\mathcal{F}$ and $\mathcal{X}$ is the smallest set containing $\mathcal{X}$ such that $f(t_1,\ldots,t_n)$ is in $\mathcal{T}(\mathcal{F},\mathcal{X})$ whenever $f \in \mathcal{F}_n$ and $t_i \in \mathcal{T}(\mathcal{F},\mathcal{X})$ for $i = 1,\ldots,n$. Variable-free terms are called *ground*; the set $\mathcal{T}(\mathcal{F},\emptyset)$ of ground terms is also written $\mathcal{G}(\mathcal{F})$.

A term t in $\mathcal{T}(\mathcal{F},\mathcal{X})$ may be viewed as a finite ordered tree, the leaves of which are labeled with variables (from $\mathcal{X}$) or constants (from $\mathcal{C}$) and the internal nodes of which are labeled with function symbols (from $\mathcal{F}-\mathcal{C}$) of positive arity, with outdegree equal to the arity of the label. A *position* within a term may be represented—in Dewey decimal notation—as a sequence of positive integers, describing the path from the outermost *root* symbol to the head of the subterm at that position. By $t|_p$, we denote the *subterm* of t rooted at position p, and by $t[s]_p$ the term obtained by replacing t_p by s in t. As a consequence, $t = t[t_p]_p$. We use Λ for the root position, $\mathcal{P}os(t)$ for the set of positions of t, and $\mathcal{FP}os(t)$ for the set of non-variable positions of t.

We use $|t|$ for the size (number of nodes) of t, and $\mathcal{V}ar(t)$ for its set of variables; we say that t is *linear* if no variable occurs more than once in t.

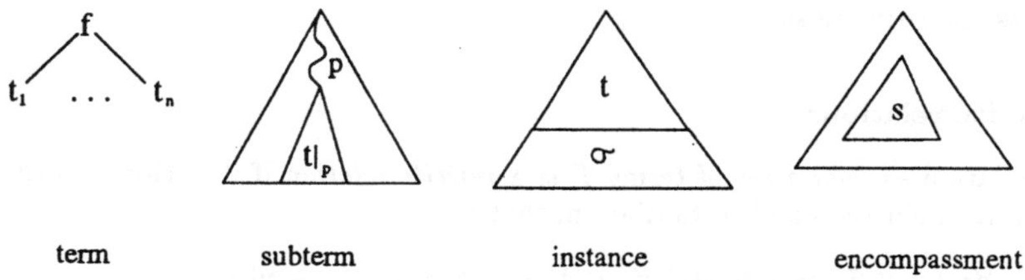

Fig. 3. Tree Representation of Terms.

3.2 Substitutions

A *substitution* is a special kind of replacement operation, uniquely defined by a mapping from variables to terms, and written out as $\{x_1 \mapsto s_1,\ldots,x_m \mapsto s_m\}$ when there are only finitely many variables x_i not mapped to themselves. Formally, a substitution σ is a function from $\mathcal{X}$ to $\mathcal{T}(\mathcal{F},\mathcal{X})$, extended to a function again written σ from $\mathcal{T}$ to itself for which we use postfix notation in such a way that $f(t_1,\ldots,t_n)\sigma = f(t_1\sigma,\ldots,t_n\sigma)$, for each $n \in N, f \in \mathcal{F}_n, i \in [1..n]$, *and* $t_i \in \mathcal{T}$. By a ground substitution, we mean a substitution γ such that $x\gamma$ is ground for all x in $\mathcal{X}$. We use postfix juxtaposition for composition of substitutions, and Σ (resp. Γ) for the set of all (resp. ground) substitutions.

A term s is *more general than* (*subsumes*) a term t, writen $s \succeq t$, if $s\sigma = t$ for some substitution σ called a *match* from s to t; we also say that t is an *instance* of s. The relation $\succeq$ is a quasi-ordering on terms, called *subsumption*, whose associated equivalence $\doteq$ is called *renaming* (*conversion*).

Subsumption and the subterm ordering are special cases of the *encompassment* quasi-ordering, in which $s \trianglelefteq t$ if a subterm of t is an instance of s, that is $s\sigma = t|_p$ for some $p \in \mathcal{P}os(t)$ and $\sigma \in \Sigma$.

We say that a substitution σ unifies two terms s and t iff $s\sigma = t\sigma$. A substitution σ is *more general* than (*subsumes*) a substitution τ, or τ is an *instance* of σ, and we write $\sigma \leq \tau$, if there exists a substitution θ such that $\sigma\theta = \tau$.

The following key result is due to Herbrand [30] and Robinson [67]:

Theorem 1. *If two terms s and t are unifiable, then they have a most general unifier mgu(s,t) which is unique up to renaming.*

3.3 Lexicography

To fix nomenclature, the letters a through h will be used for function symbols; l, r, and s through w will denote arbitrary terms; x, y, and z will be reserved for variables; p and q, for positions; lower case Greek letters, for substitutions.

Binary relations will frequently be denoted by arrows of one kind or another. If $\rightarrow$ is a binary relation, then $\leftarrow$ is its inverse, $\leftrightarrow$ is its symmetric closure ($\leftarrow \cup \rightarrow$), $\stackrel{=}{\rightarrow}$ is its reflexive closure ($\rightarrow \cup =$), $\stackrel{*}{\rightarrow}$ is its reflexive-transitive closure ($\rightarrow \circ \cdots \circ \rightarrow$) and $\stackrel{+}{\rightarrow}$ is its transitive closure ($\rightarrow \circ \stackrel{*}{\rightarrow}$).

3.4 Rewrite relations

A binary relation $\rightarrow$ over a set of terms $\mathcal{T}$ is a *rewrite relation* if it is closed both under context application and instantiation, that is

$$\forall s, t, u \in \mathcal{T}, \forall p \in \mathcal{P}os(t), \forall \sigma \in \Sigma, s \rightarrow t \Rightarrow u[s\sigma]_p \rightarrow u[t\sigma]_p.$$

The inverse, symmetric closure, reflexive closure, and transitive closure of any rewrite relation are also rewrite relations.
An ordering on terms which satisfies the above property will be called a *rewrite ordering*. An equivalence on terms which satisfies the above property is *congruence*.

Note that encompassment lies at the heart of the above definition of a rewrite relation, since a term u is rewritable iff it encompasses some rewritable term s.

Rewrite relations on terms can be generated by equations, that is unordered rewritable pairs of terms, if they are symmetric or else by rewrite rules, that is ordered rewritable pairs of terms. So, rewrite rules really differ from equations in their use.

3.5 Equations

Equation are pairs of terms written $l \simeq r$. Let E be a set of equations.

A term s rewrites to a term t at position p using an equation $l \simeq r \in E$, if there exist a substitution σ such that $s|_p = l\sigma$, and $t = s[r\sigma]_p$, or $s|_p = r\sigma$, and $t = s[l\sigma]_p$. The *equational proof step* between s and t is written $s \xleftrightarrow[l \simeq r]{p} t$ in the first case, and $s \xleftrightarrow[r \simeq l]{p} t$ in the second. We may drop the subscripts or superscripts, use E as a subscript if the actual equation is not important, as well as describe a position by a predicate if necessary. The sequence $s \xleftrightarrow[E]{*} t$ of equational proof steps between

s and t is called a proof, and $\overset{*}{\underset{E}{\longleftrightarrow}}$ is therefore the congruence relation generated by E.

For example, given the signature $0 \in \mathcal{F}_0, S \in \mathcal{F}_1, + \in \mathcal{F}_2$ for specifying the natural numbers with addition, and the usual equations for defining the addition of two natural numbers, we have the equational proof:

$$S(x)+0 \overset{\Lambda}{\underset{x+0\simeq x}{\longleftrightarrow}} S(x) \overset{1}{\underset{x+0\simeq x}{\longleftrightarrow}} S(x+0) \overset{\Lambda}{\underset{x+S(y)\simeq S(x+y)}{\longleftrightarrow}} x+S(0)$$

The semantics of equations is a particular case of the semantics of first-order predicate logic in which the only predicate symbol is interpreted as the equality. Call an *$\mathcal{F}$-Algebra $\mathcal{A}$* a pair made of a *domain* (*carrier*) A and a set of *operations* $\{f_A\}_{f\in\mathcal{F}}$ on A. A homomorphism from an *$\mathcal{F}$-Algebra $\mathcal{A}$* to another *$\mathcal{F}$-Algebra $\mathcal{B}$* is any application from A to B which preserves the operations, that is it satisfies the property $h(f_A(e_1,\ldots,e_n)) = f_B(h(e_1),\ldots,h(e_n))$. Homomorphisms from $\mathcal{A}$ to $\mathcal{A}$ are called *endomorphisms*, of which bijective ones are called *isomorphisms*.

A particular $\mathcal{F}$-algebra can be constructed from the signature $\mathcal{F}$ which plays a crucial role, and is called the *term $\mathcal{F}$-algebra* over $\mathcal{X}$. Its domain is the set of term $\mathcal{T}(\mathcal{F},\mathcal{X})$ (hereafter written $\mathcal{T}$), and its operations are defined by $f_{\mathcal{T}}(t_1,\ldots,t_n) = f(t_1,\ldots,t_n)$. Endomorphisms of the term algebra are indeed substitutions. Ground substitutions are homomorphisms from the term algebra to the ground term algebra, denoted by $\mathcal{G}(\mathcal{F})$ whose domain is the set of ground terms.

The term algebra $\mathcal{T}$ is also called the *free* algebra. It enjoys an important property: any application from $\mathcal{X}$ to the domain A of any $\mathcal{F}$-algebra $\mathcal{A}$ can be uniquely extended into a homomorphism from the free algebra $\mathcal{T}$ to $\mathcal{A}$. As a consequence, there is a unique homomorphism from the ground term algebra $\mathcal{G}$ (also called the *initial* $\mathcal{F}$-algebra) to any $\mathcal{F}$-algebra $\mathcal{A}$.

Homomorphisms allow to interpret formulae in a particular algebra. We say that $\mathcal{A}$ *satisfies* (is a *model* of) the equation $u = v$ iff $h(u) = h(v)$. We use the notation $\mathcal{A} \models u = v$, and denote by $\mathcal{M}(\mathcal{E})$ the class of models of E, that is the class of algebras which are models of each equation in E.

We now construct a particular model of E, called the *quotient algebra* and denoted $\mathcal{T}/\overset{*}{\underset{E}{\longleftrightarrow}}$, whose domain $\{[t]_E\}_{t\in\mathcal{T}}$ is made of the congruence classes defined in the term algebra by the congruence $\overset{*}{\underset{E}{\longleftrightarrow}}$ generated by the set E of equations, and whose operations are defined as expected by $f_{\mathcal{T}/\overset{*}{\underset{E}{\longleftrightarrow}}}([t_1]_E,\ldots,[t_1]_E) = [f(t_1,\ldots,t_n)]_E$. The quotient algebra is the free algebra (in the above sense) of $\mathcal{M}(\mathcal{E})$, and its subalgebra $\mathcal{G}/\overset{*}{\underset{E}{\longleftrightarrow}}$ is the initial one, also called the *standard model*. Satisfaction in the free algebra is actually equivalent to satisfaction in $\mathcal{M}(\mathcal{E})$, a result of Birkhoff [6]:

Theorem 2. $\mathcal{M}(\mathcal{E}) \models u = v$ *iff* $\mathcal{T}/\overset{*}{\underset{E}{\longleftrightarrow}} \models u = v$.

Let us again consider the previous example of natural numbers with addition, and define two algebras which are models of the axioms for addition:

For $\mathcal{A}$, we take the standard model, in which the domain is the set $\{0, 1, 2, \ldots\}$ of natural numbers, and the operations are $\{zero, succ, plus\}$ defined as expected.
For $\mathcal{B}$, we take a non-standard model, whose domain $\{0, 1, 2, \ldots, \infty, *\}$ includes the set of natural numbers, and whose operations $\{zero, succ, plus\}$ need the additional definitions $succ(\infty) = \infty, succ(*) = *, plus(x, \infty) = *, plus(x, *) = \infty, plus(\infty, y) = \infty$, and $plus(*, y) = *$.
It can be routinely checked that both $\mathcal{A}$ and $\mathcal{B}$ are models of the axioms $x + 0 \simeq x$ and $x + S(y) \simeq S(x + y)$. On the other hand, the standard model satisfies the commutativity axiom for $+$, while the other one does not. Indeed, it can be proved that the set of equations valid in the standard model is exactly the set of equations which can be proved by using replacement of equals by equals and structural induction over ground terms. We therefore call ***inductive equalities*** the equations in the following set:

$$Ind(E) = \{u = v : \mathcal{G}/\underset{E}{\overset{*}{\longleftrightarrow}} \models u = v\}$$

3.6 Rules

Rewrite rule are pairs of terms written $l \rightarrow r$. Let R be a set of rewrite rules.

A term s rewrites to a term t at position p using a rule $l \rightarrow r$ if there exist a substitution σ such that $s|_p$ is the *redex* $l\sigma$, and $t = s[r\sigma]_p$. We write $s \xrightarrow[l\rightarrow r]{p} t$ or $t \xleftarrow[l\rightarrow r]{p} s$. We may also write $s \underset{R}{\longleftrightarrow} t$ for $s \underset{R}{\longrightarrow} \cup \underset{R}{\longleftarrow} t$, hence using rules as equations.
The *derivation relation* $\xrightarrow[R]{*}$ is the reflexive transitive closure of the above rewrite relation, while $\underset{R}{\overset{*}{\longleftrightarrow}}$ is the congruence relation generated by R. The sequence of rewrite steps $s \underset{R}{\overset{*}{\longleftrightarrow}} t$ is called a *proof.*

For example, given the signature $0 \in \mathcal{F}_0, S \in \mathcal{F}_1, + \in \mathcal{F}_2$ for specifying the natural numbers with addition, and the usual rules for defining the addition of two natural numbers, we have the proofs:

$$S(x) + 0 \xrightarrow[x+0\rightarrow x]{\Lambda} S(x) \xleftarrow[x+0\rightarrow x]{1} S(x+0) \xleftarrow[x+S(y)\rightarrow S(x+y)]{\Lambda} x + S(0)$$

Definition 3. We say that s and t are *convertible* when there is a proof $s \underset{R}{\overset{*}{\longleftrightarrow}} t$. Proof shapes are depicted by appropriate names: $s \xrightarrow[R]{*} u \xleftarrow[R]{*} t$ is called a *valley* or *rewrite proof,* in which case we say that s and t are *joinable*, $s \xrightarrow[R]{=} u \xleftarrow[R]{=} t$ is a *flat* or *diamond proof,* $u \xleftarrow[R]{} s \xrightarrow[R]{} v$ is a *peak*, and $u \xleftarrow[R]{*} s \xrightarrow[R]{*} v$ is a *mountain.*

Note that a rewrite proof is simply a proof with no peak. The proof in the previous example is a rewrite proof.

We say that a term s is reducible if it contains a redex, and *irreducible* (in *normal form*) otherwise. We denote by $s\!\downarrow$ a normal form for s, that is an irreducible term t such that $s \xrightarrow[R]{*} t$, and we say that R is functional (*normalizing*) if every term has a unique computable R-normal form.

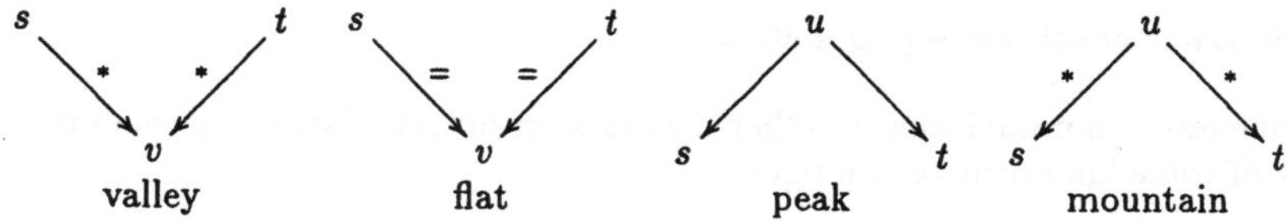

Fig. 4. Proof Shapes.

When R is functional, all terms belonging to a same class of $\overset{*}{\underset{R}{\longleftrightarrow}}$ have the same normal form, as can be easily shown by induction on the length of proofs. This normal form can be used as a representative of the class. We can therefore construct a variant of the quotient algebra, called the ***normal-form algebra*** and denoted $\mathcal{T}{\downarrow}_{\mathcal{R}}$, whose domain is the set $\{t\downarrow\}_{t\in\mathcal{T}}$ of terms in normal form, and whose operations are defined as expected by $f_{\mathcal{T}\downarrow_{\mathcal{R}}}(t_1\downarrow,\ldots,t_n\downarrow) = f(t_1,\ldots,t_n)\downarrow$. This algebra is clearly isomorphic to the quotient algebra, hence

$$\mathcal{M}(\mathcal{E}) \models u = v \; iff \; \mathcal{T}/_R \models u = v \; iff \; u\downarrow = v\downarrow .$$

Considering now the subalgebra $\mathcal{G}\downarrow_{\mathcal{R}}$ whose domain is the set of ground terms in normal form, it is therefore isomorphic to the initial algebra defined by R:

$$\mathcal{G}/_{\overset{*}{\underset{R}{\longleftrightarrow}}} \models u = v \; iff \; \mathcal{G}\downarrow_R \models u = v \; iff \; u\downarrow = v\downarrow .$$

As a consequence, rewriting is the main paradigm used for prototyping algebraic specifications.

4 Confluence Properties

Functionality appears therefore crucial. Unfortunately, it is undecidable and indeed very difficult to obtain. We will therefore approximate it, either from above by strengthening it, or from below, by relaxing it.

Approximating it from above yields the theory of terminating rewriting systems, in which every term posseses at least one normal form by assuming termination:

Definition 4. A relation $\longrightarrow$ is said to be ***terminating*** (***strongly normalizing***) if there exists no infinite sequence of related elements $t_1 \longrightarrow t_2 \longrightarrow \ldots \longrightarrow t_i \longrightarrow \ldots$.

Approximating it from below yields the theory of combinatory reduction systems, of which the lambda-calculus is a particular case. Here, we will simply require that every term that posseses a normal form possesses a unique one. But some terms may have no normal form at all, and of course, there may be non-terminating sequences of rewrites issued from a term that possesses a normal form.

In both cases, it is useful to study first abstract properties of relations, which we do now.

4.1 Abstract confluence properties

The uniqueness of normal forms, whether they exist or not, is related to a confluence property of some kind depicted on figure 5:

Definition 5. A relation $\rightarrow$ is said to be:
Church-Rosser if every proof enjoys a rewrite proof: $\forall s, t \text{ s.t. } s \overset{*}{\longleftrightarrow} t,\ \exists v \text{ s.t. } s \overset{*}{\longrightarrow} v \overset{*}{\longleftarrow} t$,
confluent if every mountain enjoys a rewrite proof: $\forall u, s, t \text{ s.t. } s \overset{*}{\longleftarrow} u \overset{*}{\longrightarrow} t,\ \exists v \text{ s.t. } s \overset{*}{\longrightarrow} v \overset{*}{\longleftarrow} t$,
locally confluent if every peak enjoys a rewrite proof: $\forall u, s, t \text{ s.t. } s \overset{*}{\longleftarrow} u \overset{*}{\longrightarrow} t,\ \exists v \text{ s.t. } s \overset{*}{\longrightarrow} v \overset{*}{\longleftarrow} t$,
strongly confluent if every peak enjoys a diamond proof: $\forall u, s, t \text{ s.t. } s \overset{*}{\longleftarrow} u \overset{*}{\longrightarrow} t,\ \exists v \text{ s.t. } s \overset{*}{\longrightarrow} v \overset{*}{\longleftarrow} t$.

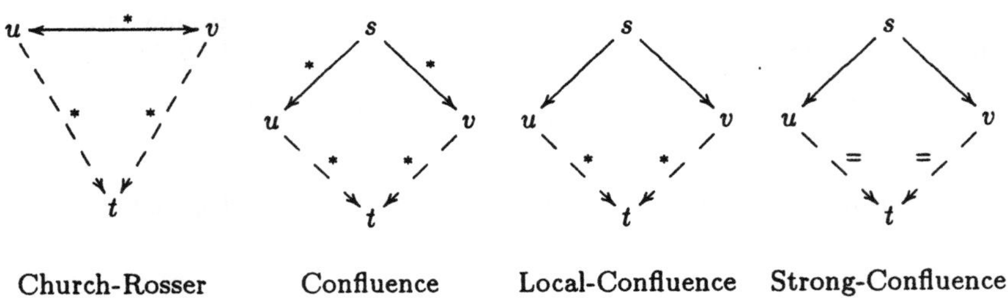

Fig. 5. Confluence Properties.

These properties are intimately related. First of all, the Church-Rosser property and confluence are equivalent as can be easily shown by induction on the length of an arbitrary proof (or better on $\rightarrow_{mul}$ which allows a single use of the induction argument). Confluence and local confluence are not equivalent in general, as shown on the picture 6. However, they are equivalent when the relation $\rightarrow$ is terminating, a result known as Newmann's lemma [59], whose proof by induction on $\rightarrow$ is pictured below.

Theorem 6. *: Assume $\rightarrow$ is terminating. Then $\rightarrow$ is confluent iff it is locally confluent.*

The alternative to termination is the use of strong confluence, a result due to Rosen [68] which can be proved by using a tiling argument as shown on picture 6.

Theorem 7. *A relation $\rightarrow$ is confluent if it is strongly confluent.*

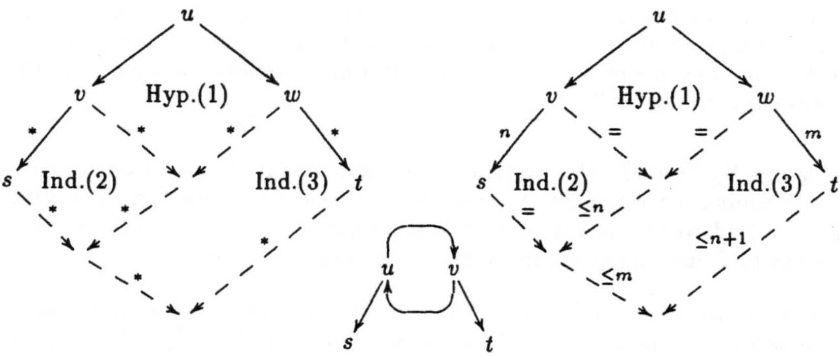

Fig. 6. Newman's and Rosen's Lemmas.

Although strong confluence appears to be very special, it is often used to prove the confluence of a relation $\longrightarrow$ by applying it to another relation $\Rightarrow$ having the same transitive closure as $\longrightarrow$. This applies in particular to the pure lambda-calculus, and more generally to combinatory reduction systems in the sense of Klop [46].

4.2 Confluence properties of rewrite relations

In practice, what really matters is to have functionality for ground terms. We therefore will distinguish two kinds of confluence properties of a rewrite relation, on $\mathcal{T}$ called *Church-Rosser* and *confluence*, and on $\mathcal{G}$ called *ground Church-Rosser* and *ground confluence*. Ground confluence, however, is undecidable even for terminating relations. In contrast, confluence can be more easily grasped, and we will see that it is indeed decidable for terminating rewrite relations.

Orthogonal Systems

Definition 8. A rewrite system R is *linear* if the left-hand side of each rule in R is linear, *overlapping* if a left-hand side unifies with a renamed non-variable subterm of any other left-hand side or with a renamed proper subterm of itself, and *orthogonal* if it is both left-linear and non-overlapping.

The importance of orthogonal systems lies in the following result:

Theorem 9 [32]. *Every orthogonal system is confluent.*

What is really shown is the strong confluence of the parallel rewrite relation, in which all redexes in a term are rewritten at once thanks to the absence of overlappings and the left-linearity of rules. Obviously, this relation has the same transitive closure as the starting one. The strong confluence proof is similar to the local confluence proof depicted on 8, but convertible terms s and t can now always be joined in at most one step from each by rewriting all redexes at once since the rewrite system is orthogonal.

Terminating Systems In this section, we restrict our attention to terminating relations. Terminating confluent systems are of course functional: any rewriting strategy will lead to the unique normal form. Besides, local confluence can itself be restricted to finitely many distinguished peaks, as shown now:

Definition 10. Let $l \rightarrow r$ and $g \rightarrow d$ be two rewrite rules (with variables renamed apart) and σ a most general unifier of g and a nonvariable subterm $l|_p$ of l. Then the peak $r\sigma \xleftarrow[l \rightarrow r]{\Lambda} l\sigma \xrightarrow[g \rightarrow d]{p} l\sigma[d\sigma]_p$ is said *critical*, and the equation $r\sigma \simeq l\sigma[d\sigma]_p$ is called a *critical pair*. We use $cp(R)$ for the set of all critical pairs between rules in R.

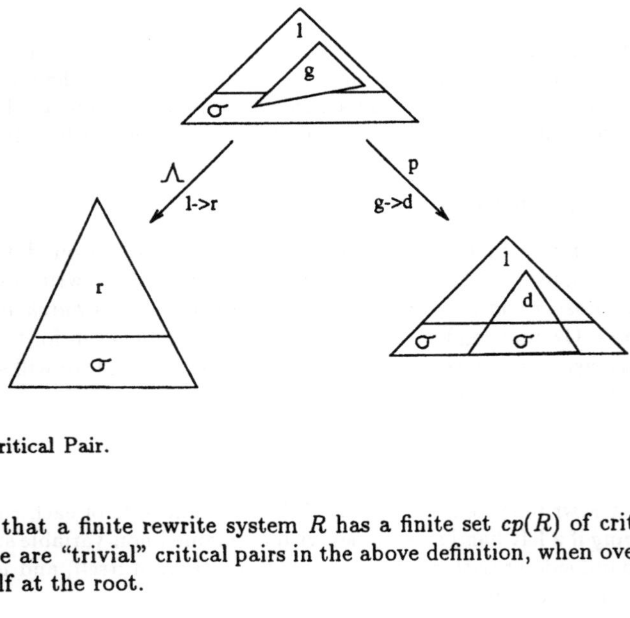

Fig. 7. Critical Pair.

Note that a finite rewrite system R has a finite set $cp(R)$ of critical pairs, and that there are "trivial" critical pairs in the above definition, when overlapping a rule with itself at the root.

Theorem 11 [47]. *A rewrite system is locally confluent if and only if all its critical pairs are joinable.*

The proof is depicted on figure 8

Therefore, a terminating system is confluent if and only if all its critical pairs are joinable, a decidable property.

We show next two examples of local-confluence proofs. For the Grecian urn which comes first figure 9, all critical pairs are joinable, hence the Grecian urn is confluent

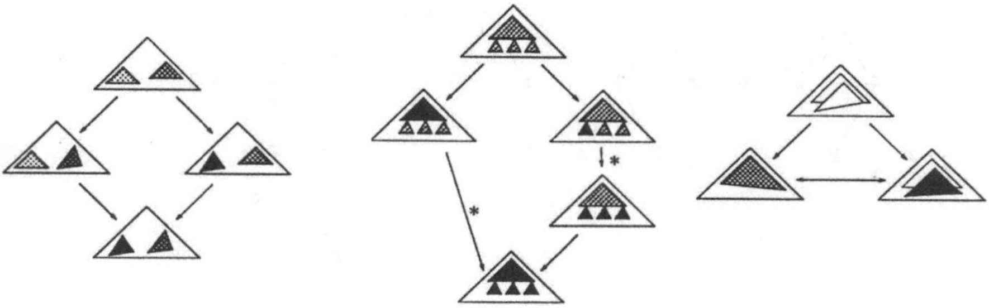

Fig. 8. Critical Pair Lemma.

since it is trivially terminating. For the group theory which comes second figure 10, there are four critical pairs of which two are confluent and the other two are not. Note how associativity overlaps itself yielding one of the two joinable critical pairs.

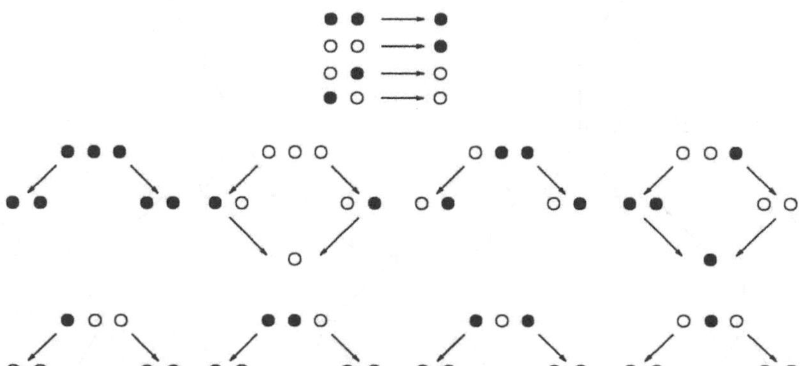

Fig. 9. Critical Pairs of the Grecian Urn

Canonical systems Rewrite systems which are both terminating and confluent are called *convergent*. We now consider an additional property in order to characterize a rewrite system under a given order.

Definition 12. A system R is *reduced* if the following terms are irreducible:

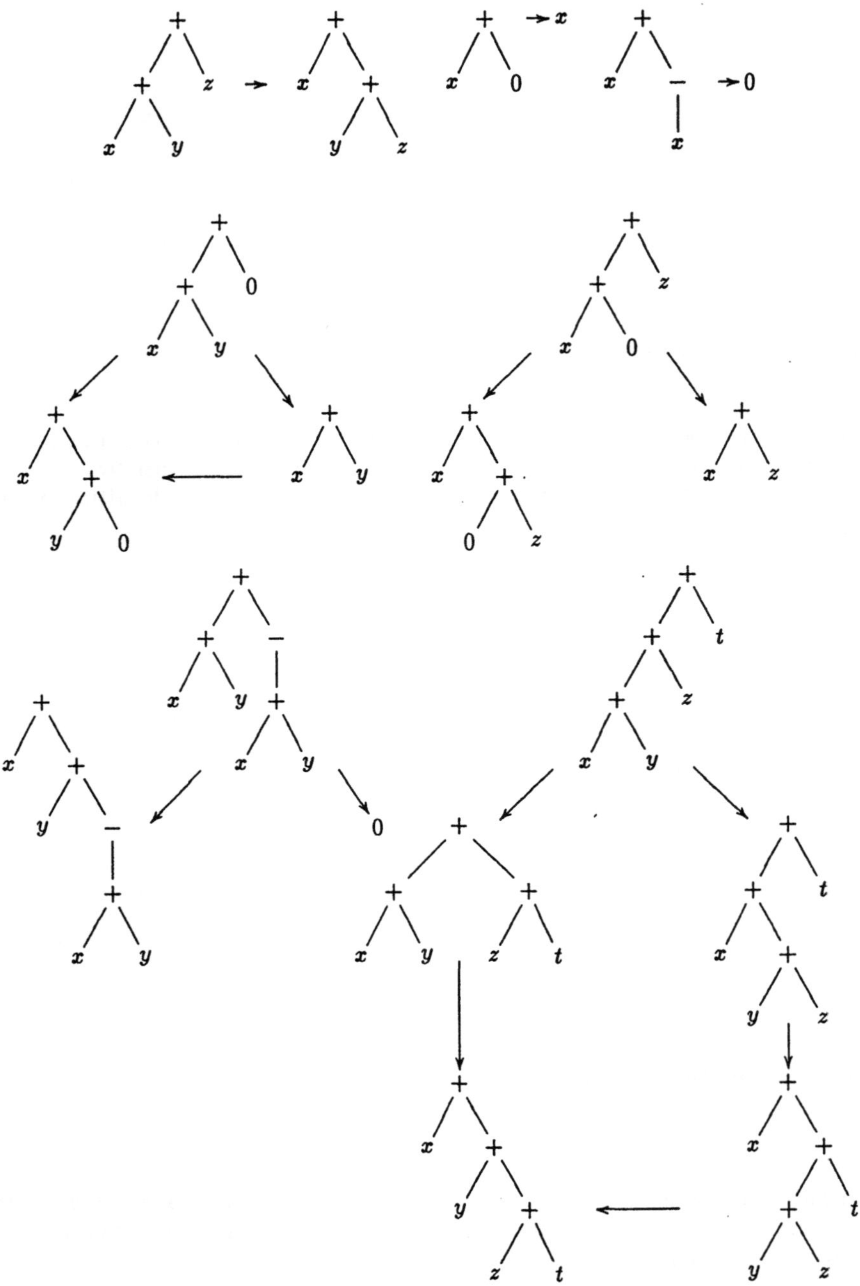

Fig. 10. Critical Pairs of Group theory.

the right-hand side r of each rule $l \rightarrow r \in R$, and
any term s properly encompassed into a left-hand side l of a rule $l \rightarrow r \in R$. A confluent, terminating, and reduced rewrite system is called *canonical.*

In particular, a strict subterm of a left-hand side must be irreducible in a reduced system. Note also that a reduced system may contain two rules $l \rightarrow r$ and $l' \rightarrow s$ such that l' is a renaming of l. In contrast, this is not possible in a canonical system due to confluence.

Canonical systems have an interesting uniqueness property, an observation first made by Lankford:

Theorem 13 [55]. *Suppose R and S are two canonical (not necessarily finite) rewrite systems having the same equational theory. Suppose further that the combined system $R \cup S$ is terminating. Then R and S must be the same (up to renaming of variables).*

This result has important consequences to be developed in section 6. Its proof is routinely obtained by expressing that each rule in R has a rewrite proof in S and vice versa.

5 Termination

In general, it is undecidable whether a system is terminating, even if both sides of all rules are monadic [35] or if it has only one left-linear rule [14]. For ground systems, however, termination is decidable [35].

In order to concentrate the termination proof on the finite number of rewrite rules rather than on the infinite number of possible rewrites, we need a special kind of ordering:

Definition 14. A *reduction ordering* on a set of terms $\mathcal{T}$ is any well-founded rewrite ordering of $\mathcal{T}$.

Termination is assured if each of the rules in R is contained in a reduction ordering; conversely, if R is terminating, then the relation $\xleftrightarrow[R]{+}$ itself is a reduction ordering.

Rewrite ordering are necessary for proving termination of rules which may apply at subterms. Rules applying always at the top, of which PROLOG is a particular case do not require well-founded orderings closed under replacement. Similarly, ground rules do not require well-founded orderings closed under instantiation. This remark is used in section 6.3.

5.1 Interpretation Orderings

We therefore need to provide with powerful enough reduction orderings that will enable us to solve the practical cases we may face with. A popular method for designing reduction orderings is by interpreting terms in some well-founded set, the set of terms being a particular case. The interpretation mapping will be a homomorphism τ from the ground terms $\mathcal{G}(\mathcal{F})$ to the well-founded set seen as an $\mathcal{F}$-algebra **A** equipped with a well-founded ordering $\succ$. The homomorphism and ordering are constrained to satisfy the following monotonicity condition: $f_\tau(\ldots x \ldots) \succ f_\tau(\ldots y \ldots)$ whenever $x \succ y$, for all f in $\mathcal{F}$ and all x, y, etc. in A. Then, the ordering $\succ_\tau$, under which $s \succ t$ if $\tau(s) \succ \tau(t)$, for s and t in $\mathcal{G}$, is well-founded.

The previous method allows us to construct reduction orderings on ground terms, which can then be lifted to open terms. By defining $s \succ t$ for arbitrary terms s and t iff $s\sigma \succ t\sigma$ for all ground substitutions σ, one obtains the finest possible such extension, but the obtained ordering may not be decidable. By adding the variables to the domain A of $\mathcal{A}$, we can map the variables in $\mathcal{X}$ to their copy in A. Then, let $s \succ_\tau t$ only if $\tau(s) \succ \tau(t)$ for all assignments of values in A to the variables in $\tau(s)$ and $\tau(t)$. A system is terminating iff such $\mathcal{A}$, τ, and $\succ$ exist.

For example, the system below, which computes the disjunctive normal form of formulae, can be shown terminating with an exponential mapping into the natural numbers:

$$\begin{aligned} not(not(x)) &\rightarrow x \\ not(or(x,y)) &\rightarrow and(not(x), not(y)) \\ not(and(x,y)) &\rightarrow or(not(x), not(y)) \\ and(x, or(y,z)) &\rightarrow or(and(x,y), and(x,z)) \\ and(or(y,z), x) &\rightarrow or(and(y,x), and(z,x)) \end{aligned}$$

The mapping $\tau : \mathcal{T} \rightarrow \{2, 3, \ldots\}$ is defined by:

$$\begin{aligned} or_\tau(x,y) &= x + y + 1 \qquad & not_\tau(x) &= 2^x \\ and_\tau(x,y) &= x \times y & c_\tau &= 2 \end{aligned}$$

for all x and y in $\{2, 3, \ldots\}$ and constant c in $\mathcal{F}_0$, with numbers compared under their natural ordering $>$. The second rule, for instance, always decreases the interpretation of a term, since $\tau(not(or(x,y))) = 2^{x+y+1}$ is greater than $\tau(and(not(x), not(y))) = 2^{x+y}$, for all x and y.

The use, in particular, of *polynomial interpretations* was developed in [51; 52]. Here a multivariate integer polynomial $f_\tau(x_1, \ldots, x_n)$ of degree n is associated with each n-ary symbol f in $\mathcal{F}_n$, for all n. The choice of coefficients must satisfy the monotonicity condition and ensure that terms are mapped into nonnegative integers only, as is the case, for example, when all coefficients are positive. Each rule $l \rightarrow r$ must be shown to be reducing; the polynomial $\tau(l) - \tau(r)$ must be positive for all values of variables greater than the minimal value of a ground term.

5.2 Extension Orderings

We will insist here in *divide and conquer* methods for building orderings, based on the use of functionals that preserve well-founded orderings. There are three major ones.

Lexicographic extension The *lexicographic extension* transforms n well-founded orderings $>_1, \ldots, >_n$ on n sets $S_1, \ldots, S_n$ into a well-founded ordering $(>_1, \ldots, >_n)_{lex}$ on the product $S_1 \times \ldots \times S_n$, by letting:

$$(a_1, \ldots, a_n)(>_1, \ldots, >_n)_{lex}(b_1, \ldots, b_n) \text{ if}$$

$$a_1 >_1 b_1 \text{ or } a_1 = b_1 \text{ and } (a_2, \ldots, a_n)(>_2, \ldots, >_n)_{lex}(b_2, \ldots, b_n)$$

We use to write $>_{lex}$ when all orderings $>_1, \ldots, >_n$ are the same ordering $>$. This definition extends to quasi-orderings, by using the equivalences associated to the quasi-orderings instead of the syntactic equality.

Example: $\gtrdot$ and $\trianglerighteq$ are well-founded orderings on terms.

To prove the claim, we interpret a term t by the pair $\langle |s|, |\{p \in \mathcal{P}os(t) : t|_p \in \mathcal{X}\} - \{\mathcal{V}ar(t)|\rangle$, and compare pairs in the lexicographic extension $(>_N)lex$.

Multiset Extension The *multiset extension* transforms any well-founded ordering $>$ on a set S into a well-founded ordering $\geq_{mul}$ on multisets on S. It is defined as the reflexive transitive closure of the relation between two multisets, say M_1 and M_2 such that M_2 is obtained from M_1 by replacing any element x by a finite number (possibly zero) of elements strictly smaller than x in the ordering on S.

This ordering is not yet powerful enough to show that Hercules will eventually defeat the Hydra. Actually, the Hydra is a kind of *nested multiset*. Defining the appropriate extension is left to the reader. Note however that a nested multiset is a term on a vocabulary of one varyadic symbol.

Term Extensions The *term extension*, also called *recursive path ordering* transforms any well-founded ordering $\succ$ (called a *precedence*) on an alphabet $\mathcal{F}$ of function symbols into a rewrite-ordering $\succ_{rpo}$ on terms in $\mathcal{T}(\mathcal{F})$. To define it, we must first assign a status, lexicographic or multiset to all function symbols in $\mathcal{F}$, therefore split into two subsets Lex and Mul. The induced ordering is as follows:

$$s = f(s_1, \ldots, s_m) \succeq_{rpo} g(t_1, \ldots, t_n) = t$$

if either of the following (1 or 2 or 3) hold:

(1) $s_i \succeq_{rpo} t$ for some s_i, $i \in [1..m]$

(2) $f = g$ and

case a: $f \in Lex$, and $\begin{cases} s \succ_{rpo} t_1, \ldots, s \succ_{rpo} t_n \\ (s_1, \ldots, s_n)(\succeq_{rpo})_{lex}(t_1, \ldots, t_n) \end{cases}$

case b: $f \in Mul$, and $(s_1, \ldots, s_n)(\succeq_{rpo})_{mul}(t_1, \ldots, t_n)$.

(3) $f \succ g$ and $s \succ_{rpo} t_i$ for all $i \in [1..n]$;

where $s \succ t$ if $s \succeq t$, but $s \not\preceq t$. When the precedence on function symbols is total, then the recursive path ordering is total. The ordering is strict if there are no multiset symbols, and its equivalence is generated by Malcev's permutative axioms for the multiset symbols otherwise.

See [20] for a more general version of this ordering mixing syntactic and semantic considerations.

As previously discussed for interpretation orderings, there are two ways to extend the recursive path ordering to a rewrite ordering on terms with variables. The finest extension turns out to be decidable [11; 42] but NP-hard. The other (polynomial) one is obtained by considering the variables as new constants which do not compare with other function symbols nor between themselves in the precedence.

For example, we can prove the termination of the Ackerman's function by using the lexicographic version of the recursive path ordering:

$$\begin{aligned} ack(0, y) &\rightarrow succ(y) \\ ack(succ(x), 0) &\rightarrow ack(x, succ(0)) \\ ack(succ(x), succ(y)) &\rightarrow ack(x, ack(succ(x), y)) \end{aligned}$$

For example, the third rule is contained in $>_{lpo}$ since x occurs in $succ(x)$ and $ack(succ(x), succ(y))$ is lexicographically greater than $ack(succ(x), y)$.

We can also prove termination of the first set of rules of section 2.9 by letting: $sort \succ insert \succ cons \succ max \succ min \succ S$. It is actually not necessary to compare $cons$, S and 0, nor is it actually necessary to compare max and min.

Trees are Ordinals This suggestive title is again due to Dershowitz [19]. Since the term ordering is total on $\mathcal{G}$ when all statuses are lexicographic, and the precedence is total, then the set of ground terms becomes totally ordered, hence is isomorphic to some initial segment of the ordinals. This correspondence between trees and ordinals is pictured in figure 11 in the case of a vocabulary of three symbols $*$ (binary), S (unary) and 0 (a constant) ordered by decreasing arity.

Although the order-type of this ordering is rather small [22], it can be made bigger by using trees as nodes, and so on ... It can be made even bigger by using infinite regular trees. However, it never gets really big[2], and hence, it is far from being huge[3].

Simplification Orderings The recursive path ordering which we just discussed is a particular case of a simplification ordering:

Definition 15. A simplification ordering is a rewrite ordering $\succ$ which has the so-called *subterm property*: $\forall s \forall p \in \mathcal{P}os(s)\ p \neq \Lambda \Rightarrow s \succ s|_p$.

Note that a term is always strictly bigger than any of its strict subterms in any reduction ordering, provided they compare. Otherwise, termination would be violated: if $s|_p \succ s$, then, since $\succ$ is a rewrite ordering, $s_p \succ s = s[s_p]_p \succ s[s]_p$, and so on ... As an important consequence, total rewrite orderings are indeed simplification orderings. There is a sort of converse due to Dershowitz [17]:

Theorem 16. *Any simplification ordering on $\mathcal{G}(\mathcal{F})$ is a reduction ordering if $\mathcal{F}$ is finite.*

[2] The notion of a big ordinal was told to me by Stan Wainer.

[3] The notion of a huge ordinal was told to me by Mitsuhiro Okada.

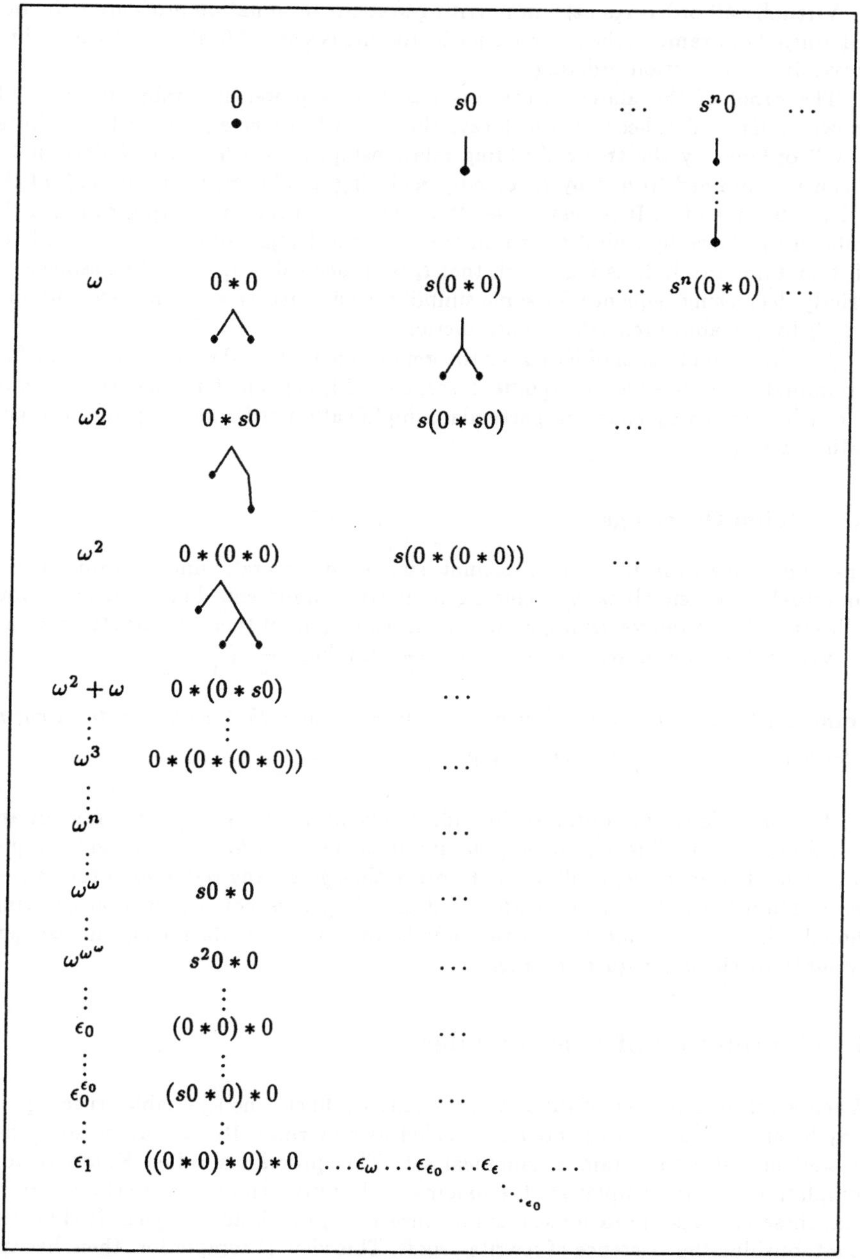

$>_{\text{rpo}}$ with $* > s > 0, * \in$ Lex

Fig. 11. Trees are ordinals [19]

Virtually all orderings used in rewriting-systems termination profs are reduction orderings. For example, the recursive path ordering is a simplification ordering, which proves it is a reduction ordering.

The proof of the above theorem is based on a powerful combinatorial result known as Kruskal's theorem, which says that the set of terms (provided $\mathcal{F}$ is finite) is well ordered by the tree-embedding relationship, in which s is *embedded* in t if it can be obtained from t by repeatedly replacing a subterm u of t by any of the strict subterms of u. It is easy to see that any simplification ordering contains the embedding. Since, by definition, an infinite sequence $\{t_i\}_{i \in N}$ of terms is well-ordered iff it contains two indices $i < j$ such that t_i is embedded in t_j, were this sequence a strictly decreasing sequence for some simplification ordering $\succ$, then we would have $t_j \succeq t_i$ by the above remark, a contradiction.

A quite elegant proof of Kruskal's theorem is due to Nash-Williams, and based on a minimal counter-sequence argument, see, e.g., [15; 17]. The finiteness condition on $\mathcal{F}$ can be relaxed by choosing particular simplification orderings, e.g. the recursive path ordering.

5.3 Union Orderings

The most powerful techniques cannot suffice in general, since termination is undecidable. When all fails, including mimicking arguments borrowed from your colleagues, it is time for trying a few tricks. Among all of them, let us stress a one.

Say that $\underset{1}{\rightarrow}$ commutes over $\underset{2}{\rightarrow}$ if $s \underset{1}{\rightarrow} t \underset{2}{\rightarrow} u$ implies $s \underset{2}{\rightarrow} \circ (\underset{1}{\rightarrow} \cup \underset{2}{\rightarrow})^* u$.

Lemma 17. *Assume that* $\underset{1}{\rightarrow}$ *commutes over* $\underset{2}{\rightarrow}$ *and that both are terminating relations. Then* $\underset{1}{\rightarrow} \cup \underset{2}{\rightarrow}$ *is terminating.*

The proof is simply obtained by inductively pushing the $\rightarrow_2$ steps first in any mixed derivation. This lemma is quite useful as the basis for a divide and conquer argument. For example, it allows us to prove that given any reduction ordering $\rightarrow$, the relation $(\rightarrow \cup \trianglerighteq)^+$ is a well-founded ordering. It is not a reduction ordering, though, since closure under substitution fails. In contrast, replacing encompassment by subterm yields a reduction ordering.

6 Completion of a set of rules

When a terminating set of rules is not locally confluent, non-joinable critical pairs may be simplified, then oriented and added as new rules. Rules may be simplified as well in order to obtain a canonical set. This process is called Knuth-Bendix completion [47]. In completion, the axioms used are therefore in a constant state of flux; these changes are expressed as inference rules, which add a dynamic character to establishing the existence of rewrite proofs. This view of completion, then, has two main components: an inference system, used in the completion process to generate new rewrite rules, and a rewrite relation on proofs, that shows how any proof can be normalized to a rewrite proof, as long as the appropriate rules have been generated.

So, adding new rules by computing critical pairs is simply a tool used to achieve our goal: to transform every proof into a rewrite proof using interreduced rules by rewriting (minimal) subproofs which do not obey the above property. According to [23], we will break the tradition by displaying the proof transformations first, and the inference rules second. Let $\succ$ be our reduction ordering used to orient rules, and $\gg$ the well-founded ordering on rules defined as follows: $s \rightarrow t \gg l \rightarrow r$ if (i) $s \trianglerighteq l$ under the encompassment ordering, or else (ii) $s \doteq l$ and $t \succ r$.

6.1 Completion Proof Rewrite Rules

$$
\begin{array}{rcl}
u \longleftarrow s \longrightarrow v & \Rightarrow & u \xrightarrow{*} \xleftarrow{*} v \\
 & & if\ u \longleftarrow s \longrightarrow v \text{ is not a critical peak} \\
u \xleftarrow[l \rightarrow r]{q} s \xrightarrow[g \rightarrow d]{q \cdot p} v & \Rightarrow & u \xleftrightarrow[r\sigma \simeq l\sigma[d\sigma]_p]{q} v \\
 & & if\ u \longleftarrow s \longrightarrow v \text{ is a critical peak} \\
u \xleftrightarrow[l \simeq r]{p} v & \Rightarrow & u \xrightarrow[l \rightarrow r]{p} v \\
 & & if\ l \succ r \\
u[l\sigma]_p \xleftrightarrow[l \simeq r]{p} u[r\sigma]_p & \Rightarrow & u[l\sigma]_p \xrightarrow[g \rightarrow d]{p \cdot q} u[l'\sigma]_p \xleftrightarrow[l' \simeq r]{p} u[r\sigma]_p \\
 & & if\ l \xrightarrow[g \rightarrow d]{q} l' \\
u \xleftrightarrow[s \simeq s]{} u & \Rightarrow & u \\
u[l\sigma]_p \xrightarrow[l \rightarrow r]{p} u[r\sigma]_p & \Rightarrow & u[l\sigma]_p \xrightarrow[l \rightarrow r']{p} u[r'\sigma]_p \xleftarrow[g \rightarrow d]{p \cdot q} u[r\sigma]_p \\
 & & if\ r \xrightarrow[g \rightarrow d]{q} r' \\
u[l\sigma]_p \xrightarrow[l \rightarrow r]{p} u[r\sigma]_p & \Rightarrow & u[l\sigma]_p \xrightarrow[g \rightarrow d]{p \cdot q} u[l'\sigma]_p \xleftrightarrow[l' \simeq r]{p} u[r\sigma]_p \\
 & & if\ l \xrightarrow[g \rightarrow d]{q} l' \ and\ l \rightarrow r \gg g \rightarrow d
\end{array}
$$

In the sequel, we will use $\Rightarrow_{KB}$ for the rewrite relation on proofs generated by the above rules. These proof rewrite rules are built up in a very natural way. The first two express the critical pair lemma. The third is the orientation of an equation into a rule. Note that the critical pair may not yet be an axiom, so it will have to be *generated.* The fourth is the simplification of equations, which again *produces* a new equation while it eliminates an old one making the underlying process of equation and rule generation non-monotonic. The fifth eliminates equations that have become trivial after some simplifications. The sixth reduces rules on the right, in order to achieve canonicity. The last reduces rules on the left, yielding new equations that must be again checked for orientation. The extra condition for the application of this rule is required for proving termination of these proof rewrite rules. So, all proof rewrite rules except the first two change the set of axioms, hence give rise to appropriate inference rules.

6.2 Completion Inference Rules

An *inference rule* (for our purposes) is a binary relation between pairs $(E; R)$, where E is a set of equations and R is a set of rewrite rules. We define the following set KB of six inference rules:

$$
\begin{array}{ll}
\text{Deduce:} & (E;R) \vdash (E \cup \{s \simeq t\}; R) \text{ if } s = t \in \mathrm{cp}(R)\\
\text{Orient:} & (E \cup \{s \simeq t\}; R) \vdash (E; R \cup \{s \rightarrow t\}) \text{ if } s \succ t\\
\text{Simplify:} & (E \cup \{s \simeq t\}; R) \vdash (E \cup \{s \simeq u\}; R) \text{ if } t \rightarrow_R u\\
\text{Delete:} & (E \cup \{s \simeq s\}; R) \vdash (E; R)\\
\text{Compose:} & (E; R \cup \{s \rightarrow t\}) \vdash (E; R \cup \{s \rightarrow u\}) \text{ if } t \rightarrow_R u\\
\text{Collapse:} & (E; R \cup \{s \rightarrow t\}) \vdash (E \cup \{u \simeq t\}; R)\\
 & \qquad \text{if } s \xrightarrow[l \rightarrow r]{p} u \text{ with } s \rightarrow t \gg l \rightarrow r
\end{array}
$$

The role of these inference rules is to make sure that if a left-hand side of a proof rewrite rule matches a particular proof, then that proof can be rewritten. This may not be the case unless E and R contain the additional rules and equations which are needed in the right-hand side of the proof rewrite rule. Adding these equations and rules is the role of the inference rules.

We write $(E; R) \vdash_{KB} (E'; R')$ if the latter may be obtained from the former by one application of a rule in KB.

6.3 Termination

Given a starting set E of equations, let us call $R*$ and $E*$ the smallest set of equations and rules which is closed under the above inference rules. To this set, we can associate the corresponding infinite set of proof rewrite rules defined in section 6.1. The role of these rules is to normalize proofs. Hence, their termination is a crucial property. To this end, let us define an ordering on proofs in which a proof is interpreted as a multiset of elementary proof steps, and a proof step is in turn interpreted by a pair of a multiset and a term as follows:

the equational proof step $s \longleftrightarrow t$ is interpreted by the pair $\langle\{s,t\}, \bot\rangle$;

the rewrite proof step $s \xrightarrow[l \rightarrow r]{} t$ is interpreted by the pair $\langle\{s\}, l \rightarrow r\rangle$.

Pairs are compared lexicographically, using the multiset ordering $\succ_{mul}$ induced by the given reduction ordering $\succ$ for the first component, and the ordering $\gg$ on rules for the second. Multisets of pairs, measuring the complexity of proofs, are compared by the multiset extension of the ordering on pairs. The ordering on proofs is therefore

$$((\succ_{mul}, \gg)_{lex})_{mul}.$$

Note how this ordering considers the justification of proofs, and not just the terms in it. Since the ordering is built up from well-founded orderings by means of extension that preserve well-founded orderings, the obtained ordering is well-founded. Checking the rules is left as a usefull exercise.

6.4 Completion Sequences

In practice, the completion rules are usually applied in some order, yielding a *completion sequence* $(E_0; \emptyset) \vdash_{KB} (E_1; R_1) \vdash_{KB} \cdots \vdash_{KB} (E_n; R_n) \cdots$ which may be infinite (completion *diverges*) or finite in case there are no possible non-redundant inferences after some finite step n. In the latter case, completion *succeeds* if $E_n = \emptyset$, and *fails* otherwise.

For example, completion succeeds when given the following fragment of group theory:

$$x \cdot 1 \simeq x, 1 \cdot x \simeq x, x^- \cdot (x \cdot y) \simeq y,$$

by generating the eight-rule canonical system

$$\begin{array}{rr} 1 \cdot x \rightarrow x & x \cdot 1 \rightarrow x \\ x^- \cdot x \rightarrow 1 & x \cdot x^- \rightarrow 1 \\ 1^- \rightarrow 1 & (x^-)^- \rightarrow x \\ x^- \cdot (x \cdot y) \rightarrow y & x \cdot (x^- \cdot y) \rightarrow y \end{array}$$

using size as the reduction ordering. A completion sequence, in which *Simplification*, *Compose* and *Collapse* are applied eagerly is shown in table 1. In contrast, completion diverges when given the single following equation $fgfx \simeq gfx$, where f and g are both unary (prefix) operator symbols, by generating the following equations and rules:

$$\begin{array}{rcll} fgfx & \underset{1}{\rightarrow} & gfx & \textit{Orient} \\ fggfx & \simeq & ggfx & \textit{Deduce}(1,1) \\ fggfx & \underset{2}{\rightarrow} & ggfx & \textit{Orient} \\ fgggfx & \simeq & gggfx & \textit{Deduce}(2,1) \\ & \ldots & & \end{array}$$

Divergence is of course undecidable [44] Note that no trick can circumvent divergence except changing the completion ordering, since by canonicity of the obtained set of rules, there cannot exist both a finite and an infinite set of rules defining the same congruence and included in the same well-founded ordering on terms. On the other hand, there are methods for circumventing failure mentioned in section 10.

6.5 Fairness

The role of the rewrite rules on proofs is to show that every proof is eventually normalized, this is why termination of these rules is crucial, and that normal form proofs are indeed rewrite proofs. For this latter property to hold, an hypothesis is needed, called *fairness.* Let us call *persisting* a rule which is never reduced on the left nor on the right once it has been generated:

$$E_\infty = \bigcup_{i \geq 0} \bigcap_{j \geq i} E_j$$

[4] The first field, i, indicates the number of completion steps. The second field is the set of rules at step i. At step 3 for example, R_3 is made of all rules in R_2 and of the oriented rule at step 4 which is displayed below R_2. Since the set of rules does not change at step 4, steps 3 and 4 are displayed between the same horizontal lines. The third field displays the set of equations in E_i on the same line(s) as the step number i. The last field indicates the name of the inference rule used at step i.

i	R_i	E_i	inference
0		$1 \cdot x \simeq x$	
		$x \cdot 1 \simeq x$	
		$x^- \cdot (x \cdot y) \simeq y$	
1	$x \cdot 1 \to x$	$1 \cdot x \simeq x$	orient
		$x^- \cdot (x \cdot y) \simeq y$	
2	R_1	$x^- \cdot (x \cdot y) \simeq y$	orient
	$1 \cdot x \to x$		
3	R_2		orient
4	$x^- \cdot (x \cdot y) \to y$	$x^- \cdot x \simeq 1$	deduce (1,3)
5	R_3		orient
6	$x^- \cdot x \to 1$	$1^- \simeq 1$	deduce (1,5)
7	R_5		orient
8	$1^- \to 1$	$(x^-)^- \cdot y \simeq x \cdot y$	deduce (3,3)
9	R_7		orient
10	$(x^-)^- \cdot y \to x \cdot y$	$1^- \cdot y \simeq 1 \cdot y$	deduce (7,9)
11	R_9		orient
	$1^- \cdot y \to 1 \cdot y$		
12	R_9		compose (11,2)
	$1^- \cdot y \to y$		
13	R_9	$1 \cdot y \simeq y$	collapse (12,7)
14		$y \simeq y$	simplify (2)
15			delete
16		$(x^-)^- \simeq x$	deduce (1,9)
17	R_9		orient
	$(x^-)^- \to x$		
18	R_7	$x \cdot y \simeq x \cdot y$	collapse (9,17)
19	$1^- \cdot y \to y$		delete
20	$(x^-)^- \to x$	$x \cdot (x^- \cdot y) \simeq y$	deduce (3,17)
21	R_{18}		orient
22	$x \cdot (x^- \cdot y) \to y$	$x \cdot x^- \simeq 1$	deduce (1,21)
23	R_{21}		orient
	$x \cdot x^- \to 1$		

Table 1. A successful completion sequence for a fragment of group theory[4]

$$R_\infty = \bigcup_{i \geq 0} \bigcap_{j \geq i} R_j$$

Then, a completion sequence is *fair* if the critical peaks formed with the persisting rules are all reducible with respect to the rewrite rules on proofs. This is in particular so, if the corresponding critical pairs have been computed, but this may not be always necessary.

If success occurs after a finite number n of steps, then the resulting system is a decision procedure for E_0. In case of divergence, a semi-decision procedure can be obtained if the completion sequence is *fair*:

Theorem 18 [33]. *Given two terms s and t such that* $s \overset{*}{\underset{E_0}{\longleftrightarrow}} t$, *and a fair completion sequence* $(E_0; \emptyset) \vdash_{KB} (E_1; R_1) \vdash_{KB} \cdots \vdash_{KB} (E_n; R_n) \cdots$ *such that* E_∞ *is empty, there exists an integer i such that s and t have the same normal form with respect to* R_j *for all* $j \geq i$.

The strategy described below, by a regular expression of the inference rules names, is fair:

$$((\textbf{Collapse}^*;\ \textbf{Compose}^*;\ \textbf{Simplify}^*;\ \textbf{Delete}^*;\textbf{Orient}^*)^*;\ \textbf{Deduce})^*$$

Efficiency of completion is strongly related to the possibility of eliminating redundant critical pair computations, especially in the context of completion modulo. See [2; 77] for various critical pair criteria and [4; 61; 53] for completion strategies which eliminate redundancy.

6.6 Limitations of rewriting techniques

There are two kinds of limitations. First, completion may fail in finding a canonical rewrite system even if one exists. This is so because orienting non-persisting equations may require an ordering more powerful than the one in use (which may be powerful enough to orient the set of persisting rules). Second, an equation may be generated, like commutativity, that cannot be oriented without loosing the termination property. These questions are briefly addressed in section 10.

Although rewriting appears as a very general and powerful technique for solving such problems in a uniform manner, it may not always apply: there are computable (with a decidable word problem) algebras, for which the word problem cannot be decided by means of (plain) rewriting in any presentation of the algebra, a result of Squier [70]. However, Squier's example can be solved in a conservative extension of the algebra, by choosing a different set of generators as recently shown by Lafont at the "Ecole de Printemps d'Informatique Théorique". This can of course always be done, since rewrite rules can encode Turing machines [54].

7 Solving queries

A convergent set R of rewrite rules for some equational theory E allows one to solve various kinds of queries. A query will be a logical formula built over the equality predicate, to be interpreted in some (possibly quotient of) term algebra ($\mathcal{T}(\mathcal{F})$ or $\mathcal{T}(\mathcal{F}, \mathcal{X})$) by the congruence generated by the set E (equivalently R) of axioms. Accordingly, the query will be solved in the quotient under consideration (we may simply say in $\mathcal{T}(\mathcal{F})$ or $\mathcal{T}(\mathcal{F}, \mathcal{X})$ in the following, leaving the quotient implicit). We will distinguish queries after their logical complexity. Since conjunctions and disjunctions do not cause particular problems, we will simplify our notations by forgetting about these two connectives, hence consider quantified atomic formulae.

7.1 Word problems and universal queries

Query: $s = t$, where s and t are ground.

This query is solved by computing the normal forms of s and t and comparing them. Such a query is usually called a word problem. Since the convergence property of the set R of rules yields a decision procedure for equality, we can also answer to negated such queries:

Query: $s \neq t$, where s and t are ground.

Word problems generalize easily to universal queries :

Query: $\forall \mathcal{V}ar(s,t) \ \ s = t$

and their negation:

Query: $\exists \mathcal{V}ar(s,t) \ \ s \neq t$

This is of course done as above by considering the variables as new constants. This can be justified by looking at the latter query, where existential variables can be first skolemized into constants, and we are back to a ground query. This remark applies to the former query, by using a double negation. Note that the equality predicate is implicitely interpreted here in $\mathcal{T}(\mathcal{F}, \mathcal{X})$, an algebra with infinitely many free constants. Adding new constants in an arbitrary algebra may not be sound: there are indeed pathological algebras with a decidable word problem for which universal queries are undecidable.

7.2 Existential queries

These queries may have free variables as well as existentially quantified ones:

Query: $\exists X \ \ s = t$, where $X \subseteq \mathcal{V}ar(s,t)$

The goal here is to find (ground) values for the free variables that make the formula true in the quotient of the Herbrand universe by the congruence generated by a convergent set of rules R. Variables in X are auxiliary variables (*parameters*) whole value is useless. The values of the free variables are supposed to be enumerated by the resolution process, or represented by means of a so-called *solved form*, see [37]. This will be achieved by a set of non-deterministic transformation rules that simplify the query until a solved form is obtained on each branch.

Due to the existence of a rewrite proof for an arbitrary valid equation, instead of solving a goal

$$\exists X \ \ s = t,$$

one solves a pair of "directed" goals,

$$\exists X \exists z \ \ s \rightarrow z \wedge t \rightarrow z, \text{ for some new variable } z.$$

The coming rules transform such a query into a substitution encoded as a formula of the form $x_1 \rightarrow u_1 \wedge \ldots \wedge x_n \rightarrow u_n$, where the x_i are distinct free variables of the query, and do not belong to any u_j.

Decompose: $f(\vec{s}) \to f(\vec{t}) \Rightarrow \bigwedge_{i=1}^{i=n} s_i \to t_i$

Delete: $s \to s \Rightarrow T$

Eliminate: $x \to s \wedge P \Rightarrow x \to s \wedge P\{x \mapsto s\}$
if $x \in \mathcal{X}$, and $x \notin \mathcal{V}ar(s)$

Mutate: $f(\vec{s}) \to t \Rightarrow s_1 \to l_1 \wedge \ldots \wedge s_n \to l_n \wedge r \to t$
if $f(l_1, \ldots, l_n) \to r$ is a copy of a rule in R with its variables renamed apart $\mathcal{V}ar(P, t, s_1, \ldots, s_n)$

Imitate: $f(\vec{s}) \to y \Rightarrow \exists y_1, \ldots, y_n \;\; y \to f(y_1, \ldots, y_n) \wedge \bigwedge_{i=1}^{i=n} s_i \to y_i$
if $y \in \mathcal{X}$, $y_i \notin \mathcal{V}ar(y, s_1, \ldots, s_n)$

Parameters: $\exists X, y \;\; y \to u \wedge P \Rightarrow \exists X \;\; P$
if $y \notin \mathcal{V}ar(P, u)$
$\exists X, y \;\; u \to y \wedge P \Rightarrow \exists X \;\; P$
if $y \notin \mathcal{V}ar(P, u)$
$\exists X, y \;\; P \Rightarrow \exists X \;\; P$
if $y \notin \mathcal{V}ar(P)$

For example, if R is $\{f(x,x) \to c(x), a \to b\}$, then the query $f(y,a) = c(y)$ can be solved by the following sequence of rewrites:

$$\exists z \;\; f(y,a) \to z \wedge c(y) \to z \Rightarrow_{Mutate}$$

$$\exists z, x \;\; y \to x \wedge a \to x \wedge c(x) \to z \wedge c(y) \to z \Rightarrow_{Eliminate}$$

$$\exists z, x \;\; y \to x \wedge a \to x \wedge c(x) \to z \Rightarrow_{Imitate}$$

$$\exists z, x \;\; y \to x \wedge x \to a \wedge c(x) \to z \Rightarrow_{Eliminate}$$

$$\exists z, x \;\; y \to a \wedge x \to a \wedge c(a) \to z \Rightarrow_{Parameters}$$

$$\exists z \;\; y \to a \wedge c(a) \to z \Rightarrow_{Parameters}$$

$$y \to a$$

The solution $\{y \mapsto b\}$ can be obtained with an application of *Mutate* replacing the application of *Imitate*:

$$\exists z, x \;\; y \to x \wedge a \to x \wedge c(x) \to z \Rightarrow_{Mutate}$$

$$\exists z, x \;\; y \to x \wedge x \to b \wedge c(x) \to z \Rightarrow_{Eliminate}$$

$$\exists z, x \;\; y \to b \wedge x \to b \wedge c(b) \to z \Rightarrow_{Parameters}$$

$$\exists z \;\; y \to b \wedge c(b) \to z \Rightarrow_{Parameters}$$

$$y \to b$$

Theorem 19 [21]. *Let R be a convergent rewrite system and Q the existential query $\exists X, z \;\; s \to z \wedge t \to z$. Then, for every solution γ of Q, there exists a sequence of rewrites from Q whose result is a substitution more general than γ.*

By exhibiting more primitive transformation rules, this set of rules subsumes *narrowing*, as used for this purpose in [25]:

Definition 20. A term s *narrows* to a term t, via substitution $\sigma = mgu(s|_p, l)$, at position p using the rule $l \rightarrow r$, symbolized $s \underset{l \rightarrow r, \sigma}{\overset{p}{\rightsquigarrow}} t$, if $t = s\sigma[r\sigma]_p$.

By encoding each narrowing step by appropriate sequences of the rules, the above theorem can now be reformulated as follows: For every solution γ of the above existential query Q, there exists a narrowing derivation from $s = t$:

$$s_0 = t_0 \underset{\sigma_0}{\rightsquigarrow} s_1 = t_1 \underset{\sigma_1}{\rightsquigarrow} \ldots \underset{\sigma_{n-1}}{\rightsquigarrow} s_n = t_n$$

such that $s = s_0$, $t = t_0$, $\theta = mgu(s_n, t_n)$ and $\sigma = \sigma_1 \ldots \sigma_{n-1}\theta$ is more general than γ.

Taking for example the rules for defining the addition over natural numbers $x + 0 \rightarrow x$, $x + S(y) \rightarrow S(x + y)$, the query $S(0) + y = S(S(x))$ originates in the narrowing tree depicted on figure 12.

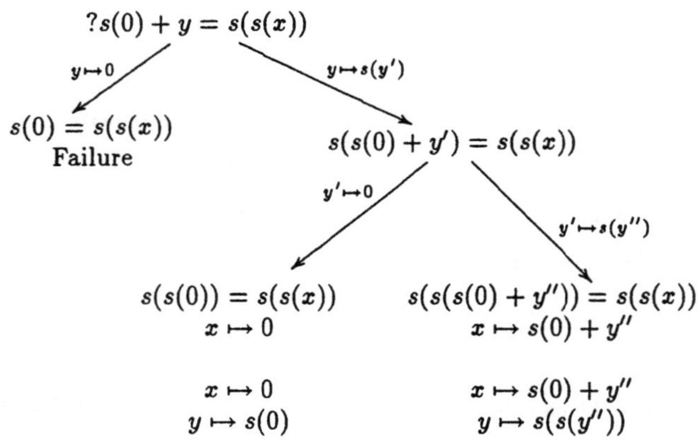

Fig. 12. Narrowing Tree.

The narrowing process resembles completion as pointed out by Dershowitz, but less critical pairs are computed. Variations on narrowing include: *normal narrowing* [25] (in which terms are normalized via $\rightarrow^!_R$ before narrowing), *basic narrowing* [36] (in which the substitution part of prior narrowings is not subsequently narrowed), and their combination [62]. Basic narrowing can be elegantly formulated by the following two inference rules operating on pairs of equations $s = t$ (assuming $=$ commutative) and substitutions θ, with $s = t, \emptyset$ as the initial pair:

Narrowing: $s = t \; ; \; \theta \;\; \Rightarrow \;\; \exists \mathcal{V}ar(l, r) s[r]_p = t \; ; \; \theta\sigma$
if σ is the most general unifier of s_p and l

Success: $s = t \; ; \; \theta \;\; \Rightarrow \;\; T \; ; \; \theta\sigma$
if σ is the most general unifier of s and t

7.3 Negative queries

The above technique extends to negative queries, by making use of ground reducibility tests:

Query: $\exists X \;\; s \neq t$, where $X \subseteq \mathcal{V}ar(s,t)$

Definition 21. An equation $s \simeq t$ is *ground reducible* iff for all ground substitutions γ, either $s\gamma = t\gamma$ or one of $s\gamma, t\gamma$ is reducible. In case $s\eta$ and $t\eta$ are different and irreducible, $s = t$ is not ground reducible *because of* η.

Narrowing: $s \neq t \; ; \; \theta \;\; \Rightarrow \;\; s' \neq t' \; ; \; \theta\sigma$
if σ is the most general unifier of s_p and l

Success: $s \neq t \; ; \; \theta \;\; \Rightarrow \;\; T \; ; \; \theta\eta$
if $s = t$ is not ground reducible because of η

Ground reducibility is indeed decidable [65; 43; 9; 7; 8], and substitutions because of which an equation is not ground reducible are enumerable [27]. The test is however NP-hard. The algorithm given in [8] has an exponential complexity with respect to the height of terms, which is proved optimal.

7.4 Inductive queries

These queries have all their variables universally quantified over ground terms. Inductive queries do not differ from universal queries at the syntactic level, but the equality predicate is interpreted in the quotient of the ground term algebra. Answering to an inductive query normally requires an induction scheme (over the term algebra). This induction scheme can actually be hidden in a completion process.

Query: $\forall X \in \mathcal{T}(\mathcal{F}) \;\; s = t$, where $X \subseteq \mathcal{V}ar(s,t)$

For convergent systems R_0, any equation between distinct ground-normal forms is considered to be inconsistent with R_0. The observation that an equation $s = t$ is valid in the initial algebra associated with R_0 iff no inconsistency follows from $R_0 \cup \{s \simeq t\}$ is the basis of the method of inductive theorem proving pioneered by Musser [58]: if there exists a ground-convergent system with the same ground normal forms as R_0, and which presents the same equational theory as $R_0 \cup \{s \simeq t\}$, then inconsistency is precluded.

Solving these queries will therefore be done by adding the equation $s = t$ to the algebra, and completing the set $R_0, \{s \simeq t\}$. To ensure that normal forms are not changed during completion, it will be enough to check the left hand sides of the added rules for ground-reducibility [16; 40].

Definition 22. A term s is *ground reducible* by R at a set of *covering positions* S iff all its ground instances by irreducible substitutions are reducible by R at a position in S..

For example, $s(s(x))$ is ground reducible by $s(s(0)) \to 0$ at the set $\{\Lambda, 1\}$ of positions.

Since we are only interested in ground-normal forms, ground-convergence will actually be enough. As a consequence, not all critical pairs are actually needed. Only those obtained by overlapping with R_0 at a set of covering positions of the generated rules are necessary. In the rules below, R denotes the set of added rules only:

$$\begin{array}{ll}
\text{Delete:} & (E \cup \{s \simeq s\}; R) \vdash (E; R) \\
\text{Compose:} & (E; R \cup \{s \to t\}) \vdash (E; R \cup \{s \to u\}) \text{ if } t \to_{R \cup R_0} u \\
\text{Simplify:} & (E \cup \{s \simeq t\}; R) \vdash (E \cup \{s \simeq u\}; R) \text{ if } t \to_{R \cup R_0} u \\
\text{Orient:} & (E \cup \{s \simeq t\}; R) \vdash (E; R \cup \{s \to t\}) \\
& \qquad \text{if } s \succ t \text{ and } s \text{ is ground-reducible by } R_0 \\
\text{Disprove:} & (E \cup \{s \simeq t\}; R) \vdash \text{False} \\
& \qquad \text{if } s \succ t \text{ and } s \text{ is not ground-reducible by } R_0 \\
\text{Collapse:} & (E; R \cup \{s \to t\}) \vdash (E \cup \{u \simeq t\}; R) \\
& \qquad \text{if } s \longrightarrow^{p}_{l \to r} u \text{ with } s \to t \gg l \to r \\
\text{Deduce:} & (E; R) \vdash (E \cup \{s \simeq t\}; R) \\
& \text{if } s \simeq t \text{ is a critical pair of } R_0 \text{ at a covering position in } R
\end{array}$$

Theorem 23. *An equation $s \simeq t$ is not an inductive consequence of a ground-convergent set R_0 of rules iff inductive completion returns false, provided it does not fail.*

The above result follows from [16; 40] and [28]. The associativity proof of the addition over natural numbers defined previously is classical. Noting that the position Λ in $(x+y)+z$ is a covering position (1 is another one), the associativity rule $(x+y)+z \to x+(y+z)$ generates two critical pairs with the starting rules whose confluence is shown on figure 13

Choosing the other covering position yields a divergent completion. This must of course be the case in general since membership to the inductive theory of a set of equations is co-semi-decidable.

8 Modularity

Assume given two modules, each one specifying a set $\mathcal{D}_i$ $(i = 1, 2)$ of *defined functions* with respect to a given set $\mathcal{C}_i$ of *constructors* by means of a set R_i of rewrite rules, the left-hand sides of which are headed by symbols in $\mathcal{D}_i$. The set $\mathcal{F}_i = \mathcal{C}_i \cup \mathcal{D}_i$ is called the *signature* of the module. Putting these two modules together results in the union of the two rewriting systems R_1 and R_2.

If $\mathcal{F}_1 \cap \mathcal{F}_2 = \emptyset$, their *disjoint union* is denoted by $R_1 \oplus R_2$.

If $\mathcal{F}_1 \cap \mathcal{F}_2 = \mathcal{C}_1 \cap \mathcal{C}_2 \neq \emptyset$, their *combined union* is denoted by $R_1 \cup R_2$.

If $\mathcal{F}_1 \cap \mathcal{D}_2 = \emptyset$, but $\mathcal{F}_2 \cap \mathcal{D}_1 \neq \emptyset$, their *hierarchical union* is denoted by $R_1 + R_2$. R_1 is the base, and R_2 the enrichment. There may be several successive enrichments in a hierarchical union.

Note that defined symbols are never shared.

$$R_0 \begin{cases} x + 0 & \rightarrow & x \\ x + s(y) & \rightarrow & s(x+y) \end{cases}$$

$$(x+y)+z \rightarrow x+(y+z)$$

↑

covering position

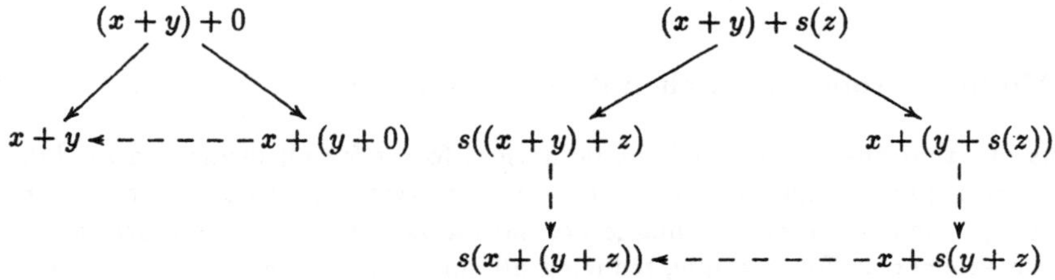

Fig. 13. Associativity of +.

Since obtaining a confluent and terminating rewriting system from a set of equations may be a hard task, it is desirable that both properties are closed under these unions. A property of term rewriting systems is *modular* if it is closed under union. We are interested in particular in the modularity of termination and confluence. Of course, neither property is closed under arbitrary union. However, it is well known that confluence is modular for disjoint union, a celebrated result of Toyama [74]. This is not true of combined unions, as pointed out by Kurihara and Ohuchi [49]:

$$R_1 = \{f(x,x) \rightarrow a,\ f(x, c(x)) \rightarrow b\}, \quad R_2 = \{g \rightarrow c(g)\}$$

Both systems are confluent, but the term $f(g,g)$ has two different normal forms, a and b, in the union. The same happens with hierarchical systems.

In contrast, termination is not modular, even for disjoint unions [73]:

$$R_1 = \{f(0,1,x) \rightarrow f(x,x,x)\}, \quad R_2 = \{g(x,y) \rightarrow x,\ g(x,y) \rightarrow y\}$$

Indeed, there is the following infinite reduction sequence:

$$f(g(0,1), g(0,1), g(0,1)) \rightarrow$$

$$f(0, g(0,1), g(0,1)) \rightarrow f(0,1,g(0,1)) \rightarrow$$

$$f(g(0,1), g(0,1), g(0,1)) \ldots$$

Contradicting hopes, the above example can be refined so as to have confluent definitions for both rewrite systems [24]:

$$R_1 = \{f(0,1,x) \to f(x,x,x),\ f(x,y,z) \to 2,\ 0 \to 2,\ 1 \to 2\}$$

$$R_2 = \{g(x,y,y) \to x,\ g(y,y,x) \to x\}$$

for which there is again an infinite reduction sequence originating from $f(g(0,2,1), g(0,2,1), g(0,2,1))$.

8.1 Modular properties of alien-decreasing unions

It turns out that the *existence* of unique normal forms for all terms is a modular property of disjoint unions [56; 57]. So, if two rewrite systems are both confluent and terminating, their union enjoys unique normal forms even when there are infinite reduction sequences. For example, reducing all subterms of the above sequence to 2 yields the normal form $f(2,2,2)$. Using implicitely a result of Rusinowitch [69], Kurihara and Ohuchi pointed out that a very simple strategy allows to compute this normal form: *shared rewriting* in which different occurrences of a same variable in a right-hand side of rule are shared. This is also true of combined unions [49], and it generalizes to even more complex situations [57; 26].

Definition 24. Let R_1 and R_2 be two rewrite systems over the respective signatures $\mathcal{F}_1 = \mathcal{C}_1 \cup \mathcal{D}_1$ and $\mathcal{F}_2 = \mathcal{C}_2 \cup \mathcal{D}_2$ such that $\mathcal{D}_1 \cap \mathcal{D}_2 = \emptyset$. A term in the union $\mathcal{T}(\mathcal{F}_1 \cup \mathcal{F}_2, \mathcal{X})$ is *homogeneous* if it uses symbols of $\mathcal{F}_1 - \mathcal{D}_2$ or $\mathcal{F}_2 - \mathcal{D}_1$ exclusively. Derivations using rules in $R_1 \cup R_2$ are *homogeneous* if all terms in the sequence are homogeneous with respect to the same signature.

Non-homogeneous derivations may be homogenized by computing *caps* and *aliens*. The cap of a term is its topmost homogeneous part obtained by abstracting different aliens by different variables, and its multiset of aliens subterms is made of its maximal subterms whose root does not belong to the signature of the cap. We then say that a relation $\to$ on terms in the union is *alien-decreasing* if $s \to t$ in a cap position implies
$aliens(t) \rhd_{mul} aliens(s)$, or
$aliens(t) =_{mul} aliens(s)$ and $cap(s) \to^{+} \cup \rhd cap(t)$.

This seemingly complex definition captures many different ones found in the literature, like conservative disjoint unions [69], in which a variable must have less occurrences in the right hand-side of a rule than in its left-hand side, conservative combined unions, shared-rewriting of combined unions [49], crosswise disjoint unions in which the function symbols occurring in the left-hand sides of one system do not occur in the right-hand sides of the other [66].

Theorem 25. *Let R_1, R_2 be terminating term rewriting systems over $\mathcal{T}(\mathcal{D}_1 \cup \mathcal{C}, \mathcal{X})$ and $\mathcal{T}(\mathcal{D}_2 \cup \mathcal{C}, \mathcal{X})$ respectively, where $\mathcal{D}_1 \cap \mathcal{D}_2 = \emptyset$. If $\to_{R_1}$, $\to_{R_2}$ are alien-decreasing then $\to_{R_1 \cup R_2}$ is terminating.*

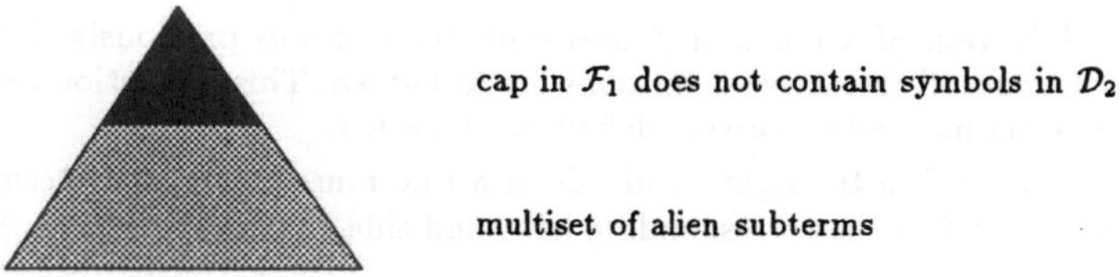

Fig. 14. Caps and Aliens

The proof is interesting by being based on a induction argument traditional in proof theory, from which we borrow some vocabulary: *For any term t and strongly normalizable substitution γ, $t\gamma$ is strongly normalizable.* The theorem follows directly, taking the identity substitution for γ. Now, γ being assumed strongly normalizable, we can use it for constructing a well-founded ordering on which the induction will operate: a term $t\gamma$ is interpreted by the pair $I(t) = \langle t, \{\gamma\}\rangle$, where $\{\gamma\}$ denotes the multiset $\{x\gamma \mid x \in Var(t)\}$. These pairs are compared in the well-founded ordering $(\rhd, (\rightarrow_{R_1 \cup R_2} \cup \rhd)_{mul})_{lex}$. The proof needs three cases, according to the position where $t\gamma$ reduces to $t\gamma'$, see [26]. The proof uses the property that a term may have several decompositions: given one, finding a lesser one allows to apply the induction hypothesis to the term.

Assume now that both rewrite systems R_1 and R_2 are *simply terminating*, that is, can be shown terminating by use of a simplification ordering. Let P_f be the rewrite system made of the projection rules $f(\ldots, x_i, \ldots) \rightarrow x_i$ for all $1 \leq i \leq n$. Since a set R of rules is simplifying iff $R \cup \bigcup_{f \in \mathcal{F}} P_f$ is terminating, it suffices to note that the obtained set of rules is alien-decreasing when the starting set R is, in order to prove that simple termination is a modular property of alien-decreasing unions, a result obtained by Kurihara and Ohuchi for combined unions [50].

8.2 Modular properties of hierarchical unions

We now investigate the case of hierarchical unions $R_0, R_1, \ldots, R_n$, *the base* R_0 of which is assumed alien-decreasing, while the *incremental developments* $R_1, \ldots, R_n$ define functions by generalized primitive recursive schema introduced by Jouannaud and Okada in the context of higher-order definitions [41]:

Definition 26. Let f a function symbol not belonging to the signature $\mathcal{F}$ of the rewrite system R, defined by a rewrite system R_f whose rules follow the so-called *general scheme*, that is are of the form: $f(t_1, \ldots, t_n) \rightarrow v$
where $t_1, \ldots, t_n \in T(\mathcal{F}, \mathcal{X})$, $v \in T(\mathcal{F} \cup \{f\}, \mathcal{X})$, and for all subterms $f(u_1, \ldots, u_n)$ of v, $\{t_1, \ldots, t_n\} \rhd_{mul} \{u_1, \ldots, u_n\}$.

The general scheme imposes two restrictions on the incremental developments:

1. Since the definition of a function f uses only the symbols previously defined and f itself, mutually recursive definitions are forbidden. This restriction can be lifted by pairing mutually recursive definitions together.
2. The arguments of f in the right-hand side of a rule must be strict subterms of the arguments of f in the corresponding left-hand side.

Termination is a modular property of hierarchical systems:

Theorem 27. *Let $R = R_0 + R_1 + \ldots + R_n$ be a hierarchical system such that R_0 is terminating and alien-decreasing, and $R_1, \ldots, R_n$ are incremental developments satisfying the general scheme. Then $\rightarrow_R$ is terminating.*

The proof of the above result uses a technique similar to the one employed in the previous section. This result, however, is not as strong as we would like it to be due to the second restriction pointed out above. When it is not satisfied, tricks can be used in order to apply the theorem above in an indirect fashion, see [26].

The above results extend to higher-order definitions like those given in section 2.9. Besides higher-order rewrite rules, the combination may contain typed-lambda terms, or even polymorphically typed lambda-terms [41], or even terms of the calculus of constructions [5]. Note that such combinations are usual in proof theory: Gödel's system T is a particular case of a higher-order rewrite system following the general schema combined with the simply typed lambda-calculus. Its most powerful generalization, the Calculus of Constructions with Inductive types, is another such combination, in which the higher-order rules may *not* follow the general schema. Since this calculus has been shown confluent and terminating [76], we may think that there is still room for improving these results.

9 Rewriting and Automata

The language of ground terms in normal form plays a key role for proving inductive theorems and evaluating negated equational queries through ground reducibility. Describing the language of ground term in normal form for a given rewrite system appears therefore to be very important. A natural tool for this purpose is provided by tree automata theory, although tree grammars have also been used [9].

Inductive reducibility is also the main concept underlying the transformation of a quotient of a ground term algebra into a free order-sorted algebra by exhibiting the underlying constructors when possible [10]. Order-sorted algebras are many sorted algebras equipped with an ordering on sorts interpreted as the subset relationship on the carrier. Comon pointed out that this is equivalent to interpret sorts as recognizable subsets of the set of terms. Indeed, many-sorted signatures and non-deterministic bottom-up tree automata are different views of the same objects in which sorts are identified to states and function declarations to transitions.

Assume given the following OBJ-like [29] specification for Boolean expressions:

sorts :	N, Z, B
subsorts :	$N < Z$
operation :	$id :\rightarrow N$
operation :	$+ : Z \times Z \rightarrow Z$
operation :	$>: Z \times Z \rightarrow B$
operation :	$\vee : B \times B \rightarrow B$
operation :	$\wedge : B \times B \rightarrow B$
operation :	$\neg : B \rightarrow B$

This specification can be seen as the tree automaton given figure 15, which can in turn be expressed by the rules given figure 16.

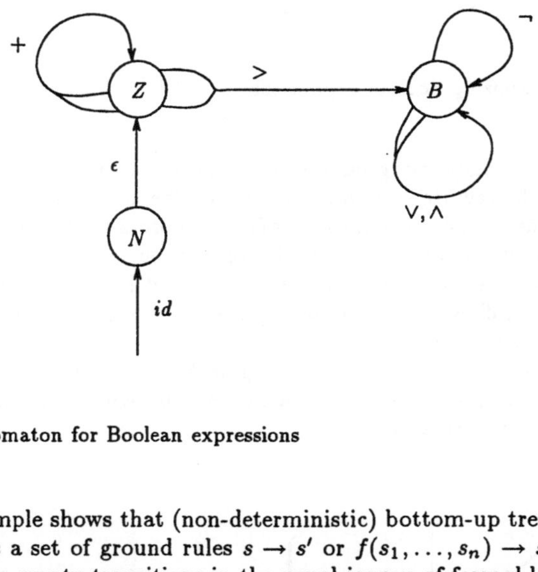

Fig. 15. Automaton for Boolean expressions

This example shows that (non-deterministic) bottom-up tree automata can also be viewed as a set of ground rules $s \rightarrow s'$ or $f(s_1, \ldots, s_n) \rightarrow s'$. The rules $s \rightarrow s'$ correspond to empty-transitions in the usual jargon of formal language theory, and to subsorts in the jargon of algebraic specifications. Non-determinism is obtained when there are critical pairs between pairs of rules of any kind.

So, bottom-up tree automata are particular rewrite systems able to type-check terms with respect to a signature. We will say that a ground term is *recognized* by an automaton if it reduces to a final state by using the rules, and that a set of terms is *regular* if it is recognized by a bottom-up tree automaton.

Bottom-up tree automata are a faithful generalization of automata on words. Non-determinism can be reduced, they can be minimized, emptiness can be decided efficiently by using cleaning techniques, pumping applies, they are closed under Boolean operations and projection, hence enjoy a logical characterization "à la Buchi", see e.g. [72] for a survey on this question.

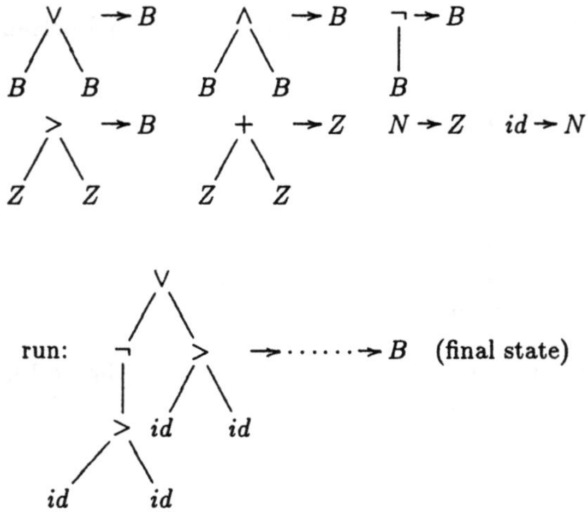

Fig. 16. Rules for recognizing Boolean expressions

Going back to the language of ground terms in normal form, it is easy to see that it is recognizable if the rewrite system is left-linear. The automaton will simply use for states the subterms of the left-hand sides of rules in order to store the information that a left-hand side of rule has been partially recognized.

For example, with the rewrite system $\{ssx \rightarrow x\}$ describing natural numbers modulo 2, we get the following automaton, whose states are $\{zero, one, reducible\}$, with $\{zero, one\}$ as final states:

In contrast, the language may not be recognizable when the rewrite system is not left-linear. For example, $\{x + x \rightarrow 0\}$ defines a non-regular language of ground normal-forms defined on figure 17.

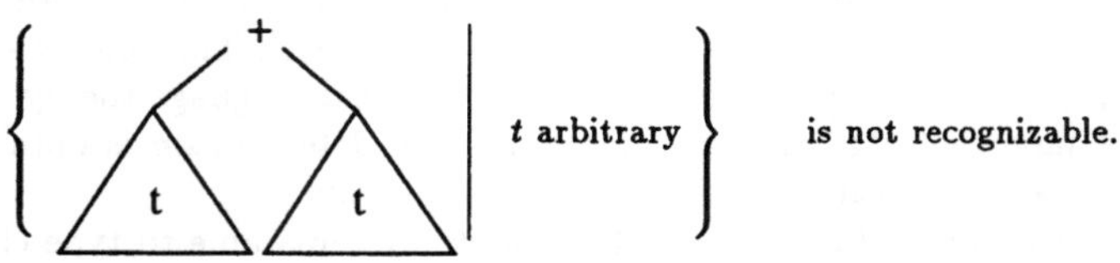

Fig. 17. A non-recognizable set of terms

The usual proof is based on a pumping lemma for regular tree languages. Adding the new rule $\{x + y \rightarrow 0\}$ makes the language regular. Fortunately, whether a finite rewrite system defines a regular laguage of ground terms in normal form is decidable [48].

We are now able to examplify the already mentioned interesting application of these automata. Consider the specification of integers as given on figure 18. This specification has an important drawback: the constructor symbols *succ* and *pred* are not free, they appear on top od left-hand sides of rules. But, since the rules are left-linear, we can indeed give an equivalent presentation obtained by computing the automaton recognizing the language of ground terms in normal form. This automaton is shown on the same figure. The state *red* in which all reducible terms are recognized is of course not final. Besides, the rules for the defined operations, if any, of the starting signature can be mechanically transformed into new rules for the transformed defined operations [10].

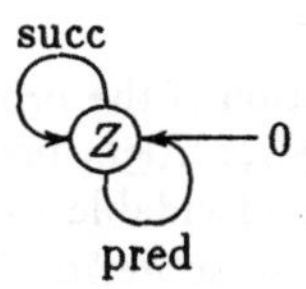

$$succ(pred(x)) \rightarrow x$$
$$pred(succ(x)) \rightarrow x$$

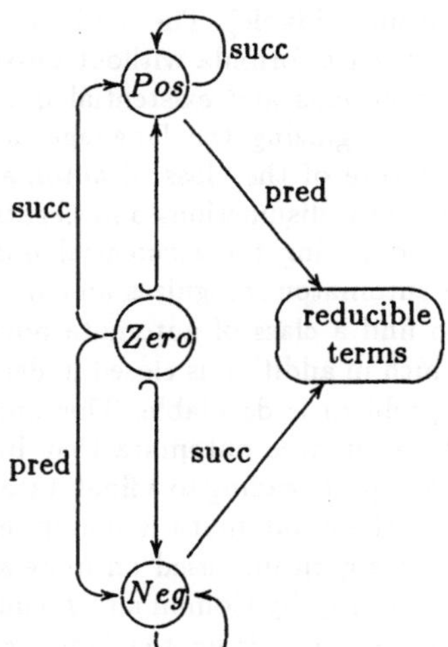

Fig. 18. Specification and automaton for the integers

Ground reducibility refers to the language of terms in normal form which are instances of a given term *t*. The entire set of ground terms in normal form is of course a particular case, taking a variable for *t*. Describing this language is the key to a decision procedure for ground reducibility.

We will first show that we can answer to our question when the rules are left-linear, even if the term t is not left-linear. By definition, t is ground-reducible iff R and $R \cup \{t \rightarrow\}$ define the same language of ground terms in normal form. If the latter is not regular, which is decidable as we have seen, then t is not ground-reducible. If it is regular, then we can compute the minimal automaton of both languages and compare them. Ground reducibility is therefore decidable when the normal form algebra has a regular carrier.

Answering the question in the general case is much more difficult. We first remark that ground reducibility can be expressed by means of a logical formula, whose set of solutions is the set of ground terms in normal form. Given $R = \{l_i \rightarrow r_i\}_{i \in [1..n]}$ such that $\mathcal{V}ar(l_i) \cap \mathcal{V}ar(l_j) = \emptyset \ if \ i \neq j$, a term t is not ground-reducible by R iff the formula

$$\exists \mathcal{V}ar(t, l_1, \ldots, l_n) \;\; x \trianglerighteq t \wedge \bigwedge_{i=1}^{i=n} x \not\trianglerighteq l_i .$$

has no solution in the term algebra. In this formulation of the problem, the formula is built-up with finitely many unary predicates $\trianglerighteq t, \trianglerighteq l_1, \ldots, \trianglerighteq l_n$ instead of the binary predicate $\trianglerighteq$, whose first-order language is indeed undecidable as a consequence of the undecidability of the Σ_3 fragment of the theory of subterm [75].

It turns out that this will be enough a restriction for decidability (the first-order theory of $\trianglerighteq$ can easily be shown undecidable). The method for proving decidability is based on Buchi's technique: given a formula without universal quantifiers (they can of course be replaced by negations and existential quantifications), we start by constructing an automaton recognizing the language defined by each atomic subformula; we then use the closure of the class of automata under the Boolean operations for handling conjunctions, disjunctions and negation; we finally use the closure under projection for eliminating the existential quantifiers; the encoded formula has no solution iff the automaton recognizes an empty language.

So, the main question is to find a class of automata powerful enough so as to encode the atomic formulae, which in addition is closed under the above operations, and for which the emptyness problem is decidable. This approach is taken in [7], in which the authors use bottom-up tree automata that include (some bounded number of) *equality tests* along any path leading to a final state. The actual algorithm based on pumping properties of these automata is doubly exponential as already mentioned. A simply exponential algorithm based on more sophisticated pumping techniques has been given very recently by Comon and Jacquemart [8].

The following is an automaton recognizing the language of ground terms in normal form for the single rule $f(x, f(a, x)) \rightarrow a$. A transition of the form $l \longrightarrow_{p \neq q} r$ means that the rule $l \rightarrow r$ applies provided $l|_p$ and $l|_q$ are different terms. The use of more complicated tests in which positions appear as arguments of terms with the intended meaning that $s[p]_q$ is the term $s[l|p]_q$ is only a matter of convenience, although it permits reduce the size of the automaton by an exponential factor.

Ground reducibility being an expensive test, it is worth avoiding it. Fortunately, this test is not necessary for (most) specifications whose set of function symbols can be split into constructors and defined symbols in the sens of Huet [34]: defined symbols generate inductive reducibility tests that are automatically satisfied as soon as the test applies to a term containing such a defined symbol.

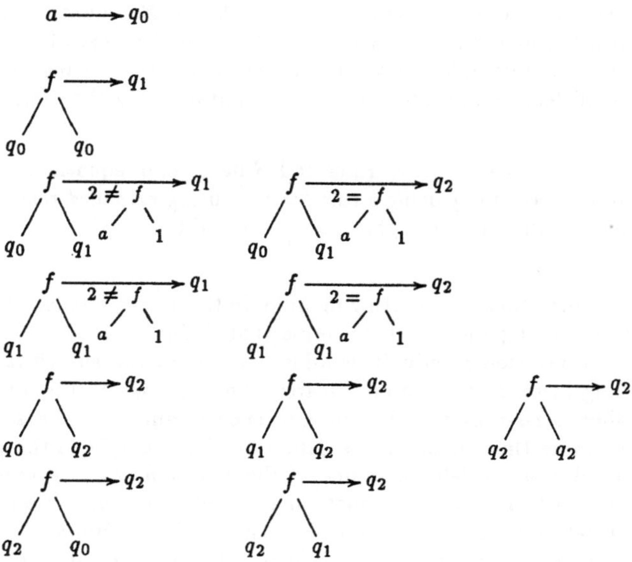

Fig. 19. Automaton with constraints

Plaisted was the first to prove the decidability of inductive reducibility [64], and Comon to consider the problem in terms of formal language theory (using grammars and reachability techniques instead of automata and pumping techniques).

10 Generalizations

The techniques developed so far require termination. We will briefly mention how to generalize them in order to accommodate more general situations, by using rewriting modulo or ordered rewriting.

Ordered rewriting uses directly the equations for rewriting, provided their instances are directed by means of an appropriate ordering.

Definition 28. A term s rewrites to a term t at position p with respect to a rewrite ordering $\succeq$ using an equation $l \simeq r \in E$, if there exists a substitution σ such that $s|_p = l\sigma, t = s[r\sigma]_p$, and $l\sigma \succ r\sigma$, or $s|_p = r\sigma, t = s[l\sigma]_p$, and $r\sigma \succ l\sigma$. We write $s \xrightarrow[l \simeq r]{p} t$ in the first case, and $s \xrightarrow[r \simeq l]{p} t$ in the second.

In the latter view of rewriting, an equation $l \simeq r$ is indeed a *rewrite rule* $l \rightarrow r$ if there are pairs s, t such that $s \xrightarrow[l \simeq r]{p} t$, but no pairs u, v such that $u \xrightarrow[r \simeq l]{p} v$.

Although commutativity can be used for ordered rewriting as defined above, for example $a+b \xrightarrow[x+y \simeq y+x]{} b+a$ if $a+b > b+a$ in the given ordering, it is in practice useful to separate the equations into two sets denoted by R and S. Equations in R will be used for rewriting. Equations in S will be used to find R-redexes by equational reasoning. This general principle known under the name of *rewriting modulo* has been proved very efficient in a number of cases by appropriately choosing R and S.

Definition 29 [63]. Let R be a set of rules and S be a set of equations. A term s rewrites to a term t at position p using a rule $l = r$ by using *extended rewriting* with S if there exist a substitution σ such that $s|_p \overset{*}{\underset{S}{\longleftrightarrow}} l\sigma$ and $t = s[r\sigma]_p$.

There are other definitions of rewriting modulo. In the above, a redex at position p is obtained by restricting the search to terms that differ from s by applying S-equalities under the position p only. This allows to obtain the substitution σ by S-pattern matching instead of equational search. The case where the whole term is searched is called *class-rewriting*, but its efficiency is dubious. The search can be restricted even more than in the above definition, for example to the positions of non-linear variables of the left hand side of the equation [23]. Although much more efficient (only standard pattern matching is used in conjunction with S-equality tests), the latter variant has not yet demonstrated its usefulness. Normalized rewriting is another variant of rewriting modulo in which terms must be normalized with respect to the S part before to be rewritten according to the R part. The S part is therefore itself a rewrite system, possibly modulo associativity and commutativity. An important example is given by symbolic computations over polynomials.

Termination for whatever variant of rewriting modulo S needs an *S-rewrite ordering*, that is a quasi ordering whose strict part is a rewrite ordering and whose equivalence contains the congruence generated by S. This is an area where research is still actively going on. On the other hand, most problems relating to confluence are basically solved [39; 1]. Orthogonal systems have received little attention, though. For ordered rewriting, joinability of an appropriate notion of critical pairs is a sufficient condition for confluence [31]. An interesting conjecture is the decidability of ground confluence for ordered rewriting when the ordering used is a total recursive path ordering with all statuses lexicographic.

The above definitions can be extended to more sophisticated notions of equations, such as conditional equations, order-sorted equations, and constrained equations. We refer to [21] for the definitions.

Completion has been extended to handle various situations, in particular to cope with associative-commutative axioms [63] which cannot be handled as rules. More generally, completion frameworks for rewriting modulo are described in [39; 3; 53], for ordered rewriting in [31; 60], and for constraint rewriting in [13]. Normalized completion embodies various completion methods, including the computation of a Gröbner basis of an polynomial ideal over a commutative ring [53]. Ordered completion, the variant based on ordered rewriting, has the important property that it always succeeds if it terminates. In addition, if the ordering used has the subterm property and a *canonical* set of rules exists for the considered theory, then ordered completion is guaranteed to find it, hence it is refutationally complete [1].

Acknowledgments: to Evelyne Contejean for her help with the drawings.

References

1. Leo Bachmair. *Canonical Equational Proofs.* Birkhäuser, Boston, 1991.
2. Leo Bachmair and Nachum Dershowitz. Critical pair criteria for completion. *Journal of Symbolic Computation*, 6(1):1–18, 1988.
3. Leo Bachmair, Nachum Dershowitz, and Jieh Hsiang. Orderings for equational proofs. In *Proc. 1st IEEE Symp. Logic in Computer Science, Cambridge, Mass.*, pages 346–357, June 1986.
4. Leo Bachmair, Harald Ganzinger, Christopher Lynch, and Wayne Snyder. Basic paramodulation and superposition. In Deepak Kapur, editor, *Proc. 11th Int. Conf. on Automated Deduction, Saratoga Springs, NY, LNCS 607.* Springer-Verlag, June 1992.
5. F. Barbanera, M. Fernández, and H. Geuvers. Modularity of strong normalization in the λ-algebraic-cube. 1993.
6. G. Birkhoff. On the structure of abstract algebras. In *Proc. Cambridge Phil. Society, 31*, 1935.
7. A-C. Caron, J.-L. Coquidé, and M. Dauchet. Encompassment properties and automata with constraints. In Claude Kirchner, editor, *Proc. 5th. Int. Conf. on Rewriting Techniques and Applications*, Lecture Notes in Computer Science, vol. 690, Montreal, Canada, 1993. Springer-Verlag.
8. H. Comon and Florent Jacquemard. Ground reducibility and automata with disequality constraints. In P. Enjalbert, editor, *Proc. 11th Symp. on Theoretical Aspects of Computer Science*, Paris, 1994. Springer-Verlag.
9. Hubert Comon. Unification et disunification: Théorie et applications. Thèse de Doctorat, Institut National Polytechnique de Grenoble, France, 1988.
10. Hubert Comon. Inductive proofs by specifications transformation. In *Proc. 3rd Rewriting Techniques and Applications, Chapel Hill, LNCS 355*, pages 76–91. Springer-Verlag, April 1989.
11. Hubert Comon. Solving symbolic ordering constraints. *International Journal of Foundations of Computer Science*, 1(4):387–411, 1990.
12. Hubert Comon. Disunification: a survey. In Jean-Louis Lassez and Gordon Plotkin, editors, *Computational Logic: Essays in Honor of Alan Robinson.* MIT Press, 1991.
13. Hubert Comon. Completion of rewrite systems with membership constraints. In W. Kuich, editor, *Proc. 19th Int. Coll. on Automata, Languages and Programming, LNCS 623*, Vienna, 1992. Springer-Verlag. An extended version is available as LRI Research Report number 699, Sept. 1991.
14. Max Dauchet. Simulation of a turing machine by a left-linear rewrite rule. In *Proc. 3rd Rewriting Techniques and Applications, Chapel Hill, LNCS 355*, 1989.
15. Nachum Dershowitz. A note on simplification orderings. *Information Processing Letters*, 9(5):212–215, November 1979.
16. Nachum Dershowitz. Applications of the Knuth-Bendix completion procedure. In *Proceedings of the Seminaire d'Informatique Theorique*, pages 95–111, Paris, France, December 1982.
17. Nachum Dershowitz. Orderings for term rewriting systems. *Theoretical Computer Science*, 17(3):279–301, March 1982.
18. Nachum Dershowitz. A taste of rewrite systems. In *Proc. Functional Programming, Concurrency, Simulation and Automated Reasonning*, 1993.
19. Nachum Dershowitz. Trees, ordinals and termination. In *Proc. CAAP 93, LNCS 668*, 1993.
20. Nachum Dershowitz and Charles Hoot. Topics in termination. In *Proc. 5th Rewriting Techniques and Applications, Montréal, LNCS 690*, 1993.

21. Nachum Dershowitz and Jean-Pierre Jouannaud. Rewrite systems. In J. van Leeuwen, editor, *Handbook of Theoretical Computer Science*, volume B, pages 243–309. North-Holland, 1990.
22. Nachum Dershowitz and Mitsuhiro Okada. Proof-theoretic techniques for term rewriting. In *Proc. 3rd IEEE Symp. Logic in Computer Science, Edinburgh*, June 1988.
23. Hervé Devie. Une approche algébrique de la réécriture de preuves équationnelles et son application à la dérivation de procédures de complétion. Thèse de Doctorat, Université de Paris-Sud, France, Octobre 1991.
24. K. Drosten. *Termerssetzungssysteme*. PhD thesis, University of Passau, Germany, 1989. In german.
25. M. Fay. First-order unification in an equational theory. In *Proc. 4th Workshop on Automated Deduction, Austin, Texas*, pages 161–167, 1979.
26. M. Fernandez and J.-P. Jouannaud. Modularity properties of term rewriting systems revisited. submitted, 1993.
27. Maribel Fernández. Narrowing based procedures for equational disunification. *Applicable Algebra in Engineering Communication and Computing*, 3:1–26, 1992.
28. Laurent Fribourg. A strong restriction of the inductive completion procedure. *Journal of Symbolic Computation*, 8:253–276, 1989.
29. J. A. Goguen, T. Winkler, José Meseguer, K. Futatsugi, and Jean-Pierre Jouannaud. *Applications of Algebraic Specifications Using OBJ*, chapter Introducing OBJ*. Cambridge University Press, 1993. D. Coleman, R. Gallimore and J. Goguen eds.
30. Jacques Herbrand. Recherches sur la théorie de la démonstration. Thèse d'Etat, Univ. Paris, 1930. Also in: Ecrits logiques de Jacques Herbrand, PUF, Paris, 1968.
31. Jieh Hsiang and Michaël Rusinowitch. On word problems in equational theories. In *Proc. in 14th ICALP Karlsruhe*, July 1987.
32. Gérard Huet. Confluent reductions: abstract properties and applications to term rewriting systems. *Journal of the ACM*, 27(4):797–821, October 1980.
33. Gérard Huet. A complete proof of correctness of the Knuth-Bendix completion algorithm. *Journal of Computer and System Sciences*, 23:11–21, 1981.
34. Gérard Huet and J.-M. Hullot. Proofs by induction in equational theories with constructors. *Journal of Computer and System Sciences*, 25(2), 1982.
35. Gérard Huet and Dallas S. Lankford. On the uniform halting problem for term rewriting systems. Research Report 283, INRIA, March 1978.
36. J.-M. Hullot. Canonical forms and unification. In *Proc. 5th Conf. on Automated Deduction, Les Arcs, France, LNCS 87*. Springer-Verlag, July 1980.
37. Jean-Pierre Jouannaud and Claude Kirchner. Solving equations in abstract algebras: A rule-based survey of unification. In Jean-Louis Lassez and Gordon Plotkin, editors, *Computational Logic: Essays in Honor of Alan Robinson*. MIT-Press, 1991.
38. Jean-Pierre Jouannaud and Hélène Kirchner. Completion of a set of rules modulo a set of equations. *SIAM Journal on Computing*, 15(4), November 1986.
39. Jean-Pierre Jouannaud and Hélène Kirchner. Completion of a set of rules modulo a set of equations. *SIAM Journal on Computing*, 15(4):1155–1194, 1986.
40. Jean-Pierre Jouannaud and Emmanuel Kounalis. Automatic proofs by induction in theories without constructors. *Information and Computation*, 82(1), July 1989.
41. Jean-Pierre Jouannaud and Mitsuhiro Okada. Executable higher-order algebraic specification languages. In *Proc. 6th IEEE Symp. Logic in Computer Science, Amsterdam*, pages 350–361, 1991.
42. Jean-Pierre Jouannaud and Mitsuhiro Okada. Satisfiability of systems of ordinal notations with the subterm property is decidable. In *Proc. 18th Int. Coll. on Automata, Languages and Programming, Madrid, LNCS 510*, 1991.

43. Deepak Kapur, Paliath Narendran, and Hantao Zhang. On sufficient completeness and related properties of term rewriting systems. *Acta Informatica*, 24(4):395–415, 1987.
44. Hélène Kirchner and M. Hermann. Meta-rule inference from crossed rewrite systems. Draft version, presented at CTRS, Montreal, 1990.
45. Jan Willem Klop. Term rewriting systems: a tutorial. *EATCS Bulletin*, 32:143–182, June 1987.
46. Jan Willem Klop. Term Rewriting Systems. In S. Abramsky, Dov.M. Gabbay, and T.S.E. Maibaum, editors, *Handbook of Logic in Computer Science*, volume 2, pages 1–116. Clarendon Press, 1992.
47. Donald E. Knuth and Peter B. Bendix. Simple word problems in universal algebras. In J. Leech, editor, *Computational Problems in Abstract Algebra*, pages 263–297. Pergamon Press, 1970.
48. G. Kucherov and M. Tajine. Decidability of regularity and related properties of ground normal form languages. Research Report 92-R-062, CRIN, Nancy, France, 1992. To appear in Information and Computation.
49. Mahahito Kurihara and Azuma Ohuchi. Non-copying term rewriting and modularity of termination. Hokkaido University.
50. Mahahito Kurihara and Azuma Ohuchi. Modularity of simple termination of term rewriting systems with shared constructors. *Theoretical Computer Science*, 103:273–282, 1992.
51. Dallas S. Lankford. Canonical inference. Memo ATP-32, University of Texas at Austin, March 1975.
52. Dallas S. Lankford. On proving term rewriting systems are noetherian. Technical Report MTP-3, Mathematics Department, Louisiana Tech. Univ., 1979.
53. Claude Marché. Réécriture modulo une théorie présentée par un système convergent et décidabilité des problèmes du mot dans certains classes de théories équationnelles. Thèse de Doctorat, Université de Paris-Sud, France, 1993.
54. José Meseguer and Joseph Goguen. Initiality, induction and computability. In M. Nivat and J. Reynolds, editors, *Algebraic Methods in Semantics*, chapter 14. Cambridge Univ. Press, 1985.
55. Y. Métivier. About the rewriting systems produced by the knuth-bendix completion algorithm. *Information Processing Letters*, 16(1):31–34, 1983.
56. A. Middeldorp. Modular aspects of properties of term rewriting systems related to normal forms. In *Proc. 3rd Rewriting Techniques and Applications, Chapel Hill, LNCS 355*, pages 263–277. Springer-Verlag, 1989.
57. Aart Middeldorp. *Modular Properties of Term Rewriting Systems.* PhD thesis, Free University of Amsterdam, Netherland, 1990.
58. D. Musser. Proving inductive properties of abstract data types. In *Proc. 7th ACM Symp. Principles of Programming Languages, Las Vegas*, 1980.
59. M. H. A. Newman. On theories with a combinatorial definition of 'equivalence'. *Ann. Math.*, 43(2):223–243, 1942.
60. Robert Nieuwenhuis and Albert Rubio. Completion of first-order clauses by basic superposition with ordering constraints. Tech. report, Dept. L.S.I., Univ. Polit. Catalunya, 1991. To appear in Proc. 11th Conf. on Automated Deduction, Saratoga Springs, 1992.
61. Robert Nieuwenhuis and Albert Rubio. Basic superposition is complete. In B. Krieg-Bruckner, editor, *Proc. European Symp. on Programming, LNCS 582*, pages 371–389, Rennes, 1992. Springer-Verlag.
62. W. Nutt, P. Réty, and Gert Smolka. Basic narrowing revisited. *Journal of Symbolic Computation*, 7(3/4):295–318, 1989.

63. Gerald E. Peterson and Mark E. Stickel. Complete sets of reductions for some equational theories. *Journal of the ACM*, 28(2):233–264, April 1981.
64. David Plaisted. Semantic confluence tests and completion methods. *Information and Control*, 65:182–215, 1985.
65. David A. Plaisted. Semantic confluence tests and completion methods. *Information and Control*, 65(2-3):182–215, May/June 1985.
66. Detlef Plump. Collapsed tree rewriting: Completeness, confluence and modularity. In *Proc. Conditional Term Rewriting Systems, LNCS 656*, 1993.
67. J. A. Robinson. A machine-oriented logic based on the resolution principle. *Journal of the ACM*, 12(1):23–41, 1965.
68. Barry K. Rosen. Tree-manipulating systems and church-rosser theorems. *J. of the Association for Computing Machinery*, 20(1):160–187, January 1973.
69. Michaël Rusinowitch. On termination of the direct sum of term rewriting systems. *Information Processing Letters*, 26:65–70, 1987.
70. C. C. Squier. Word problems and a homological finiteness condition for monoids. *Journal of Pure and Applied Algebra*, 49:201–217, 1987.
71. M. Stone. The theory of representations for Boolean algebra. *Transactions of the American Mathematical Society*, 40:37–111, 1936.
72. W. Thomas. Automata on infinite objects. In J. van Leeuwen, editor, *Handbook of Theoretical Computer Science*, pages 134–191. Elsevier, 1990.
73. Y. Toyama. Counterexamples to termination for the direct sum of term rewriting systems. *Information Processing Letters*, 25:141–143, April 1987.
74. Y. Toyama. On the Church-Rosser property for the direct sum of term rewriting systems. *Journal of the ACM*, 34(1):128–143, April 1987.
75. Ralf Treinen. A new method for undecidability proofs of first order theories. Tech. Report A-09/90, Universität des Saarladandes, Saarbrücken, May 1990.
76. Benjamin Werner. *Méta-théorie du Calcul des Constructions Inductives*. Thèse Univ. Paris VII, France, 1994.
77. H. Zhang. Automated proof of ring commutativity problems by algebraic methods. *Journal of Symbolic Computation*, 9:423–427, 1990.
78. I. I. Zhegalkin. On a technique of evaluation of propositions in symbolic logic. *Matematicheskii Sbornik*, 34(1):9–27, 1927.

Action Structures and the Pi Calculus

Robin Milner

Laboratory for Foundations of Computer Science, Computer Science Department, University of Edinburgh, The King's Buildings, Edinburgh EH9 3JZ, UK[1]

Abstract: Action structures are proposed as a variety of algebra to underlie concrete models of concurrency and interaction. An action structure is equipped with *composition* and *product* of actions, together with an indexed family of *abstractors* to allow parametrisation of actions, and a *reaction* relation to represent activity. The eight axioms of an action structure make it an enriched strict monoidal category.

In Part I the notion of action structure is developed mathematically, and examples are studied ranging from the evaluation of expressions to the statics and dynamics of Petri nets. For algebraic process calculi in particular, it is shown how they may be defined by a uniform superposition of process structure upon an action structure specific to each calculus.

The theory of action structures emphasizes the notion of *effect*; that is, the effect which any interaction among processes exerts upon its participants. Effects together with *incidents* (roughly speaking, observable actions) allow a uniform treatment of bisimulation congruence.

In Part II, the π-calculus is treated as an extended example of this uniform process theory. Three action structures for the π-calculus are examined; they are presented in a simple graphical form. The incidents for π-calculus are best characterized in terms of *reachability*, a graphical notion. This leads to a bisimulation congruence for the *synchronous* π-calculus, in which individual transitions may involve actions of arbitrary complexity.

Keywords: abstractor, action, action structure, bisimulation, concurrency, effect, incident, interaction, monoidal category, process structure, pi-calculus, reaction relation, transition.

Contents:

---

[1] This work was done with the support of a Senior Fellowship from the Science and Engineering Research Council, UK.

PART I: Action Structures

1 Introduction

The purpose of this paper is to explore action structures; they provide a new algebraic foundation for concrete models of concurrent computation, but are free from the ad hoc details which are often present in such models.

1.1 Background As far as modelling sequential computation is concerned, a classic mathematical structure is Scott's notion of *domain* [13]. It allows us to understand complete computations by examining the partial computations which approximate them. Indeed domains also help us to understand concurrent computations in this way, as shown by Nielsen, Plotkin and Winskel [25] in an important paper which relates domains both to event structures and to Petri nets [27].

Event structures and Petri nets are good examples of mathematical structures specific to concurrency. But we appear still to lack a canonical structure which is *combinational*, i.e. which explains how processes are synthesized, and which embodies the dynamics of *interaction* among processes. There are algebraic process calculi [3, 14, 19] and algebraic theories of Petri nets [7, 9]; these models have theoretical content which emphasizes combination and interaction. But what they have in common has not been distilled, so that as a family –and even individually– they can appear ad hoc.

It is therefore reasonable to look for a canonical algebraic structure for concurrency, with interaction at its heart; in this paper we propose *action structures* as a candidate. An action structure is an algebra which allows complex actions to be built from simple ones; in addition it has a dynamic component which defines the possibly complex performance of an action.

Central to action structures are two concerns. The first is that an action may be simple in one view, complex in another. Here are two examples, to suggest that this is consistent with common usage of the term "action". First, think of a sequential program accessing the fourteenth member of a linked list. At a low level this consists of many actions –chaining down the list– and this is how an implementer must treat it; at another level it is clearly a single action, an interaction between a program and the memory. As another example, suppose that you treat me to lunch because you want to borrow money from me. The lunch consists of many interactions, undoubtedly a complex pattern! Later, however, you and I will remember it as a single action: you stung me. This relativity of atomicity is well known to systems engineers; one only has to think of the difficult questions associated with transactions in a database. Studies of action refinement in process algebras, e.g. by van Glabbeek and Goltz [12] and Aceto and Hennessy [1], have begun to place relative atomicity on a theoretical footing; one purpose of action structures is to continue the task.

The second concern of action structures is with the *effect* which a shared action exerts upon its participants. Each participant suffers a different effect. In the case of the lunch you are richer, I am stung, the waiter is tired. [2] For a more formal example consider the following transition in CCS [19]:

$$\overline{p}x.P \mid p(y).Q \longrightarrow P \mid \{x/y\}Q,$$

[2] The waiter also took part, and illustrates another difficulty with concurrency. Unlike you and me, he will not remember our lunch as an atomic action; for him, it was inextricably interwoven with the serving of other lunches. This shows that the level of atomicity does not only vary from one global view of a system to another; it even varies between parts of the system.

where p is a port at which x is being output by one process and at which the other is ready to receive input, and $\{x/y\}Q$ means the substitution of x for y in Q. In action structures this will be modelled by a *reaction*, which may be written roughly as

$$\overline{p}x \otimes p(y) \searrow \mathbf{id} \otimes \{x/y\}.$$

In general, a reaction $a \searrow a'$ in an action structure represents part of the performance of an action a; the performance may continue by further reaction $a' \searrow a''$, and so on. In the example, a reaction of the product of two actions (itself an action) reduces it to a remainder action which is the product of **id** and $\{x/y\}$; these, and their product, are degenerate actions which we call *effects*. They are degenerate in that they perform no further reaction; instead they represent the effect exerted upon P and Q respectively by the performance of the original action, i.e. by the occurrence of a communication through the port p. Note that the unit action **id**, the effect on P, represents *no* effect; it is only Q that feels an effect, which in this case is a substitution.

Strong motivation for action structures came from the π-calculus [24]. A distinguishing feature of the π-calculus is that port-names, like p in the above, can be transmitted as data; this can create another kind of effect, namely a new liaison between two agents. To illustrate, in the π-calculus we may adapt the previous example as follows:

$$((\nu x)\overline{p}x.P) \mid p(y).Q \longrightarrow (\nu x)(P \mid \{x/y\}Q).$$

Here the restriction (νx) localises the port x to $\overline{p}x.P$; but an effect of the reaction is to widen the scope of x, so that it becomes a port which can carry information between P and Q. In action structures the corresponding reaction is now roughly speaking

$$((\nu x)\overline{p}x) \otimes p(y) \searrow (\nu x)(\mathbf{id} \otimes \{x/y\}),$$

and we see the newly created liaison as a component of the resulting effect. Thus the π-calculus suggested that one should try to distill a mathematical notion of effect from these instances.

The π-calculus also gave a second motivation. Through its treatment of names and restriction the π-calculus allows a complex action, such as accessing the nth element of a list, to be programmed in terms of many simple actions; so one would like a version of the π-calculus in which such a complex action can be a single transition of a process. This would correspond to the original π-calculus in the way that Synchronous CCS [18], where a transition may involve arbitrarily many "particulate" interactions, corresponds to CCS where it may only involve a single one. This motivation is traced in Part II, Section 10; it leads naturally to the three algebraic operations of an action structure. A main outcome of this paper is that process calculi like CCS, the π-calculus and their variants can be factored into two parts: at the lower level an action structure which is different for each of them, and at the upper level a process structure which is common to them all.

1.2 Outline of Part I In Section 2 we introduce action structures. We first present static action structures as an algebraic theory; then we add dynamics, a preorder preserved by the algebraic operations; finally we define a natural notion of homomorphism and substructure.

Section 3 gives several examples of action structures: functions, expression evaluation, process calculi and Petri nets. It also suggests how to treat the notion of *port*, or communication interface, uniformly in action structures. These examples lie comfortably within the action structure framework.

In Section 4, we treat the central notion of *effect*, the influence of the performance of an action on its participants. Effects take their place as a sub-actionstructure consisting of *inert* actions – those which have nothing left to perform. Some of the examples of Section 3 are re-visited to see what effect structure is appropriate in each case.

The next four sections are devoted to building process structure in a uniform way upon an action structure. Sections 5 and 6 deal respectively with the statics and with the dynamics of processes. The latter comes from two sources: from actions, especially the communications between processes, and from the process control operators.

Sections 7 and 8 present an important consequence of the theory: that the natural bisimulation equivalence over processes, presented uniformly in terms of the underlying action structure, is indeed a congruence relation. This crucially depends upon the mathematical properties of effects, and also of a new kind of actions known as *incidents*. An incident-set is a subset of the actions satisfying a simple property. There are many possible incident-sets, and each one determines a bisimulation equivalence; the larger the incident-set, the finer the equivalence. Section 7 defines incident-sets, and proves that under a simple condition there is a smallest incident-set including a given set of actions. Section 8 establishes that every bisimulation equivalence based upon an incident-set is a congruence.

Section 9 alludes to work by other authors which is related, or with which a relation should be sought, apart from papers already mentioned in the text.

1.3 Acknowledgements I am grateful to Yoram Hirshfeld, Davide Sangiorgi and David Turner for useful discussions. The presentation has been strengthened by some category-theoretic ideas with the help of Barry Jay and John Power, whom I thank for their insights.

2 Action structures

An action structure is an algebra with some extra structure. It may also be seen as strict monoidal category with two additions, one structural and one dynamic. The elements of an action structure are called *actions*; they are its morphisms, in categorical terms. The structural addition is an indexed family of functions over actions called *abstractors*; categorically they are endo-functors. Abstractors allow actions to be parametric upon certain constituents. The dynamic addition is called *reaction*; that is, a preorder over actions which represents activity.

2.1 Definition (static action structure) Let X be a set; in this context its members $x, y, \ldots$ will often be called *names*. Then a *static action structure* A over X consists of

- A monoid $(M, \otimes, \epsilon)$ of *arities*, and an assignment to each x of an arity in M. Categorically, M is the monoid of objects.

- For each $m, n \in M$ a set $A_{m,n}$ of *actions*. If $a \in A_{m,n}$ we say that a has *arity* $m \to n$, and write $a : m \to n$.

- Nullary operators $\mathbf{id}_m$ (the identities), binary operators $\cdot$ (composition) and $\otimes$ (tensor product), and a family $\{\mathbf{ab}_x \mid x \in X\}$ of unary operators (abstractors).

The operators are subject to the following rules of arity:

$$\mathbf{id}_m : m \to m \qquad\qquad \frac{a : k \to n \qquad b : m \to p}{a \otimes b : k \otimes m \to n \otimes p}$$

$$\frac{a : k \to m \qquad b : m \to p}{a \cdot b : k \to p} \qquad\qquad \frac{x : k \qquad a : m \to n}{\mathbf{ab}_x a : k \otimes m \to k \otimes n}$$

and we freely omit the explicit arity m in $\mathbf{id}_m$, meaning that it can be supplied in any way which is consistent with the context. The operators satisfy eight laws:

$$\begin{array}{llll} \text{C1.} & a \cdot \mathbf{id} = a = \mathbf{id} \cdot a & \text{C2.} & a \cdot (b \cdot c) = (a \cdot b) \cdot c \\ \text{P1.} & a \otimes \mathbf{id}_\epsilon = a = \mathbf{id}_\epsilon \otimes a & \text{P2.} & a \otimes (b \otimes c) = (a \otimes b) \otimes c \\ \text{PF1.} & \mathbf{id} \otimes \mathbf{id} = \mathbf{id} & \text{PF2.} & (a \cdot b) \otimes (c \cdot d) = (a \otimes c) \cdot (b \otimes d) \\ \text{AF1.} & \mathbf{ab}_x \mathbf{id} = \mathbf{id} & \text{AF2.} & \mathbf{ab}_x(a \cdot b) = (\mathbf{ab}_x a) \cdot (\mathbf{ab}_x b). \end{array}$$

■

Often the arity monoid M will be $(\mathsf{N}, +, 0)$, the natural numbers under addition; then the arity of each name x will be 1. But we may instead wish the arity of a name to be a *sort*, in some set S; then M will be the free monoid generated by S. More complex arities also arise in constructions of action structures.

The laws not involving the abstractors, i.e. C, P and PF, are just those of a strict monoidal category. PF asserts that $\otimes$ is a (bi)functor. Similarly, AF asserts that each $\mathbf{ab}_x$ is a functor.

We now proceed to the dynamics.

2.2 Definition (dynamic action structure) A *(dynamic) action structure* A over X consists of a static action structure together with a preorder $\searrow$ over each $A_{m,n}$ called *reaction*, such that

- Each operator preserves reaction, e.g. $a \searrow a'$ implies $a \cdot b \searrow a' \cdot b$ and $b \cdot a \searrow b \cdot a'$;
- $\mathbf{id}$ is inactive, i.e. $\mathbf{id} \searrow a$ implies $a = \mathbf{id}$. ■

Reaction is often the transitive reflexive closure of an elementary relation $\overset{1}{\searrow}$ which may be called *atomic reaction*, representing the occurrence of an elementary event. An example is the interaction between complementary actions $\alpha, \overline{\alpha}$ in CCS; in Section 3.6 we shall see that $\alpha \otimes \overline{\alpha} \overset{1}{\searrow} \tau$.

One of the aims in studying action structures is to learn how to treat the *composite* reactions in an action structure as the *elementary* steps of a higher-level process calculus. In Sections 5 and 6 we show how to build familiar process calculi in a uniform way upon an action structure; then the transition relation, already familiar in process calculi, is definable in terms of the lower-level reaction relation.

2.3 Terminology We shall usually consider dynamic, not static, action structures; so we usually omit the adjective "dynamic".

When the arity monoid M and the name-set X are understood we shall just use A to denote either the action-set or the entire action structure. At other times we may write (M, X, A) and (N, Y, B) for two distinct action structures; we may write $(M, X, A, \searrow)$ if we want to emphasise reaction; we may also write M_A, X_A, $\searrow_A$ to denote the parts of a given action structure A. Such abuses, used with care, will save heavy formality.

2.4 Definition (homomorphism of action structures) A *homomorphism* $\Phi : A \rightarrow B$ between two action structures A and B consists of

- A monoid homomorphism $\Phi : M_A \rightarrow M_B$;
- A map $\Phi : X_A \rightarrow X_B$ such that $x : m$ implies $\Phi x : \Phi m$;
- A map $\Phi : A \rightarrow B$ of actions such that

  $a : m \rightarrow n$ implies $\Phi a : \Phi m \rightarrow \Phi n$;

  Φ preserves all the operations $\mathbf{id}$, $\cdot$, $\otimes$ and $\mathbf{ab}_x$;

  $a \searrow_A a'$ implies $\Phi a \searrow_B \Phi a'$.

Φ is an *isomorphism* if there is a homomorphism $\Phi^{-1} : B \rightarrow A$ such that both $\Phi \circ \Phi^{-1}$ and $\Phi^{-1} \circ \Phi$ are identity maps on arities, names and actions. ∎

As we add further structure to the notion of action structure, we may also refine the notion of homomorphism.

2.5 Definition (sub-actionstructure) An action structure B is a *sub-actionstructure* of A if there is a homomorphism $\Phi : B \rightarrow A$ such that

1. Φ is injective on X_B, M_B and B ;
2. For all $b, b' \in B$, $b \searrow_B b'$ iff $\Phi b \searrow_A \Phi b'$. ∎

In defining substructure notions it is common to require the implication in clause 2 only from left to right, and to use the term *full* for a substructure obeying the double implication. But for the present we have no use for the weaker condition, so we adopt the stronger one to avoid tediously repeating the term "full".

2.6 Discussion The aspect of action structures which is least familiar is abstraction. The examples of Section 3 below will show that abstraction permits parametrisation of actions upon entities of some kind; but it also shows that these entities can be quite diverse in nature: ordinary values in one case, port-names in another, places and transitions in the case of Petri nets. So the notion of action structure presents, via abstraction and its axioms, something common to all these parametrisations.

But does it capture *all* that they have in common? This is really a vague question, since we clearly don't want to limit ourselves to the examples of Section 3. Certainly for those applications it will be seen that the names X stand for entities, and that the abstractors bind those names. In fact for each name x there is usually an action $\langle x \rangle$ such that $\langle x \rangle \cdot \mathbf{ab}_y a$ imposes the substitution $\{^x/_y\}$ upon a; to be precise, it represents $\langle x \rangle \otimes \{^x/_y\} a$.

We might therefore expect concrete notions such as *free name*, *substitution* and *alpha-conversion* to be captured by the formulation and axioms of action structures. But they are not, for two reasons. First, the present axioms of action structure lie easily in the mind because they have a simple and appealing pattern. Second, they are strong enough to deliver interesting properties of process calculi in a uniform way, which is a main purpose this paper. However, these concrete notions are axiomatized for a subclass of action structures called *action calculi*, which are introduced in [21].

2.7 Categorical formulation Having discussed how the notion of action structure may be *strengthened*, we now look briefly at the natural formulation in categorical terms, which turns out to be a little *weaker* (i.e. more general).

The axioms AF require the abstractors to be functors. But we have not allowed the functors $\mathbf{ab}_x$ to be quite arbitrary; the rule of arity for $\mathbf{ab}_x a$ in Definition 2.1 asserts that the operation of the functor $\mathbf{ab}_x$ upon objects is taken to be $k \otimes (-)$, where k is fixed by x. But this relationship between the abstraction functors and the monoidal operation is not essential for many properties of action structures. Thus a more abstract definition of action structure would avoid specifying how the arity $\mathbf{ab}_x m$ depends upon x. The more specific formulation of Definition 2.1 aims at a comfortable compromise between mathematical generality and the specific interesting applications.

Concerning notation: I have adopted the forward version of composition, so that for $a : k \rightarrow m$ and $b : m \rightarrow p$ I write $a \cdot b$, not $b \circ a$ which is standard for category theorists. It would be confusing use the latter in this work, since in Sections 5 and 6 we find that composition $a \cdot P$ is closely allied to the prefixing operation of $a \bullet P$ in process algebra. Some prefer to use ";" for forward composition, but the use of punctuation in an algebraic expression can be distracting.

3 Examples of action structures

We begin by fitting some widely-known ideas into the framework of action structures, and then proceed to show how these structures can accommodate less standard notions found in concurrency theory.

3.1 Functions Let D be some fixed set of values. Define $\mathsf{FUN} = (M, X, A, \searrow)$ to be the action structure in which

$$\begin{aligned} M &= (\mathsf{N}, +, 0) \\ X &= \emptyset \\ A_{m,n} &= D^m \rightarrow D^n \\ \searrow &= \text{the identity relation.} \end{aligned}$$

The operations over A are as follows, where we use $\vec{v}$ for a tuple $v_1 \ldots v_m$ of variables, and sometimes use a comma to concatenate tuples:

$$\begin{aligned} \mathbf{id}_m &\stackrel{\text{def}}{=} \lambda\vec{v}\,.\,\vec{v} \\ f \cdot g &\stackrel{\text{def}}{=} \lambda\vec{v}\,.\,g(f\vec{v}) \\ f \otimes g &\stackrel{\text{def}}{=} \lambda\vec{u}\vec{v}\,.\,(f\vec{u}, g\vec{v}) \qquad (\vec{u}, \vec{v} \text{ all distinct}). \end{aligned}$$

There are no abstractors, so we only have to verify the axioms of a monoidal category; this is simple and well-known.

3.2 Functions with abstractors Let X be a set, which we shall call the *global variables*, and let $\mathrm{Env} \stackrel{\text{def}}{=} X \rightarrow D$, the valuations of global variables in D. Define the action structure $\mathsf{FUN}[X] = (M, X, A, \searrow)$ as follows:

$$M = (\mathsf{N}, +, 0)$$

$$X = \text{as given}$$
$$A_{m,n} = \text{Env} \rightarrow (D^m \rightarrow D^n)$$
$$\searrow = \text{the identity relation.}$$

The operations are nearly as simple as in FUN; we only have to distribute the parameter $\varrho \in$ Env properly. Let us write $\varrho_{x \mapsto v}$ for the valuation which maps x to the value v, but otherwise assigns values as ϱ does. Then

$$\mathbf{id}_m \varrho \stackrel{\text{def}}{=} \lambda \vec{v} \,.\, \vec{v}$$
$$(f \cdot g)\varrho \stackrel{\text{def}}{=} \lambda \vec{v} \,.\, g\varrho(f\varrho\vec{v})$$
$$(f \otimes g)\varrho \stackrel{\text{def}}{=} \lambda \vec{u}\vec{v} \,.\, (f\varrho\vec{u}, g\varrho\vec{v}) \qquad (\vec{u}, \vec{v} \text{ all distinct})$$
$$(\mathbf{ab}_x f)\varrho \stackrel{\text{def}}{=} \lambda u\vec{v} \,.\, (u, f\varrho_{x \mapsto u}\vec{v}) \qquad (u, \vec{v} \text{ all distinct}).$$

Thus $(\mathbf{ab}_x f)\varrho$ takes one more argument than $f\varrho$; besides pushing this argument (u) through as a result, it also treats it as the value of x when evaluating f. It is instructive to check the laws AF for abstractors (2.1).

Clearly FUN is isomorphic to FUN[$\emptyset$]. More generally, whatever X is, there is an injective homomorphism from FUN to FUN[X]; it sends each $f \in D^m \rightarrow D^n$ to the function $g \in \text{Env} \rightarrow (D^m \rightarrow D^n)$ such that $g\varrho\vec{v} = f\vec{v}$ for every $\varrho \in$ Env. Thus FUN is a sub-actionstructure of FUN[X].

3.3 Evaluating expressions We have seen how functions constitute a static action structure. We now wish to consider how the expressions which denote functions constitute a *dynamic* action structure, which is given meaning by a homomorphism to FUN[X]. It must be repeated that we are doing little more than casting familiar ideas in a new form. In effect we are presenting a dynamic variant of a Lawvere algebraic theory, with equations replaced by reactions; see Barr and Wells [4]. The previous examples 3.1 and 3.2 are also Lawvere theories.

Let Exp_m be the class of expressions built from the global variables X and formal (or local) variables $y_1, \ldots, y_m$, by a fixed set of operators standing for operations over D. Thus Exp_m is the smallest set such that

- Each of $y_1, \ldots, y_m$ and each $x \in X$ is an expression in Exp_m;
- If $e_1, \ldots, e_k$ are expressions in Exp_m and $\underline{op}$ is a k-ary operator (standing for a function $op \in D^k \rightarrow D$), then $\underline{op}\langle e_1, \ldots, e_k\rangle$ is an expression in Exp_m.

An expression in Exp_m obviously denotes a function in $\text{Env} \rightarrow (D^m \rightarrow D)$, i.e. an action with arity $m \rightarrow 1$ in FUN[X]. Expressions and their evaluation can be treated as a dynamic action structure EXP[X] $= (M, X, A, \searrow)$, where

$$M = (\mathsf{N}, +, 0)$$
$$X = \text{as before}$$
$$A_{m,n} = \text{Exp}_m^n$$
$$\searrow \quad \text{defined below.}$$

Now, for $e \in \text{Exp}_m$, write $e[\vec{d}]$ to mean the result of replacing the m formal variables $\vec{y}$ simultaneously in e by the m expressions $\vec{d}$; write $\{\vec{d}/\vec{x}\}e$ to mean the result of replacing the

k global variables $\vec{x}$ simultaneously in e by the k expressions $\vec{d}$. Each member of $A_{m,n}$ is an n-tuple of expressions, which we shall write as $\langle e_1, \ldots, e_n \rangle$. Then the operations in $\mathsf{EXP}[X]$ are

$$\begin{aligned}
\mathbf{id}_m &\stackrel{\text{def}}{=} \langle y_1, \ldots, y_m \rangle \\
\langle \vec{d} \rangle \cdot \langle \vec{e} \rangle &\stackrel{\text{def}}{=} \langle \vec{e}[\vec{d}] \rangle \\
\langle \vec{d} \rangle \otimes \langle \vec{e} \rangle &\stackrel{\text{def}}{=} \langle \vec{d}, \vec{e}[y_{k+1}, \ldots, y_{k+m}] \rangle \\
& \qquad \text{where } \langle \vec{d} \rangle : k \to l,\ \langle \vec{e} \rangle : m \to n \\
\mathbf{ab}_x \langle \vec{e} \rangle &\stackrel{\text{def}}{=} \langle y_1, \{y_1/x\}(\vec{e}[y_2, \ldots, y_{m+1}]) \rangle .
\end{aligned}$$

Only the abstractors here are perhaps unfamiliar; the effect of $\mathbf{ab}_x$ is to transform x into the first formal variable, incrementing the indices of the others by 1.

We now turn to the reaction relation of EXP, which represents evaluation. First, suppose that for each value $v \in D$ there is a nullary operator $\underline{v}$; we write $\underline{v}\langle\ \rangle$ as $\underline{v}$ and call it a *constant*. For $a, a' : m \to n$, define $a \searrow^{1} a'$ to mean that a' can be gained from a by replacing a single subexpression $\underline{op}\langle \underline{v_1}, \ldots, \underline{v_k} \rangle$ in a by the constant $\underline{op(v_1, \ldots, v_k)}$. Then define reaction to be the transitive reflexive closure of $\searrow^{1}$ so that $\searrow \stackrel{\text{def}}{=} (\searrow^{1})^{*}$.

Now of course we look for a natural homomorphism $\Phi : \mathsf{EXP}[X] \to \mathsf{FUN}[X]$ sending each expression to its denotation – a function. What is this denotation? For $e \in \mathrm{Exp}_0$ (an expression with no formal variables) it is given by $[\![e]\!]$, where

$$\begin{aligned}
[\![x]\!]\varrho &\stackrel{\text{def}}{=} \varrho x \\
[\![\underline{op}\langle e_1, \ldots, e_k \rangle]\!]\varrho &\stackrel{\text{def}}{=} op([\![e_1]\!]\varrho, \ldots, [\![e_k]\!]\varrho) .
\end{aligned}$$

Then define Φ to act upon arities and names as the identity, and upon actions by

$$\Phi(\langle e_1, \ldots, e_n \rangle)\varrho \stackrel{\text{def}}{=} \lambda \vec{v} . \langle [\![e_1[\underline{\vec{v}}]]\!]\varrho, \ldots, [\![e_n[\underline{\vec{v}}]]\!]\varrho \rangle .$$

It is routine to check that Φ is a homomorphism; in particular, that it preserves reaction, i.e.

$$a \searrow_{\mathsf{EXP}[X]} a' \ \text{ implies } \ \Phi a = \Phi a'$$

(recalling that $\searrow_{\mathsf{FUN}[X]}$ is the identity).

Action structures are not limited to such familiar things as functions and expressions, as our next few examples show. Moreover, even these familiar cases yield more interest when we come to *effect structures* in the next section.

3.4 Synchronous CCS (1) Synchronous CCS [18] is a process calculus in which processes $P, Q, \ldots$ are built from actions $\alpha, \beta, \ldots$ by various constructions, among which are *prefixing* $\alpha.P$ and *product* $P \times Q$. The dynamics of processes is given by *labelled transitions* of the form $P \xrightarrow{\alpha} P'$; these are defined by transition rules such as

$$\alpha.P \xrightarrow{\alpha} P \qquad\qquad \frac{P \xrightarrow{\alpha} P' \quad Q \xrightarrow{\beta} Q'}{P \times Q \xrightarrow{\alpha \times \beta} P' \times Q'} .$$

The actions of Synchronous CCS are just elements of the free Abelian group generated by a fixed set Π of (positive) *particles* $\pi, \pi', \ldots$. Thus an instance of the product rule is

$$\frac{P \xrightarrow{\pi_1 \overline{\pi_2} \pi_3} P' \qquad Q \xrightarrow{\pi_2 \overline{\pi_3}} Q'}{P \times Q \xrightarrow{\pi_1} P' \times Q'}$$

because $\pi_1\overline{\pi_2}\pi_3 \times \pi_2\overline{\pi_3} = \pi_1$ is a product of actions.

Note that Synchronous CCS represents the interaction between two particles by using the group inverse operation. In the context of action structures we prefer to take the actions to be a free commutative monoid, treating reaction separately from the static algebra of actions. (In Part II we shall find this separation necessary in treating the π-calculus.)

We now define a simple action structure in which reaction represents for the first time the *interaction* between two actions; that is, unlike in previous examples, it is possible for $a \otimes b$ to react when neither a nor b can react independently.

Let $\overline{\Pi} \stackrel{\text{def}}{=} \{\overline{\pi} \mid \pi \in \Pi\}$, the negative particles. Then we define the action structure $\mathsf{SCCS} = (M, X, A, \searrow)$, where

$$\begin{aligned} M &= \text{a singleton} \\ X &= \emptyset \\ A &= \text{multisets over } \Pi \cup \overline{\Pi}\,; \end{aligned}$$

further, $\searrow$ is the least preorder on A such that for all $a \in A$ and $\pi \in \Pi$

$$a \cup [\,\pi, \overline{\pi}\,] \searrow a\,.$$

(Here $\cup$ means union of multisets; we write multisets in square brackets.) Then the identities, composition and tensor product of the action structure are just

$$\begin{aligned} \mathbf{id} &\stackrel{\text{def}}{=} \emptyset \\ \cdot, \otimes &\stackrel{\text{def}}{=} \cup\,, \end{aligned}$$

and there are no abstractors because the name-set X is empty. One can easily show that the action structure SCCS is freely generated by the elementary actions $[\,\pi\,]$, $[\,\overline{\pi}\,]$ ($\pi \in \Pi$) and by the single reaction rule $[\,\pi\,] \otimes [\,\overline{\pi}\,] \searrow \mathbf{id}_0$.

Of course, this action structure only explains the *actions* of synchronous CCS, not its processes or process combinators. But some of the latter can indeed be explained via action structures, as we now show.

3.5 Synchronous CCS (2) In the full calculus Synchronous CCS, processes P are subject to three operations which involve the binding of particle names. They are:

$$\begin{aligned} \text{Restriction:} &\quad P\backslash\pi \\ \text{Renaming:} &\quad P[\pi'/\pi] \\ \text{Parametrisation:} &\quad F(\pi) \stackrel{\text{def}}{=} P\,. \end{aligned}$$

Now we wish to represent all binding uniformly by the abstraction operator **ab** (which in Section 5 will be extended to operate upon processes as well as actions). Indeed, we shall find it appropriate to deal with restriction and renaming entirely at the level of actions. In the case of Synchronous CCS, this means that we now introduce a name-set X which is just the particle-set Π; thus restriction and renaming of particles will be definable in terms of the abstractors $\mathbf{ab}_x$. In part II we shall use the symbol π for a π-calculus particle, which is not just a name but something with a little more structure..

Accordingly, we now generalize SCCS to SCCS$[X] = (M, X, A, \searrow)$, where

$$M \stackrel{\text{def}}{=} (\mathsf{N}, +, 0)$$
$$X = \text{a denumerable set (the particle-names)},$$

and the actions A and reaction relation $\searrow$ are defined as follows. Each action $a : m \to n$ takes the form

$$(\vec{x})(\nu\vec{y})\, S\, \langle\vec{z}\rangle \qquad (|\vec{x}| = m,\ |\vec{z}| = n),$$

where $\vec{x}$, $\vec{y}$ and $\vec{z}$ are names in X, and S –the *body*– is a multiset over $X \cup \overline{X}$. We say that the names $\vec{x}$, $\vec{y}$ and $\vec{z}$ are respectively *imported*, *restricted* and *exported*, the first two being binding operations over the whole action. We shall assume always that the names $\vec{x}, \vec{y}$ are all distinct, and we do not distinguish a from any variant gained by change of bound names or by permutation of restricted names. Finally, reaction is defined by $\searrow \stackrel{\text{def}}{=} (\searrow^1)^*$, where $a \searrow^1 a'$ means that a' can be derived from a by deleting a single pair of complementary particles from its body.

Let us look at several simple cases of actions. Note that any of the four parts of an action may be empty, and we may then omit that part. Thus:

- The elementary interactions $[\,x\,], [\,\overline{x}\,]$ would be written in full as $(\,)(\,)[\,x\,]\langle\,\rangle$ and $(\,)(\,)[\,\overline{x}\,]\langle\,\rangle$, and have arity $0 \to 0$. The reaction relation is defined by the single rule $[\,x\,] \otimes [\,\overline{x}\,] \searrow \mathbf{id}_0$.
- Restriction is the action $\boldsymbol{\nu} : 0 \to 1 = (\nu y)\langle y\rangle$, whose full form is $(\,)(\nu y)\,\emptyset\,\langle y\rangle$.
- The name x as a datum is the action $\langle x\rangle : 0 \to 1$, whose full form is $(\,)(\,)\,\emptyset\,\langle x\rangle$.

Note that the datum $\langle x\rangle$ is distinct from the elementary action $[\,x\,]$.

The action structure operations are as follows. Let

$$a \equiv (\vec{u})(\nu\vec{v})\, R\, \langle\vec{w}\rangle \qquad (\vec{u}, \vec{v} \text{ not occurring in } b)$$
$$b \equiv (\vec{x})(\nu\vec{y})\, S\, \langle\vec{z}\rangle \qquad (\vec{x}, \vec{y} \text{ not occurring in } a).$$

Then:

$$\mathbf{id}_m \stackrel{\text{def}}{=} (\vec{x})\langle\vec{x}\rangle \qquad (|\vec{x}| = m)$$
$$a \cdot b \stackrel{\text{def}}{=} (\vec{u})(\nu\vec{v}\vec{y})\, R \cup \{\vec{w}/\vec{x}\}S\, \langle\{\vec{w}/\vec{x}\}\vec{z}\rangle$$
$$a \otimes b \stackrel{\text{def}}{=} (\vec{u}\vec{x})(\nu\vec{v}\vec{y})\, R \cup S\, \langle\vec{w}\vec{z}\rangle$$
$$\mathbf{ab}_x a \stackrel{\text{def}}{=} (x\vec{u})(\nu\vec{v})\, R\, \langle x\vec{w}\rangle \qquad (x \text{ not in } \vec{u} \text{ or } \vec{v}).$$

Thus we see that abstraction is a combination of import and export.

Recall that SCCS is freely generated by the actions $[\,\pi\,], [\,\overline{\pi}\,]$ $(\pi \in \Pi)$. Analogously, every action in SCCS$[X]$ can be expressed (using the above operations) in terms of:

$$[\,x\,], [\,\overline{x}\,] : 0 \to 0 \quad \text{– elementary interactions } (x \in X)$$
$$\langle x\rangle : 0 \to 1 \quad \text{– datum } (x \in X)$$
$$\boldsymbol{\nu} : 0 \to 1 \quad \text{– new name}$$
$$\boldsymbol{\omega} : 1 \to 0 \quad \text{– old name.}$$

These actions have all been already defined except ω, whose full form is $(x)()\emptyset\langle\rangle$, which discards a name. (Think of ν and ω as "new" and "old", or "create" and "destroy", or "gain" and "lose".) But $\mathsf{SCCS}[X]$ is not *freely* generated by these actions; certain identities hold among them, e.g. $\langle x\rangle \cdot \omega = \mathbf{id}_0$. [3]

We do not claim that this way of presenting the actions of Synchronous CCS is directly useful in applied work. Its purpose is to show how Synchronous CCS fits into the action structure framework, and it prepares for the treatment of the π-calculus in Part II.

3.6 CCS There are two differences between pure CCS [19] and Synchronous CCS which affect the way we should treat them using action structures. The first is that in the product $P \times Q$ of Synchronous CCS the processes P and Q must act in synchronization, whereas either of them may remain idle in the parallel composition $P \mid Q$ of CCS . This is reflected in the first two transition rules for CCS parallel composition:

$$\frac{P \xrightarrow{\alpha} P'}{P \mid Q \xrightarrow{\alpha} P' \mid Q} \qquad \frac{Q \xrightarrow{\alpha} Q'}{P \mid Q \xrightarrow{\alpha} P \mid Q'},$$

while interaction is permitted by the third rule:

$$\frac{P \xrightarrow{\alpha} P' \quad Q \xrightarrow{\overline{\alpha}} Q'}{P \mid Q \xrightarrow{\tau} P' \mid Q'}.$$

This third rule also reflects the other way in which CCS differs from the synchronous version; that is, the atomic action τ of CCS represents the occurrence of a single interaction between processes, while in Synchronous CCS the unit action 1 represents the simultaneous occurrence of any number of such interactions. In a sense the latter is closer to reality, for one needs a central scheduler to implement the interleaving of single interactions in CCS. But CCS was designed as an applicable theory, rather than as a programming language; the expansion theorem of CCS is easy to use just because interactions are treated singly. Moreover in [18] it was demonstrated that at the semantic level of observation equivalence it makes no difference whether single interactions of CCS are treated as interleaved or as simultaneous.

If single interactions were allowed to be simultaneous, then the action structure for CCS could be just that for Synchronous CCS; the only difference would be that the concurrent composition of CCS is treated as a derived operation in terms of product and delay (see Section 5.1 below). But to treat single interactions as interleaved only requires two slight changes in the action structure $\mathsf{SCCS}[X]$. So we define $\mathsf{CCS}[X]$ as a slight variant of $\mathsf{SCCS}[X]$, as follows. First we adjoin the special particle τ to the particles $X \cup \overline{X}$, and in a typical action $a = (\vec{x})(\nu\vec{y})\,S\,\langle\vec{z}\rangle$ we take S to be a multiset of this larger class of particles. Second, we define $a \searrow^1 a'$ iff a' can be derived from a by replacing a single pair of complementary particles in S by τ (instead of deleting them). In Section 7, we shall see how this adaptation allows us to restrict attention to actions containing at most two particles.

3.7 Ports The constructions $\mathsf{SCCS}[Z]$ and $\mathsf{CCS}[Z]$ involve the complementary actions $[\,z\,]$ and $[\,\overline{z}\,]$ $(z \in Z)$, with the reaction rule

$$[\,z\,] \otimes [\,\overline{z}\,] \searrow \mathbf{id}_0 \;(\text{or } [\,\tau\,])\,;$$

[3] In [21] a complete set of identities is given, so that $\mathsf{SCCS}[X]$ is indeed the action structure freely generated from the above actions divided by the equivalence induced by the identities. These identities characterize an important subclass of action structures called *action calculi*.

since the actions have arity $0 \to 0$ this can also be written

$$[\overline{z}] \cdot [z] \searrow \mathbf{id}_0 \text{ (or } [\tau]) .$$

But this only models the pure form of CCS, where $[z]$ or $[\overline{z}]$ is simply an indivisible action. By contrast, in CCS with value-passing we think of z as a port through which data can flow. How does this fit into the framework of action structures? Well, Example 3.2 above –the action structure $\mathsf{FUN}[X]$ for functions– is about operations upon data; we can think of a datum of arity m (an m-tuple of values) being fed into a function of arity $m \to n$. Now, how can we allow ourselves to import and export m-ary data through a port z say, as well as passing them into and out of functions? It is natural to extend $\mathsf{FUN}[X]$ by introducing a set Z (disjoint from X) of port-names, and for each port name $z \in Z$ and each appropriate arity m to introduce actions

$$\begin{aligned} \mathbf{out}_z &: m \to 0 \\ \mathbf{in}_z &: 0 \to m . \end{aligned}$$

Now the actions $[\overline{z}]$ and $[z]$ of $\mathsf{SCCS}[Z]$ and $\mathsf{CCS}[Z]$ can be regarded as special cases of the above for which $m = 0$, since no data flows in those calculi. Therefore, generalising their reaction rule also, we introduce the reaction rule

$$\mathbf{out}_z \cdot \mathbf{in}_z \searrow \mathbf{id}_m .$$

This can also be written

$$\mathbf{in}_z \otimes \mathbf{out}_z \searrow \mathbf{id}_m ,$$

for it is a simple exercise to show that if $a : m \to 0$ and $b : 0 \to n$ in a strict monoidal category, then $a \cdot b = a \otimes b = b \otimes a$. This rule exactly expresses synchronized communication at a port.

Thus we are indicating how to superpose the action structure $\mathsf{SCCS}[Z]$ or $\mathsf{CCS}[Z]$ upon an arbitrary action structure, extending the latter by adding ports through which its "data" may pass. [4] Let us describe this superposition a little more formally. Let $\mathcal{B} = (M, X, B, \searrow)$ be any action structure, and let Z (disjoint from X) be the ports we wish to add. To construct $\mathcal{B}^{(Z)} = (M', X', B', \searrow')$, the action structure $\mathcal{B}$ freely extended by addition of the ports Z, we perform the following steps.

First define

$$\begin{aligned} M' &= M \times \mathsf{N} && \text{– Cartesian product of monoids} \\ X' &= X \cup Z && \text{– a disjoint union.} \end{aligned}$$

Let $\Phi : X \to X'$ and $\Phi : M \to M'$ be the injections given by $\Phi x = x$ and $\Phi m = (m, 0)$. The arities in X' are $x : (m, 0)$ for each $x : m \in X$, and $z : (\epsilon, 1)$ for each $z \in Z$.

Next, the actions B' are as follows:

1. There is an injection $\Phi : B \to B'$; if $a : m \to n$ in B then $\Phi a : \Phi m \to \Phi n$ in B'.

2. There are actions corresponding to those of $\mathsf{SCCS}[Z]$:

$$\begin{aligned} \mathbf{in}_z &: (\epsilon,0) \to (m,0) && \text{– import data through port } z \in Z \\ \mathbf{out}_z &: (m,0) \to (\epsilon,0) && \text{– export data through port } z \in Z \\ \langle z \rangle &: (\epsilon,0) \to (\epsilon,1) && \text{– the port } z \in Z \text{ as a datum} \\ \nu &: (\epsilon,0) \to (\epsilon,1) && \text{– created port} \\ \omega &: (\epsilon,1) \to (\epsilon,0) && \text{– discarded port.} \end{aligned}$$

[4] The quotation marks are to remind us that some action structures may be abstract and not concerned with data; nonetheless our construction makes sense formally even in these cases.

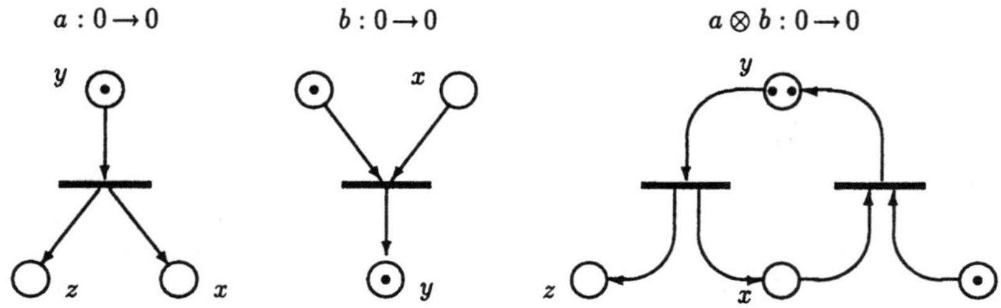

Figure 1: Two PT-nets and their product

Finally, the reaction relation is generated by the rule

$$\Phi a \searrow' \Phi b \text{ iff } a \searrow b$$

ensuring that B is a sub-actionstructure of $B^{(Z)}$, together with the rule

$$\mathbf{out}_z \cdot \mathbf{in}_z \searrow' \mathbf{id}_{(m,0)} \qquad (z \in Z, m \in M).$$

This construction remains to be detailed more precisely. Note in particular that we have *not* just taken the Cartesian product of B with $\mathsf{SCCS}[Z]$; for the port actions $\mathbf{in}_z$ and $\mathbf{out}_z$ are indexed by $z \in Z$, but their arities are drawn from B.

Note also that although we allow a port z to appear as a datum $\langle z \rangle$, it may not be exported through another port $w \in Z$; for the arities $\langle z \rangle : (\epsilon, 0) \to (\epsilon, 1)$ and $\mathbf{out}_w : (m, 0) \to (\epsilon, 0)$ forbid the composition $\langle z \rangle \cdot \mathbf{out}_w$. This is exactly the limitation of CCS with value-passing – namely, that ports cannot be passed as messages. (Indeed we expect to be able to present an action structure for CCS with value-passing just by extending $\mathsf{FUN}[X]$, for suitable data D, by adding ports in this way.) The π-calculus, for which we define action structures in Part II, corresponds exactly to the removal of this limitation.

3.8 Petri nets The monoidal structure of Petri nets was first demonstrated by Meseguer and Montanari [17]. They model each place-transition net (PT-net) as a monoidal category, whose objects are markings and whose morphisms correspond to firing sequences. It is natural to ask whether the extra ingredients in an action structure can play a rôle in the formulation of Meseguer and Montanari. But they already model the dynamics of a net N by the morphisms of the monoidal category corresponding to N, so there is no rôle for the dynamics of action structures. Nor is there an obvious rôle for abstractors at the level of a single net N.

But there is a natural action structure at a higher level, where each marked PT-net N corresponds to an action. Here the abstractors find a good rôle to play, while an atomic reaction $\searrow^1$ corresponds to the firing of a single transition; if $a \searrow^1 a'$ then a' is the same net as a with a different marking.

This action structure is treated in more detail in [21]; here we sketch it, for comparison with the previous examples. First we take $(\mathsf{N}, +, 0)$ to be the arity monoid, and take a name-set X infinite at each arity. Then an action of arity $0 \to 0$ is just a marked PT-net some of whose places bear a name from X, no two places bearing the same name. Call this a *place-named* net. Figure 1 shows two place-named nets and their product; the product is formed just by coalescing like-named places.

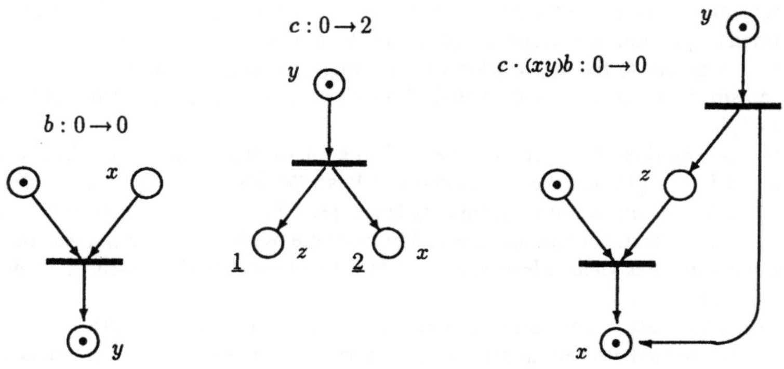

Figure 2: The composition of PT-net actions

We now turn to actions of arbitrary arity. An action $d : m \rightarrow n$ consists of a place-named net N together with a vector $\vec{x}$ of m distinct names representing "bound" or parametric places in the net, and a map (not necessarily injective) from $\{1, \ldots, n\}$ into the places of N indicating the "target" places. Typically we would write $d = (\vec{x})c$ with $c : 0 \rightarrow n$. (We allow alpha-conversion of bound names in the usual way.) Figure 2 shows $c : 0 \rightarrow 2$ formed from a by targetting two places.

To form the product of two actions $a : 0 \rightarrow m$ and $b : 0 \rightarrow n$ with zero source arity, simply increment the target labels in b by m and proceed to coalesce like-named places as in Figure 1. To form the product of two arbitrary actions $(\vec{x})a$ and $(\vec{y})b$ first ensure by alpha-conversion that the bound names in one action do not occur free or bound in the other, then form $a \otimes b$ as above; the result is $(\vec{x}\vec{y})(a \otimes b)$.

To compose two actions, e.g. to form $c \cdot (xy)b$ where b and c are as in Figure 2, first alpha-convert to disjoin bound names as for product; then coalesce each target place in the first action with the corresponding bound place in the second action, removing the target labels of the first action and the bound names of the second.

To form the abstraction $\mathbf{ab}_x d : 1{+}m \rightarrow 1{+}n$ for arbitrary $d : m \rightarrow n$, first increment the target labels in d by one; then target the place named x as 1 (adding an unconnected place x if there is none) and bind x.

Finally, the reaction relation is generated by $\searrow^{1}$ which represents the firing of a single transition. With a as in Figure 1, we have $a \searrow^{1} a'$ where both places x, z are marked in a', and a' is inactive. Moreover it is easily seen that $a \otimes b \searrow^{1} a' \otimes b$, but the latter is *not* inactive.

So our action structure contains not only place-named nets, but also nets whose places are targeted and/or bound. Because of this, many operations on nets can be represented algebraically. A simple example is hiding a place, i.e. removing its name and thus fixing its pre- and post-transitions (since it may no longer be coalesced with a like-named place). This is just our familiar restriction operator $\boldsymbol{\nu}$. In fact $\boldsymbol{\nu} : 0 \rightarrow 1$ is the action

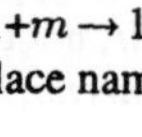

consisting of a single place – unnamed, unmarked and targeted; to hide the place z in any $c : 0 \rightarrow n$ we just form $\boldsymbol{\nu} \cdot (z)c : 0 \rightarrow n$. Other simple operations are: increasing the marking at a named place; renaming a place; coalescing two differently-named places. Also all place-named nets can be formed by product alone, from simple nets with at most one transition.

A dual action structure to the above allows the naming of transitions, instead of places. These two action structures can also be conflated, allowing naming (hence coalescing) both of transitions and of places.

Much work has been done on algebras of nets. Two good examples are the algebra AFP of Cherkasova and Kotov [9], and the box calculus of Best, Devillers and Hall [7] which is oriented towards defining concurrent programming languages. The action structure approach must be placed in that context, but this task is not addressed here; we have only been concerned to show that action structures yield at least one way of treating Petri nets algebraically. Its use of parametrisation appears to be new.

The relationship must also be examined between the present notion of action structure homomorphism, specialised to Petri nets, and the categorical treatment of nets and their morphisms by Winskel [28] and others.

4 Effects and effect structures

A central aim of action structures is to make explicit the way in which an interaction exerts an effect upon its participants. Recall from Section 1 our example in the π-calculus; the transition

$$((\boldsymbol{\nu} x)\overline{p}x.P) \mid p(y).Q \longrightarrow (\boldsymbol{\nu} x)(P \mid \{^x/_y\}Q)$$

is represented roughly speaking by the reaction

$$((\boldsymbol{\nu} x)\cdot \overline{p}x) \otimes p(y) \searrow (\boldsymbol{\nu} x)\cdot (\mathbf{id} \otimes \{^x/_y\})$$

in an appropriate action structure.

We shall refer to degenerate actions, such as the result of this reaction, as *effects*; our aim is to see how effects are related to other actions. In passing, note that one might expect the result of a reaction of a product to be itself a product, whose factors are the separate effects exerted by the reaction on each of the participants P and Q. But that is not the case here. A central idea of the π-calculus is that the effect of a communication may be a new liaison between a pair of agents, i.e. a restriction; this is an effect upon the *pair*, not upon its individual members.

What distinguishes effects from other actions? *Inertia*, the inability to contribute to further reaction, is one distinguishing property.

4.1 Definition (inactivity, inertia) An action a is *inactive* if $a \searrow a' \Rightarrow a' = a$. An action a is *inert* if, whenever $b \cdot a \searrow c$, then there exists b' such that $b \searrow b'$ and $c = b' \cdot a$. ■

Thus if a is inert then any reaction of $b \cdot a$ must be due to b alone.

4.2 Proposition If an action is inert then it is inactive.

Proof The identity $\mathbf{id}$ is inactive by assumption; so take $b = \mathbf{id}$ in the definition. ■

4.3 Proposition Each unit action $\mathbf{id}$ is inert. If a and b are inert then so is $a \cdot b$. ■

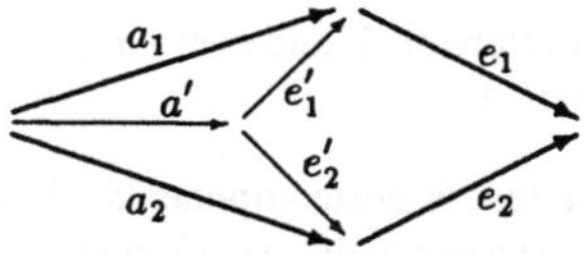

Figure 3: Decomposition for a postcomponent

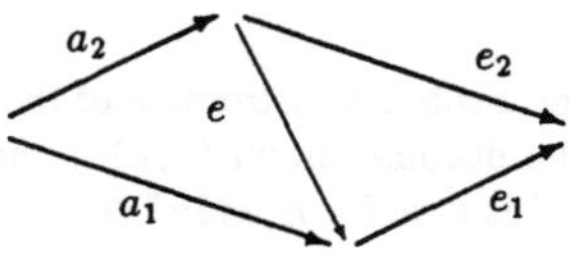

Figure 4: Preordered decompositions: $(a_1, e_1) \leq (a_2, e_2)$

Inertia is *not* symmetric with respect to composition. If a is inert then $a \cdot b$ may still have reactions not arising from b alone. For example a may be roughly speaking a substitution, which will indeed be an inert action, and the *effect* of this upon b may induce a reaction.

As well as demanding that each effect is an inert action, we shall now demand certain properties of the *set* of effects in an action structure. This will not fully determine the effects; indeed it will be useful to consider different subsets of actions as effects.

4.4 Definition (postcomponent) Let A be an action structure, and E a sub-actionstructure of A. Then E is a *postcomponent* of A if whenever $a = a_1 \cdot e_1 = a_2 \cdot e_2$, where $a_i \in A$ and $e_i \in E$ $(i = 1, 2)$, then there exist $a' \in A$ and $e'_1, e'_2 \in E$ such that (see Figure 3):

$$a_i = a' \cdot e'_i \ (i = 1, 2) \ \text{ and } \ e'_1 \cdot e_1 = e'_2 \cdot e_2.$$

Further, if all members of E are inert then E is an *effect structure* for A. ■

We continue to use a, b, c, d to range over all actions; when the effect structure E is understood we use e, f, g to range over effects.

Note that "postcomponent" is a purely static notion. The condition that E is a postcomponent is about decompositions $a = a' \cdot e$ of an action, where $e \in E$. A few more definitions will clarify the rôle played by decompositions.

4.5 Definition (decomposition) Let E be a sub-actionstructure of the action structure A. Then the pair (a', e) is a *decomposition of a for E* if $a = a' \cdot e$ and $e \in E$. We define the following preorder over decompositions for E:

$$(a_1, e_1) \leq (a_2, e_2) \text{ if } a_1 = a_2 \cdot e \text{ and } e \cdot e_1 = e_2 \text{ for some } e \in E.$$

(See Figure 4.) We say that a decomposition (a^*, e^*) of a for E is *maximal* if $(a', e) \leq (a^*, e^*)$ for any other decomposition (a', e) of a for E. ■

Thus the postcomponent condition requires that the decompositions of any action form a directed set under $\leq$. In establishing that a given sub-actionstructure of A is an effect structure, this condition is usually the hardest to check. It is helpful to look for *maximal* decompositions, since they often exist for interesting postcomponents. Their importance is as follows:

4.6 Proposition If E is sub-actionstructure of A, and every $a \in A$ has a maximal decomposition for E, then E is a postcomponent of A. ■

In fact something even stronger often holds for postcomponents which we wish to use as effect structures. In general, for given a, the decomposition (a, e) may be maximal for some effects e but not for others. But for many effect structures there is a subset of A which behaves better than this:

4.7 Definition (pure action) Let E be a sub-actionstructure of the action structure A. Then a is *pure for* E if for every $e \in E$ the decomposition (a, e) is maximal. We say that the decomposition (a^*, e^*) is *pure for* E if a^* is pure for E and $e^* \in E$. ■

In verifying that some given sub-actionstructure E of inert actions is an effect structure it is a good strategy first to identify the pure elements, and then to show that every action a has a pure decomposition, which is of course maximal. The strategy may not always work –there may not be enough pure elements– but it works in two of the cases we examine in Section 4.8 below. We shall also find in Section 6 that the concept of bisimilarity between processes is simpler when the the underlying action structure always has pure decompositions.

Let us denote by P the set of elements pure for E in A. Having identified P as an interesting set of actions, should we expect it to be a sub-actionstructure of A? It turns out to be so in the two examples just mentioned, where pure decompositions always exist. Under one or two extra conditions, which also hold in those examples, the pair (P, E) of sub-actionstructures becomes a *factorisation* of A, in the sense of Freyd and Kelly [10]. I am grateful to John Power for prompting me to look in this direction. There is a duality between the two members of a factorisation; we therefore have –at least in these examples– a duality between pure actions and effects, which may shed extra light on these concepts.

4.8 Examples What possible effect structures exist for action structures in general, and in particular for the examples in Section 3? Certainly the units $E = \{\mathbf{id}_m \mid m \in M\}$ always form an effect structure. It is not a very interesting one, since every $a : m \to n$ possesses only one effect in E, namely $\mathbf{id}_n$. Indeed, there is advantage in finding a larger effect structure; for we shall see later that this makes bisimilarity a more generous congruence relation over processes.

We turn now to the example action structures of Section 3. Both FUN and $\mathsf{FUN}[X]$ are static, i.e. reaction is the identity, so all actions are inert; therefore these action structures are effect structures for themselves. We would need to know more specific properties of the functions to determine any other effect structure here, apart from the units.

The evaluation of expressions, $\mathsf{EXP}[X]$ in Section 3.3, is immediately more interesting. First, there are many inactive actions which are not inert; $\langle y_1, y_2 \underline{+} \underline{4} \rangle$ is inactive, but if we precompose $\langle \underline{2}, \underline{3} \rangle$ to it we get $\langle \underline{2}, \underline{3} \underline{+} \underline{4} \rangle$ – which is active. However, any action $\langle e_1, \ldots, e_n \rangle$ containing only local variables and constants (i.e. no global variables or non-nullary operators) turns out to be inert, provided it contains each local variable at most once. An example is $\langle y_2, y_1, \underline{4} \rangle$. To see why each y_i should appear at most once, consider $a = \langle y_1, y_1 \rangle$; if $b = \langle \underline{2} \underline{+} \underline{3} \rangle$, we have

$$b \cdot a = \langle \underline{2} \underline{+} \underline{3}, \underline{2} \underline{+} \underline{3} \rangle \searrow \langle \underline{5}, \underline{2} \underline{+} \underline{3} \rangle$$

and we cannot find b' for which $b \searrow b'$ and $\langle \underline{5}, \underline{2} \underline{+} \underline{3} \rangle = b' \cdot a$. [5]

[5] Of course if we take $b' = \langle \underline{5} \rangle$ then we *do* have the weaker property $\langle \underline{5}, \underline{2} \underline{+} \underline{3} \rangle \searrow b' \cdot a$. This suggests that a

So do the actions containing only local variables and constants, and each y_i at most once, form an effect structure? They do indeed; but it turns out to be one in which maximal decompositions do not always exist. The reader may enjoy checking these facts, and also verifying that effects cannot contain global variables.

Now consider the subset of these actions containing each y_i *exactly* once. It is easy to see that they form a sub-actionstructure. And in contrast to the previous case we find that pure decompositions always exist, so we have an effect structure by Proposition 4.6. The pure actions are of the form $\langle d_1, \ldots, d_m \rangle$ where no d_j is a constant $\underline{v}$. For a pure decomposition (a^*, e^*) of any $a = \langle e_1, \ldots, e_n \rangle$, take a^* to be some permutation $\langle e_{i_1}, \ldots, e_{i_m} \rangle$ of the expressions e_i excluding any which is a constant.

For SCCS(Section 3.4) it's easy to see that the only effect structure is $\{\mathbf{id}_0\}$; for a non-empty multiset cannot be inert. But SCCS$[X]$ and CCS$[X]$ (Sections 3.5 and 3.6) are more interesting. Let E contain all actions $(\vec{x})\,(\nu\vec{y})\,S\,\langle\vec{z}\rangle$ in which the multiset S is empty; then all members of E are certainly inert. Also E is indeed an effect structure. Again, this can be shown by identifying the pure actions; they turn out to be all those $(\vec{x}))\,(\nu\vec{y})\,S\,\langle\vec{z}\rangle$ in which every bound name (one of $\vec{x}$ or $\vec{y}$) occurs in $\vec{z}$, and every restricted name (one of $\vec{y}$) also occurs in S.

These last two examples of effect structures are strikingly different; so it appears that the notions of effect and effect structure are quite flexible.

4.9 Inertia reconsidered According to Definition 4.1, for a to be inert means that a cannot create reactions if you *postcompose* it, i.e. if you form $c \cdot a$ for some c. But our interest in effects is based on the idea that an effect e can create reactions if you *precompose* it, i.e. if you form $e \cdot c$ for suitable c. Now it turns out in all our examples that an effect e can create reactions *only* if you precompose it, e.g. not if you juxtapose it as in $e \otimes c$. So why do we not take this stronger condition to be the definition of inertia? Does it make much difference if we do?

The answer is: There may be fewer inert actions with this stronger condition, but it does not affect the concept of effect structure at all. Let us make this claim precise.

4.10 Definition (preaction) A *preaction* φ is a unary operation on actions built from operations of the form $c \cdot (-)$, $c \otimes (-)$, $(-) \otimes c$ and $\mathbf{ab}_x(-)$ where c is any action. ■

The term "preaction" is chosen to indicate that *precomposition* $c \cdot (-)$, but not *postcomposition* $(-) \cdot c$, is allowed in forming preactions. Structural congruence and the reaction relation extend in an obvious way to preactions. We now define strong inertia using preactions, where in definingo inertia we used only precomposition. So strong inertia clearly implies inertia.

4.11 Definition (strong inertia) An action a is *strongly inert* if whenever $\varphi(a) \searrow b$, then $b = \varphi'(a)$ for some preaction φ' such that $\varphi \searrow \varphi'$. ■

Now we show that it would make no difference to the notion of effect structure, Definition 4.4, if we demanded that its members be *strongly* inert.

---

form of inertia weaker than Definition 4.1 might be useful, with the condition $c = b' \cdot a$ weakened to $c \searrow b' \cdot a$. It remains to be seen how much of the ensuing theory goes through with this weaker notion. But it is worth noting that for the other examples discussed in Section 3 the weakening does not seem to increase the set of inert actions.

4.12 Lemma Every preaction φ can be expressed as $\varphi = a \cdot \psi$, where ψ is a preaction formed by the operations $\mathbf{id} \otimes (-)$, $(-) \otimes \mathbf{id}$ and $\mathbf{ab}_x(-)$.

Proof By induction on the structure of preactions. To take just the most interesting case, consider the preaction $c \otimes \varphi$. By assumption $\varphi = a \cdot \psi$ where ψ is of the required form, so

$$\begin{aligned} c \otimes \varphi &= c \otimes (a \cdot \psi) \\ &= (c \otimes a) \cdot (\mathbf{id} \otimes \psi) \end{aligned}$$

which is of the required form. ■

4.13 Proposition Let E be an effect structure. Then all its members are strongly inert.

Proof Let $e \in E$ and let φ be a preaction such that $\varphi(e) \searrow b$. Now let $\varphi = a \cdot \psi$ as in the lemma; so $a \cdot \psi(e) \searrow b$. But $\psi(e) \in E$ is inert, so for some a' we have $a \searrow a'$ and $b = a' \cdot \psi(e)$. But $a' \cdot \psi$ is a preaction and $\varphi \searrow a' \cdot \psi$. So we have shown e to be strongly inert. ■

Thus we have lost nothing in demanding only the weaker condition of inertia in defining an effect structure; moreover this weaker condition is simpler to verify in examples.

Preactions will turn up again in Section 7, where we define the notion of *incident*.

5 Process structure: statics

We can use action structures in more than one way to model concurrent processes. In the first way, each *single transition* of a process would be modelled by an action. At the higher level of the processes, these single transitions may be considered atomic, even though the detailed performance of an action –i.e. the reaction within it– may be highly complex.

A second approach is to model a *whole process* by an action, as well as its individual transitions. This approach depends upon finding ways of building a higher-level action structure (for the processes) from a lower-level one (for the transitions). *Action calculi* [21], a subclass of action structures, make considerable progress in this direction.

Here we take the first approach. We show how certain control operations can be added to an arbitrary action structure, to yield a process calculus. This yields a generic process calculus, and we can carry its theory forward a fair way independently of the underlying action structure.

Thus we bring some unity to the study of process calculi. Specifically, we can unite the study of CCS, Synchronous CCS and the π-calculus. Other calculi such as ACP [3] and CSP [14] may also fit into the framework; this remains to be seen.

The two approaches do not conflict. In fact, the uniform treatment in this paper of bisimilarity congruence in process calculi is a preparation for studying bisimilarity in action calculi. For in contrast with actions which are arrows $a : m \to n$ with both a source and a target arity, the processes $P : m$ which we are about to introduce have only a source arity; in fact they correspond to actions $a : m \to \epsilon$ with trivial target arity. One reason for adopting this constraint is that we remain closer to existing process calculi; another is that the technical problems of bisimilarity are easier to handle.

5.1 Process syntax We study processes over an arbitrary action structure, whose actions will be denoted by $a, b, \ldots$ as before. Processes, denoted by $P, Q, \ldots$, are given by the syntax

$$\begin{aligned} P \; ::= \; & a \cdot P \;\mid\; P \otimes P \;\mid\; (x)P \\ & \mid\; \bullet P \;\mid\; \mathbf{0} \;\mid\; P + P \;\mid\; \partial P \;\mid\; !P \;. \end{aligned}$$

The first three constructions lift the operations of an action structure to process level, but allowing composition only between an action and a process rather than the composition $P \cdot Q$ of two processes. [6] Note that we use $(x)P$ rather than $\mathbf{ab}_x P$ for abstraction over processes.

The fourth construction $\bullet P$ represents what we shall call *committal*. Once we have defined reaction for processes, then we shall find that a process P may be capable of many different reactions of the form

$$P \searrow a \cdot \bullet P'$$

which we shall call *commitments*. To establish a link with CCS syntax we shall elide the "$\cdot$" before "$\bullet$", writing

$$P \searrow a \bullet P'.$$

Each commitment represents P's ability to perform a single action a, considered atomic at process level, with continuation P'. Moreover, if (a', e) is a decomposition of a then our theory will reflect the intuition that the effect e exerts an influence upon the continuation P'. The need to write down this influence explicitly, as $e \cdot P'$, is one reason for admitting the composition construction at process level.

The last five constructions from committal ($\bullet$) onwards represent control, and we call them *control terms*; equally we call $\bullet, \mathbf{0}, +, \partial$ and ! *control operators*, and refer to them respectively as committal, zero, sum, delay and replication. The last four are drawn from Synchronous CCS and the π-calculus; but they do not depend on the nature of the *actions* in those calculi, so it makes sense to use them more generally.

The set of five control operators is not claimed to be complete or fundamental; indeed the notion of action structure has greater claim to this virtue. But they work well in their parent calculi, so they serve as exemplars for adding control structure uniformly to action structure. Recall from [18] that, in the framework of Synchronous CCS, the parallel composition operator "|" of CCS is recovered by the definition

$$P \mid Q \stackrel{\text{def}}{=} P \times \partial Q + \partial P \times Q\,.$$

(The $\times$ of Synchronous CCS will be represented here by $\otimes$). The dynamics of operators is presented in Section 6; the rest of this section is devoted to statics.

5.2 Arity Processes do not constitute an action structure; in this approach they are not actions. Instead, each process has an arity $m \in M$ given by the rules

$$\frac{a : m \to n \qquad P : n}{a \cdot P : m} \qquad \frac{P : m \qquad Q : n}{P \otimes Q : m \otimes n} \qquad \frac{x : k \qquad P : m}{(x)P : k \otimes m}$$

$$\frac{P : m}{\bullet P : m} \qquad \mathbf{0} : m \qquad \frac{P : m \qquad Q : m}{P + Q : m} \qquad \frac{P : m}{\partial P : m} \qquad \frac{P : \epsilon}{!P : \epsilon}$$

These are mostly straightforward. A process P with non-ϵ arity should be thought of as parametric; its parameter would typically be provided by an effect precomposed to P. Note that $!P$ is only allowed when $P : \epsilon$. Why? Well, we expect $!P$ to be behaviourally equivalent to $(P \otimes !P) + \bullet !P$, so the arities of these two must be the same. Bearing in mind the rules for sum and committal, this implies that the two arities $P : k$ and $!P : m$ satisfy the equations $m = k \otimes m = k$, which implies $m = k = \epsilon$ (at least if the arity monoid is free).

[6] As remarked earlier, we do not here consider processes as arrows with both source and target arity; thus we avoid composition of processes. This makes the present theory simpler; in particular, it admits the simple head normal form of Definition 5.6.

5.3 Structural congruence We impose a structural congruence, $\equiv$, upon the syntax of processes. It is defined by the following laws, i.e. it is the smallest congruence satisfying them:

$$\begin{aligned} P &\equiv \mathbf{id} \cdot P \\ a \cdot (b \cdot P) &\equiv (a \cdot b) \cdot P \\ (a \cdot P) \otimes (b \cdot Q) &\equiv (a \otimes b) \cdot (P \otimes Q) \\ (x)(a \cdot P) &\equiv (\mathbf{ab}_x a) \cdot (x)P\ . \end{aligned}$$

As far as possible these are the laws of action structures lifted to process level. In particular they allow us to confuse the process-level and action-level operations with impunity.

There are no equational laws for the control operators. We could have imposed some, e.g. $P + Q \equiv Q + P, \ldots$, but we prefer to attain these equations for a *behavioural* (rather than structural) congruence. Thus the meaning of the control operators is entirely provided by the process dynamics in Section 6.

We shall need the following notion: An occurrence in P of an action a or process term Q is called a *ready* occurrence if it is not within any control term. For example a occurs readily in $P \otimes a \bullet Q$ but not in $P + a \bullet Q$.

Henceforth we shall usually omit the composition symbol, writing $a \cdot b$ as ab and $a \cdot P$ as aP. We give (elided) composition highest binding power, and all other binary operators lowest binding power; so $ab \otimes cd$ means $(ab) \otimes (cd)$, $\mathbf{ab}_x cd$ means $\mathbf{ab}_x(cd)$, $(x)P \otimes Q$ means $((x)P) \otimes Q$ and $\partial P + Q$ means $(\partial P) + Q$.

5.4 Passive processes The dynamics of processes will be a cooperation between reactions due to the underlying action structure, and control reductions. The former can only be contributed by ready occurrences of actions a. Intuitively we can say that a process is passive (cannot react) if no $a \neq \mathbf{id}$ has a ready occurrence. Thus if $a \neq \mathbf{id}$ we would say that $a \bullet P \otimes \bullet Q$ is active; on the other hand $a \bullet P + \bullet Q$ is passive, because a control reduction is needed to allow a to react. Formally:

5.5 Definition (passive process) The *passive* processes are the smallest set closed under structural congruence such that

- All control terms are passive;
- If P, Q are passive, then so are $P \otimes Q$ and $(x)P$. ∎

(The important feature here is the omission of composition, aP, from the second clause.) It will follow from Lemma 6.6 that no reaction can occur in a passive process, until a control reduction has occurred.

We shall use $H, J, K, \ldots$ to range over passive processes. The next few results show that any P can be uniquely expressed in the form aH, up to structural congruence.

5.6 Definition (head normal form) If $P \equiv aH$, where H is passive, then the latter is called a *head normal form* (*hnf*) of P. ∎

5.7 Proposition Every process possesses a head normal form.

Proof By induction on the structure of process terms. ■

This justifies our structural congruence laws (Section 5.3), since they seem to be the minimum which will yield such a result. The next proposition strengthens this claim:

5.8 Proposition Head normal forms are unique up to structural congruence; that is, if $aH \equiv bJ$ then $a = b$ and $H \equiv J$.

Proof We first show that $aH \equiv bJ$ implies $a = b$. For this purpose, we use an interpretation $\mathcal{I}$ of process terms in the underlying action structure; $\mathcal{I}$ is defined in the obvious way for composition, product and abstraction over processes, and takes all control terms to the unit action. Thus

$$\begin{aligned} \mathcal{I}(aP) &\stackrel{\text{def}}{=} a\mathcal{I}(P) \\ \mathcal{I}(P \otimes Q) &\stackrel{\text{def}}{=} \mathcal{I}(P) \otimes \mathcal{I}(Q) \\ \mathcal{I}((x)P) &\stackrel{\text{def}}{=} \mathbf{ab}_x\mathcal{I}(P) \\ \mathcal{I}(R) &\stackrel{\text{def}}{=} \mathbf{id}_m \qquad \text{if } R \text{ is a control term.} \end{aligned}$$

(In the last case m is the arity of the control term.)

Note that if $P : m$ then $\mathcal{I}(P) : m \rightarrow n$ for some n which is determined by the translation. It is easy to check that the four laws of structural congruence preserve $\mathcal{I}$, i.e. $P \equiv Q$ implies $\mathcal{I}(P) = \mathcal{I}(Q)$. From Definition 5.5 it follows that $\mathcal{I}(H) = \mathbf{id}$ for any passive process H. Hence from $aH \equiv bJ$ we infer

$$a = \mathcal{I}(aH) = \mathcal{I}(bJ) = b.$$

Next we show that $aH \equiv bJ$ implies $H \equiv J$. For any P, let $\widehat{P}$ be the result of replacing all ready occurrences of actions by the unit action. For example if P is $R \otimes a\partial(bQ)$ then $\widehat{P}$ is $\widehat{R} \otimes \partial(bQ)$. In general the arity of $\widehat{P}$ will differ from that of P, but not in the case of a passive term H – in fact, $\widehat{H} \equiv H$.

Now suppose that $P \equiv Q$ is proved by a single application of one of the four laws of structural congruence to a subterm of P. Then it can be checked that also $\widehat{P} \equiv \widehat{Q}$. Essentially this is because in each law a ready occurrence on one side corresponds uniquely to a ready occurrence on the other side. So in general $P \equiv Q$ implies $\widehat{P} \equiv \widehat{Q}$. Thus from $aH \equiv bJ$ we infer, as required,

$$H \equiv \widehat{aH} \equiv \widehat{bJ} \equiv J\,. \qquad ■$$

Finally, we shall need the following simple proposition about committal:

5.9 Proposition If $\centerdot P \equiv \centerdot Q$ then $P \equiv Q$.

Proof For any R, let $\widehat{R}$ be the result of simultaneously replacing every ready occurrence of a commitment $\centerdot S$ in R by S itself. By means similar to the second half of the proof of Proposition 5.8, we see that $R_1 \equiv R_2$ implies $\widehat{R_1} \equiv \widehat{R_2}$. But $\widehat{\centerdot P}$ is P, so the result follows. ■

6 Process structure: dynamics

Our presentation of process dynamics is best introduced by contrast with what was done for CCS and Synchronous CCS. Consider the following transition of Synchronous CCS:

$$(\pi.P + \pi'.Q) \times \overline{\pi}.R \xrightarrow{1} P \times R\,;$$

according to [18] it is derived by three transition rules, whose general form is:

$$\alpha.P \xrightarrow{\alpha} P \qquad\qquad \frac{P \xrightarrow{\alpha} P'}{P + Q \xrightarrow{\alpha} P'}$$

$$\frac{P \xrightarrow{\alpha} P' \qquad Q \xrightarrow{\beta} Q'}{P \times Q \xrightarrow{\alpha\times\beta} P' \times Q'}.$$

Here we wish to derive such transitions in another way; we wish to distinguish the contributions made to any transition by *control* from those made by *action*. We shall use the relations $\searrow_c$ and $\searrow_a$ for these respective contributions. So taking SCCS as the underlying action structure, the above transition will be derived in three stages using the rules given below in 6.2 and 6.3:

$$\begin{array}{rlll} (\pi \bullet P + \pi' \bullet Q) \otimes \overline{\pi} \bullet R & \searrow_c & \pi \bullet P \otimes \overline{\pi} \bullet R & \text{by a rule for } + \\ & \equiv & (\pi \otimes \overline{\pi})(\bullet P \otimes \bullet R) & \\ & \searrow_c & (\pi \otimes \overline{\pi}) \bullet (P \otimes R) & \text{by a rule for } \bullet \\ & \searrow_a & \mathbf{id} \bullet (P \otimes R) & \text{by the rule for } \searrow_a. \end{array}$$

where we have written π for the action $[\,\pi\,]$ etc. Thus a single process-transition is represented by a sequence of reductions. We now proceed to the formal definitions.

6.1 Definition (context, ready context) A *context* C is a process term with a single hole, denoted by [ ], to be filled by a process term. We write $C[P]$ for result of placing P in the context C. If the hole occurs readily then C is a *ready* context. ■

$[\,] \otimes Q$ is an example of a ready context; $[\,] + Q$ is not ready. Structural congruence extends to contexts, and preserves readiness.

6.2 Definition (action preorder) The *action preorder* $\searrow_a$ over processes is the least preorder closed under structural congruence, such that for all ready C

$$a \searrow a' \text{ implies } C[aP] \searrow_a C[a'P].$$

We use $\searrow_a^1$ to represent a single use of this rule, so that $\searrow_a \;=\; (\searrow_a^1)^*$. ■

6.3 Definition (control preorder) The *control preorder* $\searrow_c$ over processes is the least preorder closed under structural congruence satisfying the following *control rules*:

$$\begin{array}{llll} \bullet P \otimes \bullet Q \searrow_c \bullet(P \otimes Q) & P + Q \searrow_c P & \partial P \searrow_c P & !P \searrow_c P \otimes\, !P \\ (x)\bullet P \searrow_c \bullet(x)P & P + Q \searrow_c Q & \partial P \searrow_c \bullet\partial P & !P \searrow_c \bullet\, !P \end{array}$$

and such that for all ready C

$$P \searrow_c P' \text{ implies } C[P] \searrow_c C[P'].$$

We use $\searrow^1_c$ to represent a single use of one of the control rules in a ready context; then $\searrow_c = (\searrow^1_c)^*$. ■

Each of the six rules for sum, delay and replication either makes some process ready for action, or makes a commitment. But the two rules for committal (the first column) have a different purpose; they lift commitments to the top level, as this example shows:

$$a \bullet P \otimes (x)b \bullet Q \searrow (a \otimes \mathbf{ab}_x b) \bullet (P \otimes (x)Q)\,.$$

6.4 Definition (reduction preorder) The *reduction preorder* $\searrow$ over processes is the least preorder which includes $\searrow_a$ and $\searrow_c$. Thus $\searrow = (\searrow_a \cup \searrow_c)^*$. ■

Of course $\searrow = (\searrow^1)^*$ where $\searrow^1 = (\searrow^1_a \cup \searrow^1_c)^*$; so $\searrow$ consists just of a sequence of action and control steps.

A moment's thought shows that an action step creates no new possible control steps; so all the control steps can be done first. Hence

6.5 Proposition $\searrow = \searrow_c \searrow_a$. ■

One illustration of this result is the example from Synchronous CCS with which we began this section.

To prepare for the introduction of simulation and bisimulation over processes, we now prove two propositions which represent the basic theory of reduction. This theory rests conveniently on the notion of head normal form introduced in the previous section.

6.6 Lemma Let $aH \searrow_a bJ$. Then $a \searrow b$ and $H \equiv J$. (Compare Corollary 5.9.)

Proof It is enough to prove the result for $\searrow^1_a$. In fact it is easiest to prove first that $P \searrow^1_a P'$ implies

$$P \equiv cR,\ \ P' \equiv c'R \text{ and } c \searrow c' \text{ for some } c, c' \text{ and } R.$$

We prove this by a straightforward induction on the structure of the ready context C for which $P \equiv C[dS]$, $P' \equiv C[d'S]$ and $d \searrow d'$.

Now take P and P' to be aH and bJ. Then for some c, c', R we have

$$aH \equiv cR,\ \ bJ \equiv c'R \text{ and } c \searrow c'.$$

Now let R have hnf $c''K$. Then from the uniqueness of hnfs it follows that

$$H \equiv K \equiv J \text{ and } a = cc'' \searrow c'c'' = b.$$ ■

The corresponding lemma for $\searrow_c$ is easier:

6.7 Lemma Let $aH \searrow_c bJ$. Then there exists c such that $H \searrow_c cJ$ and $ac = b$.

Proof Again, it is enough to prove the result for $\searrow^1_c$. There is some ready context C and some K, Q such that

$$H \equiv C[K],\ \ K \searrow^1_c Q \text{ and } bJ \equiv aC[Q].$$

Now let $\mathcal{C}[Q]$ have hnf cJ'. Then $H \searrow_{\mathrm{c}}^{1} cJ'$, and by the uniqueness of hnfs it follows that $b = ac$ and $J \equiv J'$. ∎

Putting these two lemmas together, with the help of Proposition 6.5 we have proved a characterisation of reduction in terms of hnfs:

6.8 Proposition Let $aH \searrow bJ$. Then for some c, $H \searrow_{\mathrm{c}} cJ$ and $ac \searrow b$. ∎

Our second basic result about dynamics concerns commitments $R \searrow c \centerdot R'$. It asserts that commitments of a composite process arise from commitments of its components.

6.9 Proposition Let $P \searrow a \centerdot P'$. Then

1. If P is bQ then for some c
$$Q \searrow_{\mathrm{c}} c \centerdot P' \quad \text{and} \quad bc \searrow a.$$

2. If P is $Q \otimes R$, then for some b, c, Q' and R'
$$Q \searrow_{\mathrm{c}} b \centerdot Q', \quad R \searrow_{\mathrm{c}} c \centerdot R', \quad b \otimes c \searrow a \quad \text{and} \quad Q' \otimes R' \equiv P'.$$

3. If P is $(x)Q$ then for some b and Q'
$$Q \searrow_{\mathrm{c}} b \centerdot Q', \quad \mathbf{ab}_x b \searrow a \quad \text{and} \quad (x)Q' \equiv P'.$$

Proof For clause 1 consider the hnf of P, hence of R, and apply Proposition 6.8. It will be enough to outline the proof of clause 3; clause 2 is similar. From the assumption, we have $(x)Q \searrow_{\mathrm{c}} a' \centerdot P'$ where $a' \searrow a$. Now by inspection the control reduction must take the form of $n \geq 0$ steps of the form
$$(x)Q_i \searrow_{\mathrm{c}}^{1} (x)Q_{i+1}$$
where $Q \equiv Q_0$, $Q_i \searrow_{\mathrm{c}}^{1} Q_{i+1}$ $(0 \leq i < n)$ and $Q_n \equiv b \centerdot Q'$, followed by a last step of the form
$$(x)Q_n \equiv (\mathbf{ab}_x b)\,(x) \centerdot Q' \searrow_{\mathrm{c}}^{1} (\mathbf{ab}_x b) \centerdot (x)Q'.$$
(Note that no further control reduction is possible after this step.) Thus we have $Q \searrow_{\mathrm{c}} b \centerdot Q'$ where $\mathbf{ab}_x b = a' \searrow a$ and $(x)Q' \equiv P'$ as required. ∎

6.10 Commitment and transition No further control reduction is possible for a commitment
$$P \searrow a \centerdot P'\,;$$
if also a is inactive, then no further reduction at all can occur.

At this point recall that we intend to model the single transitions of processes as actions in an action structure. (See the beginning of Section 5.) Commitments are the key to this. As a first approximation we can think of the above commitment as a single transition of P, and write it as $P \xrightarrow{\mathrm{a}} P'$, recalling that process calculi like CCS are defined in terms of transitions labelled with actions. But this is not quite enough; for we declared that we should make explicit the way in which an action exerts an effect $\mathrm{e} \in E$ upon the continuation of a process. Moreover the transition labels in CCS, except for τ, represent that part of an action which may interact with or be observed by the environment; but we saw at the end of Section 4 that effects are strongly inert, which implies that they cannot react with the environment. So in proceeding from $P \searrow a \bullet P'$ to $P \xrightarrow{a} P'$ we somehow want to absorb any effect of a into P'. How shall we do this? Since an action does not possess an unique effect, though its effects are directed under precomposition (see after Definition 4.5), the following definition is appropriate:

6.11 Definition (transition) The *transition relations* $\{\xrightarrow{a} \mid a \in A\}$ over processes are defined by

$$P \xrightarrow{a} P' \text{ iff, for some } e \text{ and } P'',\ \ P \searrow ae \bullet P'' \text{ and } eP'' \equiv P'. \qquad \blacksquare$$

These relations behave well. We now show that they are preserved by composition, abstraction and product, by reaction, and by factorisation of an action through an effect:

6.12 Proposition Let $P \xrightarrow{a} P'$. Then

1. $cP \xrightarrow{ca} P'$;
2. $(x)P \xrightarrow{\mathbf{ab}_x a} (x)P'$;
3. if also $Q \xrightarrow{b} Q'$ then $P \otimes Q \xrightarrow{a \otimes b} P' \otimes Q'$;
4. if $a \searrow a'$ then $P \xrightarrow{a'} P'$;
5. if $a = a'e$ then $P \xrightarrow{a'} eP'$. ■

The next step is to define a notion of (bi)simulation over processes, which uses these transitions. It takes into account that the rôle of an effect e is not to be observed, but to modify the successor process; thus an action with one effect may be simulated by another with different effect, provided the two actions have the same observable content.

6.13 Definition (simulation) A binary relation $\mathcal{S}$ over processes is a *simulation* if, whenever $(P, Q) \in \mathcal{S}$ and $P \xrightarrow{a} P'$, then there exist a', e and Q' such that

$$a = a'e,\ \ Q \xrightarrow{a'} Q' \text{ and } (eP', Q') \in \mathcal{S}.$$

If $\mathcal{S}$ is also symmetric then it is called a *bisimulation*.

If the pair (P, Q) is in some simulation then we say that Q *simulates* P and write $P \lesssim Q$. If (P, Q) is in some bisimulation then we say that P and Q are *bisimilar* and write $P \sim Q$. ■

We shall hold off the theory of $\lesssim$ and $\sim$ until Section 8, because it is a special case of what is done there. But it is worth recording here a corollary of the main result:

6.14 Theorem (congruence) $\lesssim$ is a precongruence, and $\sim$ a congruence, for the algebra of processes; that is, these relations are respectively a preorder and an equivalence, and are preserved by all the operations introduced in Section 5.1.

Proof This follows from Theorem 8.7 by taking K, the incident-set, to be the entire action structure. ■

It is also worth observing that a simpler notion of simulation suffices, in the case that every action has a pure decomposition (see Definition 4.7):

6.15 Definition (pure simulation) $\mathcal{S}$ is a *pure simulation* if, whenever $(P, Q) \in \mathcal{S}$ and $P \xrightarrow{a} P'$ for some pure a, then there exists Q' such that $Q \xrightarrow{a} Q'$ and $(P', Q') \in \mathcal{S}$. If $\mathcal{S}$ is also symmetric, then it is a *pure bisimulation*. ∎

Then under the stated condition this gives rise to exactly the same theory:

6.16 Proposition If every action has a pure decomposition, then $P \lesssim Q$ (*resp.* $P \sim Q$) iff (P, Q) is in some pure simulation (*resp.* bisimulation). ∎

This theory is quite pleasant. But even if we choose a natural effect structure –and whether or not pure decompositions exist– it turns out that this notion of bisimulation does not yield the familiar equivalence in familiar cases; it is still too strong. We may try to weaken it by restricting the range of a in Definitions 6.13 and 6.15; but this does not always yield a congruence. The remedy lies in the concept of an *incident*, which is the subject of the following section.

7 Incident-sets

We begin this section by showing why bisimilarity, as defined in Definition 6.13, is a stronger equivalence than we would wish.

Recall from Section 3.4 the action structure $\mathsf{SCCS}[Z]$. Choose the effect structure found in Section 4.8, which consists of all actions $(\vec{x})\,(\nu\vec{y})\,S\,\langle\vec{z}\rangle$ whose multiset S is empty. Now $[\,z, \overline{z}\,]$ is a different action from $[\,z', \overline{z'}\,]$. So when we consider processes over this action structure we find

$$[\,z\,]\centerdot P \otimes [\,\overline{z}\,]\centerdot Q \not\sim [\,z'\,]\centerdot P \otimes [\,\overline{z'}\,]\centerdot Q,$$

since the former has the transition $\cdots \xrightarrow{[z,\overline{z}]} P \otimes Q$ which cannot be matched by the latter in the sense required for a simulation (6.13). But in Synchronous CCS as originally defined the two processes *are* bisimilar (they are both bisimilar to $\mathbf{id}\centerdot(P \otimes Q)$), and we would wish them so.

It is tempting to relax the definition of simulation by restricting consideration to transitions $\xrightarrow{a}$ in which a is inactive, thus excluding things like $[\,z, \overline{z}\,]$. This works for Synchronous CCS, but in general we have to be more careful if we are to retain the congruential property of bisimulation. Indeed, for the synchronous π-calculus (see Part II) it does not work to consider only the inactive actions in simulation.

Throughout this section we assume some fixed action structure A and effect structure E.

7.1 Definition (incident-set) An *incident-set* K is a subset of the actions of A such that

1. For all $a \in A$ and $e \in E$ such that ae is defined, $ae \in K$ iff $a \in K$;

2. For any preaction φ, if $\varphi(a) \searrow \ell \in K$ then there exists $k \in K$ for which $a \searrow k$ and $\varphi(k) \searrow \ell$. ∎

Henceforward, when an incident-set K is understood we shall call its members *incidents*, and use k, ℓ, m to range over them. Here, as in the definition of an effect structure, we have imposed the weakest conditions which yield interesting results. For example K need not be a sub-actionstructure.

Clause 2 is the main condition which will ensure congruence. Clause 1 is auxiliary; an important consequence of it is that for a given incident k, its decompositions (k', e) –where k' is also an incident– form a directed set (see after Definitions 4.4 and 4.5):

7.2 Proposition Let $k_1e_1 = k_2e_2$. Then there exist k, f_1 and f_2 such that

$$k_1 = kf_1, \ k_2 = kf_2 \text{ and } f_1e_1 = f_2e_2. \qquad \blacksquare$$

Note that this result is only concerned with the statics of incidents.

7.3 Examples (1) It is easy to see that in general, in any action structure A, the entire set A of actions is an incident-set. But we look for smaller incident-sets, because we wish to restrict a in the definition of simulation (6.13) to range over an incident-set. These smaller incident-sets can differ strikingly in different cases.

In both EXP[X] and SCCS[X] (Sections 3.3 and 3.5) it turns out that the inactive actions constitute an incident-set. (This appears to be mainly because the reaction relation is confluent in both cases.) For SCCS[X] these are just the actions $(\vec{x})(\nu\vec{y})\,S\,\langle\vec{z}\rangle$ in which S contains no complementary pair $x, \overline{x}$. A smaller incident-set K consists of the inactive actions which also have no restricted particles; that is, if $v \in S$ or $\overline{v} \in S$ then $v \notin \vec{y}$. This is a more appropriate incident-set, since no restricted particle in an action a can ever participate in a reaction of $\varphi(a)$, for any preaction φ. For this K we conjecture that the K-bisimilarity defined in 7.5 below coincides with the original strong congruence for Synchronous CCS [18].

7.4 Examples (2) The case of CCS[X] is interesting. Recall from Section 3.6 that its reaction rule replaces a complementary pair of particles by the special particle τ, instead of just deleting them. What difference does this make to the possible choice of incident-set? We may take exactly the inactive actions as an incident-set, as for SCCS[X]; this allows an incident to consist of many simultaneous interactions. But due to the presence of the special particle τ we can make the transitions $P \xrightarrow{k} P'$ $(k \in K)$ correspond much more closely to the original transitions of CCS [19] which were labelled by a single particle. We outline how to achieve this.

We have referred to the particle x as *positive* and $\overline{x}$ as *negative*; for now, let us say that τ is *both* positive *and* negative. Define the *weight* of an action to be the pair (p, n), where p (resp. n) is the number of its positive (resp. negative) particles; so each τ counts both in p and in n. Then reaction in CCS[X] preserves the weight of an action; it follows quite easily from this that the set of inactive actions whose weight does not exceed some fixed bound $(p_{\max}, n_{\max})$ constitutes an incident-set. In particular those with weight at most $(1,1)$ constitute an incident-set; these are the inactive actions with bodies $\emptyset$, $[\tau]$, $[x]$, $[\overline{x}]$ or $[x, \overline{y}]$ for $x \neq y$. Again, we get a more appropriate incident-set K if we also exclude actions containing restricted particles. This K is in fact the smallest incident-set containing the action $[\tau]$ (see Theorem 7.7), and the corresponding K-bisimilarity is close to the original strong congruence for CCS; but it is a *stronger* congruence, since it is also preserved by instantiation of port-names. [7]

In our examples so far the natural incidents have been inactive. In general, and especially in the case of the π-calculus and Petri nets, the situation is more delicate; in these cases the inactive actions do not constitute an incident-set, and there are important incident-sets which contain active actions. Incidents for the π-calculus are examined in Part II.

Having looked at some particular cases, we naturally wish to know what non-empty incident-sets to expect in a given action structure A. Can there be none, or many? If there are many, then what can we say about the family of incident-sets in A?

---

[7] In the standard presentation of CCS [19] instantiation of port-names is not admitted; in fact the strong congruence $p.\mathbf{0} \mid \overline{q}.\mathbf{0} \sim p.\overline{q}.\mathbf{0} + \overline{q}.p.\mathbf{0}$ holds if p and q are different port-names, but fails if they are identical.

We begin with a simple observation. Recall that an incident-set is relative to a given effect structure E. It turns out that, under a modest condition, an incident-set must include the whole of E:

Exercise Let K be an incident-set. Assume that for each $m \in M$ there is a member of K with target arity m. Show that $E \subseteq K$. (*Hint*: use inertia.) ■

Next recall that A is always an incident-set for itself. So in view of our examples, there is often more that one non-empty incident-set. In that case, is there in some sense a best one? We shall show that, given any set B of actions, under a mild condition there is a smallest incident-set which includes B. This is significant for the bisimilarity congruences $\sim_K$ (one for each incident-set K) which we establish in Section 8. For since $\sim_K$ increases as K decreases, we then have the most generous congruential bisimilarity in which two bisimilar processes must at least "agree" with respect to the incidents in B.

For the rest of the section we continue to keep the action structure (A, E) fixed. We shall also assume that $\searrow$ in A is generated by $\overset{1}{\searrow}$ so that $\searrow\ = (\overset{1}{\searrow})^*$.

7.5 Definition (convergence) A set $S \subseteq A$ is *convergent* if no infinite reaction sequence $a_0 \overset{1}{\searrow} a_1 \overset{1}{\searrow} a_2 \overset{1}{\searrow} \cdots$ has infinitely many members in S. ■

Now A may not itself be convergent, but it may possess a convergent incident-set. Actually, in all our examples except for Petri nets, A itself is convergent.

If there were a smallest incident-set including B, it should be the intersection of all incident-sets including B. Unfortunately this class is not in general closed under intersection. But:

7.6 Proposition Let $\mathcal{C}$ be any set of incident-sets which contains a convergent member. Then $\bigcap \mathcal{C}$ is also an incident-set.

Proof Clearly $\bigcap \mathcal{C}$ satisfies 7.1(1), so we must show that it also satisfies 7.1(2). Suppose not. Then for some preaction φ, some $k \in \bigcap \mathcal{C}$ and some $b \in A$ we have

$$\phi(b) \searrow k \text{ but there is no } \ell \in \bigcap \mathcal{C} \text{ for which } b \searrow \ell \text{ and } \varphi(\ell) \searrow k.$$

Take any $K \in \mathcal{C}$. We get a contradiction by finding a reaction sequence

$$b = b_0 \overset{+}{\searrow} b_1 \overset{+}{\searrow} \cdots$$

which passes infinitely often through K. (Here $\overset{+}{\searrow} = (\overset{1}{\searrow})^+$, the transitive closure of $\overset{1}{\searrow}$.) We also ensure that $\varphi(b_i) \searrow k$ for all i.

Suppose $b_0, b_1, \ldots, b_j$ are already determined. By assumption, $b_j \notin \bigcap \mathcal{C}$. There are two cases:

1. $b_j \in K$. Then for some other $K' \in \mathcal{C}$, $b_j \notin K'$. Then choose $b_{j+1} \in K'$ such that $b_j \searrow b_{j+1}$ and $\varphi(b_{j+1}) \searrow k$.
2. $b_j \notin K$. Then choose $b_{j+1} \in K$ with the same conditions.

Clearly adjacent members of this sequence are distinct, so $b_j \overset{+}{\searrow} b_{j+1}$ as required; also the sequence passes infinitely often through K by construction. ■

Now, by taking $\mathcal{C}$ in Prop 7.6 to be all incident-sets which include B, we have established the following.

7.7 Theorem (minimum incident-set) Let $B \subseteq A$. If there is a convergent incident-set including B, then there is a smallest incident-set including B. ∎

7.8 Discussion The main topic of this section was the structure of the class of incident-sets. The main result here, the existence of a smallest incident-set under certain conditions, does not depend at all upon process structures and the uniform way (defined in Section 5) in which they can be built from action structures. In fact the notion of incident-set appears to be intrinsic to action structure theory. The only part of its definition which is partly motivated by process structure is the notion of preaction, Definition 4.10. However, preactions also seem to have intrinsic interest, being closely connected with inertia.

8 Bisimulation and congruence

We are now ready to introduce a natural class of preorders and equivalences, one for each incident-set. The properties of an incident-set will ensure that they are congruential; this is the main goal of our uniform study of bisimulation.

8.1 Definition (K-simulation) Let K be an incident-set. A binary relation $\mathcal{S}$ over processes is a K-*simulation* if, whenever $(P, Q) \in \mathcal{S}$ and $P \xrightarrow{k} P'$, then there exist k', e and Q' such that

$$k = k'e, \quad Q \xrightarrow{k'} Q' \text{ and } (eP', Q') \in \mathcal{S}.$$

If $\mathcal{S}$ is also symmetric, then it is called a K-*bisimulation*.

If (P, Q) is in some K-simulation we say that Q K-*simulates* P, and write $P \lesssim_K Q$. If (P, Q) is in some K-bisimulation we say that P and Q are K-*bisimilar*, and write $P \sim_K Q$. ∎

To establish that $P \lesssim_K Q$, or that $P \sim_K Q$, it is convenient to have a more generous class of relations than K-simulations. The following notion is useful:

8.2 Definition (K-simulation up to composition) If $\mathcal{S}$ is a relation over processes, then $\mathcal{S}^{\cdot}$, the *composition closure* of $\mathcal{S}$, is defined by

$$\mathcal{S}^{\cdot} \stackrel{\text{def}}{=} \{(aP, aQ) \mid a \text{ an action and } (P, Q) \in \mathcal{S}\}.$$

$\mathcal{S}$ is a K-*simulation up to composition* if, whenever $(P, Q) \in \mathcal{S}$ and $P \xrightarrow{k} P'$, then there exist k', e, Q' such that

$$k = k'e, \quad Q \xrightarrow{k'} Q' \text{ and } (eP', Q') \in \mathcal{S}^{\cdot}.$$

If $\mathcal{S}$ is also symmetric then it is called a K-*bisimulation up to composition*. ∎

The importance of this definition is that, to establish (bi)similarity, we only have to exhibit a (bi)simulation up to composition. To be precise:

8.3 Proposition Let $\mathcal{S}$ be a K-simulation up to composition. Then $\mathcal{S}^{\cdot}$ is a K-simulation. Hence if $(P, Q) \in \mathcal{S}$ then $P \lesssim_K Q$, and if in addition $\mathcal{S}$ is symmetric then $P \sim_K Q$.

Proof It is enough to show the first part. Let $\mathcal{S}$ be a K-simulation up to composition and (aP, aQ) be a typical member of $\mathcal{S}^{\cdot}$ with $(P, Q) \in \mathcal{S}$. Let $aP \xrightarrow{k} P'$. Then By 7.1 there exist ℓ, e and P'' such that

$$P \xrightarrow{\ell} P'', \quad a\ell \searrow ke \text{ and } eP' \equiv P''.$$

Since $(P, Q) \in \mathcal{S}$, there exist ℓ', f and Q'' such that

$$Q \xrightarrow{\ell'} Q'', \quad \ell' f = \ell \text{ and } (fP'', Q'') \in \mathcal{S}^{\cdot}.$$

It follows from Proposition 6.12 that $aQ \xrightarrow{a\ell'} Q''$. But $a\ell' f \searrow ke$ and e is inert, so there exists k' such that

$$a\ell' \searrow k' \text{ and } k'f = ke.$$

Now from Proposition 7.2 there exist k'', f' and e' such that

$$k' = k''f', \quad k = k''e' \text{ and } f'f = e'e.$$

So again from Proposition 6.12 we have $aQ \xrightarrow{k'} Q''$, hence

$$aQ \xrightarrow{k''} f'Q'',$$

and since $k = k''e'$ it only remains to show that $(e'P', f'Q'') \in \mathcal{S}^{\cdot}$. But

$$(e'P', f'Q'') \equiv (e'eP'', f'Q'') \equiv (f'fP'', f'Q'')$$

and we are done, since $\mathcal{S}^{\cdot}$ is composition-closed and $(fP'', Q'') \in \mathcal{S}^{\cdot}$. ■

As a simple corollary, we find that $\lesssim$ and $\sim$ are preserved by composition:

8.4 Proposition $P \lesssim_K Q$ implies $aP \lesssim_K aQ$, and $P \sim_K Q$ implies $aP \sim_K aQ$.

Proof If $P \lesssim_K Q$ then (P, Q) is in some K-simulation $\mathcal{S}$, which is *a fortiori* a K-simulation up to composition. But $(aP, aQ) \in \mathcal{S}^{\cdot}$ which is itself a K-simulation by the Proposition. Hence $aP \lesssim_K aQ$. ■

With the help of this we can now establish

8.5 Proposition $\lesssim_K$ is a preorder and $\sim_K$ is an equivalence.

Proof It will be enough to show that $\lesssim_K$ (hence similarly $\sim_K$) is transitive. For this, we show that $\mathcal{S} = \{(P, R) \mid \exists Q.\ P \lesssim_K Q \lesssim_K R\}$ is a K-simulation. So let $P \lesssim_K Q \lesssim_K R$ and $P \xrightarrow{k} P'$. It is enough to find k'', f'' and R' for which $k = k''f''$, $R \xrightarrow{k''} R'$ and $(f''P', R') \in \mathcal{S}$.

First, there exist k', f and Q' for which

$$k = k'f, \quad Q \xrightarrow{k'} Q' \text{ and } fP' \lesssim_K Q'.$$

Further, there exist k'', f' and R' for which

$$k' = k''f', \quad R \xrightarrow{k''} R' \text{ and } f'Q' \lesssim_K R'.$$

Now from Proposition 8.4 we know that $f'fP' \lesssim_K f'Q'$; then the required conditions hold if we take $f'' = f'f$. ■

In final preparation for the proof that these relations are congruential, we shall need a development of Proposition 6.9. It shows how an incident-transition of a composite process must arise from incident-transitions of its components, as follows:

8.6 Proposition Let $P \xrightarrow{k} P'$. Then

1. If P is bQ then for some ℓ and e

$$Q \xrightarrow{\ell} Q', \ b\ell \searrow ke \text{ and } eQ' \equiv P'.$$

2. If P is $Q \otimes R$, then for some ℓ, m, e, Q' and R'

$$Q \xrightarrow{\ell} Q', \ R \xrightarrow{m} R', \ \ell \otimes m \searrow ke \text{ and } e(Q' \otimes R') \equiv P'.$$

3. If P is $(x)Q$ then for some ℓ, e and Q'

$$Q \xrightarrow{\ell} Q', \ \mathbf{ab}_x\ell \searrow ke \text{ and } e\,(x)Q' \equiv P'.$$

Proof It will be enough to do clause 2. We have $P \searrow ke \bullet P''$ for some e and P'' with $eP'' \equiv P'$. So by Proposition 6.9(2) there exist b, c, Q' and R' such that

$$Q \searrow_{\partial} b \bullet Q', \ R \searrow_{\partial} c \bullet R', \ b \otimes c \searrow ke \text{ and } Q' \otimes R' \equiv P''.$$

The rest follows by applying the second clause of Definition 7.1, first to the preaction $(-) \otimes c$ to find ℓ, then to the preaction $\ell \otimes (-)$ to find m. ■

We are now ready for the main result.

8.7 Theorem (incident congruence) K-similarity $\lesssim_K$ is a precongruence, and K-bisimilarity $\sim_K$ is a congruence.

Proof We proved in Proposition 8.4 that composition preserves these relations. We must now do the same for all other process constructions. In many cases this amounts to proving that an appropriate relation is a (bi)simulation up to composition; the detailed reasoning often follows the pattern of the proof of Proposition 8.3.

To show that product preserves $\lesssim_K$ it will be enough to show that

$$\mathcal{S} \stackrel{\text{def}}{=} \{(P \otimes R, \ Q \otimes R) \mid P \lesssim_K Q\}$$

is a K-simulation up to composition.

Let $P \otimes R \xrightarrow{k} S$. Assuming $P \lesssim_K Q$, we shall find k'', e' and T such that $Q \otimes R \xrightarrow{k''} T$, $k''e' = k$ and $(e'S, T) \in \mathcal{S}^{\cdot}$.

First, from Proposition 8.6(2) there exist ℓ, m, e, P' and R' for which

$$P \xrightarrow{\ell} P', \ R \xrightarrow{m} R', \ \ell \otimes m \searrow ke \text{ and } e(P' \otimes R') \equiv S.$$

Now $P \lesssim_K Q$, so there exist ℓ', f and Q' such that

$$Q \xrightarrow{\ell'} Q', \ \ell' f = \ell \text{ and } fP' \lesssim_K Q'.$$

We deduce $(\ell' \otimes m)(f \otimes \mathbf{id}) = \ell \otimes m \searrow ke$; but $f \otimes \mathbf{id}$ is inert so there exists k' for which

$$\ell' \otimes m \searrow k' \text{ and } k'(f \otimes \mathbf{id}) = ke.$$

From the last equation, by Proposition 7.2 there exist k'', e'' and e' for which

$$k' = k''e'', \quad k = k''e' \text{ and } e''(f \otimes \mathbf{id}) = e'e.$$

From the above transitions for Q and R, using Proposition 6.12 we deduce

$$Q \otimes R \xrightarrow{k''} e''(Q' \otimes R'),$$

so we shall take T to be the result of this transition. Then from the above equations we find

$$(e'S, T) \equiv (e''(fP' \otimes R'),\ e''(Q' \otimes R'))$$

which is in $\mathcal{S}^{\cdot}$ since $fP' \lesssim_K Q'$.

This completes the proof that product preserves incident-simulation, and a similar proof works for abstraction. It remains to show that the control operators preserve it. We consider only commitment and replication; summation and delay are no harder.

For commitment, let $P \lesssim_K Q$ and consider $(\bullet P,\ \bullet Q)$. Let $\bullet P \xrightarrow{k} P'$; we shall show $\bullet Q \xrightarrow{k} Q'$ where $P' \lesssim_K Q'$.

We have

$$\bullet P \searrow ke \bullet P'' \text{ where } eP'' \equiv P'.$$

By Proposition 6.8 it follows that for some c

$$\bullet P \searrow_a c \bullet P'' \text{ and } c \searrow ke.$$

But $\bullet P \searrow_a^1 \ldots$ is impossible, so $c = \mathbf{id}$ and $\bullet P \equiv \bullet P''$, whence $P \equiv P''$ by Proposition 5.9. But $\mathbf{id}$ is inactive, so $ke = \mathbf{id}$. Now $\bullet Q \xrightarrow{1} Q$, i.e. $\bullet Q \xrightarrow{ke} Q$, so applying Proposition 6.12(5) we get

$$\bullet Q \xrightarrow{k} eQ \text{ and } P' \equiv eP \lesssim_K eQ.$$

This completes the proof that committal preserves incident-simulation.

For replication, the main task is to show that

$$\mathcal{S} \stackrel{\text{def}}{=} \{(R \otimes\ !P,\ S \otimes\ !Q) \mid R \lesssim_K S,\ P \lesssim_K Q\}$$

is a simulation up to composition – and hence so also is $\mathcal{S} \cup \{(!P,\ !Q) \mid P \lesssim_K Q\}$ by a similar argument.

So assume $R \lesssim_K S$ and $P \lesssim_K Q$, and let $R \otimes\ !P \xrightarrow{k} U$. We shall find k'', e' and V for which $S \otimes\ !Q \xrightarrow{k''} V$, $k = k''e'$ and $(e'U, V) \in \mathcal{S}^{\cdot}$. W.l.o.g. we can assume that (for some $n \geq 0$) the first n+1 atomic reductions of the transition are an unfolding of $!P$ into n copies of P followed by a commitment:

$$R \otimes\ !P \searrow_a^1 R \otimes P \otimes\ !P \searrow_a^1 \cdots \searrow_a^1 R \otimes P^n \otimes\ !P \searrow_a^1 R \otimes P^n \otimes \bullet\ !P,$$

where P^n means $\overbrace{P \otimes \cdots \otimes P}^{n\ times}$ (or the term is absent if $n = 0$).

So $R \otimes P^n \otimes \bullet\ !P \xrightarrow{k} U$; then by Propositions 8.6(2) there exist ℓ, m, e, R' and P' such that

$$R \otimes P^n \xrightarrow{\ell} R',\quad !P \xrightarrow{m} P',\quad \ell \otimes m \searrow ke \text{ and } e(R' \otimes P'') \equiv U. \tag{1}$$

Considering the first of these transitions, we know that $R \otimes P^n \lesssim_K S \otimes Q^n$, so there exist ℓ', f and S' for which

$$S \otimes Q^n \xrightarrow{\ell'} S', \;\; \ell' f = \ell \;\text{ and }\; fR' \lesssim_K S'. \qquad (2)$$

Since $(\ell' \otimes m)(f \otimes \mathbf{id}) \searrow ke$ and $f \otimes \mathbf{id}$ is inert, there exists k' for which

$$\ell' \otimes m \searrow k' \;\text{ and }\; k'(f \otimes \mathbf{id}) = ke,$$

whence by Proposition 7.2 there also exist k'', e'' and e' for which

$$k' = k''e'', \;\; k = k''e' \;\text{ and }\; e''(f \otimes \mathbf{id}) = e'e.$$

Considering the second transition of (1), by reasoning similar to the argument for commitment above we find that there exists g such that

$$mg = \mathbf{id} \;\text{ and }\; P' \equiv g\,!P.$$

Now let us consider a transition for $S \otimes\, !Q$; from (2) we easily find that

$$S \otimes\, !Q \xrightarrow{\ell' \otimes 1} S' \otimes\, !Q;$$

also $\ell' \otimes \mathbf{id} = (\ell' \otimes m)(\mathbf{id} \otimes g) \searrow k''e''(\mathbf{id} \otimes g)$, so by Prop 6.12(4),(5) we get

$$S \otimes\, !Q \xrightarrow{k''} e''(\mathbf{id} \otimes g)(S' \otimes\, !Q).$$

Take V to be the result of this transition. Since $k = k''e'$, it only remains to prove that $(e'U, V) \in \mathcal{S}^{\cdot}$. But

$$e'U \equiv e'e(R' \otimes P') \equiv e''(f \otimes \mathbf{id})(R' \otimes g\,!P) \equiv e''(\mathbf{id} \otimes g)(fR' \otimes\, !P),$$

and we are done since $fR' \lesssim_K S'$.

This completes the proof that $\lesssim_K$ is a precongruence; an analogous proof shows that $\sim_K$ is a congruence. ■

Before leaving the topic of K-simulation, we note that when every action has a pure decomposition, then the notion of pure simulation (Definition 6.15) can be extended in an obvious way to *pure K-simulation* and a corresponding equivalence.

9 Related work

This paper is one of many attempts to find a common theoretical framework for concurrent computation. It would take a long essay to classify these attempts, or even to compare them all to this one. I shall mention only three alternative lines.

The *chemical abstract machine* (CHAM) of Berry and Boudol [6] has common aspects with action structures. Indeed, in particular cases such as the π-calculus action structure the actions, with a multiset of particles as their principal component, are close to what the natural CHAM for the π-calculus looks like. Crudely, one can see action structures as an attempt to make an algebra of CHAMs, with parametrisation. Other differences, e.g. in the dynamics, arise from treating reaction as algebraically as possible. Whatever the difference, the CHAM has been a guiding influence in the present work.

Meseguer proposes conditional rewriting logic [16] as a general framework for unifying the study of concurrency. Action structures are not about logic, but they should be compared with the categories, so-called $\mathcal{R}$-systems, which are models –in the technical sense– of a Meseguer rewrite theory. This is a rather general class, and contains the monoidal categories which represent individual Petri nets [17], already referred to in Section 3.8 above. One may contrast this work with action structures as follows: in an $\mathcal{R}$-system the carrier of the algebra consists of the computations of an individual system (e.g. Petri net), while in an action structure the carrier consists of the systems themselves. Direct comparison is therefore difficult; nonetheless it would be of interest to explore a combination of action structures with Meseguer's logical approach.

Recently Nielsen and Winskel have proposed asynchronous transition systems as a general model of concurrency [26]. These are essentially transition systems with additional structure recording the independence among events. This is one difference from action structures, whose definition makes no mention of independence among reactions, though independence can be examined in particular action structures – e.g. in Petri nets. But other differences are perhaps deeper. A good point to examine is the rôle played by morphisms. A morphism in the category of asynchronous transition systems describes, crudely speaking, how one process (e.g. Petri net) may implement or simulate another. In contrast a homomorphism of action structures describes how one *class* of agents, e.g. π-calculus processes, may implement another, e.g. arithmetic expressions. Both notions deserve study; once again comparison is difficult because the models work at different levels.

One may find this disparity among models bewildering. Are we doing something wrong? I think not; concurrent computing can be viewed in so many ways. This has long been apparent at the level of programming languages, and of concrete calculi; it is no surprise to find it so at the more abstract level of general models. Each model must be analysed deeply in its own terms, and be developed with awareness of the others; we must hope that this leads us to unifying concepts, but we should not expect the search to be quick and easy.

PART II: Actions for Pi Calculus

10 Review

In Part I of this paper the theory of action structures was developed, and it was shown how to extend an arbitrary action structure to a process calculus. Part II is devoted to showing how, in particular, an action structure may underlie the π-calculus. We begin by explaining how action structures indeed arose from the need for such a foundation.

10.1 How π-calculus suggests action structures In this subsection we assume nothing about action structures, but rather motivate their introduction. Consider the atomic communications which occur in process calculi, beginning with CCS [19]. The CCS prefix construction $p(y).P$, means "receive any value at port p, bind it to the variable y and proceed to P". The π-calculus has the same construction; but here values, variables and ports are all the same thing: names. Therefore ports can be bound, or parametric; this is what allows the π-calculus to model dynamically changing structure.

Consider the parallel composition

$$S \equiv x(y).y(u).P \mid \overline{x}z.Q \mid \overline{z}v.R\,;$$

here a communication at port x can occur, yielding the transition $S \longrightarrow S'$ where

$$S' \equiv z(u).P \mid Q \mid \overline{z}v.R\,.$$

(For simplicity, we assume y not free in P.) At this point a communication at z can occur, which could not occur before; this yields the transition $S' \longrightarrow S''$ where

$$S'' \equiv \{v/u\}P \mid Q \mid R\,.$$

We can say that the communication at x has *enabled* the one at z, through the instantiation of y to z. This enablement is a phenomenon of the π-calculus, not of CCS.

Prefixing imposes *precedence*, i.e. order of occurrence, among actions in the π-calculus. But since *enablement* is also present, it is reasonable to expect some benefit from separating the two notions, instead of uniting them in a single construction.

Let us look at one way of doing this. Use π to stand for a *particle*, i.e. a simple prefix of the form $x(y)$ or $\overline{x}y$, and consider two cases of the construction $\pi.\pi'.R$:

$$\begin{aligned} P &\equiv x(y).\overline{z}w.P' \\ Q &\equiv x(y).\overline{y}w.Q'\,. \end{aligned}$$

In both cases π *precedes* π', while only in Q does π *enable* π'. But suppose that we allow compound prefixes as follows:

$$\begin{aligned} P &\equiv [\,x(y), \overline{z}w\,].P' \\ Q &\equiv [\,x(y), \overline{y}w\,].Q' \end{aligned} \tag{3}$$

where the binding of a variable like y is still to the right, but the commutation $\pi, \pi' \equiv \pi', \pi$ is allowed provided it does not make a free variable bound or vice versa. Thus the comma represents enablement.

Such a step has been suggested by several people. Joachim Parrow, who with David Walker and myself first introduced the π-calculus, mooted the idea some years ago. Samson Abramsky found that proof terms for linear logic could be written in π-calculus more accurately with enablement; Gianluigi Bellin and Philip Scott [5] have pursued this idea further. I am also grateful to Bob Constable for pointing out a natural property of concurrent logic programming which can be loosely paraphrased as follows: A variable becomes accessible for a certain purpose just when it is sufficiently instantiated for that purpose. We may think of instantiation as provision of data; in general terms, the idea of data flow as a form of control is an old and natural one, likely to reappear in different contexts.

We now carry our π-calculus example forward a short way, leading to the operations of action structures. Let us suppose that we extend the π-calculus by allowing a prefix to be any sequence of particles, allowing commutations as described above. We may call this the *synchronous* π-calculus, since it corresponds to the π-calculus just as Synchronous CCS [18] corresponds to CCS. Now recall the multiple prefixes in (3); they are

$$a = [\,x(y), \overline{z}w\,] \text{ and } b = [\,x(y), \overline{y}w\,]\,.$$

It is natural to think of a as the juxtaposition, or *product*, of its parts:

$$a = \pi \otimes \pi_z, \text{ where } \pi = x(y) \text{ and } \pi_z = \overline{z}w \ .$$

On the other hand b is not simply a juxtaposition, because one particle enables the other; so we prefer to think of it as a *composition* of its parts:

$$(?) \qquad b = \pi \cdot \pi_y, \text{ where } \pi_y = \overline{y}w.$$

But this formulation pays little attention to how π binds y in π_y, which is the essence of enablement. The most natural thing, rather than introduce a family of binding operators $x(y)$ –one for each pair of names x and y– is to introduce a singly-indexed *abstraction* operator $\mathbf{ab}_y$. This allows a more accurate expression for b:

$$b = b_1 \cdot b_2, \text{ where } b_1 = \mathbf{in}_x \text{ and } b_2 = \mathbf{ab}_y \pi_y.$$

Here $\mathbf{in}_x$ is an action which inputs a name at port x; then the composition ensures that this single name is bound to y in π_y.

This short exercise leaves some loose ends, but it illustrates the rôle played in an action structure by its three operations: *product*, *composition* and *abstraction*. The dynamical ingredient of action structures, *reaction*, is what allows the particles in the action $[\, x(y), \overline{x}z \,]$ to annihilate one another, leaving in their place the substitution $\{{}^{z}/_{y}\}$, which is itself treated as a degenerate kind of action: an *effect*.

10.2 What π-calculus needs from an action structure A uniform way of building a process structure upon an action structure was developed in Sections 5–8 of Part I. We shall now determine what is needed in an action structure, to recover essentially the polyadic π-calculus defined in [20]. The same will suffice also for the synchronous π-calculus.

There are three process constructions in the polyadic π-calculus which are not among those in Section 5.1; they are

$P \mid Q$	parallel composition
$(\boldsymbol{\nu} x)P$	restriction
$\pi.P$	prefixing

where π may be an input particle $x(\vec{y})$, an output particle $\overline{x}\langle\vec{y}\rangle$ or the τ particle (silent action). The names $\vec{y}$ must be distinct in the case of input since they are binding occurrences.

Parallel composition may be defined in terms of other operations, as was done in [18] to derive CCS from synchronous CCS, as follows:

$$P \mid Q \stackrel{\text{def}}{=} (P \otimes \partial Q) + (\partial P \otimes Q) \,.$$

For the rest, we need only require that the π-calculus action structure [8] contain the following five elementary kinds of action:

τ	$: 0 \to 0$	– the silent action
$\mathbf{out}_x$	$: m \to 0$	– output m names through port x
$\mathbf{in}_x$	$: 0 \to m$	– input m names through port x
$\langle x \rangle$	$: 0 \to 1$	– the name x as a datum
$\boldsymbol{\nu}$	$: 0 \to 1$	– a new name as a datum.

[8] We assume now that the arity monoid in this action structure is the natural numbers under addition, and that each name has arity 1.

We shall use the following two abbreviations (where m is the length of $\vec{y}$):

$$\langle\vec{y}\rangle \stackrel{\text{def}}{=} \langle y_1\rangle \otimes \cdots \otimes \langle y_m\rangle$$
$$(\vec{y})(-) \stackrel{\text{def}}{=} (y_1)\cdots(y_m)(-)\,.$$

We can now define prefixing and restriction upon processes as follows, adopting the device of [18] that prefixing incorporates a delay (∂):

$$\begin{aligned} \tau.P &\stackrel{\text{def}}{=} \boldsymbol{\tau}\cdot\bullet\partial P \\ \overline{x}\langle\vec{y}\rangle.P &\stackrel{\text{def}}{=} \langle\vec{y}\rangle\cdot\mathbf{out}_x\cdot\bullet\partial P \\ x(\vec{y}).P &\stackrel{\text{def}}{=} \mathbf{in}_x\cdot(\vec{y})\bullet\partial P \\ (\boldsymbol{\nu}x)P &\stackrel{\text{def}}{=} \boldsymbol{\nu}\cdot(x)P \end{aligned}$$

where for clarity, just for now, we have written composition $\cdot$ explicitly. What is more, the reaction relation $\searrow$ will be the least for which

$$\mathbf{out}_x \otimes \mathbf{in}_x \searrow \boldsymbol{\tau}\otimes\mathbf{id}_m\,.$$

In fact, for the synchronous π-calculus $\boldsymbol{\tau}$ is not needed; for in that case –as for synchronous CCS in Section 3.5– the reaction rule is just

$$\mathbf{out}_x \otimes \mathbf{in}_x \searrow \mathbf{id}_m\,.$$

10.3 Outline of Part II Section 11 presents the action structure PIC_τ for the original π-calculus, together with its effect structure; it also presents PIC, which is a candidate to underlie the synchronous π-calculus. Section 11 finishes with an illustration of synchronous activity, in accessing compound data structures.

In Section 12 an incident-set for PIC_τ is presented; via the general theory of Sections 7–8, this provides a bisimilarity congruence which is satisfactory for the original π-calculus. An incident-set is also presented for the *monadic* PIC. But –surprisingly– it is seen that the analogous set for the full *polyadic* PIC is *not* an incident-set; there is an anomaly.

Section 13 gives a graphical presentation of PIC. This graphical form is a new way of thinking about process mobility, and supplies intuitions which are not seen so clearly in the syntactic form. But it does more; for it suggests a natural way to overcome the anomaly of the previous section.

Section 14 takes up this suggestion; it defines the *reflexive* π-calculus, in terms of the action structure RPIC, which rests upon a new technical notion of *reflexive substitution.* (This notion has the intriguing virtue that it provides a new characterisation for the ubiquitous restriction operator $\boldsymbol{\nu}$; restriction is just the reflexive substitution of a name for itself!) RPIC seems the best candidate to underlie the synchronous π-calculus; it is actually *simpler* than PIC, since it has a purer graphical form; RPIC also yields a natural and elegant incident-set, thereby providing a bisimilarity congruence. This incident-set, and the proof that it is one, rest on a simple graph-theoretic notion of *reachability*.

11 The action structures PIC and PIC_τ

In the previous section we found that the prefixing and restriction operations of the polyadic π-calculus can be defined in terms of a few more elementary kinds of action. We are now ready to examine action structures generated by these primitives (and one other).

The most natural action structure is actually one which underlies synchronous π-calculus. ("Synchronous" here means that arbitrarily many interactions *may* be synchronized; it does not imply that synchronization is *necessary*.) This action structure, which we shall call PIC, needs no silent τ action. We shall also define PIC_τ, differing from PIC only in the presence of τ and a corresponding adjustment to the reaction rule; by this means the interleaving treatment of actions in the original π-calculus is faithfully recovered.

11.1 Actions In Section 3.4 the action structure SCCS was defined. The body of each action is a multiset over $\Pi \cup \overline{\Pi}$, where Π and $\overline{\Pi} = \{\overline{\pi} \mid \pi \in \Pi\}$ are respectively the positive and negative particles. Reaction consists simply in cancelling a complementary pair $\pi, \overline{\pi}$. SCCS also allows some particles in an action to be restricted.

To get PIC from SCCS, all we need to do is to allow information, i.e. names, to flow between complementary particles in a reaction. It is also convenient to think of a restriction $\boldsymbol{\nu} x$ as a particle. So particles π of PIC are now given by

$$\pi \ ::= \ x(\vec{y}) \ \mid \ \overline{x}\langle\vec{y}\rangle \ \mid \ \boldsymbol{\nu} x .$$

The first two kinds are *input* and *output* (collectively *i/o*) particles, of positive and negative sign respectively. In $x(\vec{y})$ the names $\vec{y}$ must be distinct names, since they are binding. The name x in $\boldsymbol{\nu} x$ is also binding. In PIC_τ, τ is admitted as a fourth kind of particle.

Because of binding, we cannot take the body of an action to be simply a multiset. We wish to allow an action to contain a sequence such as

$$\ldots \ x(y), \ \overline{y}\langle z\rangle \ \ldots$$

representing the receipt of a name y followed by its use as a port. At the process level we may think of the successive reactions of these particles as simultaneous; but we must still admit that the reaction of the second depends upon the receipt of the datum y via the reaction of the first.

We therefore declare the body S of an action to be a *partial sequence*

$$[\pi_1, \ \ldots, \ \pi_n]$$

of particles; that is, a sequence in which we allow the commutation $\pi, \pi' = \pi', \pi$ of any neighbouring pair neither of which binds a name occurring in the other. The general form of an action $a : m \rightarrow n$ is then

$$a \ = \ (\vec{x}) S \langle\vec{y}\rangle,$$

where

- $\vec{x}$ is an m-vector of distinct names, the *imported* names;
- S is a partial sequence of particles, the *body*;
- $\vec{y}$ is an n-vector of names, not necessarily distinct, the *exported* names.

The imported names are bound. The scope of each bound name extends to the right of its binding occurrence, and includes the exported names. We identify actions which only differ by alpha-conversion (change of bound names). We tacitly assume in what follows that by suitable alpha-conversion, a name bound in an action is distinct from any other name in that action or in any other action under discussion.

As examples, we immediately define the kinds of action (five for PIC, six for PIC_τ) in terms of which all others may be expressed. They are all listed in Section 10, except ω.

$$\begin{array}{rrcll} & \mathbf{out}_x & : m \to 0 & \stackrel{\text{def}}{=} & (\vec{y})[\,\overline{x}\langle\vec{y}\rangle\,]\langle\,\rangle \qquad (|\vec{y}| = m) \\ & \mathbf{in}_x & : 0 \to m & \stackrel{\text{def}}{=} & (\,)[\,x(\vec{y})\,]\langle\vec{y}\rangle \qquad (|\vec{y}| = m) \\ & \langle x\rangle & : 0 \to 1 & \stackrel{\text{def}}{=} & (\,)[\;]\langle x\rangle \\ & \boldsymbol{\nu} & : 0 \to 1 & \stackrel{\text{def}}{=} & (\,)[\,\boldsymbol{\nu}x\,]\langle x\rangle \\ & \omega & : 1 \to 0 & \stackrel{\text{def}}{=} & (x)[\;]\langle\,\rangle \\ (\mathrm{PIC}_\tau \text{ only}) & \tau & : 0 \to 0 & \stackrel{\text{def}}{=} & (\,)[\,\tau\,]\langle\,\rangle\,. \end{array}$$

We shall often allow ourselves to omit any empty component of an action; e.g. we may write $(\vec{y})[\,\overline{x}\langle\vec{y}\rangle\,]$ for $(\vec{y})[\,\overline{x}\langle\vec{y}\rangle\,]\langle\,\rangle$. We shall also write $\boldsymbol{\nu}\vec{x}$ for a sequence $\boldsymbol{\nu}x_1,\ \boldsymbol{\nu}x_2, \cdots$ of particles.

11.2 Operations The action structure operations are defined identically for PIC and PIC_τ. First, the identities are given by

$$\mathbf{id}_m \stackrel{\text{def}}{=} (\vec{x})\langle\vec{x}\rangle \qquad (|\vec{x}| = m).$$

To define the other operations we use substitutions like $\sigma = \{\vec{u}/\vec{v}\}$, where $\vec{v}$ is a vector of distinct names; $\sigma(-)$ is the result of simultaneously replacing each free occurrence of u_i in $(-)$ by the corresponding v_i, first alpha-converting $(-)$ to change any bound uses of $\vec{v}$. We shall also use a comma to concatenate partial sequences. Suppose $a = (\vec{u})S\langle\vec{v}\rangle$ and $b = (\vec{x})T\langle\vec{y}\rangle$; then composition, product and abstraction are given as follows:

$$\begin{array}{rcll} a \cdot b & \stackrel{\text{def}}{=} & (\vec{u})S, \sigma T\langle\sigma\vec{y}\rangle & \qquad (\sigma = \{\vec{v}/\vec{x}\}) \\ a \otimes b & \stackrel{\text{def}}{=} & (\vec{u}\vec{x})S, T\langle\vec{v}\vec{y}\rangle & \\ \mathbf{ab}_x a & \stackrel{\text{def}}{=} & (x\vec{u})S\langle x\vec{v}\rangle. & \end{array}$$

Recall our convention that a bound name differs from all names used outside its scope. Thus $a \otimes b$ is simply the juxtaposition of the two actions; no order is dictated by the concatenation, since the convention ensures that $S, T = T, S$ in this case. But in $a \cdot b$ the substitution σ may replace some names free in T by names bound in S, and then $S, \sigma T \neq \sigma T, S$.

It is a routine matter to verify the action structure axioms. Note in particular that the export of x in $\mathbf{ab}_x a$ ensures that $\mathbf{ab}_x$ is functorial.

Exercise Show that ω cannot be expressed in terms of the other basic actions and the above-defined operations. ■

For future convenience we define the following:

$$\begin{array}{rcll} (x)a & \stackrel{\text{def}}{=} & (x\vec{u})S\langle\vec{v}\rangle & \qquad (\text{where } a = (\vec{u})S\langle\vec{v}\rangle,\ x \notin \vec{u}) \\ (\boldsymbol{\nu}x)a & \stackrel{\text{def}}{=} & \boldsymbol{\nu} \cdot (x)a & \end{array}$$

11.3 Dynamics The atomic reaction relation $\searrow^1$ of PIC is defined as follows. Whenever a takes the form

$$(\vec{x})\, S, [\,\overline{u}\langle\vec{v}\rangle,\ u(\vec{w})\,], T\,\langle\vec{y}\rangle$$

(after suitable commutations), then

$$a \searrow^1 (\vec{x})\, S, \sigma T\, \langle \sigma\vec{y} \rangle,$$

where σ is the substitution $\{\vec{v}/\vec{w}\}$. We call $\overline{u}\langle\vec{v}\rangle, u(\vec{w})$ a *redex* of a.

Example Let $a = (x)[\,\overline{u}\langle x\rangle,\ u(w),\ \overline{w}\langle y\rangle,\ x(z)\,]\langle wz\rangle$. Then

$$\begin{aligned} a \ &\searrow^1\ (x)[\,\overline{x}\langle y\rangle,\ x(z)\,]\langle xz\rangle \\ &\searrow^1\ (x)\langle xy\rangle. \end{aligned}$$

■

The port of a redex (in the example, first u then x) may be free or bound. Note that $\searrow^1$ is indeed preserved by the operations; for example $a \searrow^1 a'$ implies $b \cdot a \searrow^1 b \cdot a'$. The reaction relation $\searrow$ is defined to be $(\searrow^1)^*$, the transitive reflexive closure of the atomic relation.

The only difference in PIC_τ is that a redex is not just deleted, but replaced by τ. So in PIC_τ the above example becomes

$$a \searrow^1 \cdots \searrow^1 (x)[\,\tau,\ \tau\,]\langle xy\rangle\ .$$

Exercise If e is inactive, or even inert, $e \cdot a$ may still have reactions not arising solely from a. Consider the reactions of $\langle u\rangle \cdot a$ in the above example. ■

11.4 Discussion In the original π-calculus every prefix, and therefore every action, consisted of just a single particle. By introducing the action structure PIC or PIC_τ, have we not gone much further than we need to capture the behaviour of the π-calculus?

In fact these action structures reflect the real performance of distributed systems rather more closely than does the original formulation of π-calculus. For the transition system of the latter explicitly prevents more than a single interaction to occur in a single transition, and to implement this would actually require a global scheduler – something which is absent in a distributed system by definition! But the π-calculus was designed to be a tractable theoretical tool, not to be a faithful model of implementation; its interleaving of actions is what makes the expansion law of the π-calculus [24] easy to use, since the number of terms in the expansion does not explode exponentially. The same point was made for CCS in Section 3.6. Moreover, the interleaving was justified at least for CCS [18] (and the same should hold for π-calculus) by a theorem that at the semantic level of observation equivalence it makes no difference whether single observations are treated as simultaneous or interleaved.

In Section 7.4 an appropriate incident-set was defined to recapture the interleaving of actions for CCS; later we shall similarly use PIC_τ to recapture interleaving in the π-calculus.

11.5 Effects and transitions What should be the effect of performing an action in PIC or PIC_τ, i.e. what effect structure should we adopt? There is no single answer; to simplify the decision, we shall consider only actions whose body contains no i/o or τ particles (so they are certainly inert), i.e. actions of the form

$$(\vec{x})[\,\nu\vec{u}\,]\langle\vec{y}\rangle.$$

Into this class fall the following: $\langle x\rangle$, a free datum; $(x)\langle xx\rangle$, which forms a link by identifying two names; and $[\,\nu x\,]\langle xx\rangle$, which forms a private link. These are important cases; we shall see the last illustrated in Section 11.6. But the class contains some undesirables; one such is ω, which discards a name. It is undesirable because discarding can be made into a postcomponent of *any* action; for example if $a : 1 \to 1$ then we can make $(xy)\langle x\rangle$ (which discards its second argument) into an effect of a by writing

$$a = a \cdot (x)\langle xx\rangle \cdot (xy)\langle x\rangle.$$

To exclude the discarding of names, we propose that the effect structure E for PIC and PIC$_\tau$ consist of all actions

$$(\vec{x})[\,\nu\vec{u}\,]\langle\vec{y}\rangle, \qquad \vec{x} \cup \vec{u} \subseteq \vec{y}.$$

It is routine to verify that E is indeed an effect structure. Furthermore, every action has a pure decomposition for E. (See Definition 4.7.) In fact the pure actions P are those actions $(\vec{x})S\langle\vec{y}\rangle$ which satisfy two conditions:

- $\vec{y}$ contains no free names, and no repetitions;
- every restricted name in S occurs in at least one i/o particle.

Exercise Show that every a in PIC or PIC$_\tau$ can be expressed as $a = p \cdot e$, where $p \in P$ and $e \in E$. Show further that this is unique up to isomorphism; that is, if also $a = p' \cdot e'$ ($p' \in P$ and $e' \in E$) then there exists $i = (\vec{u})\langle\vec{v}\rangle$, where $\vec{v}$ is a permutation of $\vec{u}$, making the following diagram commute:

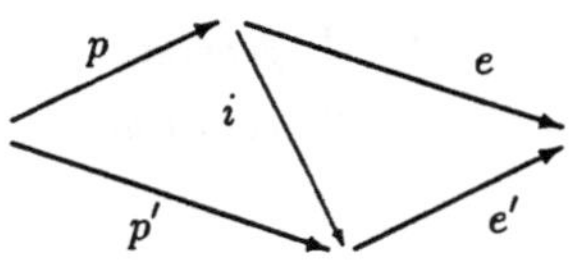

This makes (P, E) into a Freyd-Kelly factorization in PIC and in PIC$_\tau$ [10], considered as categories. ■

Transitions were introduced in Section 6.11. Recall from there that if P has a commitment $P \searrow a \bullet P'$, then any decomposition (b, e) of a for E can be used to provide a transition

$$P \xrightarrow{\ b\ } eP'.$$

In the next section we focus on the case when b is a special kind of action known as an *incident*, as defined in Section 7. In such actions, all internal reaction which could not affect external interaction has been carried out.

11.6 Illustration Let us now illustrate the power of composite action in PIC, when extended by process structure as described in Section 5. We shall use an encoding of data structures as processes, close to that described in [20]. Specifically, we consider lists. A single list-cell located at ℓ, containing a value-name v and pointing to a successor cell located at ℓ', may be represented by the action

$$[\,\ell(cn),\ \bar{c}\langle v\ell'\rangle\,] : 0 \to 0.$$

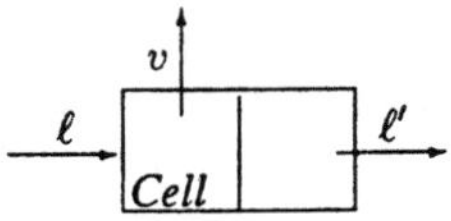

We shall represent it by the process *Cell* of arity 3, parametric upon ℓ, v and ℓ':

$$Cell \stackrel{\text{def}}{=} (\ell v \ell')\,\partial[\,\ell(cn),\ \overline{c}\langle v\ell'\rangle\,]\,\boldsymbol{.}\,\mathbf{0}.$$

Similarly, the empty list-cell *Nil* of arity 1, parametric upon its location, is given by

$$Nil \stackrel{\text{def}}{=} (\ell)\,\partial[\,\ell(cn),\ \overline{n}\langle\,\rangle\,]\,\boldsymbol{.}\,\mathbf{0}.$$

For the present discussion we shall understand a *value*, i.e. a list-member, to be a process *Val* of arity 1, of the form $(v)\,!V$; it is parametric upon its location, and the replication allows its repeated "use". Thus $\langle v\rangle \cdot Val$ is a value located at v. So what is the list containing k values $Val_1, \ldots, Val_k$? It is itself a value $L : 1 \to 0$, given by

$$\begin{aligned} L \;=\; (\ell_0)\,!(\nu v_1 \ell_1 \cdots v_k \ell_k)\Big(\, &\langle \ell_0 v_1 \ell_1\rangle Cell \otimes \cdots \otimes \langle \ell_{k-1} v_k \ell_k\rangle Cell \otimes \langle \ell_k\rangle Nil \\ &\otimes \langle v_1\rangle Val_1 \otimes \cdots \otimes \langle v_k\rangle Val_k \Big). \end{aligned}$$

Note how each value Val_i has been located for inclusion in the list.

Let us define the single actions *head* and *tail*, which get hold of the first element and the remainder of an arbitrary list value. They are

$$\begin{aligned} head : 1 \to 1 &\stackrel{\text{def}}{=} get\,first \\ tail : 1 \to 1 &\stackrel{\text{def}}{=} get\,second\,, \end{aligned}$$

where

$$\begin{aligned} get : 1 \to 2 &\stackrel{\text{def}}{=} (\ell)[\,\nu cn,\ \overline{\ell}\langle cn\rangle,\ c(v\ell')\,]\langle v\ell'\rangle \\ first : 2 \to 1 &\stackrel{\text{def}}{=} (v\ell)\langle v\rangle \\ second : 2 \to 1 &\stackrel{\text{def}}{=} (v\ell)\langle \ell\rangle\,. \end{aligned}$$

Now let us see how to define processes, given these definitions. Suppose that we have L located at ℓ_0, and we wish to provide its first member as parameter to a process $P : 1$. This can be done by prefixing the action $\langle l_0\rangle head$ to P; then in a single transition this action gains access to the head of L, as follows:

$$\langle \ell_0\rangle head\,\boldsymbol{.}\,P \otimes \langle \ell_0\rangle L \xrightarrow{\;b\;} (\nu v \cdots)\Big(\langle v\rangle P \otimes \langle v\rangle Val_1 \otimes \langle \ell_0\rangle L\ \cdots\Big).$$

We see that P has indeed "received" v, the location of L's first member Val_1. Thanks to replication, the located list is still available for further access. The action b will normally be an incident (see Section 5); the left-hand side performs a commitment, and the action of this commitment decomposes into (b, e), where the effect e in this case produces the private link v between P and the value Val_1.

Steps down the list L need not be performed one-by-one; the following single transition provides access to the fourteenth member of L, where $tail^n$ means the composition of n *tails*:

$$\langle \ell_0\rangle tail^{13} head\,\boldsymbol{.}\,P \otimes \langle \ell_0\rangle L \xrightarrow{\;b'\;} (\nu v \cdots)\Big(\langle v\rangle P \otimes \langle v\rangle Val_{14} \otimes \langle \ell_0\rangle L\ \cdots\Big).$$

Let us briefly consider this illustration in broader terms. It naturally prompts the question of why it should be necessary to distinguish processes from actions. Why should we not add the control operations $\bullet$, $\mathbf{0}, +, \partial$ and ! to action structures, combining the notions of process structure and action structure?

One answer is that by distinguishing these notions we get a treatment of processes closer to existing process calculi. However, it may be that process calculi themselves would be enriched by treating them as action structures. The second answer is that there are some interesting technical problems to resolve before the extension of action structures (by adding control operations) is well understood. Work in this direction will be reported elsewhere.

12 Incident-sets for π-calculus

In this section we provide incident-sets, as defined in Section 7, both for PIC_τ and for the monadic version of PIC.

12.1 The nature of incidents In many action structures the reaction relation $\searrow$ is confluent. When that occurs, it seems that the interesting incident-sets contain only inactive actions. All the examples of action structures treated in Section 3 were confluent except that for Petri nets, for which an incident-set was not defined. In particular, readers may check that the incident-set given there for CCS contains only inactive actions. (It is also worth noting that it is not a sub-actionstructure; the product of two incidents may indeed be active.)

Now PIC is not confluent, as can be easily seen as follows. Let $a = [\,\overline{x}\langle w\rangle,\ \overline{x}\langle z\rangle,\ x(y)\,]$; then a has two distinct one-step reactions neither of which can be continued:

$$a \searrow [\,\overline{x}\langle w\rangle\,] \not\searrow$$
$$a \searrow [\,\overline{x}\langle z\rangle\,] \not\searrow$$

Indeed we can use this to show that *any* non-empty incident-set for PIC must contain active members. For let K be such an incident-set, and suppose it contains the one-particle action $\ell = [\,\overline{x}\langle z\rangle\,]$. (The exercise below shows that indeed if $K \neq \emptyset$ then $\ell \in K$.) Let $a = [\,\overline{x}\langle z\rangle,\ x(y)\,]$; a is active, and we show $a \in K$. Let φ be the preaction $[\,\overline{x}\langle w\rangle\,] \otimes (-)$. Then $\varphi(a) \searrow \ell \in K$; so there exists $k \in K$ for which $a \searrow k$ and $\varphi(k) \searrow \ell$. But the only action $a' \neq a$ such that $a \searrow a'$ is the unit action $\mathbf{id}_0$, and clearly $\varphi(\mathbf{id}_0) \not\searrow \ell$. Hence $a \in K$, as we claimed.

Exercise If K is a non-empty incident-set for PIC, show that $[\,\overline{x}\langle z\rangle\,] \in K$. (*Hint*: Let k be any member of K; consider the preaction $k \otimes [\,x(y)\,] \otimes (-)$. ∎

A main aim of this paper is to give interesting and natural incident-sets for PIC. First we shall tackle the easier task of finding a very modest incident-set for PIC_τ.

12.2 Restricted particles and incidents Recall that in a PIC_τ reaction the redex is not merely cancelled, but replaced by τ.

Now we define the *weight* $|a| = (p, n)$ of an action a exactly as for CCS (Section 7.4); p is the number of positive particles, n is the number of negative particles, restriction particles νx do not count, and a τ particle is positive *and* negative, so counts towards both p and n. We see that $|a \cdot b| = |a \otimes b| = |a| + |b|$ and $|\mathbf{ab}_x a| = |a|$. Also, reaction preserves weight; this is the important difference between PIC_τ and PIC, and is what allows PIC_τ to have a rather small incident-set, as we now see.

Call an i/o particle in a *restricted* if its port is restricted, *free* otherwise. Define the set K_τ^{PIC} to contain every action in PIC_τ with weight at most $(1,1)$ and no restricted i/o particles. We can now establish

12.3 Theorem (incidents for PIC_τ) K_τ^{PIC} is an incident-set for PIC_τ.

Proof It is easy to verify the first clause of Definition 7.1. To verify the second clause, it is enough to verify it for the four basic forms of preaction: $c\cdot(-)$, $c\otimes(-)$, $(-)\otimes c$ and $\mathbf{ab}_x(-)$.

Suppose then that $\varphi(a)\searrow\ell \in K_\tau^{\mathsf{PIC}}$, where φ takes one of these four forms. Note that this reaction takes at most one step. We must find $k \in K_\tau^{\mathsf{PIC}}$ such that $a\searrow k$ and $\varphi(k)\searrow\ell$. We know $|a| \le |\ell| \le (1,1)$, so we consider the possible cases for $|a|$.

If $|a| = (0,0)$ then already $a \in K_\tau^{\mathsf{PIC}}$ and we are done. If $|a| = (1,0)$ or $(0,1)$ then a has a single i/o particle. If it is free then $a \in K_\tau^{\mathsf{PIC}}$ and we are done. If not, then $\varphi(a)$ also has this restricted particle, and no other i/o particle with the same port; but then $\varphi(a) \notin K_\tau^{\mathsf{PIC}}$, and a one-step reaction cannot reduce it to a member of K_τ^{PIC}, contrary to hypothesis.

If $|a| = (1,1)$ and $a \notin K_\tau^{\mathsf{PIC}}$ (else we are done) there are two cases:

- a has one free and one restricted particle. Then $\varphi(a)$ also has this restricted particle, and we proceed as in the previous case.
- a has two restricted i/o particles. Then $\varphi(a)$ has the same, and they must form a redex which is reduced in the reaction $\varphi(a)\searrow\ell$. But this reduction can also be done in a, yielding $k \in K_\tau^{\mathsf{PIC}}$ (with a single τ particle) for which $\varphi(k)\searrow\ell$. ■

This argument would fail if φ could take the form of a postcomposition $(-)\cdot c$; hence the asymmetry in the notion of preaction is important.

We may ask whether this incident-set has special status among those possible for PIC_τ. It was shown in Theorem 7.7 that under certain conditions (satisfied by PIC_τ) there is a smallest incident-set which contains certain given actions. Is K_τ^{PIC} the smallest containing τ? For the full PIC_τ, it is not quite, but nearly so; the discrepancy has to do with i/o particles which carry an empty vector of names. But the assiduous reader may enjoy the following

Exercise Show that in the monadic PIC_τ (where every i/o particle carries a single datum name) K_τ^{PIC} as defined above is the smallest incident-set. *Hint*: By similar methods to the previous exercise, first show that if $\tau \in K$ then every action of weight 1 is in K; then proceed to consider those of weight 2. ■

How does the bisimilarity induced by K_τ^{PIC} relate to the congruences defined in terms of bisimilarity for the original π-calculus [24]? It turns out to be a little stronger; but it is also more natural in one sense. For it is obtained by a direct definition, whereas the original bisimilarities were not quite congruential; the congruences were defined as "bisimilarity under all substitutions". The relative merits of these congruences remain to be studied; one also hopes that action structures can throw some light upon the structure of the family of strong bisimilarities, not just for π-calculus but more generally.

12.4 Reachability and incidents We now turn to the more challenging question of incident-sets for PIC. Here, the absence of τ means that we no longer have a way to bound the weight of actions in an incident. Instead we rely on another intuition, to reduce the class of possible incidents. Suppose a is in the context of some preaction φ and we are interested in the performance of $\varphi(a)$, e.g. its multistep reduction to an incident. Now a may contribute in two ways to this reduction; it may react internally, or it may interact with φ. Looking at a in isolation, it has two kinds of internal reaction; those which may preclude (or be precluded by) a's possible interactions with its environment, and those which neither preclude nor are precluded by such external interactions.

For the latter kind of internal reaction, nothing (i.e. no external interaction) is lost by letting them be performed earlier rather than later. By confining our attention (in forming an incident-set) to actions which have carried out *all* such internal reactions, we limit the size of the incident-set without endangering the essential second clause of Definition 7.1. A further consideration is that an incident should contain no particle which cannot eventually take part in some reaction.

We are thus led to the notion of *reachability*. In defining it, we assume as usual that all bound names are distinct from each other and from all names occurring free.

Definition The set of names *reachable* in a given action $a \in$ PIC is the smallest such that

- all the imported and free names of a are reachable;
- if z is reachable and $z(\vec{w})$ or $\overline{z}\langle\vec{w}\rangle$ occurs in a then the names $\vec{w}$ are reachable.

An i/o particle is *reachable* in a if its port is reachable; an action is *reachable* if all its i/o particles are reachable. ∎

Example Let

$$a = (x)[\,\nu u,\ x(y),\ u(v),\ \overline{u}\langle y\rangle,\ \overline{v}\langle z\rangle\,]\langle yu\rangle;$$

then only the names x, y and the particle $x(y)$ are reachable. Note that a contains a redex $u(v)\,\overline{u}\langle y\rangle$ which has an unreachable port. If we reduce it we have

$$a \searrow^{1} (x)[\,\nu u,\ x(y),\ \overline{y}\langle z\rangle\,]\langle yu\rangle;$$

the reaction has created a reachable particle $\overline{y}\langle z\rangle$ from one which was previously unreachable, and has reduced the unreachable action a to a reachable one. ∎

Surprisingly enough, the status of reachable actions differs according to whether we are considering full PIC, or only monadic PIC. We can prove the following:

12.5 Theorem (incidents for monadic PIC) The set K^{PIC} of reachable actions is an incident-set in monadic PIC. ∎

The proof of the theorem follows similar lines to that of Theorem 14.4, concerning an incident-set for the reflexive π-calculus. I conjecture, but have not proved, that K^{PIC} is the smallest incident-set in monadic PIC which includes the effects.

In Section 12.6 below we see why K^{PIC} is *not* an incident-set for full PIC. This anomaly arises from a phenomenon of the binding convention in PIC. It may seem puzzling, if we recall from [20] that the polyadic π-calculus is definable from the monadic calculus. But that definition relies upon the prefixing construction of the π-calculus, and prefixing is absent in PIC! In Section 14 below we enrich PIC to remove the anomaly.

Some insight into incident-sets can be gained from the following exercise, which shows clearly how the notion of incident separates proper from improper transitions.

Exercise Consider again the list-processing illustration in Section 11.6. Consider the commitments of $(\nu\ell_0)(\langle\ell_0\rangle head \bullet P \otimes \langle\ell_0\rangle L)$, where now the access to the list is restricted. Show that there is a large number of possible commitments of the form

$$(\nu\ell_0)\big(\langle\ell_0\rangle head \bullet P \otimes \langle\ell_0\rangle L\big) \searrow a \bullet Q,$$

but only one in which a an incident. Show that this incident decomposes into two actions; the incident $[\,\nu cn\,]$ which discards some "used" names, and the effect

$$[\,\nu\ell_0,\ \nu v_1\ell_1 \cdots v_k\ell_k\,]\langle \ell_0 v_1 v_1 \ell_1 \cdots v_k \ell_k\rangle$$

which transmits private names to Q. ∎

12.6 Anomaly In PIC, though not in monadic PIC, the interplay between binding and reaction has a surprising consequence which can be illustrated by a simple example. Let

$$b \;=\; [\,v(x),\ \underline{\overline{u}\langle xy\rangle,\ u(zw)},\ \overline{v}\langle w\rangle\,]\langle x\rangle,$$

or by rearrangement

$$b \;=\; [\,u(zw),\ \underline{\overline{v}\langle w\rangle,\ v(x)},\ \overline{u}\langle xy\rangle\,]\langle x\rangle,$$

where the two redexes in b are shown by underlining. These two redexes are disjoint (i.e. they involve distinct particles), so one may expect that they can be reduced in arbitrary order. But here are the two reaction sequences for b:

$$b \;\searrow\; [\,\underline{v(x),\ \overline{v}\langle y\rangle}\,]\langle x\rangle \searrow \langle y\rangle$$
$$b \;\searrow\; [\,u(zw),\ \overline{u}\langle wy\rangle\,]\langle w\rangle\ \not\searrow^{1}.$$

The second sequence can go no further! There is no redex, due to PIC's binding convention. Yet if we allowed a reaction, then the result should again be $\{y/w\}\langle w\rangle = \langle y\rangle$. In forbidding such redexes, PIC denies certain reactions which are intuitively acceptable.[9]

Our example can now be used to show that K^{PIC} is indeed not an incident-set for full PIC. Let c be a slight variant of b, as follows:

$$c \;=\; [\,\nu v,\ v(x),\ \underline{\overline{u}\langle xy\rangle,\ u'(zw)},\ \overline{v}\langle w\rangle\,]\ .$$

Now let φ be the preaction $\langle u\rangle \mathbf{ab}_{u'}(-)$; then (as in the first reaction sequence for b)

$$\varphi(c) \searrow [\,\nu v\,]\langle u\rangle\ .$$

Since $[\,\nu v\,]\langle u\rangle \in K^{\mathrm{PIC}}$, we look for $k \in K^{\mathrm{PIC}}$ such that

$$c \searrow k \ \text{ and }\ \varphi(k) \searrow [\,\nu v\,]\langle u\rangle.$$

But $c \notin K^{\mathrm{PIC}}$, and its only reaction is

$$c \searrow c'\ \not\searrow^{1},\quad \text{where } c' \;=\; [\,\nu v,\ u'(zw),\ \overline{u}\langle wy\rangle\,]\ ;$$

furthermore $\varphi(c') = [\,\nu v,\ u(zw),\ \overline{u}\langle wy\rangle\,]\langle u\rangle$ has no redex, so cannot reduce to $[\,\nu v\,]\langle u\rangle$. So K^{PIC} is not an incident-set for full PIC.

This anomaly of resolved in Section 14.2 by the notion of *reflexive substitution.*

---

[9] Even in monadic PIC, an action may have two disjoint redexes, but after one is reduced the other is no longer a redex; consider $[\,v(x), \overline{u}\langle x\rangle, u(y), \overline{v}\langle y\rangle\,]$. But in monadic PIC it turns out that if both can be reduced in *one* order, then both can be reduced in the other order. This is what lies behind Theorem 12.5, and fails for polyadic PIC.

13 A graphical action structure

The action structure PIC has been presented in a very syntactic way. This helps formal manipulation, but some essential aspects are still hidden. Here we show that both the statics and the dynamics of PIC correspond to simple graphical ideas. This graphical presentation also helps us to understand the notions of effect and incident. Further, it suggests an enrichment of PIC which has appealing properties.

13.1 Bare π-nets We begin with a simple kind of graph called a *bare π-net.* It has *nodes* and *directed arcs.* Each node is drawn as a pointed oval with a *head,* a *waist* and a *tail* as shown.

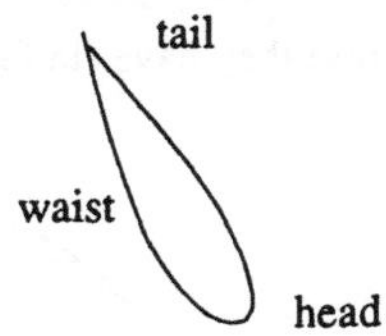

Each directed arc is either a *source* or a *target* arc. We first consider *monadic* arcs. A monadic source arc is drawn from the head of a node (the *source* of the arc) to the waist of a node (the *port*); a monadic target arc is drawn from the waist of a node (the *port* of the arc) to the tail of a node (the *target*).

In general, an arc may be *polyadic*; a source arc may have an ordered sequence of k source nodes, and a target arc may have an ordered sequence of k distinct target nodes ($k \geq 0$):

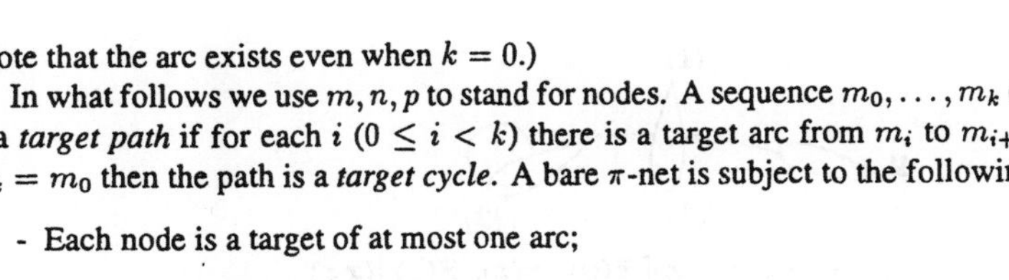

(Note that the arc exists even when $k = 0$.)

In what follows we use m, n, p to stand for nodes. A sequence $m_0, \ldots, m_k$ of nodes ($k \geq 0$) is a *target path* if for each i ($0 \leq i < k$) there is a target arc from m_i to m_{i+1}. If $k > 0$ and $m_k = m_0$ then the path is a *target cycle.* A bare π-net is subject to the following constraints:

- Each node is a target of at most one arc;
- There are no target cycles.

We say a (bare) π-net is *monadic* if all its arcs are monadic. We shall use only monadic π-nets in our illustrations, but the work of this section and the next applies to all π-nets.

Here are two admissible bare monadic π-nets:

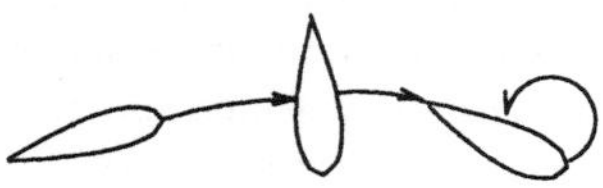

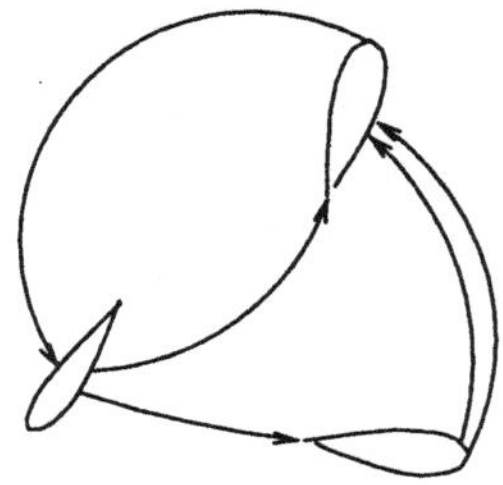

Here are two which are inadmissible, because they have (in fact they are) target cycles:

If a node m is the port of a target arc then we say it *carries* the arc and its target nodes; similarly, if m is the port of a source arc then it carries the arc and its source nodes.

13.2 Actions and π-nets We now define π-nets, and show how to represent the actions of PIC as π-nets.

Suppose that Z is the set of names free in an action $a : h \rightarrow k$ of PIC. Then the π-net representing a consists of a certain bare π-net together with

- an injective map from $Z \cup \{1, \ldots, h\}$ into the non-target nodes;
- a map (not necessarily injective) from $\{1, \ldots, k\}$ into the nodes.

Here are two π-nets, obtained by adding appropriate maps to our two admissible bare π-nets. The first map is indicated by *tail-labels* such as $\boldsymbol{y}$ or **(1)**, the second by *head-labels* such as $\langle \mathbf{1} \rangle$. Note that our definition ensures that a node may not bear a tail-label if it is a target of an arc.

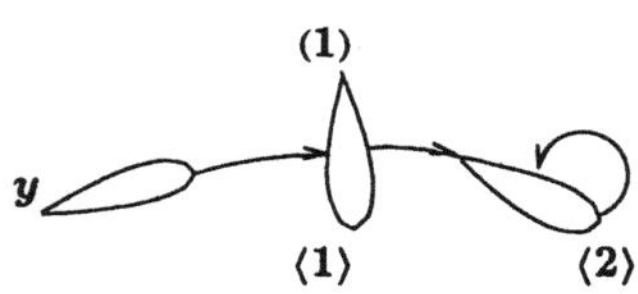

$$\boldsymbol{a : 1 \rightarrow 2 \; = \; (x)[\, \overline{x}\langle y \rangle,\ x(z),\ \overline{z}\langle z \rangle \,]\langle xz \rangle}$$

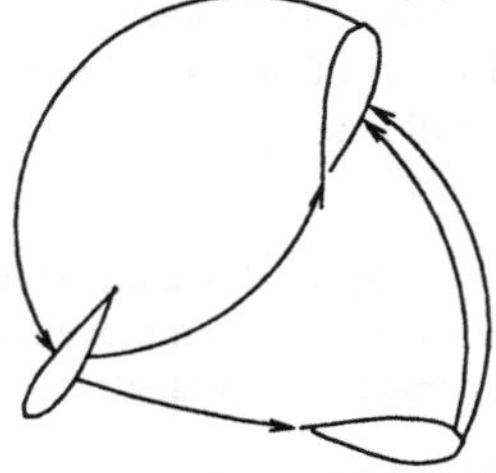

$$b : 0 \rightarrow 3 \; = \; [\nu x,\; x(y),\; \overline{x}\langle y\rangle,\; x(z),\; \overline{y}\langle z\rangle,\; \overline{y}\langle z\rangle\,]\langle yzz\rangle$$

For each of these π-nets we have shown the PIC action which it stands for. To translate an arbitrary PIC action a to a π-net is quite simple. Let $a = (\vec{x})S\langle\vec{y}\rangle : h \rightarrow k$. Then:

- create a node $\boldsymbol{z}$ (i.e. having this tail-label) for each name z occurring free or bound in a;
- for $1 \leq j \leq k$ give node $\boldsymbol{y}_j$ the head-label $\langle \boldsymbol{j}\rangle$;
- for each particle $\overline{u}\langle\vec{v}\rangle$ in S draw a source arc with source nodes $\vec{\boldsymbol{v}}$ and port node $\boldsymbol{u}$;
- for each particle $u(\vec{v})$ in S draw a target arc with port node $\boldsymbol{u}$ and target nodes $\vec{\boldsymbol{v}}$, removing the latter tail-labels;
- for each particle νu in S remove the tail-label of node $\boldsymbol{u}$;
- for $1 \leq i \leq h$, replace the tail-label of node $\boldsymbol{x}_i$ by $\langle \boldsymbol{i}\rangle$.

The result, which we call *draw*(a), is clearly a π-net on the names free in a. In particular it has no target cycles because a contains no cycle of input particles each binding the next.

It is also easy to define an inverse translation

$$\textit{write} : \pi\text{-nets} \rightarrow \mathsf{PIC};$$

to form *write*(N), first introduce a new name z for every unnamed node in N; then construct the particles – one i/o particle for each arc, and one restriction particle for each non-target node with no tail-label. We omit the remaining details, since they are obvious. The absence of target cycles ensures that the particles in *write*(N) can be written in an appropriate linear order.

Now it is easy to show that *write*(*draw*$(a)) = a$. It is nearly true that *draw*(*write*$(N)) = N$; the only discrepancy is when N contains a node which has a name $\boldsymbol{z}$ (say) as tail-label, but no head-label and no incoming or outgoing arcs. Let us call such a node *vacuous*. A vacuous node has no significance; it affects neither the algebraic operations nor the dynamics (both defined below). So we shall not distinguish two π-nets differing only in their vacuous nodes. Then indeed we have a bijection

$$\mathsf{PIC} \;\underset{\overset{\longleftarrow}{\textit{write}}}{\overset{\textit{draw}}{\longrightarrow}}\; \pi\text{-nets}\,.$$

13.3 Operations and dynamics The operations of composition, product and abstraction on π-nets are simply described. The first two involve the *coalescing* of two nodes; this means forming a single node bearing the arcs and labels of both the given nodes. Whenever this is done, it is ensured that the resulting π-net is still legal.

Let $M_1 : h_1 \rightarrow k_1$ and $M_2 : h_2 \rightarrow k_2$ be π-nets.

- To form $M_1 \cdot M_2$, for each i ($1 \le i \le k_1 = h_2$) coalesce the node in M_1 whose head-label is $\langle i \rangle$ with the node in M_2 whose tail-label is (i), and remove these labels. Also coalesce nodes in M_1 and in M_2 which share a name z as tail-label.
- To form $M_1 \otimes M_2$, first increment all tail-labels (i) and head-labels $\langle j \rangle$ in M_2 by h_1 and k_1 respectively. Also, as above, coalesce any pair of nodes with the same tail-label z.
- To form $\mathbf{ab}_x M_1$, first increment all tail-labels (i) and head-labels $\langle j \rangle$ in M_1 by 1. Then assign head-label $\langle 1 \rangle$ to the node with tail-label x, and replace its tail-label by (1).

We now define reaction. In the monadic case a *redex* of M consists of a configuration

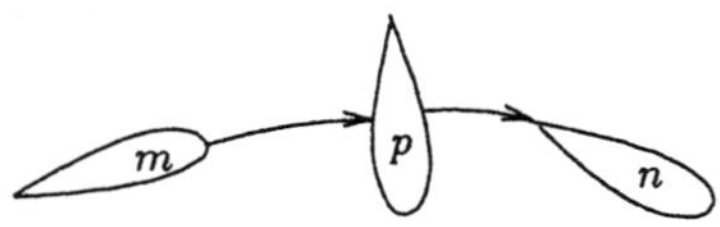

where there is no target path from n to m. (This implies $m \neq n$.) Then there is a one-step reduction $M \searrow^1 M'$, where M' is derived from M just by removing the two arcs shown and coalescing m with n. We call m, n and p respectively the *source*, the *target* and the *port* of the redex. The reaction relation $\searrow$ π-nets is defined to be the transitive reflexive closure of $\searrow^1$. Note that reaction does not depend at all upon labelling; in effect we have defined it for bare π-nets.

Here is an illustration of a two-step reaction:

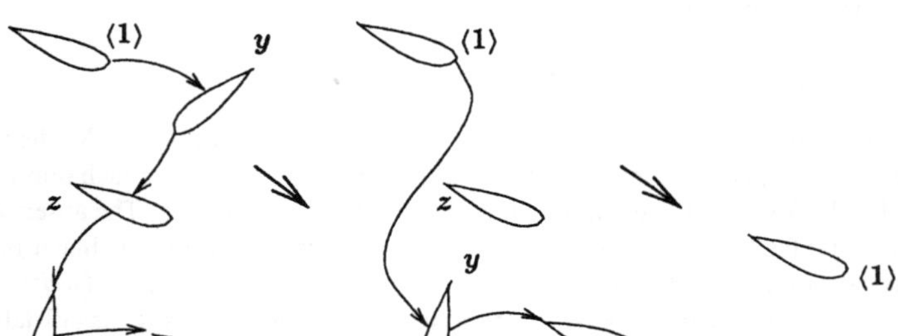

$$[\nu x,\ \overline{y}\langle x\rangle,\ \overline{z}\langle y\rangle,\ z(u),\ u(v)]\langle x\rangle \searrow [\nu x,\ \overline{y}\langle x\rangle,\ y(v)]\langle x\rangle \searrow [\nu x]\langle x\rangle$$

In the polyadic case a redex is of the following form, with no target path from any n_j to any m_i:

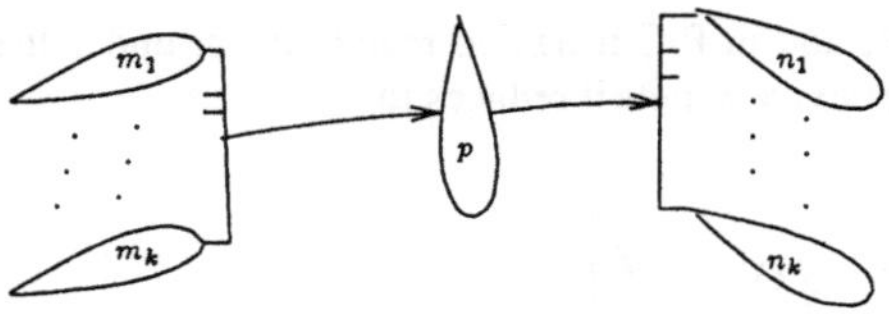

The reduction again consists of removing the two arcs and coalescing m_i with n_i, for each i. (Note that reduction is not vacuous when $k = 0$, since the arcs are removed.)

This completes the definition of π-nets as an action structure. It is now a routine matter to check that the bijection consisting of the two translations *draw* and *write* is actually an isomorphism of action structures; that is, it preserves both the operations and the dynamics in both directions. Thus π-nets are an alternative presentation of PIC.

Recall that the effect structure of PIC consists of all actions of the form $(\vec{x})[\,\nu\vec{y}\,]\langle\vec{z}\rangle$ where $\vec{y} \cup \vec{z} \subseteq \vec{x}$. In terms of π-nets, these are just arc-free nets where every node has a head-label.

13.4 Reflexive π-nets Having seen that the action structure of π-nets is isomorphic to PIC, one naturally asks what is gained by this new presentation. One answer is that it gives a less formal, more geometric, view of reaction. We may compare it with the geometry of interaction of Lafont [15], following Girard [11]; it differs notably in that reaction in PIC is not a confluent reduction relation. Another answer is that the geometric view may suggest generalisations or variations of PIC. We now examine one generalisation.

Two related constraints have been placed upon π-nets:

C1 A π-net must have no target cycles.

C2. In a redex there must be no target path from a target node to a source node.

C1 arises from the binding convention in PIC. C2 may also be considered to arise from binding. First, it disallows $x(\vec{y}), \overline{x}\langle\vec{z}\rangle$ as a redex if $\vec{y}$ and $\vec{z}$ have any name in common. In reducing a redex such as $x(y), \overline{x}\langle y\rangle$ apparently the two particles would simply cancel each other, causing the name y – previously bound – to become free; at first sight it is unclear what this could mean. Second, C2 prevents a reaction from creating a target cycle and thus violating C1.

We now define *reflexive* π-nets to be the more general class of nets for which constraints C1 and C2 are removed. The operations and reaction relation for reflexive π-nets are defined exactly as before, but without the constraints. It is routine to check that this gives us another action structure, of which the (non-reflexive) π-nets form a sub-actionstructure. We shall shortly give it a syntactic formulation, and it will be called RPIC. But first let us focus on its most interesting new aspect. Consider the following (non-reflexive) π-net:

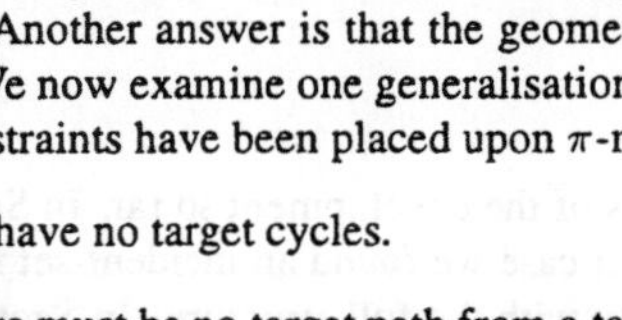

It is the action $[\,x(y),\ \overline{x}\langle y\rangle\,]\langle y\rangle$, and in PIC it has no reaction; but under the reaction relation which we have defined over reflexive π-nets it reduces to

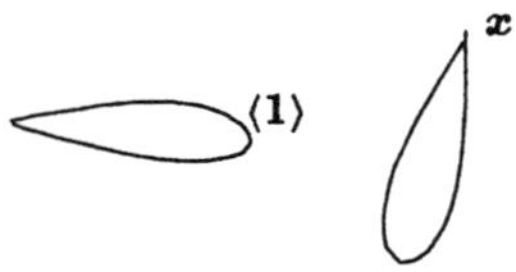

The port node x has become vacuous, so the result is the action $[\,\nu y\,]\langle y\rangle$, or just ν. So apparently in RPIC, just as the reaction of $x(y), \overline{x}\langle z\rangle$ produces the substitution $\{^z/_y\}$ when $z \neq y$, so the reaction of $x(y)\,\overline{x}\langle y\rangle$ produces the restriction νy. This reaction is geometrically sensible, since it arises simply from removing a constraint previously placed upon the reaction rule; but its implication is remarkable, for it suggests that restriction is formally a kind of *self*-substitution (but a kind which is unfamiliar). The substitution $\{^z/_y\}$ means that y should be taken to denote z; so if we follow the geometry's suggestion then the self-substitution $\{^y/_y\}$ should mean that y denotes itself, i.e. *denotes nothing else*. Though fanciful, this aligns with the intuition that a restricted name is indeed distinct from every other name. It explains why we have used the term *reflexive* for the wider class of nets.

In the next section we set out to make self-substitution not only geometrically sensible but also formally sensible.

14 Reflexive π-calculus

In this section we put together two strands of the development so far. In Section 12 we studied incident-sets for PIC_τ and PIC; in the latter case we found an incident-set for the monadic sub-actionstructure of PIC, but met a difficulty with the full structure. In Section 13 the graphical characterisation of PIC suggested a richer action structure, reflexive PIC or RPIC, which does not arise from the formal presentation.

Here we present RPIC formally, a surprisingly easy task once we establish the idea of reflexive substitution. Then we find a natural and elegant incident-set for RPIC.

14.1 The action structure RPIC: statics An action of RPIC takes the form

$$(\vec{x})M\langle\vec{y}\rangle,$$

where the body M is a *multiset* of particles, in contrast to PIC where the body is a *partial sequence* of particles. (The particles are unchanged.) Associated with this difference is a difference of binding convention. In RPIC, as in PIC, there is just one binding occurrence of each bound name x in an action a; but the scope of x is now the whole of a, whereas in PIC it extends only to the right of the binding occurrence.

The effect is to admit cyclic binding dependency among particles (i.e. target cycles). Thus the actions

$$[\,x(y),\ y(z),\ z(x)\,] \text{ and } [\,x(x)\,]$$

are admissible in RPIC. (Note that neither of them has any free names!)

At present we seek no direct significance for these target cycles; they should be treated for now as a formal extension of PIC which allows simple technical development. For one thing, multisets are easier to handle that partial sequences; for another, there will be (as we said) a natural incident-set – and this arises because RPIC does not suffer from the anomaly of PIC's binding rule which was pointed out in Section 12.6.

The operations of composition, product and abstraction are defined as for PIC, but with multiset union in place of the concatenation of partial sequences. If $a = (\vec{u})M\langle\vec{v}\rangle$ and $b = (\vec{x})N\langle\vec{y}\rangle$ then

$$\begin{aligned} a \cdot b &\stackrel{\text{def}}{=} (\vec{u})\, M \cup \sigma N\, \langle\sigma\vec{y}\rangle \qquad (\sigma = \{\vec{v}/\vec{x}\}) \\ a \otimes b &\stackrel{\text{def}}{=} (\vec{u}\vec{x})\, M \cup N\, \langle\vec{v}\vec{y}\rangle \\ \mathbf{ab}_x a &\stackrel{\text{def}}{=} (x\vec{u})\, M\, \langle x\vec{v}\rangle . \end{aligned}$$

14.2 The action structure RPIC: **dynamics** In RPIC any complementary pair of particles

$$\overline{u}\langle\vec{v}\rangle,\ u(\vec{w})$$

in the body of an action a is a redex of a; the contrast with PIC is that members of $\vec{v}$ may be among the bound names $\vec{w}$. By analogy with PIC we allow the one-step reduction

$$a \searrow^{1} \{\vec{v}/\vec{w}\} * \widehat{a},$$

where $*$ represents the action of *reflexive substitution* (of $\vec{v}$ for $\vec{w}$) and $\widehat{a}$ stands for a with the redex deleted. Informally, reflexive substitution performs literal replacement of names, but introduces a restriction particle whenever a name is required (via a cycle) to "replace itself".

In defining the operation $\{\vec{y}/\vec{x}\} * a$ we assume always that the names $\vec{x}$ are distinct and not bound in a. But $\vec{y}$ may include members of $\vec{x}$ and even bound names of a; reflexive substitution operates upon a fixed presentation of an action, with no alpha-conversion.

First, define $^{y}/_{x}(-)$ to mean the literal replacement of the name y for the name x; for example $^{y}/_{x}((y)\langle x\rangle) = (y)\langle y\rangle$. Then reflexive substitution is given recursively by

$$\begin{aligned} \{\,\} * a &\stackrel{\text{def}}{=} a \\ \{z/w, \vec{y}/\vec{x}\} * a &\stackrel{\text{def}}{=} \begin{cases} \{^{z}/_{w}\vec{y}/\vec{x}\} * {}^{z}/_{w}a & (z \neq w) \\ \{\vec{y}/\vec{x}\} * (\nu w)a & (z = w) \end{cases} \end{aligned}$$

To ensure that this is a proper definition, it is important to check that it yields the same result (up to alpha-conversion) no matter in which order the elements of a substitution $\{\vec{y}/\vec{x}\}$ are applied. This can be proved by induction on the size of the substitution. The reader may check, for example, that

$$\{y/x, x/y\} * [\,\overline{y}\langle z\rangle\,]\langle x\rangle = \{y/y\} * [\,\overline{y}\langle z\rangle\,]\langle y\rangle = [\,\nu y,\, \overline{y}\langle z\rangle\,]\langle y\rangle,$$

and that a different substitution order yields the same up to alpha-conversion:

$$\{x/y, y/x\} * [\,\overline{y}\langle z\rangle\,]\langle x\rangle = \{x/x\} * [\,\overline{x}\langle z\rangle\,]\langle x\rangle = [\,\nu x,\, \overline{x}\langle z\rangle\,]\langle x\rangle\,.$$

14.3 Discussion It is a routine matter now to verify that the action structure of reflexive π-nets and the action structure RPIC are isomorphic. The isomorphism is just as described in Section 13.2 for the non-reflexive case. It remains to be seen, by experimenting with applications, whether modelling with RPIC is significantly different from PIC in cases where the richer constructions of RPIC are not needed. Indeed, it is not even clear when and where this extra richness will be useful in practice. What then are the advantages of RPIC?

First, though its significance is not clear, we have found a new characterization of restriction; the reduction

$$[\,u(x),\ \overline{u}\langle x\rangle\,](x) \searrow \boldsymbol{\nu}$$

shows it arising from an interaction, rather than as an independent notion.

Second, the theory becomes easier in at least one important matter. For not only do we now find a natural incident-set where we failed for PIC, but we also find this easier to verify than in the case of monadic PIC because we can use graph-theoretic methods without being hampered by awkward constraints.

We now set about this. First, we rephrase the notion of reachability (Section 12.4) in graphical terms. The *reachable* nodes and arcs of a reflexive π-net are the least set such that

- every node with a target-label is reachable;
- every node and arc carried by a reachable node is reachable.

(This is a special case of a more general definition given later.) Further, a reflexive π-net is *reachable* if all its arcs are reachable.

The remainder of this section is devoted to a proof of the following

14.4 Theorem (incidents for RPIC**)** The set K^{RPIC} of reachable reflexive π-nets is an incident-set for RPIC. ■

We need some definitions and lemmas to prove this result.

14.5 Redexes and residuals A *redex* of $a \in$ RPIC consists of a source arc and a target arc with a common port, and their associated source and target nodes. This is just as in Section 6.3 for π-nets, but without constraints. Two redexes are *disjoint* if their source arcs differ and their target arcs differ.

We use r, s, t to stand for redexes. The one-step reaction or *firing* corresponding to a redex r is written

$$a \searrow^{r} a^{r},$$

where the *reductum* a^r is gained from a by removing the source and target arcs of r and coalescing each source node with the corresponding target node.

Given a redex r of a, each node n in a corresponds in an obvious way to a node n^r in a^r called its *residual* (under r). Thus if m and n are a corresponding source and target of the redex r we have $m^r = n^r$, but otherwise distinct nodes have distinct residuals. Similarly each arc π in a except the source and target arcs of r has a residual arc π^r in a^r; distinct arcs always have distinct residuals. A collection of arcs and nodes in a, such as a redex, also has a residual under r (the collection of residuals of its members). The following is easy to prove in RPIC (but is false in PIC, as shown in Section 12.6):

14.6 Lemma If s is a redex disjoint from r then its residual s^r is a redex of a^r. Moreover the two reductions commute, i.e. $(a^r)^{s^r} = (a^s)^{r^s}$. This result is shown as b in the diagram:

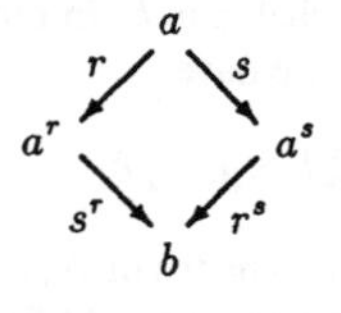

■

14.7 Reachability Let P be a subset of the nodes of an action a. The set of *P-reachable* nodes of a is the smallest set which includes P and all target-labelled nodes, and such that if m is P-reachable and m carries n then n is P-reachable.

When $P = \{p\}$ we say *p-reachable*, and also talk of *reachable from* P or p, or (when P is empty) just *reachable*. It is clear that m is P-reachable iff it is reachable from some $p \in P$. We say an arc is *P-reachable* in a if it is carried by a P-reachable node. A node or arc which is not P-reachable is *P-unreachable*.

Intuitively, a reachable (i.e. $\emptyset$-reachable) arc of a is a particle which can eventually react with its environment without a performing any internal reactions. The number of external reactions needed to "reach" it is equal to the length of a path of (reachable) arcs leading to it from a target-labelled node.

We say a redex is *P-reachable* in a if its port is P-reachable.

There is a delicate interplay between reachable and unreachable redexes and nodes. If n is unreachable, then the firing of an *unreachable* redex can make n reachable; for the reduction may coalesce an unreachable node with a reachable one. In contrast, however, the firing of a *reachable* redex cannot improve reachability:

14.8 Lemma Let r be a P-reachable redex of a, and n be P-unreachable. Then n^r is P^r-unreachable in a^r.

Proof Prove the contrapositive; suppose n^r is P^r-reachable in a^r, and construct a path reaching n from P in a. ■

For our present purposes we also need the idea of a set of nodes which can only be reached from each other. We say that a set N of nodes in a is *isolated* if whenever $n \in N$ is P-reachable then $P \cap N \neq \emptyset$. In other words, every $n \in N$ is unreachable, and if it is p-reachable then $p \in N$.

As an example, for any P and a the set of P-unreachable nodes is isolated in a.

Let us say that r is an *N-redex* if its port is in N, and π is an *N-arc* if it is carried by a member of N. The importance of isolation lies in the following:

14.9 Lemma Let $a \searrow \ell$ where ℓ is a reachable action. Suppose that N is isolated in a, and that there is at least one N-arc. Then there is an N-redex r such that

$$a \overset{r}{\searrow} a^r \searrow \ell.$$

Proof Since N is isolated, the given N-arc is unreachable. So a is not reachable, and the reduction must be of length at least 1. The proof is by induction on the length of the reduction

$$a \overset{s}{\searrow} b \searrow \ell.$$

If the first redex s is an N-redex we are done. If not then it is P-reachable, where P is the complement of N in the node-set of a. But every node of N is P-unreachable in a, so by Lemma 14.8 every node of N^s is P^s-unreachable in b. In other words N^s is isolated in b; also there exists an N^s-arc. Hence by induction we have

$$a \searrow^{s} b \searrow^{t} c \searrow \ell,$$

where t is an N^s-redex in b. But s coalesces no nodes of N, so t is the residual r^s of an N-redex r in a. But r and s are disjoint redexes of a, so Lemma 14.8 yields, as required,

$$a \searrow^{r} a^r \searrow^{s^r} c \searrow \ell . \qquad \blacksquare$$

14.10 Preactions We need to establish some properties of arbitrary preactions. In each case it will be enough to verify them for the four basic forms of preaction given in Section 4.10.

Let φ be any preaction and a any action. For each node n of a we define its *image* a^φ in $\varphi(a)$ in the obvious way. Note that if φ is a precomposition then distinct nodes of a may have the same image. Similarly each arc π of a has an image π^φ; distinct arcs always have distinct images.

Any collection of arcs and nodes in a has an image in $\varphi(a)$. In particular

14.11 Lemma Let r be a redex in a. Then its image r^φ is a redex in $\varphi(a)$, and moreover $\varphi(a) \searrow^{r^\varphi} \varphi(a^r)$. ■

This property happens to hold even for *post*composition, not just for preactions, though we shall not need that fact. The next property is more crucial, because it holds for preactions but not for postcomposition:

14.12 Lemma Let N be isolated in a. Then N^φ is isolated in $\varphi(a)$. Further, if s is an N^φ-redex in $\varphi(a)$ then s is the image of an N-redex in a. ■

The proof of this is straightforward for the four basic preaction forms, and hence by an inductive argument for all preactions. Notice that it fails if φ is a postcomposition. Take for example $a = [\,\boldsymbol{\nu} x,\ \overline{x}\langle y\rangle\,]\langle x\rangle$, and $\varphi(-) = (-) \cdot b$ where $b = (x)[\,\overline{z}\langle x\rangle,\ x(w)\,]$. Now $\{x\}$ is isolated in a, but it is not isolated in

$$\varphi(a) \ = \ [\,\boldsymbol{\nu} x,\ \overline{x}\langle y\rangle,\ \overline{z}\langle x\rangle,\ x(w)\,]\,,$$

because z is free and carries x. Furthermore, $\varphi(a)$ has an x-redex which is the image of no redex in a. The postcomposition of b allows it to "gain access" to the isolated nodes (names) of a.

At last we are ready to prove the main theorem.

14.13 Proof of theorem 14.4 It is easy to verify that K^{RPIC} satisfies the first clause of Definition 7.1 (Incident-sets). For the second clause, we assume that $\varphi(a)\searrow\ell \in K^{\mathsf{RPIC}}$, and must find $k \in K^{\mathsf{RPIC}}$ for which $a\searrow k$ and $\varphi(k)\searrow\ell$.

We proceed by induction on the size (number of arcs) of a. If $a \in K^{\mathsf{RPIC}}$ we are done. If not then a has an unreachable arc π. Now let N be the unreachable nodes in a, so that π is an N-arc. Since N is isolated in a, from Lemma 14.12 we deduce that N^φ is isolated in $\varphi(a)$, and π^φ is an N^φ-arc.

By Lemma 14.9 there is an N^{φ}-redex s for which

$$\varphi(a) \searrow^{s} \varphi(a)^{s} \searrow \ell.$$

By Lemma 14.12, $s = r^{\varphi}$ where r is an N-redex in a. So let

$$a \searrow^{r} a';$$

then by Lemma 14.8 again $\varphi(a)^{s} = \varphi(a')$, so

$$\varphi(a') \searrow \ell.$$

Hence by induction there exists $k \in K^{\mathsf{RPIC}}$ such that

$$a' \searrow k \text{ and } \varphi(k) \searrow \ell,$$

and we are done since $a \searrow^{r} a'$. ∎

15 Further work

The main contributions of this paper have been

- The definition of *action structures*, and the development of their intrinsic theory including the notions *effect* and *incident*;
- The uniform construction of a *process calculus* upon an arbitrary action structure, and a uniform theory of *bisimulation*;
- The exploration of particular action structures for the π*-calculus*, as an instance of the general theory.

In another paper [21] the notion of *action calculus*, a kind of syntactic action structure, has been defined. A natural further step, therefore, is to adapt the uniform construction of process calculi so that its result is indeed an action calculus; one must also extend the uniform theory of bisimilarity. As remarked at the start of Section 5 this is not a trivial step, but I believe it is possible; and if we claim that action structures provide a general framework then we should indeed carry out these constructions *within* it rather than *upon* it.

Already the examples in Section 3, as well as the π-calculus, show the action structure framework to be quite wide. It is therefore open to study at different levels. At one level we may explore the family of action structures for a particular model, e.g. Petri nets; the general theory may provide some insight for the particular model (e.g. a new kind of algebra of Petri nets). At a higher level, we may try to isolate some aspects of certain action structures to see whether they appear more widely. An example is the notion of *restriction*; it has already appeared in action structures both for π-calculus and for Petri nets, and it thus attains a more general significance. A more ambitious example (but linked with restriction) is the notion of *reflexion*. Reflexion arises naturally from the graphical presentation of π-nets (Section 13.4), yet is rather independent of the π-calculus; one may try to define the subclass of *reflexive action calculi*. It is also tempting to try to generalise the graphical presentation of PIC in Section 14; bearing in mind that graphical presentation of concurrent systems originated with Petri nets, which themselves form an action structure, one should surely try to isolate graphical notions which are common to a wide class of action structures.

Thus a main line of future work is to use action structures as a means of classification for models of concurrency; in this way, to echo the first sentence of this paper, we can hope to separate their ad hoc details from their essentials.

References

[1] Aceto, L. and Hennessy, M.C., Towards action-refinement in process algebras, to appear in Journal of Information and Computation.

[2] Austry, D. and Boudol, G., Algèbre de processus et synchronisation, Journal of Theor. Comp. Science, Vol 30, pp90–131, 1984.

[3] Baeten, J.C.M. and Weijland, W.P., Process Algebra, Cambridge University Press 1990.

[4] Barr, M. and Wells, C., Toposes, Triples and Theories, Springer Verlag, 1985.

[5] Bellin, G. and Scott, P.J. (with introduction by S. Abramsky), On the π-calculus and linear logic, to appear in Theoretical Computer Science, (Proc. MFPS Conference, Oxford 91, ed. M. Mislove and A.W. Roscoe), 1993.

[6] Berry, G. and Boudol, G., The chemical abstract machine, Journal of Theoretical Computer Science, Vol 96, pp217–248, 1992.

[7] Best, E., Devillers, R. and Hall, J.G., The box calculus: a new causal algebra with multi-label communication, in Advances in Petri Nets '92, ed. G.Rozenberg, Lecture Notes in Computer Science, Springer Verlag, Vol.609, pp21–69, 1992.

[8] Brookes, S.D., Hoare, C.A.R. and Roscoe, A.W., A theory of communicating sequential processes, Journal of ACM, Vol 31, pp560–599, 1984.

[9] Cherkasova, L.A. and Kotov, V.E., Descriptive and analytical process algebras, in Advances in Petri Nets '89, ed. G Rozenberg, Lecture Notes in Computer Science, Springer Verlag, Vol 424, pp77–104, 1989.

[10] Freyd, P.J. and Kelly, G.M., Categories of continuous functors, I, Journal of Pure and Applied Algebra, Vol 2, pp169–191, 1972.

[11] Girard, J.-Y., Linear logic, J. Theoretical Computer Science, Vol 50, pp1–102, 1987.

[12] van Glabbeek, R. and Goltz, U., Equivalence notions for concurrent systems and refinement of actions, Proc. 4th Conference on Mathematical Foundations of Computer Science, Lecture Notes in Computer Science, Springer Verlag, Vol 379, pp237–248, 1988.

[13] Gunter, C.A. and Scott, D.S., Semantic domains, in Handbook of Theoretical Computer Science, Vol A, pp633–674, Elsevier, 1990.

[14] Hoare, C.A.R., Communicating Sequential Processes, Prentice Hall, 1985.

[15] Lafont, Y., Interaction nets, Proc. 17th ACM Symposium on Principles of Programming Languages, 1990.

[16] Meseguer, J., Conditional rewriting logic as a unified model of concurrency, Journal of Theoretical Computer Science, Vol 96, pp73–155, 1992.

[17] Meseguer, J. and Montanari, U., Petri nets are monoids, Journal of Information and Computation, Vol 88, pp105–155, 1990.

[18] Milner, R., Calculi for synchrony and asynchrony, Journal of Theoretical Computer Science Vol 25, pp267–310, 1983.

[19] Milner, R., Communication and Concurrency, Prentice Hall, 1989.

[20] Milner, R., The polyadic π-calculus: a tutorial, in Logic and Algebra of Specification, ed. F.L. Bauer, W. Brauer and H. Schwichtenberg, Springer Verlag, 1993.

[21] Milner, R., Action calculi, or concrete action structures, Proc. MFCS Conference, Gdansk, Poland, Lecture Notes in Computer Science, Springer Verlag, Vol 711, pp105–121, 1993. (Also titled Action calculi I: axioms and applications, the first of a series on action calculi.)

[22] Milner, R., Action calculi II: π-nets with boxes and replication, Internal report, Laboratory for Foundations of Computer Science, Dept. of Computer Science, University of Edinburgh, 1993.

[23] Milner, R., Action calculi III: higher-order, submitted for publication. Internal report, Laboratory for Foundations of Computer Science, Dept. of Computer Science, University of Edinburgh, 1993.

[24] Milner, R., Parrow, J. and Walker D., A calculus of mobile processes, I and II, Journal of Information and Computation, Vol 100, pp1–40 and pp41–77, 1992.

[25] Nielsen, M., Plotkin, G.D. and Winskel, G., Petri nets, event structures and domains, Journal of Theoretical Computer Science, Vol 13, No 1, 1981.

[26] Nielsen, M., and Winskel, G., Models for concurrency, to appear in Volume IV of the Handbook of Logic and the Foundations of Computer Science, ed. D.Gabbay, Oxford University Press, 1993.

[27] Reisig, W., Petri Nets, EATCS Monographs on Theoretical Computer Science, ed. W.Brauer, G.Rozenberg, A.Salomaa, Springer Verlag, 1983.

[28] Winskel, G., Petri nets, algebras, morphisms and compositionality, Journal of Information and Computation, March 1987.

Linear Logic and Computation: A Survey

Andre Scedrov

Department of Mathematics, University of Pennsylvania
Philadelphia, PA 19104-6395, USA
andre@cis.upenn.edu

Abstract. This is a survey of computational aspects of linear logic related to proof search.

Keywords. Linear logic, cut free proof search, logic programming, complexity.

1 Introduction

Linear logic, introduced by Girard [14, 36, 32], is a refinement of classical logic. While the central notions of *truth* (emphasized in classical logic) and *proof construction* (emphasized in intuitionistic logic) remain important in linear logic, it might be said that the emphasis in linear logic is on *state*. Linear logic is sometimes described as being *resource sensitive* because it provides an intrinsic and natural accounting of process states, events, and resources. Linear logic also sheds new light on classical logic and its relationship to intuitionistic logic, see Girard [15, 16] and Danos *et al.* [11]. An evocative semantic paradigm for linear logic by means of games is proposed by Blass [7] and by Abramsky and Jagadeesan [2].

As an intuitive motivation, let us consider reading logical deductions so that assumptions are understood as resources, and conclusions as requirements to be met by spending the given resources. In this reading the formula *A implies A*, for instance, means that the resource A is spent to meet the requirement A. This is an axiom in linear logic. However, in this reading two standard "structural" rules, *contraction* and *weakening* are open to criticism. Contraction rule stipulates that any assumption, once it is stated, may then be reused as often as desired. The weakening rule makes it possible to carry out a deduction without using all of the assumptions. Together these rules make it possible to discard or duplicate the assumptions at will and are thus responsible for a loss of control over resources in classical (and in intuitionistic) logic. This insight is the starting point of linear logic. Removing the rules of contraction and weakening produces a *linear* system in which each assumption must be spent *exactly once*. In this system, formulas indicate finite resources that cannot necessarily be discarded or duplicated without effort.

A more subtle feature of linear logic is the distinction between two natural forms of conjunction and disjunction, one form called "multiplicative", and the other "additive". For instance, considering the hypotheses as resources, a proof of a multiplicative conjunction $\otimes$ as a conclusion forbids any sharing between the resources used to establish each conjunct, whereas the additive conjunction & requires the sharing of all of the resources. (The terminology originates from a semantic interpretation in so-called coherence spaces [14], where multiplicative conjunction $\otimes$ is seen as a product of certain undirected graphs, while additive conjunction & is seen as a disjoint sum.)

Linear logic also involves a kind of modality: a "storage" or "reuse" operator, ! , and its dual , ? . Intuitively, $!A$ provides unlimited creation of A, and $?B$ allows the unlimited consumption of B. In effect, the structural rules of contraction and weakening are replaced in linear logic by the explicit logical rules involving the modalities. For instance, the usual function type (or: intuitionistic implication) $A \Rightarrow B$ can be recovered as $!A \multimap B$, where $\multimap$ is *linear implication*, which provides the type of functions that "use" their argument exactly once. A detailed description of linear logic rules is given in Section 2 below.

Current work on computational aspects of linear logic generally falls within either of the following two paradigms. In what may be broadly called functional programming (often involving certain aspects of parallelism), computation is seen as term reduction corresponding to *proof reduction* (*i.e.,* to the *process* of cut elimination). Recent work in this area includes Abramsky [1], Benton *et al.* [6], Gonthier *et al.* [17, 18], Chirimar *et al.* [9, 10], and Danos and Regnier [12]. On the other hand, in what may be broadly called logic programming (again, often involving concurrency), computation is seen as cut free *proof search* in certain linear logic theories. From this latter perspective, the cut elimination property is used simply to allow one to concentrate on cut free proofs without loss of generality.

This survey is intended as a companion to [32]. Here we consider certain computational aspects of linear logic related to proof search. An overview of this area is given in Section 3. An illustrative example relating computational complexity and linear logic proof search is discussed in Section 4. The reader is referred to [32] for a broader overview and an extensive bibliography. Partial support from NSF Grant CCR-91-02753, from ONR Grant N00014-92-J-1916, and from an American Mathematical Society Centennial Research Fellowship is gratefully acknowledged.

2 Linear Logic Proof Rules

The formal logical framework considered in this paper is Gentzen-style sequent calculus. For us it is convenient to consider sequents of the form $\vdash \Gamma$, where Γ is a finite multiset of formulas. The reader should note that in standard presentations of sequent calculi, sequents are often built from sets of formulas, whereas we use multisets here. This difference is crucial. We maintain usual distinctions among propositional, predicate, first order, and second order calculi. We assume a set of function symbols and predicate symbols, along with their associated arities. Atomic formulas (other than constants $1, \perp, \top, 0$) are defined as usual. The equality symbol is not considered here. The inference rules for the linear logic sequent calculus are given in Figure 1.

We use the following notational conventions:

a, b, p, q	Atomic formulas
$a^\perp, b^\perp, p^\perp, q^\perp$	Negated atomic formulas
A, B, C	Arbitrary formulas
$\Gamma, \Delta, \Sigma, \Theta$	Arbitrary finite multisets of formulas

As usual, in the $\forall$ rule in Figure 1 it is assumed that X is not free in Γ. We follow notation from logic programming, where the variables are written in capital letters. The notation $?\Gamma$ is used to denote a multiset of formulas which all begin with ?. The English names for the rules given in Figure 1 are identity, cut, tensor, par, plus, with, weakening, contraction, dereliction, promotion (or storage), bottom, one, top, universal, and existential, respectively. $\otimes$, $⅋$, and $^\perp$ are *multiplicative* connectives; 1 and $\perp$ are *multiplicative* propositional constants. $\oplus$

and & are *additive* connectives; 0 and $\top$ are *additive* propositional constants. Observe the "additive" nature of the quantifiers and that there is no rule for the constant 0.

A proof is represented as a tree rooted at its conclusion sequent at the bottom and with the leaves at the top. Given this orientation, the notion of a rule occuring above or below another rule is clear.

The reader will note that linear negation is a defined concept on composite formulas, not a basic connective. One may define negation by recursion on the structure of formulas: $(p^\perp)^\perp$ is p, $(A \otimes B)^\perp$ is $A^\perp \parr B^\perp$, $(A \parr B)^\perp$ is $A^\perp \otimes B^\perp$, $(A \& B)^\perp$ is $A^\perp \oplus B^\perp$, $(A \oplus B)^\perp$ is $A^\perp \& B^\perp$, $(\forall X.A)^\perp$ is $\exists X.A^\perp$, $(\exists X.A)^\perp$ is $\forall X.A^\perp$, $1^\perp$ is $\perp$, $\perp^\perp$ is 1, $\top^\perp$ is 0, and $0^\perp$ is $\top$. If Γ consists of $A_1, \ldots, A_n$, then $\Gamma^\perp$ denotes $A_1^\perp, \ldots, A_n^\perp$.

Although the identity rule, *i.e.*, the first rule in Figure 1 is restricted to atomic formulas, the analogous identity rule for arbitrary formula A instead of p is derivable. The reader will also observe that *linear implication*, $\multimap$, may be defined by letting $A \multimap B$ be $A^\perp \parr B$. Linear implication is a multiplicative connective.

It should be noted that our presentation of sequent calculus is equivalent to a more traditional, two-sided presentation of sequent calculus (which has twice as many rules). For instance, in two-sided presentation derived rules for linear implication take the form described in Figure 2. It is readily shown that a sequent $\Gamma \vdash \Delta$ is provable in two-sided presentation (without cut) iff the sequent $\vdash \Gamma^\perp, \Delta$ is provable in our presentation (without cut, respectively).

It follows from [14] that if a sequent is provable in linear logic, then it is provable in linear logic without using the Cut rule.

3 Linear Logic Proof Search as Computation

Proof search is an important computational aspect of linear logic. This computational perspective is based on the work of Miller *et al.* [31], which introduced a paradigm for logic programming based on cut free proof search in Gentzen-style sequent calculi for intuitionistic logic. Each step in a proof search is a reverse application of a logical inference rule (except the Cut rule) to a sequent matching the conclusion of the rule in order to generate the corresponding premises. Rules for

I $$\frac{}{\vdash p, p^{\perp}}$$

Cut $$\frac{\vdash \Gamma, A \qquad \vdash A^{\perp}, \Delta}{\vdash \Gamma, \Delta}$$

$\otimes$ $$\frac{\vdash \Gamma, A \qquad \vdash B, \Delta}{\vdash \Gamma, (A \otimes B), \Delta}$$

$⅋$ $$\frac{\vdash \Gamma, A, B}{\vdash \Gamma, (A ⅋ B)}$$

$\oplus$ $$\frac{\vdash \Gamma, A}{\vdash \Gamma, (A \oplus B)} \qquad \frac{\vdash \Gamma, B}{\vdash \Gamma, (A \oplus B)}$$

$\&$ $$\frac{\vdash \Gamma, A \qquad \vdash \Gamma, B}{\vdash \Gamma, (A \& B)}$$

?W $$\frac{\vdash \Gamma}{\vdash \Gamma, ?A}$$

?C $$\frac{\vdash \Gamma, ?A, ?A}{\vdash \Gamma, ?A}$$

?D $$\frac{\vdash \Gamma, A}{\vdash \Gamma, ?A}$$

!S $$\frac{\vdash ?\Gamma, A}{\vdash ?\Gamma, !A}$$

$\perp$ $$\frac{\vdash \Gamma}{\vdash \Gamma, \perp}$$

1 $$\frac{}{\vdash 1}$$

$\top$ $$\frac{}{\vdash \Gamma, \top}$$

$\forall$ $$\frac{\vdash \Gamma, A}{\vdash \Gamma, \forall X.A}$$

$\exists$ $$\frac{\vdash \Gamma, A[t/X]}{\vdash \Gamma, \exists X.A}$$

Figure 1: Linear logic inference rules

$$\multimap\mathbf{L}\quad \frac{\Sigma \vdash A, \Theta \qquad \Gamma, B \vdash \Delta}{\Sigma, \Gamma, (A \multimap B) \vdash \Theta, \Delta} \qquad\qquad \frac{\Gamma, A \vdash B, \Delta}{\Gamma \vdash (A \multimap B), \Delta}\quad \multimap\mathbf{R}$$

Figure 2: Derived rules for linear implication

matching again formulas in the hypothesis are used only when the conclusion is atomic. It is desirable to omit the Cut rule because its premises may be more complicated than its conclusion and the construction of useful cut formulas is far from being a computational primitive around which one builds a programming language. It is a fundamental property of logical calculi that the Cut rule may be omitted without loss of completeness (see above in the case of linear logic). An important consequence of this Cut Elimination Property in the quantifier free case and in the first order case is the *Subformula Property*, that if a formula A is provable, then there exists a proof of A in which only the subformulas of A occur. The Subformula Property is a theoretical basis of proof search.

An equally important role in the proof search paradigm is played by certain *Permutability Properties*, which yield optimized presentations of cut free proofs. Permutability properties ensure that if a formula A is provable, then there exists a proof of A of a certain permuted form. That is, one may permute the order of application of inferences in a proof to achieve a kind of normal form. Permutability properties are used to give control over the shape of linear logic proofs, thus enabling a more direct computational reading, see Lincoln [28], Andreoli [5], Hodas and Miller [20], and Lincoln and Scedrov [24].

The expressiveness of linear logic proof search as a computational paradigm is brought to light in a body of research over the past several years. Previous work of Miller and others [29] had developed several examples in data bases, theorem proving, and natural language semantics where the intuitionistic logic proof search as a basis of conventional logic programming languages was inadequate. In each of these cases, a logic programming language based on linear logic proof search provided correct specifications.

In a number of papers Andreoli and Pareschi develop a declarative treatment of object communication and concurrent object-oriented computation [3, 4, 5]. As part of this work, Andreoli and Pareschi develop a concurrent specification language, LO [4].

The work of Miller and Hodas [20] describes a linear logic programming language that refines the logics behind both Prolog and λProlog. A detailed analysis of this linear logic programming language is given in [20], including a semantics based on Kripke models. Hodas [19] develops this linear logic programming language further, and explains the natural language parsing issue of gap-threading parsers within this logic programming setting. As part of this work, an implementation of this linear logic programming language was built in Standard ML. This system, called *Lolli*, is available by anonymous ftp from host ftp.cis.upenn.edu and directory pub/Lolli. All the operators of λProlog except the higher order unification are supported by Lolli , but with the addition of their linear variations. A Lolli program thus distinguishes between clauses which can be used as many, or as few, times as desired, and those that must be used exactly once. Using these features, Lolli provides a perspicuous, logical implementation of a number of sample problems, including object-oriented programming with mutable state, database update, and term rewriting.

Miller [30] develops a connection between concurrency theory and linear logic. In particular, Milner's elaboration of CCS, the π-calculus, is expressed as a simple theory in linear logic. This setting is both immediate, because reduction is identified with logical consequence, and productive, because it provides a new approach to understanding the Hennessy-Milner logic and bisimulation in CCS. Recently, Miller has been exploring how to extend the framework of logic programming from the single conclusion sequent calculus in which it is often defined to the more general and flexible, multiple conclusion setting, such as the linear logic sequent calculus. As it turns out, several concurrent specification languages, in particular CCS, the π-calculus (with non-deterministic +), and the language LO developed by Andreoli and Pareschi [4], can be thought of as multiple conclusion generalizations of logic programming.

The striking expressiveness of linear logic proof search computational paradigm is also indicated by the work on complexity and undecidability of (provability in) the natural fragments of linear logic. This work focuses on direct, lockstep representations of various generic computational frameworks by cut free proof search in fragments of linear logic. The representations reveal a structural relationship between the natural fragments of linear logic on the one hand and the standard computational complexity classes on the other. It is shown in Lincoln *et al.* [23] that the full propositional (*i.e.*, quantifier free) linear logic is undecidable. This is obtained by faithfully representing computations on counter machines by proof search in propositional linear logic, so that the representing search successfully ends in the axioms iff the counter machine computation terminates. On the other hand, Kanovich [22] shows that the multiplicative fragment alone is NP-complete. In fact, in a noticeable contrast to the standard NP-completeness of the satisfiability of propositional formulas in classical logic, in linear logic even the decision problem for *constant-only* multiplicative propositional formulas is NP-complete. This result, obtained by Lincoln and Winkler [27], shows that a simple minded, efficient, truth-table style characterization of provability in the multiplicative fragment of linear logic would imply that P = NP.

Lincoln *et al.* [23] study the fragment of propositional linear logic that consists only of the multiplicative and the additive connectives, *i.e.*, the nonmodal propositional fragment, MALL. Provability in this fragment is shown to be polynomial space (PSPACE) complete. This result provides insight into the additive connectives of linear logic, which stipulate *sharing* of resources. Indeed, it is not *a priori* clear that the sequential order of classical boolean propositional quantifiers, involved in the standard PSPACE-complete problem of quantified boolean formula validity, can be simulated by the commutative propositional connectives alone. The method introduced in [23] to express the order of boolean quantifiers is of independent interest. It may be described as a "lock-and-key" method of using a certain combination of multiplicative and additive connectives in propositional linear logic to force a particular execution strategy in successful proof searches, see Section 4 below.

The "lock-and-key" method is used in Lincoln *et al.* [25] to construct an embedding of the implicational propositional intuitionistic logic IIL into MALL. The embedding preserves cut free proofs in a formal proof system IIL* that is an optimization of IIL. In IIL* principal formulas of inference rules are never copied, and the context is shared between the premises. The embedding into MALL is efficient and provides an alternative proof of the PSPACE-hardness of MALL. (The embedding is exponential in the implication depth, but polynomial on formulas of bounded implication depth. It is well-known that implications of depth at most 2 suffice.) The embedding exploits several proof-theoretic properties of intuitionistic implication that analyze the use of resources in IIL proofs. Furthermore, the result addresses the issue of replacing copying and reuse by sharing. The optimized calculus IIL* is of independent interest. Although the optimized calculus IIL* is implicit in the work of Russian logician Vorobjev in the 1950's, IIL* has been formulated explicitly only in the past several years independently by Lincoln *et al.* [25], by Dyckhoff [13], and by Hudelmaier [21]. Versions of IIL* are being used in several theorem provers, built by Miller and Hodas (see above), by Paulson, by N. Tennant, and by Dyckhoff, Leslie, and Read.

Recent work of Lincoln and Scedrov [24] is concerned with the expressiveness of first order linear logic without modalities, MALL1, the first order analog of MALL. As in MALL, cut free MALL1 proof trees have linear depth in the length of the concluding formula (sequent). This follows from the fact that in the absence of the reuse modalities a linear logic formula can be analyzed as a principal formula of a logical rule only once, and in this analysis step the formula is decomposed into proper subformulas and spent. In fact, this *is* a technical aspect of *linearity* in linear logic. However, in contrast to MALL, where in the course of cut free proof search the size of sequents does not increase, in MALL1 proof search sequents may be very large due to the instantiations of first order quantifiers by terms that may involve function symbols. In fact, in [24] it is shown that MALL1 is nondeterministic single exponential time (NEXPTIME) hard. (Membership in NEXPTIME, and hence NEXPTIME-completeness of MALL1, follows from current work of Lincoln and Shankar [26], see below.) The hardness proof in [24] combines E. Shapiro's logic programming simulation of nondeterministic Turing machines [34] with the standard proof of the PSPACE-hardness of quantified boolean formula validity as in, *e.g.*, Stockmeyer [35], and utilizing some of the remarkable plasticity of linear logic. The hardness result is achieved through a direct encoding of Turing machine transitions, which are shared via the additives, and the existential quantification over intermediate Turing machine configurations. The linear depth bound on cut free MALL1 proofs gives an immediate single exponential upper bound on the number of Turing machine transitions that can be used in the entire proof using this encoding. In contrast to previous work on linear logic proof search as a computational paradigm [23, 20, 30, 4], in which a computation proceeds "upwards" along a proof tree, in [24] the computation steps are applied "horizontally across" the leaves of a proof tree.

Lincoln and Shankar [26] consider the first order linear logic analog of the dynamic herbrandization (skolemization) technique earlier employed for intuitionistic logic in [33]. In this technique unification is postponed to the leaves of the search tree. [26] develops such a proof search procedure for first order linear logic. In the case of MALL1 this proof search procedure is in NEXPTIME when applied to MALL1 formulas. Together with the hardness result in [24] just discussed, this shows that MALL1 is NEXPTIME-complete. In the case of the first order multiplicative fragment of linear logic, the same proof search procedure is in NP when applied to formulas from this fragment. Thus it follows from earlier propositional NP-hardness results [22] (see above) that first order multiplicative linear logic is NP-complete.

4 Example: Simulating Validity

Recall the nonmodal fragment of propositional linear logic, MALL. This fragment contains multiplicative connectives $\otimes$, $⅋$, and $^{\perp}$, additive connectives & and $\oplus$, and constants 0, 1, $\top$, and $\perp$, but no modalities !, ?, and no quantifiers. Because of the Subformula Property, it suffices to consider only the rules in the Section 2 that are associated with multiplicative and additive connectives and constants.

The PSPACE-hardness of MALL provability is obtained in [23] by a lockstep simulation of the validity problem for quantified boolean formulas (QBF). Let us briefly discuss the key aspects of the argument in [23].

A *quantified boolean formula* has the (prenex) form:

$$Q_m X_m \ldots Q_1 X_1 \colon M,$$

where M is a quantifier free boolean *matrix* and each Q_i is either $\forall$ or $\exists$ quantifying boolean variables in M. In our notation G ranges over QBF's, M ranges quantifier free boolean formulas, and X ranges over boolean variables. An *assignment* I for G maps the free variables in G to truth values T or F. The validity of G under I is written as $I \models G$. The QBF validity problem is: *Given a closed QBF G, is $\models G$?*

[23] exhibits a succinct encoding of a closed QBF G as a MALL sequent that is provable *exactly* when G is valid. One part encodes the boolean matrix, and a second part encodes the quantifier prefix. Let us describe the encoding on an example from [23].

Let G be the *valid* QBF

$$\forall X_2 \exists X_1 \colon \neg(\neg X_1 \wedge X_2) \wedge \neg(\neg X_2 \wedge X_1).$$

G is equivalent to $\forall X_2 \exists X_1 \colon (X_1 \iff X_2)$. Notice that reversing the quantifier prefix results in a *falsifiable* QBF $\exists X_1 \forall X_2 \colon (X_1 \iff X_2)$.

The encoding of the boolean matrix M of G describes the formula as a circuit with signals labeled by MALL literals. If the assignment I is encoded by a sequence of MALL formulas $\langle I \rangle$, and $[M]_a$ is the MALL formula encoding M with output labeled by the literal a, then $I \models M$ is encoded by the sequent $\vdash \langle I \rangle, [M]_a, a$, whereas $I \not\models M$ is encoded by $\vdash \langle I \rangle, [M]_a, a^\perp$.

The assignment $\langle \{X_1 \leftarrow T, X_2 \leftarrow F\} \rangle$ is encoded by the sequence of formulas $b_1^\perp, b_2$. These literals are the input signals to the encoding of the boolean formula.

The encoding $[\neg X_1]_a$ of $\neg X_1$ with output labeled a is the formula $\text{NOT}(b_1, a)$ expressing the truth table for negation within MALL:

$$\text{NOT}(b_1, a) = (b_1 \otimes a) \oplus (b_1^\perp \otimes a^\perp), \tag{1}$$

which is the negation of the formula $(b_1 \multimap a^\perp) \& (b_1^\perp \multimap a)$. The sequent

$$\vdash b_1, \text{NOT}(b_1, a), a \tag{2}$$

encodes the situation where the input X_1 is F, and asserts (correctly) that the output $\neg X_1$ is T. The sequent (2) is easily seen to have the proof

$$\cfrac{\cfrac{\cfrac{}{\vdash b_1, b_1^\perp} \quad \cfrac{}{\vdash a^\perp, a}}{\vdash b_1, (b_1^\perp \otimes a^\perp), a}}{\vdash b_1, (b_1 \otimes a) \oplus (b_1^\perp \otimes a^\perp), a}$$

Similarly, the sequent (3) representing $\{X_1 \leftarrow T\} \not\models \neg X_1$ is also provable:

$$\vdash b_1^\perp, \text{NOT}(b_1, a), a^\perp. \tag{3}$$

On the other hand, the sequent

$$\vdash b_1^\perp, \text{NOT}(b_1, a), a \tag{4}$$

$$
\dfrac{\dfrac{}{\vdash q_2, q_2^\perp}\,\mathrm{I} \qquad \dfrac{\dfrac{\dfrac{?}{\vdash q_1^\perp, q_1, b_1}}{\vdash q_1^\perp, q_1 \parr b_1}\,\parr}{\vdash q_1^\perp, ((q_1 \parr b_1) \oplus (q_1 \parr b_1^\perp))}\,\oplus}{\dfrac{\vdash q_2, q_1^\perp, q_2^\perp \otimes ((q_1 \parr b_1) \oplus (q_1 \parr b_1^\perp)) \qquad \overset{\vdots}{\vdash B, A_a, a}}{\vdash q_2, q_2^\perp \otimes ((q_1 \parr b_1) \oplus (q_1 \parr b_1^\perp)), q_1^\perp \otimes B, A_a, a}\,\otimes}\,\otimes
$$

Figure 3: A failed proof search attempt

$$
\dfrac{\dfrac{}{\vdash q_2, q_2^\perp}\,\mathrm{I} \qquad \dfrac{\dfrac{\overset{\vdots}{\vdash q_1, b_1, q_1^\perp \otimes B, A_a, a}}{\vdash q_1 \parr b_1, q_1^\perp \otimes B, A_a, a}\,\parr}{\vdash ((q_1 \parr b_1) \oplus (q_1 \parr b_1^\perp)), q_1^\perp \otimes B, A_a, a}\,\oplus}{\vdash q_2, q_2^\perp \otimes ((q_1 \parr b_1) \oplus (q_1 \parr b_1^\perp)), q_1^\perp \otimes B, A_a, a}\,\otimes
$$

Figure 4: A proof search respecting quantifier order

asserts (falsely) that $\{X_1 \leftarrow T\} \models \neg X_1$. Sequent (4) is not provable because MALL is a refinement of classical logic in which only the valid sequents of ordinary propositional logic are provable (when reading $^\perp$ as ordinary negation, $\otimes$ and $\&$ as ordinary conjunction, and $\parr$ and $\oplus$ as ordinary disjunction).

The truth tables for the other boolean connectives can similarly be expressed as MALL formulas and the entire matrix M encoded as above. The only subtlety is that when the fanout of a signal exceeds one, the encoding must provide explicit means for duplicating the corresponding MALL literal, because MALL lacks a contraction rule.

In order to encode boolean quantification, one must encode individual quantifiers as well as the dependencies between quantifiers. Given the above encoding for assignments and the boolean connectives, a first attempt would be to encode $\exists X_1$ by the formula $(b_1 \oplus b_1^\perp)$ and to encode $\forall X_2$ by $(b_2 \& b_2^\perp)$. The reason the formula $(b_1 \oplus b_1^\perp)$ behaves like existential quantification in proof search is that a nondeterministic choice can be made between

$$\frac{\vdash b_1^\perp, \Gamma}{\vdash (b_1 \oplus b_1^\perp), \Gamma} \quad \text{and} \quad \frac{\vdash b_1, \Gamma}{\vdash (b_1 \oplus b_1^\perp), \Gamma}$$

according to the assignment (T or F, respectively) to X_1 which makes $\exists X_1\colon M$ valid. Similarly, the formula $(b_2 \& b_2^\perp)$ behaves like universal quantification requiring proofs of both $\vdash b_2^\perp, \Gamma$ and $\vdash b_2, \Gamma$:

$$\frac{\vdash b_2^\perp, \Gamma \qquad \vdash b_2, \Gamma}{\vdash (b_2 \& b_2^\perp), \Gamma}$$

However, with this representation of quantifiers, the MALL encoding of $\exists X_1 \forall X_2\colon M$ would be the same as that of $\forall X_2 \exists X_1\colon M$, but only the latter formula is valid since M expresses $(X_1 \iff X_2)$.

In order to guarantee that $\exists X_1 \forall X_2\colon M$ is correctly encoded, one needs to ensure that if the encoding is provable, then there is a proof in which the choice of witness for X_1 does not depend on whether X_2 is T or F. For instance, one needs to exclude a dependency of X_1 on X_2 as shown in the following deduction:

$$\frac{\dfrac{\vdash b_1, b_2, \Gamma}{\vdash (b_1 \oplus b_1^\perp), b_2, \Gamma} \qquad \dfrac{\vdash b_1^\perp, b_2^\perp, \Gamma}{\vdash (b_1 \oplus b_1^\perp), b_2^\perp, \Gamma}}{\vdash (b_1 \oplus b_1^\perp), (b_2 \& b_2^\perp), \Gamma}$$

where each branch can be successfully completed to a proof when Γ is $[X_1 \iff X_2]_a, a$.

The solution in [23] is to encode the quantifier order in a way that forces $(b_1 \oplus b_1^\perp)$ to be introduced *below* $(b_2 \& b_2^\perp)$ in any cut free proof. For this [23] introduces new MALL atoms q_0, q_1, q_2, and encodes $\exists X_1 \forall X_2\colon M$ as:

$$\begin{array}{ll} \vdash & q_2, \\ & q_2^\perp \otimes ((q_1 \mathbin{⅋} b_1) \oplus (q_1 \mathbin{⅋} b_1^\perp)), \\ & q_1^\perp \otimes ((q_0 \mathbin{⅋} b_2) \& (q_0 \mathbin{⅋} b_2^\perp)), \\ & q_0^\perp \otimes [M]_a, a \end{array}$$

The quantifier encoding for $\exists X_1$ now hides the "key" q_1 that is needed to unlock the quantifier encoding for $\forall X_2$. In Figures 3 and 4 let B be $(q_0 \mathbin{⅋} b_2) \& (q_0 \mathbin{⅋} b_2^\perp)$, and let A_a be $q_0^\perp \otimes [M]_a$. One attempt to violate the quantifier ordering as before, is shown in Figure 3, where the subgoal $\vdash q_1^{\perp}, q_1, b_1$ is unprovable in MALL because of the absence of an applicable weakening rule. All other proof search attempts that violate the quantifier ordering also fail.

A proof search that does respect the quantifier order is shown in Figure 4. When M is $(X_1 \iff X_2)$ any such proof search ultimately fails because for any assignment I, the sequent $\vdash \langle I \rangle, [M]_a, a$ is provable iff $I \models M$ (see above).

References

[1] S. Abramsky. Computational interpretations of linear logic. *Theoretical Computer Science*, 111:3–57, 1993.

[2] S. Abramsky and R. Jagadeesan. Games and full completeness for multiplicative linear logic. Accepted for publication in Journal of Symbolic Logic, 1994. Draft available using anonymous ftp from the host theory.doc.ic.ac.uk and the file theory/papers/Abramsky/gfc.dvi.

[3] J.-M. Andreoli and R. Pareschi. Logic programming with sequent systems: a linear logic approach. In *Proc. Workshop on Extensions of Logic Programming, Tuebingen.* Lecture Notes in Artificial Intelligence, Springer-Verlag, Berlin, 1990.

[4] J.-M. Andreoli and R. Pareschi. Linear objects: Logical processes with built-in inheritance. *New Generation Computing*, 9, 1991.

[5] Jean-Marc Andreoli. Logic programming with focusing proofs in linear logic. *Journal of Logic and Computation*, 2(3), 1992.

[6] N. Benton, G. Bierman, V. de Paiva, and J.M.E. Hyland. Linear lambda-calculus and categorical models revisited. In *Computer Science Logic '92, Selected Papers*, pages 61–84. Springer LNCS, vol. 702, 1993.

[7] A. Blass. A game semantics for linear logic. *Annals of Pure and Applied Logic*, 56:183–220, 1992. Special Volume dedicated to the memory of John Myhill.

[8] S. Cerrito. A linear axiomatization of negation as failure. *Journal of Logic Programming*, 12:1–24, 1992.

[9] J. Chirimar, C. Gunter, and J. Riecke. Linear ML. In *Proc. ACM Symposium on Lisp and Functional Programming, San Francisco.* ACM Press, June 1992.

[10] J. Chirimar, C. Gunter, and J. Riecke. Reference counting as a computational interpretation of linear logic. Manuscript, 1993. Available by anonymous ftp from the machine research.att.com in the file dist/riecke/linear-logic-journal.ps.

[11] V. Danos, J.-B. Joinet, and H. Schellinx. LKQ and LKT: sequent calculi for second order logic based upon dual linear decomposition of classical implication. Manuscript, October 1993.

[12] V. Danos and L. Regnier. Local and asynchronous beta reduction. In *Proc. 8-th IEEE Symp. on Logic in Computer Science, Montreal*, pages 296–306. IEEE Computer Society Press, Los Alamitos, California, June 1993.

[13] R. Dyckhoff. Contraction-free sequent calculi for intuitionistic logic. *Journal of Symbolic Logic*, 57:795–807, 1992.

[14] J.-Y. Girard. Linear logic. *Theoretical Computer Science*, 50:1–102, 1987.

[15] J.-Y. Girard. A new constructive logic: classical logic. *Mathematical Structures in Computer Science*, 1:255–296, 1991.

[16] J.-Y. Girard. On the unity of logic. *Annals of Pure and Applied Logic*, 59:201–217, 1993.

[17] G. Gonthier, M. Abadi, and J.-J. Levy. The geometry of optimal lambda reduction. In *Proc. 19-th Annual ACM Symposium on Principles of Programming Languages, Albuquerque, New Mexico.* ACM Press, New York, NY, January 1992.

[18] G. Gonthier, M. Abadi, and J.-J. Levy. Linear logic without boxes. In *Proc. 7-th Annual IEEE Symposium on Logic in Computer Science, Santa Cruz, California*, pages 223–234. IEEE Computer Society Press, Los Alamitos, California, June 1992.

[19] J.S. Hodas. Specifying filler-gap dependency parsers in a linear logic programming language. In K. Apt, editor, *Proc. Joint International Conference and Symposium on Logic Programming*, pages 622–636, 1992.

[20] J.S. Hodas and D. Miller. Logic programming in a fragment of intuitionistic linear logic. In *Proc. 6-th Annual IEEE Symposium on Logic in Computer Science, Amsterdam*, pages 32–42. IEEE Computer Society Press, Los Alamitos, California, July 1991. Full paper to appear in *Information and Computation*. Draft available using anonymous ftp from host ftp.cis.upenn.edu and the file pub/papers/miller/ic92.dvi.Z.

[21] J. Hudelmaier. *Bounds for Cut Elimination in Intuitionistic Propositional Logic*. PhD thesis, Universität Tübingen, 1989.

[22] M. Kanovich. Horn programming in linear logic is NP-complete. In *Proc. 7-th Annual IEEE Symposium on Logic in Computer Science, Santa Cruz, California*, pages 200–210. IEEE Computer Society Press, Los Alamitos, California, June 1992. Full paper to appear in *Annals of Pure and Applied Logic*.

[23] P. Lincoln, J. Mitchell, A. Scedrov, and N. Shankar. Decision problems for propositional linear logic. *Annals of Pure and Applied Logic*, 56:239–311, 1992. Special Volume dedicated to the memory of John Myhill.

[24] P. Lincoln and A. Scedrov. First order linear logic without modalities is NEXPTIME-hard. Accepted for publication in *Theoretical Computer Science*, 1994. Available by anonymous ftp from host ftp.cis.upenn.edu and the file pub/papers/scedrov/mall1.dvi.

[25] P. Lincoln, A. Scedrov, and N. Shankar. Linearizing intuitionistic implication. *Annals of Pure and Applied Logic*, 60:151–177, 1993.

[26] P. Lincoln and N. Shankar. Proof search in first order linear logic. Manuscript, October 1993.

[27] P. Lincoln and T. Winkler. Constant-Only Multiplicative Linear Logic is NP-Complete. Manuscript, September 1992. Available using anonymous ftp from host ftp.csl.sri.com and the file pub/lincoln/comult-npc.dvi.

[28] P.D. Lincoln. *Computational Aspects of Linear Logic*. PhD thesis, Stanford University, 1992.

[29] D. Miller. Abstractions in logic programming. In P. Odifreddi, editor, *Logic and Computer Science*, pages 329–359. APIC Studies in Data Processing, Vol. 31, Academic Press, 1990.

[30] D. Miller. The π-calculus as a theory in linear logic: Preliminary results. In E. Lamma and P. Mello, editors, *Proceedings of the 1992 Workshop on Extensions to Logic Programming*, number 660 in Lecture Notes in Computer Science, pages 242–265. Springer-Verlag, 1993. Available using anonymous ftp from host ftp.cis.upenn.edu and the file pub/papers/miller/pic.dvi.Z.

[31] D. Miller, G. Nadathur, F. Pfenning, and A. Scedrov. Uniform proofs as a foundation for logic programming. *Annals of Pure and Applied Logic*, 51:125–157, 1991. Special Issue on the 2-nd Annual IEEE Symposium on Logic in Computer Science, 1987.

[32] A. Scedrov. A brief guide to linear logic. In G. Rozenberg and A. Salomaa, editors, *Current Trends in Theoretical Computer Science*, pages 377–394. World Scientific Publishing Co., 1993.

[33] N. Shankar. Proof search in the intuitionistic sequent calculus. In *Proc. of the 11th International Conference on Automated Deduction*, pages 522–536. Springer LNAI, vol. 607, 1992.

[34] E.Y. Shapiro. Alternation and the computational complexity of logic programs. *Journal of Logic Programming*, 1:19–33, 1984.

[35] L. Stockmeyer. Classifying the computational complexity of problems. *Journal of Symbolic Logic*, 52:1–43, 1987.

[36] A.S. Troelstra. *Lectures on Linear Logic.* CSLI Lecture Notes No. 29. Center for the Study of Language and Information, Stanford University, 1992.

Computable Functions on Stream Algebras

J.V. Tucker[1]
Department of Computer Science,
University College of Swansea, Swansea SA2 8PP, Wales

J.I. Zucker[2]
Department of Computer Science and Systems,
McMaster University, Hamilton, Ontario L8S 4K1, Canada

Abstract. Let A be a many-sorted algebra. A stream over A is an infinite sequence of data from A. We study classes of computable functions on streams over A. We show how models of deterministic parallel computation on A can be adapted to provide new models of computation on stream algebras over A. The primary models considered are simultaneous primitive recursion schemes, with and without the least number operator, over A. We also look at an imperative model involving concurrent assignments.

1 Introduction

We will address the problem of establishing the scope and limits of deterministic computing on many-sorted algebras with streams.

1.1 The problem

For simplicity, suppose that A is a single sorted algebra, say

$$A = (A; f_1, \ldots, f_k)$$

where $f_i : A^{n(i)} \to A$ is an operation, $n(i) \geq 0$ and $1 \leq i \leq k$. A *stream* is an infinite sequence $\xi(0), \xi(1), \ldots$ of data from A, represented by a function $\xi : \mathbb{N} \to A$. Let $[\mathbb{N} \to A]$ be the set of all streams. We consider computing a mapping of the form

$$f : [\mathbb{N} \to A]^m \to [\mathbb{N} \to A]$$

which may be called a *stream transformer*; also, we consider such functions in their uncurried or cartesian form

$$\mathit{cart}(f) : [\mathbb{N} \to A]^m \times \mathbb{N} \to A$$

[1] Research partially supported by UK SERC Research Contract SO/103/91 and the ESPRIT Basic Research Action NADA.

[2] Research partially supported by grants from the Natural Sciences and Engineering Research Council of Canada (NSERC), and by the Telecommunications Research Institute of Ontario (TRIO).

defined by $\boldsymbol{cart}(f)(\xi, n) = f(\xi)(n)$ for $\xi \in [\mathsf{N} \rightarrow A]$ and $n \in \mathsf{N}$. Note that $\boldsymbol{cart}$ is a bijection between function spaces.

We ask the following questions.

For any algebra A, what are the computable stream transformations over A?

What is their relation to the computable functions on A?

1.2 An algebraic approach

To answer these questions, we extend the algebra A to a new algebra $\bar{A}$ by adding N and $[\mathsf{N} \rightarrow A]$ and some appropriate operations. There are many choices for these operations but we will work with the simplest: let

$$\bar{A} = (A,\ \mathsf{N},\ [\mathsf{N} \rightarrow A];\ f_1, \ldots, f_k,\ 0,\ \mathsf{S},\ \mathsf{eval})$$

where $0 \in \mathsf{N}$, $\mathsf{S} : \mathsf{N} \rightarrow \mathsf{N}$ is the successor $\mathsf{S}(n) = n + 1$ for $n \in \mathsf{N}$, and $\mathsf{eval} : [\mathsf{N} \rightarrow A] \times \mathsf{N} \rightarrow A$ is defined by $\mathsf{eval}(\xi, n) = \xi(n)$ for $\xi \in [\mathsf{N} \rightarrow A]$ and $n \in \mathsf{N}$. We call $\bar{A}$ the ***stream algebra*** over A.

Consider any model of computation MC that can be applied to an arbitrary algebra A to define a set MC(A) of computable functions on A. Then MC can be applied to $\bar{A}$ to obtain a model of stream computation, and a set

$$\mathrm{MC}(\bar{A})$$

of computable functions on $\bar{A}$, including computable stream transformations. Furthermore, we consider the construct that performs the following operation which is unique to streams: from any function

$$g : D \times \mathsf{N} \rightarrow A,$$

where D is any cartesian product of carriers of $\bar{A}$, construct the function

$$\boldsymbol{\lambda abs}(g) : D \rightarrow [\mathsf{N} \rightarrow A],$$

defined by

$$(\boldsymbol{\lambda abs}(g))(d)(n) = g(d, n)$$

for $n \in \mathsf{N}$ and $d \in D$. We will call this operation ***stream abstraction***. The addition of this construct to the model of computation MC leads to the set

$$\lambda\mathrm{MC}(\bar{A})$$

of computable functions on $\bar{A}$.

The theory of computable functions on algebras contains many models of computation MC for defining functions on an algebra A. For example, see the imperative, functional, equational, logical, set theoretic models surveyed in Greibach [1975], Tucker [1980], Fitting [1981], Shepherdson [1985], Engeler [1993] and (especially) Tucker &

Zucker [1988]. Many of these models define the same distinguished class of computable functions on any A, for which a generalised Church-Turing Thesis can be assumed. With varying difficulties surrounding the construct for stream abstraction, these models may be extended to define functions on $\bar{A}$.

In computing on $\bar{A}$, three types of function must be considered: *stream valued functions*, typically of the form

$$f : [\mathrm{N} \to A]^m \times A^p \to [\mathrm{N} \to A];$$

simple valued functions of the form

$$f : [\mathrm{N} \to A]^m \times A^p \to A;$$

and functions on A with *simple values and arguments* of the form

$$f : A^p \to A.$$

As we will see, these functions have rather different technical properties.

For any appropriate model MC and stream abstraction construct we may ask questions of the form:

(*i*) ***Conservativity:*** For any function f on A, is it the case that

$$f \in \mathrm{MC}(\bar{A}) \iff f \in \mathrm{MC}(A)?$$

(*ii*) ***Invariance under cartesian form:*** For any function f on $\bar{A}$, is it the case that

(*a*) $$f \in \mathrm{MC}(\bar{A}) \iff \mathbf{cart}(f) \in \mathrm{MC}(\bar{A})?$$

(*b*) $$f \in \lambda\mathrm{MC}(\bar{A}) \iff \mathbf{cart}(f) \in \lambda\mathrm{MC}(\bar{A})?$$

(*iii*) ***λ-Elimination using cartesian forms:*** For any function f on $\bar{A}$, is it the case that

$$f \in \lambda\mathrm{MC}(\bar{A}) \iff f \in \mathrm{MC}(\bar{A})?$$

(*iv*) ***λ-Elimination + Cartesian form:*** For any function f on $\bar{A}$, is it the case that

$$f \in \lambda\mathrm{MC}(\bar{A}) \iff \mathbf{cart}(f) \in \mathrm{MC}(\bar{A})?$$

(*v*) ***Equivalence of models:*** Given two models of computation MC_1 and MC_2, is it the case that

$$\mathrm{MC}_1(\bar{A}) = \mathrm{MC}_2(\bar{A}) \quad \text{or} \quad \lambda\mathrm{MC}_1(\bar{A}) = \lambda\mathrm{MC}_2(\bar{A})?$$

(*vi*) ***Completeness:*** Is there a canonical class of stream transformations for which a *generalised Church-Turing Thesis for deterministic stream computation* may be proposed?

The investigation of each question involves a translation of stream processing programs.

1.3 Recursive functions on stream algebras

We will examine the above questions for a small range of models of computation: functional models based on generalised Kleene schemes and, in a secondary role, imperative models based on 'for' programs. The primary models are

PR	simultaneous primitive recursive functions,		
PR*	simultaneous primitive recursive functions	+	finite sequences,
μPR	simultaneous primitive recursive functions	+	least number search,
μPR*	simultaneous primitive recursive functions	+	least number search
		+	finite sequences.

For any algebra A, we have in general

$$\mathrm{PR}(A) \subsetneqq \mathrm{PR}^*(A) \qquad \text{and} \qquad \mu\mathrm{PR}(A) \subsetneqq \mu\mathrm{PR}^*(A).$$

The μPR* computable functions on A are the functions that characterise the limit of deterministic computation on A (see Tucker & Zucker [1988]).

The above classes are trivially extended to make the corresponding new stream models

$$\lambda\mathrm{PR}, \quad \lambda\mathrm{PR}^*, \quad \lambda\mu\mathrm{PR} \quad \text{and} \quad \lambda\mu\mathrm{PR}^*.$$

First, we check that the addition of streams to A to make $\bar{A}$ does not increase the computing power of the models of computation on A, *i.e.*, we give positive answers to some of the conservativity questions (*i*) above: for any function f on A,

$$f \in \mathrm{PR}(\bar{A}) \iff f \in \mathrm{PR}(A);$$

and similarly for PR*, μPR and μPR* (§5.1). Conservativity can however be violated for weaker models by adding further operations to $\bar{A}$ (§5.2).

The simple operations in the signature of $\bar{A}$ limit the power of the models in computing stream transformations on $\bar{A}$. In Thompson [1987] it is observed that for any

$$f : [\mathbb{N} \to A]^m \to [\mathbb{N} \to A]$$

if $f \in \mathrm{PR}(\bar{A})$ then f is essentially a projection function, and hence there are many such f such that

$$f \notin \mathrm{PR}(\bar{A}) \qquad \text{but} \qquad \boldsymbol{cart}(f) \in \mathrm{PR}(\bar{A}).$$

This observation applies to PR^*, μPR and μPR^* (Remark in §5.3). Thus (*ii*) is answered negatively, and computability is not invariant with respect to cartesian forms.

Thirdly, we consider the addition of stream abstraction, which is designed to allow more stream valued functions to be defined by models of computation. Thus we have, for example,

$$f \in \lambda PR(\bar{A}) \iff \textit{cart}(f) \in \lambda PR(\bar{A}),$$

and similarly for λPR^*, $\lambda\mu PR$ and $\lambda\mu PR^*$ (§5.3, Proposition 1). However, we have an example of a function f on an algebra A such that

$$f \in \lambda PR(\bar{A}) \quad \text{but} \quad f \notin PR(\bar{A}), \tag{1}$$

answering (*iii*) and also (*iv*) negatively (Theorem 3 in §5.3).

The case of λPR warrants close examination. We define four submodels of PR and λPR for stream computation.

Let $PR_0(\bar{A})$ be the set of functions on $\bar{A}$ defined by restricting all the schemes of PR to simple valued functions. We also define the set $PR_0^\lambda(\bar{A})$ of functions on $\bar{A}$ by adding to PR_0 a form of stream abstraction hidden inside composition. We show that for simple valued functions

$$PR_0^\lambda(\bar{A}) = PR_0(\bar{A}) = PR(\bar{A}) \tag{2}$$

(Section 7). In particular, these models are weaker than λPR (already over A, by the example in (1)).

Next, let $PR_-(\bar{A})$ be the set of functions on $\bar{A}$ defined by restricting only the primitive recursion scheme of PR to simple valued functions. Let $\lambda PR_-(\bar{A})$ be the set of functions on $\bar{A}$ defined by adding stream abstraction to PR_-. We show that for simple valued functions

$$\lambda PR_-(\bar{A}) = PR_-(\bar{A}) = PR(\bar{A})$$

and so these are also equal to the other models named in (2) (Sections 6 and 7). In particular, λPR_- is weaker, in general, than λPR. So *the enhancement of power of PR by stream abstraction is the result of the combination of stream abstraction and unrestricted primitive recursion over streams.*

Since the computable functions on A are those in $\mu PR^*(A)$, we are particularly interested in $\mu PR^*(\bar{A})$ and $\lambda\mu PR^*(\bar{A})$. Among a number of results we show

$$\lambda\mu PR(\bar{A}) = \lambda\mu PR^*(\bar{A})$$

(Theorem 5 in §5.5).

Sections 2,3 and 4 contain preliminaries concerning algebras, stream algebras and computable functions. Sections 5, 6 and 7 prove the results (but not in the order presented above). Section 8 contains concluding remarks.

Some of the results were announced in Tucker & Zucker [1992a]. We are grateful to the Directors of the NATO Summer School on Proof and Computation, Marktoberdorf, August 1993, for the opportunity to present a full account of these results. We also thank B C Thompson and R Stephens for useful discussions: they have an independent proof of $\mathrm{PR}_0^\lambda(\bar{A}) = \mathrm{PR}_0(\bar{A})$ from the perspective of composition of stream transformations (Stephens & Thompson [1995]).

2 Standard signatures and algebras

In this and the next section we briefly review some concepts defined and discussed in Tucker & Zucker [1988, 1989, 1991, 1992a,b,c, 1993, 1994] and Tucker, Wainer & Zucker [1990]. Detailed treatments can be found in Chapter 4 of Tucker & Zucker [1988] and Part A of Tucker & Zucker [1992a]. Background information on universal algebra can be found in Ehrig & Mahr [1985], Wechler [1991] and Meinke & Tucker [1992].

2.1 Standard signatures; Standard algebras

A *standard signature* Σ specifies (1) a finite set ***Sort***(Σ) of *sorts*, including the *numerical sort* N and *boolean sort* B; and (2) a finite set ***Func***(Σ) of *function symbols* F, each having a *type* $s_1 \times \cdots \times s_m \to s$, where $m \geq 0$ is the arity of F, $s_1, \ldots, s_m$ are the *domain sorts* and s is the *range sort* (including the case $m = 0$ for *constant symbols*). These include symbols for certain *standard operations* associated with the sorts N and B: (*a*) *arithmetical operations*, namely the constant '0', successor operation S and order relation '$<$' on the natural numbers; and (*b*) *boolean operations*, including a complete set of propositional connectives, the constants true and false, and an *equality* operator $\mathbf{eq}_s$ at some sorts s , including (at least) $s = \mathsf{N}$ and $s = \mathsf{B}$. We call those sorts s with the equality operator $\mathbf{eq}_s$, *equality sorts*.

We will only consider standard signatures Σ, and we will sometimes make one further assumption on Σ:

INSTANTIATION ASSUMPTION. *Each sort of* Σ *is* instantiated, *i.e. there is a closed term of each sort.*

A Σ-algebra A has, for each sort s of Σ, a carrier set A_s, and for each function symbol F of type $s_1 \times \cdots \times s_m \to s$, a function $F^A : A_{s_1} \times \cdots \times A_{s_m} \to A_s$. The algebra A is *standard* if $A_{\mathsf{N}} = \mathbb{N}$, the set of natural numbers, $A_{\mathsf{B}} = \mathbb{B} = \{\mathtt{tt}, \mathtt{ff}\}$, the set of truth values, and the standard operations have their *standard interpretations* on $\mathbb{N}$ and $\mathbb{B}$; in particular, the equality symbol $\mathbf{eq}_s$ is interpreted as identity on each equality sort s.

Let ***StdAlg***(Σ) denote the class of standard Σ-algebras.

We only consider standard Σ-algebras. Note however that any many-sorted algebra B can be *standardised* to such an algebra A by the adjunction of the sets $\mathbb{N}$ and $\mathbb{B}$, together with their standard operations, so that B is a reduct of A to the signature of B.

This notion of *reduct* will be important in what follows, so we define it now.

Definitions. Let Σ and Σ' be standard signatures.

(1) We write $\Sigma \subseteq \Sigma'$ to mean $\mathit{Sort}(\Sigma) \subseteq \mathit{Sort}(\Sigma')$ and $\mathit{Func}(\Sigma) \subseteq \mathit{Func}(\Sigma')$

(2) Suppose $\Sigma \subseteq \Sigma'$. Let A and A' be standard algebras over Σ and Σ' respectively.

(*i*) The *Σ-reduct $A'|_\Sigma$ of A, is the standard algebra of signature Σ, consisting of the carriers of A' named by the sorts of Σ and equipped with the functions of A' named by the function symbols of Σ.*

(*ii*) *A' is a Σ'-expansion of A iff A is the Σ-reduct of A'.*

2.2 The unspecified value u; Structures A^u of signature Σ^u

Let A be a standard Σ-algebra satisfying the Instantiation Assumption. For each Σ-sort s, let $\mathbf{u}_s$ be a new object, representing an "unspecified value", and let $A_s^u = A_s \cup \{\mathbf{u}_s\}$. For each function symbol F of Σ of type $s_1 \times \cdots \times s_m \to s$, extend its interpretation F^A on A to a function

$$F^{A,u} : A_{s_1}^u \times \cdots \times A_{s_m}^u \to A_s^u$$

by *strictness* — *i.e.*, the value is defined as $\mathbf{u}$ whenever any argument is $\mathbf{u}$. Then the algebra A^u with signature Σ^u, contains:

(*i*) the original carriers A_s of sort s, and functions F^A on them;

(*ii*) the new carriers A_s^u of sort s^u, and functions $F^{A,u}$ on them;

(*iii*) a constant unspec_s of sort s^u to denote $\mathbf{u}$ as a distinguished element of A_s^u;

(*iv*) a boolean-valued function Unspec_s of type $s^u \to \mathsf{B}$, the characteristic function of $\mathbf{u}_s$ (*i.e.*, with value $\mathbf{tt}$ on $\mathbf{u}_s$ and $\mathbf{ff}$ everywhere else on A_s^u);

(*v*) an *embedding function* i_s of type $s \to s^u$, to denote the embedding of A_s into A_s^u, and an inverse function j_s of type $s^u \to s$, which maps $\mathbf{u}_s$ to the denotation of some closed term in A_s (because of the Instantiation Assumption) for each sort s; and finally

(*vi*) an *equality* operator on A_s^u for each equality sort s.

Remarks. (1) The algebra A^u is a *Σ^u-expansion* of A. If Σ has $r+2$ sorts ($r \geq 0$), then Σ^u has $2r+4$ sorts.

(2) A^u can be *effectively coded within* A. Each element y of A_s^u is represented by the *pair* $(b, \mathsf{j}_s(y)) \in \mathbb{B} \times A_s$, where $b = \mathbf{tt}$ if $y \neq \mathbf{u}_s$ and $b = \mathbf{ff}$ otherwise. This induces, in an obvious way, a coding of the operations on A^u by operations on A. (See later, §5.5.)

2.3 Structures A^* of signature Σ^*

Define, for each Σ-sort s, the carrier A_s^* to be the set of pairs $a^* = (\xi, l)$ where $\xi : \mathsf{N} \to A_s^u$, $l \in \mathsf{N}$ and for all $n \geq l$, $\xi(n) = \mathbf{u}_s$. So l is a witness to the "finiteness" of ξ, or an "effective upper bound" for a^*. The elements of A_s^* have "starred sort" s^*, and can be considered as finite sequences or *arrays*. The resulting algebras A^* have signature Σ^*, which extends Σ^{u} by including, for each sort s, the new "starred sorts" s^* as well as s^{u}, and also the following new function symbols:

(i) the Lgth_s operator, of type $s^* \to \mathsf{N}$, where

$$\mathsf{Lgth}_s^A((\xi,l)) = l,$$

(ii) the application operator Ap_s of type $s^* \times \mathsf{N} \to s^{\mathrm{u}}$, where

$$\mathsf{Ap}_s^A((\xi,l),n) = \xi(n),$$

(iii) the null array Null_s of type s^*, where

$$\mathsf{Null}_s^A = (\lambda n \cdot \mathbf{u}_s, 0) \in A_s^*,$$

(iv) the Update_s operator of type $s^* \times \mathsf{N} \times s^{\mathrm{u}} \to s^*$, where for $(\xi,l) \in A_s^*$, $n \in \mathbb{N}$ and $x \in A_s^{\mathrm{u}}$, $\mathsf{Update}_s^A((\xi,l),n,x)$ is the array $(\eta,l) \in A_s^*$ such that for all $k \in \mathbb{N}$,

$$\eta(k) = \begin{cases} \xi(k) & \text{if } k < l, k \neq n \\ x & \text{if } k < l, k = n \\ \mathbf{u}_s & \text{otherwise,} \end{cases}$$

(v) the $\mathsf{Newlength}_s$ operator of type $s^* \times \mathsf{N} \to s^*$, where $\mathsf{Newlength}_s^A((\xi,l),m)$ is the array (η,m) such that for all k,

$$\eta(k) = \begin{cases} \xi(k) & \text{if } k < m \\ \mathbf{u}_s & \text{if } k \geq m, \end{cases}$$

and (vi) the *equality* operator on s^*, for each *equality sort* s.

(The justification for this is that if a sort s has "computable" equality, then clearly so has the sort s^*, since it amounts to testing equality of finitely many pairs of objects of sort s, up to a "computable" length.)

For $a^* \in A_s^*$ and $n \in \mathbb{N}$, we write $a^*[n]$ for $\mathsf{j}_s^A(\mathsf{Ap}_s^A(a^*,n))$. (So $a^*[n]$ is the element of A_s "corresponding to" $\mathsf{Ap}(a^*,n) \in A_s^{\mathrm{u}}$.)

REMARKS. (1) The algebra A^* is a Σ^*-*expansion* of A and of A^{u}. If Σ has $r+2$ sorts ($r \geq 0$), then Σ^* has $3r+6$ sorts.

(2) The "doubly starred" algebra $(A^*)^*$ can be effectively coded in A^*, since we can effectively code a finite sequence of starred objects of a given sort as a single starred object of the same sort, thanks to the explicit Lgth operation. More precisely, a sequence $x_0^*, \ldots, x_{k-1}^*$ of elements of A_h^* (for some sort h) can be coded as a *pair* $(y^*, n^*) \in A_h^* \times \mathsf{N}^*$, where $\mathsf{Lgth}(n^*) = k$, and for $0 \leq j < k$, $n^*[j] = \mathsf{Lgth}(x_j^*)$, and $\mathsf{Lgth}(y^*) = n^*[0]+\cdots+n^*[k]$, and for $1 \leq j \leq k$ and $0 \leq i < n^*[j]$, $y^*[n^*[0]+\cdots+n^*[j-1]+i] = x_j^*[i]$. (See later, §5.5.)

2.4 Homomorphisms and isomorphisms

Let Σ be a standard signature. The notions of *Σ-homomorphism* and *Σ-isomorphism* are defined as usual (see, *e.g.*, Meinke & Tucker [1992, §3.4]). We need a more sophisticated notion.

DEFINITION. Let Σ and Σ' be standard signatures with $\Sigma \subseteq \Sigma'$. Let B_1 and B_2 be two standard Σ'-algebras such that

$$B_2 \,|\, _\Sigma = B_1 \,|\, _\Sigma.$$

(1) A *Σ'/Σ-homomorphism* from B_1 to B_2 is a $\boldsymbol{Sort}(\Sigma')$-indexed family of mappings

$$\phi = \langle \phi_s : B_{1s} \to B_{2s} \mid s \in \boldsymbol{Sort}(\Sigma') \rangle$$

which is a Σ'-homomorphism from B_1 to B_2, such that for all $s \in \boldsymbol{Sort}(\Sigma)$, ϕ_s is the *identity* on B_{1s}.

(2) A *Σ'/Σ-isomorphism* from B_1 to B_2 is a Σ'/Σ-homomorphism which is also a Σ'-isomorphism from B_1 to B_2.

(3) B_1 and B_2 are *Σ'/Σ-isomorphic*, written $B_1' \cong_{\Sigma'/\Sigma} B_2'$, if there is a Σ'/Σ-isomorphism from B_1' to B_2'.

2.5 Product types and function types

We collect some definitions. Consider a standard signature Σ.

(1) A *product type* over Σ, or *Σ-product type*, is a symbol of the form $s_1 \times \cdots \times s_m$ ($m \geq 0$), where $s_1, \ldots, s_m$ are sorts of Σ. Its *component sorts* are $s_1, \ldots, s_m$. We define $\boldsymbol{Prod}(\Sigma)$ to be the set of Σ-product types, with elements $u, v, w, \ldots$.

(2) If $u = s_1 \times \cdots \times s_m$, we put $m = \boldsymbol{lgth}(u)$, the *length* of u. When $\boldsymbol{lgth}(u) = 1$, we identify u with its component sort. When $\boldsymbol{lgth}(u) = 0$, u is the *empty product sort*.

(3) For a Σ-product type u and Σ-sort s, let $\boldsymbol{Func}(\Sigma)_{u \to s}$ denote the set of all Σ-function symbols of type $u \to s$.

Consider an algebra A of signature Σ.

(4) For $u = s_1 \times \cdots \times s_m$, we write

$$A^u =_{df} A_{s_1} \times \cdots \times A_{s_m}.$$

Thus $x \in A^u$ iff $x = (x_1, \ldots, x_m)$, where $x_i \in A_{s_i}$ for $i = 1, \ldots, m$. Also, each Σ-function symbol F of type $u \to s$ has an interpretation $F^A : A^u \to A_s$. (If u is empty, *i.e.*, F is a constant symbol, then F^A is an element of A_s.)

(5) A *function type* over Σ, or *Σ-function type*, is a symbol of the form $u \to v$, with *domain type* u and *range type* v (both Σ-product types).

(6) For any Σ-function type $u \to v$, a *function of type $u \to v$ over A* is a function

$$f : A^u \to A^v. \qquad (*)$$

Its *component functions* (putting $v = s_1 \times \cdots \times s_n$) are $f_1, \ldots, f_n$, where

$$f_j : A^u \to A_{s_j} \qquad (**)$$

for $j = 1, \ldots, n$, and for all $x \in A^u$,

$$f(x) \simeq (f_1(x), \ldots, f_n(x)). \qquad (***)$$

(We will explain the '$\simeq$' in (7) below.) Conversely, given n functions f_j as in (**), all with the same domain type u, and with range types (or sorts) $s_1, \ldots, s_n$ respectively, we can form their *vectorisation* as a function f satisfying (*) and (***).

We will investigate computable *vector valued functions* (*) over A.

(7) Although all the functions of our signatures Σ are *total*, the computable functions on the Σ-algebras may very well be *partial* (see, *e.g.*, §3.3). We use the following notation: if $f : A^u \to A_s$ and $x \in A^u$, then $f(x)\uparrow$ ("$f(x)$ diverges") means that $x \notin \mathit{dom}(()f)$; $f(x)\downarrow$ ("$f(x)$ converges") means that $x \in \mathit{dom}(()f)$; and $f(x)\downarrow y$ ("$f(x)$ converges to y") means that $x \in \mathit{dom}(()f)$ and $f(x) = y$.

We also make the following convention for convergence of vector valued functions: in the notation of (*) and (**), for any $x \in A^u$, we say that $f(x)\downarrow$ if and only if $f_j(x)\downarrow$ for *every component function* f_j of f, in which case $f(x) = (f_1(x), \ldots, f_n(x))$. Otherwise (*i.e.*, if $f_j(x)\uparrow$ for any j with $1 \le j \le n$), we say that $f(x)\uparrow$. (That is the meaning of the symbol '$\simeq$' in (***) above.)

3 Computation on standard algebras

We will consider four notions of computability on standard many-sorted algebras. They will be formalised by *schemes*, which apply uniformly to *all* algebras of some fixed standard signature Σ. We write $\alpha, \beta, \ldots$ for schemes.

Each scheme α will have a fixed type $u \to v$, with *domain type* u and *range type* v, both product types over Σ; we will also write $\alpha : u \to v$. The semantics of such a scheme, for each $A \in \mathbf{StdAlg}(\Sigma)$, will then be a function

$$[\![\alpha]\!]^A : A^u \to A^v.$$

We will usually write α^A for $[\![\alpha]\!]^A$.

3.1 PR computability

Given a standard signature Σ, we will define the family

$$\mathrm{PR}(\Sigma) = \langle \mathrm{PR}(\Sigma)_{u \to v} \mid u, v \in \mathbf{Prod}(\Sigma) \rangle.$$

where $\mathrm{PR}(\Sigma)_{u \to v}$ is the set of schemes of type $u \to v$ over Σ. Thus for any scheme

$\alpha \in \mathrm{PR}(\Sigma)_{u\to v}$ and any $A \in \boldsymbol{StdAlg}(\Sigma)$, α defines a function on A:

$$\alpha^A : A^u \to A^v$$

of type $u \to v$. These schemes generalise the schemes for primitive recursive functions over $\mathbb{N}$ in Kleene [1952]. They are generated as follows.

Basic schemes.

(*i*) *Initial functions and constants.* For each function symbol $F \in \boldsymbol{Func}(\Sigma)_{u\to s}$, there is a scheme $F \in \mathrm{PR}(\Sigma)_{u\to s}$. On each $A \in \boldsymbol{StdAlg}(\Sigma)$, it defines the function

$$F^A : A^u \to A_s.$$

(*ii*) *Projection.* For all $m > 0$, $u = s_1 \times \cdots \times s_m$ and i with $1 \le i \le m$, there is a scheme $\boldsymbol{U}_{u,i} \in \mathrm{PR}(\Sigma)_{u\to s_i}$. It defines the function $\boldsymbol{U}_{u,i}{}^A : A^u \to A_{s_i}$ on each $A \in \boldsymbol{StdAlg}(\Sigma)$, where

$$\boldsymbol{U}^A_{u,i}(x_1,\dots,x_m) \;=\; x_i$$

for all $(x_1,\dots,x_m) \in A^u$.

(*iii*) *Definition by Cases.* For every Σ-sort s there is a scheme $\boldsymbol{dc} \in \mathrm{PR}(\Sigma)_{\mathsf{B}\times s\times s\to s}$. It defines the function $\boldsymbol{dc}^A : \mathbb{B} \times A_s{}^2 \to A_s$ on each $A \in \boldsymbol{StdAlg}(\Sigma)$, where

$$\boldsymbol{dc}^A(b,x,y) = \begin{cases} x & \text{if } b = \mathbf{tt} \\ y & \text{if } b = \mathbf{ff} \end{cases}$$

for all $x, y \in A_s$.

Induction: Building new schemes from old

(*iv*) *Vectorisation.* For all Σ-product types u, v, where $v = s_1 \times \cdots \times s_n$, and for all schemes $\beta_1,\dots,\beta_n$, where $\beta_i \in \mathrm{PR}(\Sigma)_{u\to s_i}$ for $i = 1,\dots,n$, there is a scheme $\alpha \equiv \boldsymbol{vec}_{u,v}(\beta_1,\dots,\beta_n) \in \mathrm{PR}(\Sigma)_{u\to v}$. It defines the function $\alpha^A : A^u \to A^v$ on each $A \in \boldsymbol{StdAlg}(\Sigma)$, where

$$\alpha^A(x) \;=\; (\beta_1^A(x),\dots,\beta_n^A(x))$$

for all $x \in A^u$.

(*v*) *Composition.* For all Σ-product types u, v, w, and for all schemes $\beta \in \mathrm{PR}(\Sigma)_{u\to v}$ and $\gamma \in \mathrm{PR}(\Sigma)_{v\to w}$ there is a scheme $\alpha \equiv \boldsymbol{com}_{u,v,w}(\beta,\gamma) \in \mathrm{PR}(\Sigma)_{u\to w}$. It defines the function $\alpha^A \;=\; \gamma^A \circ \beta^A : A^u \to A^w$ on each $A \in \boldsymbol{StdAlg}(\Sigma)$, where

$$\alpha^A(x) \;=\; \gamma^A(\beta^A(x))$$

for all $x \in A^u$.

(*vi*) *Simultaneous Primitive Recursion.* For all Σ-product types u, v, w, and for all schemes $\beta \in \mathrm{PR}(\Sigma)_{u\to v}$ and $\gamma \in \mathrm{PR}(\Sigma)_{\mathsf{N}\times u\times v\to v}$ there is a scheme $\alpha \equiv \boldsymbol{rec}_{u,v}(\beta,\gamma) \in \mathrm{PR}(\Sigma)_{\mathsf{N}\times u\to v}$. It defines the function $\alpha^A : \mathbb{N}\times A^u \to A^v$ on each $A \in \boldsymbol{StdAlg}(\Sigma)$, where

$$\begin{aligned} \alpha^A(0,x) &= \beta^A(x) \\ \alpha^A(z+1,x) &= \gamma^A(z,x,\alpha^A(z,x)) \end{aligned}$$

for all $x \in A^u$ and $z \in \mathbb{N}$.

Now, for any $A \in \boldsymbol{StdAlg}(\Sigma)$, we define

$$\mathrm{PR}(A) = \langle \mathrm{PR}(A)_{u\to v} \mid u,v \in \boldsymbol{Prod}(\Sigma)\rangle$$

where $$\mathrm{PR}(A)_{u\to v} = \{\alpha^A \mid \alpha \in \mathrm{PR}(\Sigma)_{u\to v}\}.$$

3.2 PR* computability

It turns out that a broader class of functions provides a closer generalisation of the notion of primitive recursiveness, namely *PR*(Σ) computability*. We define $\mathrm{PR}^*(\Sigma)$ to be the class of $\mathrm{PR}(\Sigma^*)$ schemes for which the domain and range types are in Σ, *i.e.*,

$$\mathrm{PR}^*(\Sigma) =_{df} \langle \mathrm{PR}(\Sigma^*)_{u\to v} \mid u,v \in \boldsymbol{Prod}(\Sigma)\rangle \subsetneq \mathrm{PR}(\Sigma^*).$$

Then any such scheme $\alpha \in \mathrm{PR}^*(\Sigma)_{u\to v}$ defines a function $\alpha^A : A^u \to A^v$ on each $A \in \boldsymbol{StdAlg}(\Sigma)$.

Also $\mathrm{PR}^*(A)$ is the set of $\mathrm{PR}^*(\Sigma)$-computable functions on A.

3.3 μPR computability

The class of μPR *schemes* over Σ:

$$\mu\mathrm{PR}(\Sigma) = \langle \mu\mathrm{PR}(\Sigma)_{u\to v} \mid u,v \in \boldsymbol{Prod}(\Sigma)\rangle$$

are formed by adding to the PR schemes of §3.1 the schemes generated by the following:

(*vii*) *Least number* or *μ operator.* For all Σ-product types u and for all schemes $\beta \in \mu\mathrm{PR}_{u\times\mathsf{N}\to\mathsf{B}}$ there is a scheme $\alpha \equiv \boldsymbol{min}_u(\beta) \in \mu\mathrm{PR}(\Sigma)_{u\to\mathsf{N}}$. It defines the function $\alpha^A : A^u \to \mathbb{N}$ on each $A \in \boldsymbol{StdAlg}(\Sigma)$, where for all $x \in A^u$,

$$\alpha^A(x) \simeq \mu z[\beta^A(x,z) = \mathbf{tt}].$$

That is, $\alpha^A(x) \downarrow z$ if, and only if, $\beta^A(x,y) \downarrow \mathbf{ff}$ for each $y < z$ and $\beta^A(x,z) \downarrow \mathbf{tt}$.

Also $\mu\mathrm{PR}(A)$ is the set of $\mu\mathrm{PR}(\Sigma)$-computable functions on A.

Note, this scheme (as well as that for simultaneous primitive recursion) uses the standardness of the algebra. Also, μPR computable functions are, in general, *partial*.

The above schemes generalise the schemes for partial recursive functions over N given in Kleene [1952].

3.4 μPR* computability

Again, a broader class turns out to be more useful, namely *μPR* computability.* The class $\mu\text{PR}^*(\Sigma)$ consists of those $\mu\text{PR}(\Sigma^*)$ schemes for which the domain and range types are in Σ, *i.e.*,

$$\mu\text{PR}^*(\Sigma) =_{df} \langle \mu\text{PR}(\Sigma^*)_{u\to v} \mid u, v \in \textbf{\textit{Prod}}(\Sigma) \rangle \subsetneqq \mu\text{PR}(\Sigma^*).$$

Also for any $A \in \textbf{\textit{StdAlg}}(\Sigma)$, $\mu\text{PR}^*(A)$ is the set of $\mu\text{PR}^*(\Sigma)$-computable functions on A.

3.5 Generalised Church-Turing Thesis for deterministic computation

There are a number of other models of computability, due to Moschovakis, Friedman, Shepherdson and others, which turn out to be equivalent to μPR* computability: see Tucker and Zucker [1988, §4.10]. All these equivalences have led to the postulation of a *Generalised Church-Turing Thesis for deterministic computation of functions*, which can be formulated as follows:

> *Any function on a many-sorted algebra that is computable by a deterministic algorithm is computable by a μPR* scheme.*

3.6 Models of computability with imperative programming languages

It is interesting to compare the above notions of computability involving schemes with notions of computability involving imperative programming languages. Consider programming languages for $\textbf{\textit{StdAlg}}(\Sigma)$ in which the main control structure is either a 'for' or a 'while' loop. Also, the language may or may not contain *arrays*. Thus there are four possible languages, and four corresponding notions of computability: 'for'-computability, 'for'-array-computability, 'while'-computability, and 'while'-array-computability. These can be shown to match up with the above models as follows. For any standard Σ-algebra A,

(1) PR(A)-computability $\iff$ 'for'-computability on A;

(2) PR*(A)-computability $\iff$ 'for'-array-computability on A;

(3) μPR(A)-computability $\iff$ 'while'-computability on A;

(4) μPR*(A)-computability $\iff$ 'while'-array-computability on A.

These equivalences all hold uniformly for $A \in \textbf{\textit{StdAlg}}(\Sigma)$.

They can be proved by the techniques of Tucker & Zucker [1988], or the revised version in Thompson [1987]. For a proof of (1) in the classical case over the naturals, see, *e.g.*, Davis & Weyuker [1983, Chapter 13] or Zucker & Pretorius [1993, §13].

Precise formulations and applications of results of the kind (1) above, in the context of stream algebras, are given in Section 7.

4 Stream algebras; Computation on stream algebras

We will define algebras containing streams, and show how to apply our previous models of computation on general algebras to these stream algebras. In §4.1 the general form of a stream algebra is defined. In §4.2 it is shown how to construct a stream algebra $\bar{A}$ from a given standard algebra A, and in §4.3 the models of computability are applied.

4.1 Definitions and examples

Let Σ be a standard signature. Σ is a (standard) *stream signature* if there are two designated subsets $\boldsymbol{PreStr}(\Sigma)$ and $\boldsymbol{Str}(\Sigma)$ of $\boldsymbol{Sort}(\Sigma)$, the set of *pre-stream sorts* and *stream sorts* of Σ respectively, for which the following holds:

(1) There is a bijection $s \mapsto \bar{s}$ of $\boldsymbol{PreStr}(\Sigma)$ with $\boldsymbol{Str}(\Sigma)$.

(2) For all $\bar{s} \in \boldsymbol{Str}(\Sigma)$, the only Σ-function with $\bar{s}$ in its domain is eval_s of type $\bar{s} \times \mathsf{N} \to s$ (where $\mathsf{eval}_s(\xi, n)$ means: evaluate stream ξ at argument n); and the only possible Σ-function with $\bar{s}$ in its range is $\mathsf{eval}_{\bar{s}}$ (in the case that $\bar{s}$ is also in $\boldsymbol{PreStr}(\Sigma)$).

Note that $\boldsymbol{PreStr}(\Sigma)$ and $\boldsymbol{Str}(\Sigma)$ need not be disjoint — a stream sort $\bar{s}$ can in turn be a pre-stream sort for another stream sort $\bar{\bar{s}}$.

EXAMPLES. (1) $\boldsymbol{PreStr}(\Sigma) = \boldsymbol{Str}(\Sigma) = \emptyset$: all our standard signatures up to now!

(2) S is *partitioned* into $\boldsymbol{PreStr}(\Sigma)$ and $\boldsymbol{Str}(\Sigma)$: so each sort in S is either a pre-stream sort or a stream sort — and only one of these.

DEFINITION (*Stream algebra*). Let Σ be a standard stream signature. A *standard Σ-stream algebra* A is a standard Σ-algebra, with the further conditions that for each pre-stream sort s:

(*i*) the carrier set for $\bar{s}$ is the set

$$A_{\bar{s}} \;=\; \bar{A}_s \;=\; [\mathsf{N} \to A_s]$$

of all *streams on A i.e.*, functions $\xi : \mathsf{N} \to A$;

(*ii*) the interpretation of eval_s on A is a function $\mathsf{eval}_s^A : [\mathsf{N} \to A_s] \times \mathsf{N} \to A_s$ which *evaluates* a stream at an index, *i.e.*,

$$\mathsf{eval}_s^A(\xi, n) = \xi(n).$$

REMARKS. (1) Because we have taken $\bar{A}_s$ to be the set of *all* streams on A_s, we call A the *full* standard stream algebra (with respect to its stream sorts). Note that for a full standard stream algebra A, if A_s has cardinality > 1 for some $s \in \boldsymbol{PreStr}(\Sigma)$, then $\bar{A}_s$, and hence A, is uncountable.

(2) Note that the Instantiation Assumption does not hold in general on stream algebras.

4.2 Extending signatures and algebras to stream signatures and stream algebras

Let Σ be an ordinary standard signature, and let $S \subseteq Sort(\Sigma)$. We can extend Σ to a stream signature $\overline{\Sigma}^S$ in the following way.

(1) With each $s \in S$, we associate a new *stream sort* $\bar{s}$. Then $Sort(\overline{\Sigma}^S) = Sort(\Sigma) \cup \overline{S}$, where $\overline{S} =_{df} \{\bar{s} \mid s \in S\}$. So $PreStr(\overline{\Sigma}^S) = S$, and $Str(\overline{\Sigma}^S) = \overline{S}$ (according to the definitions in §4.1).

(2) $Func(\overline{\Sigma}^S)$ consists of $Func(\Sigma)$, together with $\mathbf{eval}_s$ for each $s \in S$.

Now let A be a Σ-algebra. We can *expand* A to a standard $\overline{\Sigma}^S$-stream algebra $\bar{A}^S$, uniquely (up to $\overline{\Sigma}^S/\Sigma$-isomorphism) according to the definition in §4.1, by adding carriers $\bar{A}_s = A_{\bar{s}}$, and the operator $\mathbf{eval}_s^A$ for each $s \in S$. The algebra $\bar{A}^S$ is the *(full standard) stream algebra over A with respect to* §.

REMARKS. (1) If for any $s \in S$, A_s has cardinality > 1, then $\bar{A}^S$ is uncountable. (See Remark 1 in §4.1.)

(2) In the case that $S = Sort(\Sigma)$, this is an instance of the second example given in §4.1.

4.3 Computing on stream algebras

Let Σ be a standard signature, let $S \subseteq Sort(\Sigma)$, and let $A \in StdAlg(\Sigma)$. We will define various $\overline{\Sigma}^S$*-computation schemes*, which will generate computability classes on $\bar{A}^S$, uniformly for $A \in StdAlg(\overline{\Sigma}^S)$.

TERMINOLOGY AND NOTATION. (1) Below we will write $\overline{\Sigma}$ for $\overline{\Sigma}^S$, and $\bar{A}$ for $\bar{A}^S$.

(2) A sort of $\overline{\Sigma}$ is *simple* if it is a sort of Σ. So every sort of $\overline{\Sigma}$ is either a simple sort or a stream sort (and not both).

(3) A product type u over $\overline{\Sigma}$ is *simple* if every component sort of u is simple, *i.e.*, if it is a product type over Σ.

(4) If f is any (vector valued) mapping on $\bar{A}$, then f is *simple valued* if is range type is simple, and *stream valued* otherwise (*i.e.*, at least one sort in its range type is a stream sort). Similarly, a computation scheme on $\overline{\Sigma}$ is *simple valued* or *stream valued* according as its range type is simple or not.

(5) We extend the notation of §2.5(1): For a Σ-product type $u = s_1 \times \cdots \times s_m$, we write

$$[\mathbb{N} \to A]^u =_{df} [\mathbb{N} \to A_{s_1}] \times \cdots \times [\mathbb{N} \to A_{s_m}].$$

(6) Below, we will use x for tuples of (stream or simple) elements of $\bar{A}$, and (more specifically) z or t for elements of $\mathbb{N}$.

Note that the four scheme models given Section 3 (PR, PR*, μPR and μPR*) can be applied to $\overline{\Sigma}$, and hence to $\bar{A}$. In addition, we have other scheme models MC($\overline{\Sigma}$), and hence MC($\bar{A}$) for $A \in \boldsymbol{StdAlg}(\Sigma)$, which apply specifically to stream algebras.

DEFINITIONS OF SYSTEMS. (1) $\mathrm{PR}_-(\overline{\Sigma})$ is like $\mathrm{PR}(\overline{\Sigma})$, except that the scheme for *simultaneous primitive recursion* is restricted to have *simple range type*.

(2) $\mathrm{PR}_0(\overline{\Sigma})$ is similar, except that *all* schemes are restricted to have *simple range type*. Note that this includes all the *initial operations*, including eval. Note also that in the scheme for *composition* $\alpha \equiv \boldsymbol{com}_{u,v,w}(\beta,\gamma)$ (see §3.1(v)), the domain type v of γ, as well as its range type w, is simple (since v is also the range type of β).

(3) $\lambda\mathrm{PR}(\overline{\Sigma})$ consists of $\mathrm{PR}(\overline{\Sigma})$ plus a scheme for *stream abstraction*: For all $\overline{\Sigma}$-product types u and $s \in S$, and for all schemes $\beta \in \lambda\mathrm{PR}(\overline{\Sigma})_{u\times\mathsf{N}\to s}$, there is a scheme $\alpha \equiv \boldsymbol{\lambda abs}_u(\beta) \in \lambda\mathrm{PR}(\overline{\Sigma})_{u\to\bar{s}}$. It defines the function $\alpha^{\bar{A}} : A^u \to [\mathsf{N} \to A_s]$ on $\bar{A}$ for each $A \in \boldsymbol{StdAlg}(\Sigma)$, where

$$\alpha^{\bar{A}}(x) = \lambda z \cdot \beta^{\bar{A}}(x,z)$$

for all $x \in \bar{A}^u$.

(4) $\lambda\mathrm{PR}_-(\overline{\Sigma})$ is $\mathrm{PR}_-(\overline{\Sigma})$ plus the scheme $\boldsymbol{\lambda abs}$.

(5) $\mathrm{PR}_0^\lambda(\overline{\Sigma})$ is $\mathrm{PR}_0(\overline{\Sigma})$ plus a scheme for *stream abstraction*, which (because of the nature of PR_0) must be embedded inside a special form of the *composition* scheme: For all $\overline{\Sigma}$-product types u, v and $s \in S$, and for all schemes $\beta \in \mathrm{PR}_0^\lambda(\overline{\Sigma})_{u\times\mathsf{N}\to s}$ and $\gamma \in \mathrm{PR}_0^\lambda(\overline{\Sigma})_{u\times\bar{s}\to v}$, there is a scheme $\alpha \equiv \boldsymbol{com\text{-}\lambda abs}_{u,v}(\beta,\gamma) \in \mathrm{PR}_0^\lambda(\overline{\Sigma})_{u\to v}$. It defines the function $\alpha^{\bar{A}} : A^u \to A^v$ on $\bar{A}$ for each $A \in \boldsymbol{StdAlg}(\Sigma)$, where

$$\alpha^{\bar{A}}(x) = \gamma^{\bar{A}}(x,\ \lambda z \cdot \beta^{\bar{A}}(x,z))$$

for all $x \in \bar{A}^u$.

Note, finally, that each of the above models can be modified to incorporate the μ operator, thus: $\mu\mathrm{PR}_0(\overline{\Sigma})$, $\mu\mathrm{PR}_-(\overline{\Sigma})$, $\lambda\mu\mathrm{PR}(\overline{\Sigma})$, $\lambda\mu\mathrm{PR}_-(\overline{\Sigma})$ and $\mu\mathrm{PR}_0^\lambda(\overline{\Sigma})$.

5 Computational conservativity

5.1 General models applied to stream algebras

Here we examine the effect of applying our general models of computation (PR, PR*, μPR and μPR*) on any standard Σ-algebra A to stream algebras $\bar{A}$. Three classes of functions must be considered for these four models of computation. We begin by establishing the effect on functions on A:

$$f : A^u \to A,$$

where u is simple. First we formulate the idea that an expansion may not effect the computational power of a model of computation.

Let Σ' be a standard signature with $\Sigma \subseteq \Sigma'$, and let B be a standard Σ'-expansion of A. Let MC be a model of computation that defines functions MC(A) and MC(B) on A and B. Then B is an *MC computationally conservative expansion of A* if for every function f on A,

$$f \in \mathrm{MC}(A) \iff f \in \mathrm{MC}(B).$$

THEOREM 1. *$\bar{A}$ is a computationally conservative expansion of A with respect to PR, PR*, μPR and μPR*.*

PROOF: The four arguments involve a close inspection of the syntax of schemes. Recall from §4.3(3) that a sort tuple is *simple* if it contains only simple sorts.

Consider first PR. Suppose $\alpha \in \mathrm{PR}(\overline{\Sigma})_{u \to v}$, where $u = s_1 \times \cdots \times s_m$ and $v = s'_1 \times \cdots \times s'_n$.

We prove: If u is simple, then so is v, and further, $\alpha \in \mathrm{PR}(\Sigma)$.

The proof is by induction on $\mathbf{compl}(\alpha)$. Let us consider two cases.

(*i*) *Initial functions.* The only new initial function in $\overline{\Sigma}$ is eval, for which the domain sorts are not all simple, so the induction assertion is trivially satisfied.

(*ii*) *Composition.* Let $\alpha = \mathbf{com}(\beta, \gamma)$ where $\beta : u \to w$ and $\gamma : w \to v$, and suppose u has simple sorts. By the induction hypothesis, w is simple and $\beta \in \mathrm{PR}(\Sigma)$. Then, again by the induction hypothesis, v is simple and $\gamma \in \mathrm{PR}(\Sigma)$. Hence $\alpha \in \mathrm{PR}(\Sigma)$.

Similarly with the other cases.

For PR*: Note that the sorts of $\mathrm{PR}^*(\overline{\Sigma})$ may have any of the four forms: s, s^*, $\bar{s}$ and $\bar{s}^*$. We call the former two "unbarred" and the latter two "barred". Then we show (as in the previous case) that if u does not contain a barred sort, then neither does v, and $\alpha \in \mathrm{PR}^*(\Sigma)$.

For μPR and μPR*: The proof is similar. There is an extra scheme to check, the μ-operator. □

The second class of functions to be considered is that of *simple-valued* functions

$$f : [\mathbb{N} \to A]^u \times A^v \to A.$$

There are many non-trivial computable functions of this class in PR($\bar{A}$) and our other models. We will consider these later.

The third and final class is that of *stream valued functions*

$$f : [\mathbb{N} \to A]^u \times A^v \to [\mathbb{N} \to A].$$

DEFINITION. Let $u = s_1 \times \cdots \times s_m$ and $v = s'_1 \times \cdots \times s'_p$. A stream valued function $f : [\mathbb{N} \to A]^u \times A^v \to [\mathbb{N} \to A]$ is an *input dependent stream projection* if for some simple-valued function $f_0 : [\mathbb{N} \to A]^u \times A^v \to \mathbb{N}$,

$$f(\xi_1, \ldots, \xi_m, x_1, \ldots, x_p) = \xi_{f_0(\xi_1, \ldots, \xi_m, x_1, \ldots, x_p)}$$

We call f_0 a *representative* of f.

THEOREM 2. *If f is a stream valued function in $PR(\bar{A})$, then f is an input dependent projection.*

PROOF: The proof follows from the *subalgebra property*: $f(\xi_1, \ldots, \xi_m, x_1, \ldots, x_p)$ must lie in the subalgebra generated by $\xi_1, \ldots, \xi_m, x_1, \ldots, x_p$ (Tucker & Zucker [1988]). This subalgebra, on the stream sorts, consists only of the input streams $\xi_1, \ldots, \xi_m$ themselves! □

Note that we are saying nothing here about the computability of the representative f_0 of f. Later (Theorem 2 of Section 6) we will prove an appropriate result concerning this.

5.2 Two simple extensions of $\overline{\Sigma}$

We consider two simple extensions of $\overline{\Sigma}$ which increase its computational power with respect to PR, and, in particular, allow more stream valued functions to be PR-computable. Let A be a standard Σ-algebra.

First, adjoin to $\overline{\Sigma}$ the stream operation $\mathsf{const}_s : s \to \bar{s}$ for all Σ-sorts s, where for all $x \in A_s$, $\mathsf{const}_s(x)$ is the stream with constant value x. Then

$$\mathrm{PR}(\bar{A}, \mathsf{const}) \supsetneq \mathrm{PR}(\bar{A})$$

since trivially, $\mathrm{PR}(\bar{A}, \mathsf{const})$ includes the function const, but by the proof of Theorem 1, PR(A) does not contain *any* function (such as const) whose domain is simple, but whose range is a stream sort.

Similar results hold for PR*, μPR and μPR*.

Next, further adjoin the stream operation $\mathsf{update}_s : \bar{s} \times \mathsf{N} \times s \to \bar{s}$ for all Σ-sorts s, where for $\xi \in \bar{A}$, $n \in \mathbb{N}$, and $x \in A_s$, $\mathsf{update}_s(\xi, n, x)$ gives the "updated stream" which is like ξ except that its value at n is x (rather like the **Update** operation on arrays: see §2.3). Then we can show that for any f on A,

$$f \in \mathrm{PR}(\bar{A}, \mathsf{const}, \mathsf{update}) \iff f \in \mathrm{PR}^*(A).$$

5.3 Stream abstraction is strong

Stream abstraction allows us to compute many stream transformations.

Let Σ be a standard signature, let A be a standard Σ-algebra, and (for some $S \subseteq$ ***Sort***(Σ)) let $\bar{A}$ be the corresponding standard stream algebra, of signature $\overline{\Sigma}$.

We define a "cartesian" or "uncurrying" operation on stream-valued functions on $\bar{A}$. If $f : \bar{A}^u \to [\mathsf{N} \to A_s]$, then $\boldsymbol{cart}(f) : \bar{A}^u \times \mathsf{N} \to A_s$ is defined by

$$\boldsymbol{cart}(f)(x, z) = f(x)(z).$$

PROPOSITION 1. $f \in \lambda PR(\bar{A}) \iff \boldsymbol{cart}(f) \in \lambda PR(\bar{A})$.

PROOF: The functions f and $\boldsymbol{cart}(f)$ are easily inter-definable, by the use of eval and $\boldsymbol{\lambda abs}$. □

Note that the same result holds for PR*, μPR and μPR* in place of PR.

PROPOSITION 2. $\lambda PR(\bar{A}) \supsetneqq PR(\bar{A})$.

PROOF: *This follows from the result on* const *in §5.2, since* const *can easily be defined in* $\lambda PR(\overline{\Sigma})$. □

REMARK. Although, as we have seen, $\mathsf{const}^A \notin \mathrm{PR}(\bar{A})$, note that $\boldsymbol{cart}(\mathsf{const}^A)$, which is just a projection function, *is* in $\mathrm{PR}(\bar{A})$. The same remark applies to PR*, μPR and μPR*.

Proposition 2 can be strengthened: PR is computationally weaker than λPR, *even for functions on A*. To show this, consider the standard algebra $\mathcal{N} = (\mathsf{N}; 0, \mathsf{S})$, its stream algebra $\overline{\mathcal{N}} = (\mathsf{N}, [\mathsf{N} \to \mathsf{N}]; 0, \mathsf{S}, \mathsf{eval})$ and the iterator functional

$$I : [\mathsf{N} \to \mathsf{N}] \times \mathsf{N} \to \mathsf{N}$$

defined by

$$I(\phi, 0) = 1$$
$$I(\phi, n+1) = \mathsf{eval}(\phi, I(\phi, n))$$

for $\phi \in [\mathsf{N} \to \mathsf{N}]$ and $n \in \mathsf{N}$.

Clearly, $I(\phi, n) = \phi^n(1)$ and $I \in \mathrm{PR}(\overline{\mathcal{N}})$, and so

$$\boldsymbol{\lambda abs}(I) : [\mathsf{N} \to \mathsf{N}] \to [\mathsf{N} \to \mathsf{N}]$$

is in $\lambda\mathrm{PR}(\overline{\mathcal{N}})$. Next we define $J : [\mathsf{N} \to \mathsf{N}] \times \mathsf{N} \to [\mathsf{N} \to \mathsf{N}]$ by

$$J(\phi, 0) = \phi$$
$$J(\phi, n+1) = (\boldsymbol{\lambda abs}(I))(J(\phi, n))$$

for $\phi \in [\mathsf{N} \to \mathsf{N}]$ and $n \in \mathsf{N}$. Clearly,

$$J(\phi, n) = (\boldsymbol{\lambda abs}(I))^n(\phi).$$

COROLLARY. *The algebra* $\overline{\mathcal{N}}$ *is not* λPR *computationally conservative over* $\mathcal{N}$.

PROOF: The Ackermann function is in $\lambda\mathrm{PR}(\overline{\mathcal{N}})$. □

We note without proof:

THEOREM 3. *The class of functions on* $\mathbb{N}$ *which are in* $\lambda PR(\overline{\mathcal{N}})$ *is the class* T_1 *of functions up to level 1 in the Gödel Dialectica interpretation.*

5.4 Representations

In order to prove other computational conservativity results, we investigate the notion of a *representation* of an algebra in a retract of itself.

Let A and B be standard algebras of signature Σ and Σ', where $\Sigma \subseteq \Sigma'$ and B is a Σ'-expansion of A. A *representation* (ϕ, ψ) of B *in* A consists of the following:

(*i*) a Σ'-algebra $R_A(B)$ that is an expansion of A, in which for each additional sort $s \in \boldsymbol{Sort}(\Sigma') \setminus \boldsymbol{Sort}(\Sigma)$ there is a product type over Σ

$$w(s) = w(s,1) \times w(s,2) .... \times w(s, r(s))$$

such that the carrier of sort s in $R_A(B)$ is the set

$$A^{w(s)} =_{df} A_{w(s,1)} \times A_{w(s,2)} \times \cdots \times A_{w(s,r(s))};$$

(*ii*) a surjective Σ'/Σ-homomorphism

$$\psi : R_A(B) \to B;$$

(*iii*) some section of ψ, *i.e.*, a map

$$\phi : B \to R_A(B)$$

such that $\psi \circ \phi = \boldsymbol{id}_B$. Note that ψ is *injective*.

This formalises the idea that each new carrier B_s of B of sort $s \in \boldsymbol{Sort}(\Sigma') \setminus \boldsymbol{Sort}(\Sigma)$ is coded by a product $A^{w(s)}$ of carriers of A, using the surjective map $\psi_s : A^{w(s)} \to B_s$. Because ψ is a homomorphism, for each function

$$F^B : B_{s_1} \times \cdots \times B_{s_m} \to B_s$$

of B, the corresponding function

$$F^{R_A(B)} : A^{w(s_1)} \times \cdots \times A^{w(s_m)} \to A^{w(s)}$$

of $R_A(B)$ is a "tracking function" on the carriers of A for this coding. The injective maps $\phi_s : B_s \to A^{w(s)}$, where $\psi_s \circ \phi_s = \boldsymbol{id}_{B_s}$, are called "decoding functions".

We define the *kernel* $\equiv_\psi$ of ψ in the usual way: for all Σ'-sorts s and all $x, y \in R_A(B)_s$,

$$x \equiv_{\psi_s} y \iff \psi_s(x) = \psi_s(y).$$

Note for future reference that

$$x \equiv_{\psi_s} y \iff \psi_s(x) = \psi_s(y) \iff \phi_s\psi_s(x) = \phi_s\psi_s(y), \tag{1}$$

since ϕ_s is injective. Also, for $s \in \boldsymbol{Sort}(\Sigma)$ the congruence $\equiv_{\psi_s}$ is the *identity*, since ψ is a Σ'/Σ-homomorphism. We have

$$B \cong_{\Sigma'/\Sigma} R_A(B)/\equiv_\psi$$

by the Homomorphism Theorem (Meinke & Tucker [1992, Theorem 3.4.18]).

Note that, by cardinality considerations, $\bar{A}$ does not have a representation over A.

5.5 Computable representations

We refine the notion of representation given in §5.4 as follows. Let, again, A and B be standard algebras of signature Σ and Σ', where $\Sigma \subseteq \Sigma'$. Let MC be a model of computation on A. A representation (ϕ, ψ) of B in A is *MC-computable over A* if

(*i*) each operation of $R_A(B)$ is in MC(A); and

(*ii*) for all Σ-sorts s, the functions $\phi_s \circ \psi_s : A^{w(s)} \to A^{w(s)}$ are in MC(A).

REMARKS. (1) Condition (ii) implies (by (1) of §5.3) that if the sorts of $w(s)$ are all equality sorts, then $\equiv_\psi$ is MC-computable on $R_A(B)_s$.

(2) For all Σ-sorts s, $\phi_s \circ \psi_s$ is the identity on $A^{w(s)}$, since ψ is a Σ'/Σ-homomorphism.

We say that B is MC-*computably representable in A* if B has an MC-computable representation in A.

EXAMPLES. We showed in Tucker & Zucker [1988, §1.1.7 and §3.5.3] that for standard algebras A:

(*i*) A^{u} is PR representable in A (see §2.2, Remark 2);

(*ii*) $(A^*)^*$ is PR representable in A^* (see §2.3, Remark 2).

THEOREM 4. *Let A and B be any standard algebras, and let B be an expansion of A. Let MC be any one of:*

$$PR,\ \mu PR,\ \lambda PR,\ \lambda\mu PR,\ PR^*,\ \mu PR^*,\ \lambda PR^*,\ \lambda\mu PR^*.$$

If B has an MC-computable representation over A, then B is computationally conservative over A.

PROOF: (Outline.) Let (ϕ, ψ) be an MC-computable representation of B on A. First, for any Σ'-product type $u = s_1 \times \cdots \times s_m$, put $w(u) =_{df} w(u_1) \ldots w(u_m)$, and define the maps

$$\phi^u : B^u \to A^{w(u)}$$

and

$$\psi^u : A^{w(u)} \to B^u$$

by

$$\phi^u(x_1,\ldots,x_m) = (\phi(x_1),\ldots,\phi(x_m))$$

for $(x_1,\ldots,x_m) \in B^u$, and similarly for ψ^u.

Now, for any function f on B, define f^A to be the tracking operation of f on A w.r.t. (ϕ,ψ), *i.e.*,

$$f^A = \phi \circ f \circ \psi$$

or, more precisely, if

$$f : B^u \to B^v,$$

then

$$f^A = \phi^v \circ f \circ \psi^u : A^{w(u)} \to A^{w(v)}.$$

In other words, f^A is defined to make the following diagram commute:

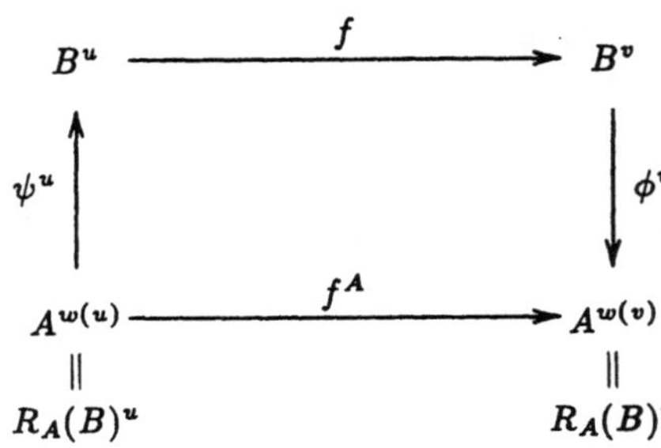

We must then show that

$$\text{if } f \in \text{MC}(B) \text{ then } f^A \in \text{MC}(A). \qquad (*)$$

The proof of $(*)$ is by induction on the scheme for f (taking each MC in turn). As an example, consider the case of composition for PR: $f = g \circ h$. Then (omitting sort superscripts):

$$\begin{aligned} f^A &= \phi \circ f \circ \psi \\ &= \phi \circ g \circ h \circ \psi \\ &= \phi \circ g \circ (\psi \circ \phi) \circ h \circ \psi \qquad (\text{since } \psi \circ \phi = \boldsymbol{id}) \\ &= (\phi \circ g \circ \psi) \circ (\phi \circ h \circ \psi) \\ &= g^A \circ h^A \end{aligned}$$

which is in PR(A), since g^A and h^A are, by the induction hypothesis.

Next we note that if f is on A, *i.e.*, if u, v are Σ-product types, then $f^A = f$, since ϕ^v and ψ^u are identities, by Remark 2 above.

The result then follows from (∗).

Note that in order to prove (∗) for those MC involving '∗' or 'λ' or both (namely PR^*, $\mu\mathrm{PR}^*$, $\lambda\mathrm{PR}$, $\lambda\mu\mathrm{PR}$, $\lambda\mathrm{PR}^*$, $\lambda\mu\mathrm{PR}^*$), we use the following

LEMMA. *Suppose A is a reduct of B, and B is computably representable over A by (ϕ, ψ). Then (for the relevant MC's):*

(i) A^* is a reduct of B^*, and (ϕ, ψ) can be extended to a computable representation (ϕ^*, ψ^*) of B^* over A^*;

(ii) $\bar{A}$ is a reduct of $\bar{B}$, and (ϕ, ψ) can be extended to a computable representation $(\bar{\phi}, \bar{\psi})$ of $\bar{B}$ over $\bar{A}$. □

For an application of Theorem 4, we use the following

LEMMA. *The algebras $\bar{\bar{A}}$, $\bar{A}^*$ and $\overline{A^*}$ have λPR representations over the stream algebra $\bar{A}$.*

PROOF: The representations are easy (but tedious) to construct directly. □

By combining this lemma with Theorem 4, we obtain:

THEOREM 5. *The algebras $\bar{\bar{A}}$, $\bar{A}^*$ and $\overline{A^*}$ are computationally conservative over $\bar{A}$ with respect to the models of computation*

$$\lambda PR, \quad \lambda\mu PR, \quad \lambda PR^*, \quad \lambda\mu PR^*.$$

Among the many equivalences this generates we are particularly interested in the following.

COROLLARY. *For any standard A, and for functions on $\bar{A}$:*

(a) the following are equivalent:

$$\lambda PR(\bar{A}), \quad \lambda PR(\bar{\bar{A}}), \quad \lambda PR^*(\bar{A}), \quad \lambda PR(\overline{A^*});$$

(b) the following are equivalent:

$$\lambda\mu PR(\bar{A}), \quad \lambda\mu PR(\bar{\bar{A}}), \quad \lambda\mu PR^*(\bar{A}), \quad \lambda\mu PR(\overline{A^*}).$$

The definitions of *representation* and *computable representation* given above are generalisations of the notion of *computable algebra* used in Bergstra & Tucker [1987]. Related notions have been used in work on compiler correctness in Thompson [1987].

6 Restricted primitive recursion scheme

In this section, we consider the effect of restricting the range type, and hence the power, of the ***rec*** scheme on stream algebras. The computation models defined in Section 4 are relevant to this: $\mathrm{PR}_-(\overline{\Sigma})$, $\mathrm{PR}_0(\overline{\Sigma})$, $\lambda\mathrm{PR}_-(\overline{\Sigma})$ and $\mathrm{PR}_0^\lambda(\overline{\Sigma})$, as well as these models with μ. First we investigate the relationship between $\lambda\mathrm{PR}_-(\overline{\Sigma})$ and $\mathrm{PR}_0^\lambda(\overline{\Sigma})$. First we must extend the definition of the ***cart*** operation appropriately to vector valued functions.

As before, let Σ be a standard signature, let A be a standard Σ-algebra, and (for some $\boldsymbol{S} \subseteq \boldsymbol{Sort}(\Sigma)$) let $\bar{A}$ $(= \bar{A}^{\boldsymbol{S}})$ be the corresponding standard stream algebra, of signature $\overline{\Sigma}(= \overline{\Sigma}^{\boldsymbol{S}})$.

For a $\overline{\Sigma}$-product type $v = s_1 \times \cdots \times s_n$, we define the *simplification* of v to be the simple type $v^0 = s_1^0 \times \cdots \times s_n^0$ where, if s_i is simple, then $s_i^0 = s_i$, and if s_i is a stream sort $\bar{s}$, then $s_i^0 = s$.

DEFINITION 1. Let f be a function on $\bar{A}$ of type $u \to v$. We will define a function $\boldsymbol{cart}(f)$ of type $u \times \mathsf{N} \to v^0$.

(*a*) If f is stream valued, then $\boldsymbol{cart}(f)$ is as defined in §5.3.

(*b*) If f is simple valued, then $\boldsymbol{cart}(f)(x,t) = f(x)$ for all $x \in \bar{A}^u$ and $t \in \mathrm{N}$.

(*c*) If f is vector valued, with components $f_1, \ldots, f_n$, then $\boldsymbol{cart}(f)$ is the vector valued function defined by

$$\boldsymbol{cart}(f)(x) \;=\; (\boldsymbol{cart}(f_1)(x), \ldots, \boldsymbol{cart}(f_n)(x))$$

for all $x \in \bar{A}^u$.

LEMMA 1. *(a)* $PR_0(\bar{A}) \subseteq PR_-(\bar{A})$

(b) $PR_0^\lambda(\bar{A}) \subseteq \lambda PR_-(\bar{A})$.

LEMMA 2. *(a) If f is stream valued, then* $f = \boldsymbol{\lambda abs}(\boldsymbol{cart}(f))$.

(b) If f is simple valued, then $f(x) = \boldsymbol{cart}(f)(x,0)$ *for all* $x \in A^u$.

THEOREM 1. $f \in \lambda PR_-(\bar{A}) \iff \boldsymbol{cart}(f) \in PR_0^\lambda(\bar{A})$.

PROOF: ($\Longleftarrow$:) This direction is immediate, since for any stream valued f, if $\boldsymbol{cart}(f) \in \mathrm{PR}_0^\lambda(\bar{A})$, then by Lemma 1, $\boldsymbol{cart}(f) \in \lambda\mathrm{PR}_-(\bar{A})$. Hence by Lemma 2, $f = \boldsymbol{\lambda abs}(\boldsymbol{cart}(f)) \in \lambda\mathrm{PR}_-(\bar{A})$.

($\Longrightarrow$:) This is proved by structural induction on the $\lambda\mathrm{PR}_-(\overline{\Sigma})$ scheme for f. Suppose, for example, that f is defined from functions g and h by the scheme for simultaneous primitive recursion:

$$\begin{aligned} f(0,x) &= g(x) \\ f(z+1,x) &= h(z,x,f(z,x)) \end{aligned}$$

for all $z \in \mathbb{N}$, $x \in A^u$. Because of the restriction on this scheme in λPR_-, the range type of f must be simple. But that is also the range type of g and h. So $\boldsymbol{cart}(f)$ is related to $\boldsymbol{cart}(g)$ and $\boldsymbol{cart}(h)$ by the equations

$$\begin{aligned} \boldsymbol{cart}(f)(0,x,t) &= \boldsymbol{cart}(g)(x,t) \\ \boldsymbol{cart}(f)(z+1,x,t) &= \boldsymbol{cart}(h)(z,x,f(z,x),t). \end{aligned}$$

for all $z \in \mathbb{N}$, $x \in A^u$ and $t \in \mathbb{N}$. But this is an instance of the primitive recursion scheme, and by the induction hypothesis $\boldsymbol{cart}(g)$ and $\boldsymbol{cart}(h)$ are in $\mathrm{PR}_0^\lambda(\bar{A})$, and hence so is $\boldsymbol{cart}(f)$. □

COROLLARY 1. *If f is simple valued, then*

$$f \in \lambda PR_-(\bar{A}) \iff f \in PR_0^\lambda(\bar{A}).$$

PROOF: If f is simple valued, then by Lemma 2, $f(x) = \boldsymbol{cart}(f)(x,0)$. Hence

$$f \in \mathrm{PR}_0^\lambda(\bar{A}) \iff \boldsymbol{cart}(f) \in \mathrm{PR}_0^\lambda(\bar{A}).$$ □

The relation between the other models named above and these two, will be clarified after the results in Section 7.

Next we show that, in the absence of stream abstraction, full primitive recursion is not more powerful than restricted primitive recursion!

As before, let Σ be a standard signature, let A be a standard Σ-algebra, and (for some $S \subseteq Sort(\Sigma)$) let $\bar{A}$ be the corresponding standard stream algebra, of signature $\overline{\Sigma}$.

For a $\overline{\Sigma}$-product type $v = s_1 \times \cdots \times s_n$, we define the N-*simplification* of v to be the simple type $v^{\mathsf{N}} = s_1^{\mathsf{N}} \times \cdots \times s_n^{\mathsf{N}}$, where, if s_i is simple, then $s_i^{\mathsf{N}} = s_i$, and if s_i is a stream sort, then $s_i^{\mathsf{N}} = \mathsf{N}$.

Recall the definition of *input dependent stream projection* and its *representative* in §5.1. We generalise the latter notion to vector valued functions.

DEFINITION 2. Let f and f^{N} be vector valued functions on $\bar{A}$ of types $u \to v$ and $u \to v^{\mathsf{N}}$ respectively, with component functions $f_1, \ldots, f_n$ and $f_1^{\mathsf{N}}, \ldots, f_n^{\mathsf{N}}$ respectively. Suppose every stream valued component of f is an input dependent stream projection. We say that f^{N} is a *representative* of f iff for $j = 1, \ldots, n$ the following hold:

(*a*) if f_j is stream valued, then f_j^{N} is a representative of f_j according to the definition in §5.1;

(*b*) if f is simple valued, then $f^{\mathsf{N}} = f$.

REMARK 1. If f is a simple valued function, then f trivially satisfies the condition that every stream valued component is an input independent stream projection, and is its own representative.

THEOREM 2. *Suppose $f \in PR(\bar{A})$. Then:*

(i) every stream valued component of f is an input dependent stream projection;

(ii) f has a representative in $PR_0(\bar{A})$.

PROOF: By structural induction on the complexity of the PR($\bar{A}$) scheme for f. Consider one case: the scheme for *composition*. The idea will be clear from a typical case: suppose $f = h \circ g$, with $g, h \in \mathrm{PR}(\bar{A})$, where for some pre-stream sort s,

$$g : \bar{A}_s \times \bar{A}_s \to \bar{A}_s \times \bar{A}_s,$$

and

$$h : \bar{A}_s \times \bar{A}_s \to A_s.$$

Then g is vector valued with components $g_1, g_2 : \bar{A}_s \times \bar{A}_s \to \bar{A}_s$, and h is simple valued. By induction hypothesis, g_1, g_2 and h are input dependent stream projections, with representatives g_1^{N}, g_2^{N} and h^{N}, respectively, all in $\mathrm{PR}_0(\bar{A})$. Thus, for all $\xi_1, \xi_2 \in \bar{A}_s$,

$$g_j^{\mathsf{N}}(\xi_1, \xi_2) = \begin{cases} 1 & \text{if } g_j(\xi_1, \xi_2) = \xi_1 \\ 2 & \text{if } g_j(\xi_1, \xi_2) = \xi_2 \end{cases}$$

for $j = 1, 2$. Also, since h is simple valued, $h = h^{\mathsf{N}} \in \mathrm{PR}_0(\bar{A})$. So $f : \bar{A}_s \times \bar{A}_s \to A_s$ is defined by

$$f(\xi_1, \xi_2) = \begin{cases} h(\xi_1, \xi_1) & \text{if } g_1^{\mathsf{N}}(\xi_1, \xi_2) = 1 \text{ and } g_2^{\mathsf{N}}(\xi_1, \xi_2) = 1 \\ h(\xi_1, \xi_2) & \text{if } g_1^{\mathsf{N}}(\xi_1, \xi_2) = 1 \text{ and } g_2^{\mathsf{N}}(\xi_1, \xi_2) = 2 \\ h(\xi_2, \xi_1) & \text{if } g_1^{\mathsf{N}}(\xi_1, \xi_2) = 2 \text{ and } g_2^{\mathsf{N}}(\xi_1, \xi_2) = 1 \\ h(\xi_2, \xi_2) & \text{if } g_1^{\mathsf{N}}(\xi_1, \xi_2) = 2 \text{ and } g_2^{\mathsf{N}}(\xi_1, \xi_2) = 2. \end{cases}$$

Hence $f \in \mathrm{PR}_0(\bar{A})$, by definition by cases. Also, since f is simple valued, by Remark 1 f trivially satisfies (i) in the statement of the Theorem, and is its own representative. □

COROLLARY 2. *If f is simple valued, then*

$$f \in PR(\bar{A}) \iff f \in PR_0(\bar{A}).$$

PROOF: From the Theorem and Remark 1. □

REMARKS. (2) Theorem 2, and hence Corollary 2, also hold with the μ-operator added throughout, *i.e.*, with μPR and $\mu\mathrm{PR}_0$ instead of PR and PR_0.

(3) However, it follows from Theorem 3 in Section 5 that Theorem 2 does *not* hold with stream abstraction added.

7 Eliminating λ-abstraction

7.1 Outline of results

Let Σ be a standard signature, and A a standard Σ-algebra. We will prove the following result (Corollary to Theorem 4 below):

$$\mathrm{PR}_0^\lambda(\bar{A}) = \mathrm{PR}_0(\bar{A}).$$

The proof depends on translating these two models of computation into imperative models based on 'for' programs, incorporating functional procedures with stream parameters. Details are given in the rest of Section 7.

As a further corollary, we have:

COROLLARY. *For simple valued functions on $\bar{A}$, the following are equivalent:*

$$PR_0(\bar{A}), \quad PR_-(\bar{A}), \quad PR_0^\lambda(\bar{A}), \quad \lambda PR_-(\bar{A}), \quad PR(\bar{A}).$$

PROOF: From the above result, and Theorem 2 in Section 6. □

It follows from Theorem 3 in Section 5 that for certain standard stream algebras $\bar{A}$, $\lambda\mathrm{PR}(\bar{A})$ contains more functions than the models listed above.

To summarise: *unrestricted primitive recursion plus λ-abstraction* is stronger than either of these alone.

In the next three subsections we will define a number of imperative programming languages, incorporating assignment, serial composition and the conditional, and, as the main control construct, a 'for' command.

In each language, functions are defined by means of *functional procedures*. Such a procedure will have a type $u \to v$, according to its lists of input and output variables, so as to define a function of that type.

7.2 The programming language $\boldsymbol{For}(\Sigma)$

We begin with the syntax of the language $\boldsymbol{For}(\Sigma)$. First, for each Σ-sort s, there are *variables* $\mathbf{a}^s, \mathbf{b}^s, \ldots, \mathbf{x}^s, \mathbf{y}^s \ldots$ of sort s. (We will usually omit the type superscript s.)

Let $\boldsymbol{Var} = \boldsymbol{Var}(\Sigma) = \bigcup_{s \in \boldsymbol{Sort}(\Sigma)} \boldsymbol{Var}_s$ be the class of variables of Σ, where $\boldsymbol{Var}_s$ is the class of variables of sort s.

We now define three syntactic classes: *terms*, *statements* and *procedures*, by *simultaneous induction*.

(1) $\boldsymbol{Term} = \boldsymbol{Term}(\Sigma)$ is the class of terms $t, \ldots,$ and for each product type u over Σ, $\boldsymbol{Term}_u$ is the class of terms of type u. These are generated by the following rules.

(*i*) A *variable* $\mathbf{x}$ of sort s is in $\boldsymbol{Term}_s$.

(*ii*) (*Signature.*) If $F \in \boldsymbol{Func}(\Sigma)_{u \to s}$ and $t \in \boldsymbol{Term}_u$ then $F(t) \in \boldsymbol{Term}_s$.

(*iii*) (*Vectorisation.*) If $t_i \in \mathbf{Term}_{s_i}$ for $i = 1, \ldots, m$, then $(t_1, \ldots, t_m) \in \mathbf{Term}_u$, where $u = s_1 \times \cdots \times s_m$. The t_i are the *components* of the *vector* $(t_1, \ldots, t_m)$.

(*iv*) (*Projection.*) If $t \in \mathbf{Term}_u$ where $u = s_1 \times \cdots \times s_m$, then $U_{u,i}(t) \in \mathbf{Term}_{s_i}$ for $i = 1, \ldots, m$,

(*v*) (*Conditional.*) If $t^{\mathsf{B}} \in \mathbf{Term}_{\mathsf{B}}$ and $t_1, t_2 \in \mathbf{Term}_u$ then

$$\textsf{if } t^{\mathsf{B}} \textsf{ then } t_1 \textsf{ else } t_2 \textsf{ fi} \in \mathbf{Term}_u.$$

(*vi*) (*Procedure call.*) If P is a procedure of type $u \to v$, and $t \in \mathbf{Term}_u$, then $P(t) \in \mathbf{Term}_v$. Here t is the *argument* of the procedure call $P(t)$. If t is a vector $(t_1, \ldots, t_m)$, then $P(t)$ is written $P(t_1, \ldots, t_m)$, and the t_i are called the *actual parameters* of the call.

NOTATION. Suppose $u = s_1 \times \cdots \times s_m$. We write $t : u$ (or sometimes t^u) to mean $t \in \mathbf{Term}_u$, and $\mathtt{x} : u$ to mean $\mathtt{x}$ is a tuple of *distinct variables* of sorts $s_1, \ldots, s_m$ respectively.

(2) $\mathbf{Stmt} = \mathbf{Stmt}(\Sigma)$ is the class of statements $S, \ldots$. The *atomic statements* are 'skip' and the *simultaneous assignment* $\mathtt{x} := t$ where for some product type u, $\mathtt{x} : u$ and $t : u$.

Statements are then generated by the rules

$$S \ ::= \ \textsf{skip} \mid \mathtt{x} := t \mid \ S_1; S_2 \mid \textsf{if } t^{\mathsf{B}} \textsf{ then } S_1 \textsf{ else } S_2 \textsf{ fi} \mid \textsf{for } t^{\mathsf{N}} \textsf{ do } S \textsf{ od}.$$

(3) $\mathbf{Proc} = \mathbf{Proc}(\Sigma)$ is the class of procedures $P, Q, R, \ldots$. These have the form

$$P \ \equiv \ \textsf{proc } D \textsf{ begin } S \textsf{ end}$$

where D is the *variable declaration* and S is the *body*. Here D has the form

$$D \ \equiv \ \textsf{in a, out b, aux c}$$

where a, b and c are lists of *input variables*, *output variables* and *auxiliary* (or *local*) *variables* respectively. Further, we stipulate:

- a, b and c each consist of distinct variables, and they are pairwise disjoint,
- every variable occurring in the body S must be declared in D (among a, b or c),
- the *input variables* a can occur only on the *rhs* of an assignment in S,
- *initialisation condition*: S has the form $S_{init}; S'$, where S_{init} consists of an *initialisation* of all the *output* and *auxiliary variables*, *i.e.*, an assignments to each of them of a closed term of the same sort, whose existence is guaranteed by the Instantiation Assumption on Σ.

The input variables are also called *formal parameters*.

Each variable occurring in the declaration of a procedure *binds* all free occurrences of that variable in the body.

If $\mathrm{a} : u$ and $\mathrm{b} : v$, then P has *type* $u \to v$, written $P : u \to v$.

The *arity* of a procedure $P : u \to v$ is the number of its input variables. Note that it is possible that the arity is 0, *i.e.*, the list of input variables is empty, and u is the empty product sort.

REMARKS. (1) The 'for' loop ('for t^{N} do S od') has the meaning: "evaluate the term t^{N} (as a natural number), and iterate S that number of times".

(2) According to these semantics, all $\boldsymbol{For}(\Sigma)$ statements must terminate.

(3) The functional procedures are constituents of the terms. So if our programming language were modified to include nonterminating control constants such as 'while', it would contain nonterminating procedures, which would materially complicate (even) the semantics of terms.

NOTATION. We will use $E, \ldots$ to denote syntactic expressions of any of the three classes.

7.3 Semantics of $\boldsymbol{For}(\Sigma)$

DEFINITION 1. Let $A \in \boldsymbol{StdAlg}(\Sigma)$. A *state* on A is a function σ from $\boldsymbol{Var}(\Sigma)$ to the carriers of A such that for all $s \in \boldsymbol{S}$,

$$\sigma \restriction \boldsymbol{Var}_s : \ \boldsymbol{Var}_s \to A_s.$$

Let $\boldsymbol{State}(A)$ be the set of states on A.

DEFINITION 2. For two states σ_1 and σ_2 and a set V of variables, $\sigma_1 \approx \sigma_2$ (rel V) means that $\sigma_1 \restriction V = \sigma_2 \restriction V$.

We could now give the semantics of each of the three syntactic classes defined above, relative to $A \in \boldsymbol{StdAlg}(\Sigma)$. For an expression E in any of these classes, we define a function $[\![E]\!]^A$, as follows.

(1) For $t \in \boldsymbol{Term}_u$,

$$[\![t]\!]^A : \ \boldsymbol{State}(A) \to A^u.$$

(2) For $S \in \boldsymbol{Stmt}$,

$$[\![S]\!]^A : \ \boldsymbol{State}(A) \to \boldsymbol{State}(A).$$

(3) For $P \in \boldsymbol{Proc}$, of type $u \to v$,

$$[\![P]\!]^A : \ A^u \to A^v.$$

(If the arity of P is 0, then $[\![P]\!]^A$ is an element of A^v.)

We often write P^A for $[\![P]\!]^A$.

REMARKS. (1) These three semantic functions are defined by *simultaneous recursion*, following the simultaneous inductive definition of the three classes ***Term***, ***Stmt*** and ***Proc*** ("structural recursion"). We omit details. It should be clear how to proceed from the semantics of 'while' programs given in Tucker & Zucker [1988], and from the informal explanation of the 'for' construct given above.

(2) For P^A to be well defined, we need the fact that the procedure P is *functional*, which means the following. Suppose

$$P \;\equiv\; \textsf{proc in a, out b, aux c begin } S \textsf{ end}.$$

Then for all σ_1, σ_2:

$$\sigma_1 \approx \sigma_2 \text{ (rel } \mathsf{a}) \implies [\![S]\!]^A\sigma_1 \approx [\![S]\!]^A\sigma_2 \text{ (rel } \mathsf{b}).$$

This is guaranteed by the *initialisation condition* for procedures (§7.2).

Functionality of procedures amounts to saying that there are *no side effects* from the output variables or auxiliary variables.

(3) The initialisation condition is a *sufficient (but not necessary) syntactic condition for functionality.* A more general syntactic condition ensuring functionality was given in Jervis [1988] and Thompson [1987]. (A semantic approach to functionality was taken in Tucker & Zucker [1988, §4.3.2].)

DEFINITION 3. Two expressions E_1 and E_2 from any one of the three syntactic classes are said to be *semantically equivalent over Σ* if for all $A \in \boldsymbol{StdAlg}(\Sigma)$, $[\![E_1]\!]^A = [\![E_2]\!]^A$.

REMARKS. (4) Throughout this Section we will describe a number of *transformations* of terms or procedures. In all cases, the transformations are effective, and preserve semantic equivalence over Σ.

(5) (proc-*conversion.*) Given a procedure

$$P \;\equiv\; \textsf{proc in } \mathsf{a}_1,\ldots,\mathsf{a}_m, \textsf{ out b, aux c begin } S \textsf{ end}$$

of type $u \to v$, and a vector of terms $(t_1,\ldots,t_m) : u$, we can define a transformation, which is a form of *simultaneous β-conversion*, to the call $P(t_1,\ldots,t_m)$ w.r.t. some (or all) of the input variables, say (for notational convenience) the first k input variables $\mathsf{a}_1,\ldots,\mathsf{a}_k$ $(1 \le k \le m)$, as follows:

$$P(t_1,\ldots,t_m) \;\mapsto\; P'(t_{k+1},\ldots,t_m),$$

where

$$P' \;\equiv\; \textsf{proc in } \mathsf{a}_{k+1},\ldots,\mathsf{a}_m, \textsf{ out b, aux c begin } S\langle \mathsf{a}_1,\ldots,\mathsf{a}_k/t_1,\ldots,t_k\rangle \textsf{ end}$$

and $S\langle \mathsf{a}_1,\ldots,\mathsf{a}_k/t_1,\ldots,t_k\rangle$ means the *simultaneous substitution* of $t_1,\ldots,t_k$ for $\mathsf{a}_1,\ldots,\mathsf{a}_k$. Note that this substitution is unproblematic, by our requirement that the a_i do not appear on the *lhs* of any assignment in S. (In this substitution, all variables in the declaration in P which occur free in any of $t_1, \ldots, t_k$ must be systematically renamed.) This transformation, which is a form of *simultaneous β-conversion*, is called proc-conversion.

Let FOR(A) be the class of functions defined by $\boldsymbol{For}(\Sigma)$ procedures on A.

THEOREM 1. *We have*

$$PR(A) = FOR(A).$$

Moreover, this equivalence is uniformly effective over Σ, in the sense that we can effectively transform any $PR(\Sigma)$ scheme α into a $\boldsymbol{For}(\Sigma)$ procedure P, and vice versa, such that for all $A \in \boldsymbol{StdAlg}(\Sigma)$,

$$\alpha^A = P^A.$$

This can be proved by an extension of the techniques used in proving the equivalence (1) in §3.6.

7.4 The programming language $\boldsymbol{For}_0(\overline{\Sigma})$

As before, let $S \subseteq \boldsymbol{Sort}(\Sigma)$, and let $\bar{A}$ be the corresponding standard stream algebra, of signature $\overline{\Sigma}(= \overline{\Sigma}^S)$.

The language $\boldsymbol{For}(\overline{\Sigma})$ is (of course) just the $\boldsymbol{For}$ language over the stream signature $\overline{\Sigma}$. More interesting is a sublanguage, $\boldsymbol{For}_0(\overline{\Sigma})$, defined below. First we need

TERMINOLOGY. (1) Variables of simple sort are called *simple variables*, and those of stream sort are called *stream variables*.

(2) Terms of simple (product) type are called *simple terms*.

Now the language $\boldsymbol{For}_0(\overline{\Sigma})$ is a sublanguage of $\boldsymbol{For}(\overline{\Sigma})$, in which the simultaneous inductive definitions of $\boldsymbol{Term}(\overline{\Sigma})$, So if $\boldsymbol{Stmt}(\overline{\Sigma})$ and $\boldsymbol{Proc}(\overline{\Sigma})$ (§7.2) are modified according to the following restrictions:

(*i*) Assignments are to (tuples of) *simple variables* only.

(*ii*) In a procedure, the lists of *output variables* and *auxiliary variables* consist of *simple variables* only. (But the *input variables* may include stream variables.) So the *range type* v of a procedure $P : u \to v$ is *simple*, but the *domain type* u may contain stream sorts.

(*iii*) The projections $U_{u,i}$ and the 'if – then – else – fi' construct for terms applies to *simple terms* only.

REMARK. Thus the only terms of stream sort are *stream variables* $\xi, \ldots$, and these can occur free in the body of a procedure only in one of the two contexts:

- $Q(\ldots, \xi, \ldots)$ (that is, as an actual parameter in a procedure call),
- $\mathsf{eval}(\xi, t^{\mathsf{N}})$.

Let $\mathrm{FOR}_0(\bar{A})$ be the class of functions defined by $\boldsymbol{For}_0(\overline{\Sigma})$ procedures on $\bar{A}$. By a modification of the proof of Theorem 1 we can show:

THEOREM 2. *We have*

$$FOR_0(\bar{A}) \;=\; PR_0(\bar{A})$$

uniformly effectively over Σ.

7.5 The programming language $\boldsymbol{For}_0^{\lambda}(\overline{\Sigma})$

This extends the language $\boldsymbol{For}_0(\overline{\Sigma})$ by adding terms of *stream sort*, by means of *stream abstraction over procedures*, or (more specifically) λ-abstraction over an input variable of sort N. Thus the simultaneous inductive definitions of $\boldsymbol{Term}(\overline{\Sigma})$, $\boldsymbol{Stmt}(\overline{\Sigma})$ and $\boldsymbol{Proc}(\overline{\Sigma})$ are further modified by adding a clause in the definition of $\boldsymbol{Term}(\overline{\Sigma})$:

(*vii*) If P is a procedure of type $u \times \mathsf{N} \to s$, where $s \in S$, and $t \in \boldsymbol{Term}_u$, then

$$\lambda P(t) \in \boldsymbol{Term}_{\bar{s}}.$$

We may write $\lambda P(t)$ more perspicuously as $\lambda \mathbf{z} P(t, \mathbf{z})$, where $\mathbf{z}$ is a variable of type N, which does not occur free in P or t, and which is bound by the '$\lambda \mathbf{z}$'. Such a term is called *stream abstraction term*. Note that it is possible that u is empty, *i.e.*, $P : \mathsf{N} \to s$, in which case the stream abstraction term is $\lambda \mathbf{z} P(\mathbf{z})$.

We call a term of stream sort a *stream term*. So now we have two kinds of stream terms: *stream variables* and *stream abstraction terms*.

REMARKS. (1) Abstraction terms can occur in procedures in two contexts (compare the Remark in §7.4.):

- as actual parameters in procedure calls:

$$Q(\ldots, \lambda \mathbf{z} P(t, \mathbf{z}), \ldots) \qquad \text{or} \qquad \lambda \mathbf{z}' Q(\ldots, \lambda \mathbf{z} P(t, \mathbf{z}), \ldots, \mathbf{z}')$$

  (so stream abstraction can be *nested*!)
- in the context $\mathsf{eval}(\lambda \mathbf{z} P(t, \mathbf{z}),\ t^{\mathsf{N}})$.

(2) (eval-*conversion*.) However the above eval-term can always be simplified to $P(t, t^{\mathsf{N}})$! This is another kind of β-conversion, which we call eval-conversion. (See Remark 5 of §7.3.)

(3) Suppose u is a product type $s_1 \times \cdots \times s_m$ $(m \geq 2)$. If u is simple (*i.e.*, if all the s_i are simple sorts), then a term of type u need not be a vector. It could be a procedure call $P(t)$ where the range type of P is u. However, if u is not simple (*i.e.*, if at least one of the s_i is a stream sort), then (since the range type of a procedure in $\boldsymbol{For}_0^{\lambda}(\overline{\Sigma})$ must be simple) the only possible term of type u is a vector $(t_1, \ldots, t_m)$ with $t_i : s_i$.

(4) An (unimportant, though annoying) syntactic detail in connection with eval-conversion is the following. If t is a vector $(t_1, \ldots, t_{m-1})$, then we interpret $P(t, t^{\mathsf{N}})$ as

$P(t_1, \ldots, t_{m-1}, t^{\mathsf{N}})$. But if t is of product type $s_1 \times \cdots \times s_{m-1}$, but not a vector, then we interpret $P(t, t^{\mathsf{N}})$ as $P(\,U_{u,1}(t), \ldots, U_{u,m-1}(t), t^{\mathsf{N}})$.

(5) The ***semantics*** of $\boldsymbol{For}_0^{\lambda}(\overline{\Sigma})$ is defined by the semantics of $\boldsymbol{For}(\overline{\Sigma})$, together with the stipulation that the meaning of ***stream abstraction terms*** is such that eval-conversion (Remark 2) preserves semantic equivalence.

Let $\mathrm{FOR}_0^{\lambda}(\bar{A})$ be the class of functions defined by $\boldsymbol{For}_0^{\lambda}(\overline{\Sigma})$ procedures on $\bar{A}$. By a modification of the proof of Theorem 2 we can show:

THEOREM 3. *We have*

$$FOR_0^{\lambda}(\bar{A}) \;=\; PR_0^{\lambda}(\bar{A})$$

uniformly effectively over Σ.

7.6 Equivalence of $\boldsymbol{For}_0^{\lambda}(\overline{\Sigma})$ and $\boldsymbol{For}_0(\overline{\Sigma})$

The main result of this section is

THEOREM 4.

$$FOR_0^{\lambda}(\bar{A}) \;=\; FOR_0(\bar{A})$$

uniformly effectively over Σ.

From this and Theorems 2 and 3 follow

COROLLARY.

$$PR_0^{\lambda}(\bar{A}) \;=\; PR_0(\bar{A})$$

uniformly effectively over Σ.

We will call a term, program or procedure of $\boldsymbol{For}_0^{\lambda}(\overline{\Sigma})$ *λ-free* if it contains no stream abstraction terms, *i.e.*, if it is in $\boldsymbol{For}_0(\overline{\Sigma})$.

The proof of Theorem 4, which occupies the rest of this subsection, amounts to defining a "λ-elimination" for $\boldsymbol{For}_0^{\lambda}(\overline{\Sigma})$ procedures, *i.e.*, a ***transformation*** $P \mapsto P'$ of such procedures to λ-free procedures. This essentially involves eliminating stream abstraction terms by "pushing them back through the calls", using proc-conversion, until they disappear by eval-conversion.

The essence of this transformation is given in the proof of the main lemma below, which says that if procedures P and $R_1, R_2, \ldots$, and terms $t_1, t_2, \ldots$ are all λ-free, then the procedure call

$$P(\ldots, \lambda \mathbf{z} R_1(t_1, \mathbf{z}), \ldots, \lambda \mathbf{z} R_2(t_2, \mathbf{z}), \ldots).$$

can be converted to a λ-free call. Stated more precisely:

Lemma. *Suppose*

$$P \;\equiv\; \textsf{proc in } \mathbf{a}_1, \ldots, \mathbf{a}_m, \textsf{ out } \mathbf{b}, \textsf{ aux } \mathbf{c} \textsf{ begin } S \textsf{ end}$$

is a λ-free procedure, and $t_1, \ldots, t_m$ are terms (of the correct types) such that for $i = 1, \ldots, m$: either (i) t_i is a simple term which is λ-free, or (ii) t_i is a stream variable, or (iii) t_i is a stream abstraction term $\lambda \mathbf{z} R(t, \mathbf{z})$, where R and t are λ-free. Then the call

$$P(t_1, \ldots, t_m) \tag{$*$}$$

can be transformed to a λ-free call.

Proof of Lemma: The proof is by structural induction on P. So assume the lemma holds for all procedures occurring in S.

Consider the call $(*)$. If none of the terms $t_1, \ldots, t_m$ is a stream abstraction term, then $(*)$ is λ-free and there is nothing to prove. Otherwise, let us suppose (for notational convenience) that the first k terms $t_1, \ldots, t_k$ (and only these) are the stream abstraction terms. (So $\mathbf{a}_1, \ldots, \mathbf{a}_k$ are stream variables.) Apply simultaneous proc-conversion (§7.3, Remark 5) to $(*)$ w.r.t. $\mathbf{a}_1, \ldots, \mathbf{a}_k$. This results in the call

$$P'(t_{k+1}, \ldots, t_m)$$

where

$$P' \;\equiv\; \textsf{proc in } \mathbf{a}_{k+1}, \ldots, \mathbf{a}_m, \textsf{ out } \mathbf{b}, \textsf{ aux } \mathbf{c} \textsf{ begin } S' \textsf{ end}$$

with

$$S' \;\equiv\; S\langle \mathbf{a}_1, \ldots, \mathbf{a}_k / t_1, \ldots, t_k \rangle.$$

Recall that, by the Remark in §7.4, there are two possible contexts for free occurrences of the stream variables $\mathbf{a}_1, \ldots, \mathbf{a}_k$ in S:

- In a term of the form $\textsf{eval}(\mathbf{a}_j, t^{\mathsf{N}})$. But then $\textsf{eval}(t_j, t^{\mathsf{N}})$ can be simplified by eval-conversion to remove the λ-abstraction (§7.5, Remark 2).
- As actual parameters in some procedure call:

  $$Q(\ldots, \mathbf{a}_{j_1}, \ldots, \mathbf{a}_{j_2}, \ldots).$$

  Then the substitution results in the call

  $$Q(\ldots, t_{j_1}, \ldots, t_{j_2}, \ldots) \tag{$**$}$$

  which satisfies the assumption of the lemma with Q in place of P. Hence, by the induction hypothesis, $(**)$ can be transformed to a λ-free call.

The above transformations result in a λ-free procedure body S'', and hence a λ-free call $P''(t_{k+1}, \ldots, t_m)$. □

Remark. The use of this lemma in the proof of Theorem 4 is reminiscent of the use of the following lemma by Gentzen in the proof of his *Hauptsatz* (cut elimination theorem): *"A derivation with a mix for its lowest inference figure, and not containing any other mix, may be transformed into a derivation (with the same endsequent) in which no mix occurs"* (Gentzen [1969, p.89]).

The proof of Theorem 4 now follows easily.

Proof of Theorem 4: We will actually prove the following three assertions:

(*a*) A procedure P can be transformed to a λ-free procedure P';

(*b*) A statement S can be transformed to a λ-free statement S';

(*c*) A term t can be transformed to a term t' such that:

(*i*) if t is simple, then t' is λ-free,

(*ii*) if t is a stream variable, then $t' \equiv t$,

(*iii*) if t is a stream abstraction term $\lambda \mathbf{z} P(t, \mathbf{z})$, then $t' \equiv \lambda \mathbf{z} P'(t', \mathbf{z})$, where P' and t' are λ-free.

(*iv*) if t is a vector $(t_1, \ldots, t_m)$, then $t' \equiv (t'_1, \ldots, t'_m)$.

(Note that if t is a vector of simple type, then it falls under both (*i*) and (*iv*). This is a harmless overlap of cases.)

All these assertions are proved by simultaneous structural induction, or more precisely, all these transformations $E \to E'$ (for E a procedure, statement or term) are defined by simultaneous structural recursion on E. We give a few of the interesting cases.

• $t \equiv P(t_1, \ldots, t_m)$.
By the induction hypothesis, there are transformations $P \to P'$ and $t_i \to t'_i$, where P' is λ-free, and t'_i are as stated above. Let $t' \equiv P'(t'_1, \ldots, t'_m)$. This satisfies the hypothesis of the lemma, and so it can be further transformed to a λ-free call.

• $t \equiv \lambda \mathbf{z} P(t_1, \ldots, t_{m-1}, \mathbf{z})$.
This similarly uses the lemma.

• $S \equiv \mathbf{x} := t$.
Notice that t is a simple term. So by the induction hypothesis, there is a transformation $t \to t'$, where t' is λ-free. Let $S' \equiv \mathbf{x} := t'$.

• $P \equiv$ proc D begin S end.
By the induction hypothesis, there is a transformation $S \to S'$, with S' λ-free. Let $P \equiv$ proc D begin S' end. □

8 Concluding remarks

Many computing systems process streams of data, and stream transformations occur in many areas of computer science. Stream transformations are used extensively in the semantics of data flow networks and related functional programming languages (see Wadge & Ashcroft [1985], for example); such mappings are seen in the theories of processes, automata and formal languages; and general stream processing functions are also used in general purpose specification and design methodologies (such as FOCUS; see Broy et al. [1993]). In the study of stream processing, there seems to be an emphasis on nondeterministic and higher order aspects of computation, and on general fixed point methods. There are many and varied mathematical formulations of streams and stream processing functions.

In contrast, the approach pursued here is conservative, being designed to study deterministic stream computation in some detail and to emphasise and exploit connections with abstract data type theory and methods from classical recursion theory, proof theory and model theory. Furthermore, this work is designed to support a mathematical theory of parallel deterministic computing systems called synchronous concurrent algorithms (SCAs). We comment on these concerns.

8.1 Computability theory on many sorted algebras

Our research on stream computation and its applications is part of a programme that aims at the precise formulation and analysis of the many interesting notions of computability, specifiability, and verifiability for many sorted algebras. Many sorted universal algebra provides a general theory of data, and so can be used in different areas of computer science and mathematics. Recent themes for computability on A are:

1. classes of models of computation and specification: Tucker and Zucker [1988, 1989, 1991, 1992a];
2. real and complex number computation: Tucker and Zucker [1992b];
3. verifiable computable functions: Tucker and Zucker [1990, 1993];
4. parameterised algebraic specifications and computable functions: Tucker & Zucker [1994].

The extension to streams suggests new directions for research, for example, involving higher order algebraic methods (Meinke [1992],) in the case of 1, 3 and 4, and topological algebras in the case of 2. It is interesting to note that work on dynamical systems is requiring models of computation on algebras of real and complex numbers: see Blum, Shub & Smale [1989].

8.2 SCA computation, hardware algebra and dynamical systems

A synchronous concurrent algorithm (SCA) is a network of processors and channels that compute and communicate in parallel, and are synchronised by a global clock. In particular, such algorithms compute on infinite streams of data, and are deterministic because of the clock. Most hardware is constructed from components that are examples of SCAs. The theory of SCAs was developed independently to unify many disparate models of parallel deterministic computation found in hardware design (*e.g.*, systolic arrays) and in dynamical system theory (*e.g.*, neural nets; cellular automata (Wolfram [1989]); and coupled map lattice dynamical systems (Crutchfield and Kaneko [1987], Kaneko [1993]).

In its form and techniques, the research programme has been closely related to that of the theory of computable functions on algebras. SCA computation is a form of $\lambda\mu$PR computation, and the primary types of SCA computation are forms of λPR computation. Some discussion of this is in Tucker & Zucker [1992a]. However the subject is independently driven by case studies in hardware design and dynamical systems, practical formal methods, and software tools.

This work began in earnest in 1983 and is recorded in several papers and theses, starting with Thompson & Tucker [1985] and Thompson [1987], respectively. A basic reference is Thompson & Tucker [1991], and some general theory is in Meinke & Tucker [1992]. Examples of work representing the two areas are:

Hardware: Hobley, Thompson & Tucker [1988], Harman & Tucker [1988, 1990, 1993], Eker and Tucker [1988, 1989], Eker, Stavridou & Tucker [1991], Martin & Tucker [1988], McConnell & Tucker [1992].

Dynamical systems: Holden, Thompson & Tucker [1990, 1991], Holden et al. [1992].

The early research on SCAs and its connections with computability theory was more fully surveyed in Tucker [1991], and its context in the theory of hardware described in McEvoy & Tucker [1990]. Recent work is able to exploit more fully both the methods of universal algebra and computablility theory. The topics 1–4 in §8.1 are directly relevant to studies of SCAs. For example, the theory of computation on real and complex numbers and stream algebras underlies the theory of neural networks and coupled map lattices.

References

J. A. Bergstra & J. V. Tucker [1987], "Algebraic specifications of computable and semi-computable data types," Theoretical Computer Science 50, 137–181.

L. Blum, M. Shub & S. Smale [1989], "On a theory of computation and complexity over the real numbers: NP-completeness, recursive functions and universal machines," Bulletin American Mathematical Society (N.S.) 21, 1–46.

M. Broy, F. Dederichs, C. Dendorfer, M. Fuchs, T. F. Gritzner & R. Weber [1993], "The design of distributed sytems: An introduction to FOCUS," Inst. f. Informatik, Technical University of Munich, Technical Report TUM-19202-2, January 1993.

J. P. Crutchfield & K. Kaneko [1987], "Phenomenology of spatio-temporal chaos," in Directions in Chaos, H. Bai-lin, ed., World Scientific.

M. D. Davis & E. P. Weyuker [1983], Computability, Complexity and Languages, Academic Press.

H. Ehrig & B. Mahr [1985], Fundamentals of Algebraic Specification 1, EATCS Monographs #6, Springer-Verlag.

S. M. Eker, V. Stavridou & J. V. Tucker [1991], "Verification of synchronous concurrent algorithms using OBJ3. A case study of the Pixel Planes architecture," in Designing Correct Circuits, G. Jones & M. Sheeran, eds., Springer-Verlag, 231–252.

S. M. Eker & J. V. Tucker [1988], "Specification, derivation and verification of concurrent line drawing algorithms and architectures," in Theoretical Foundations of Computer Graphics and CAD, R.A. Earnshaw, ed., 449–516.

S. M. Eker & J. V. Tucker [1989], "Specification and verification of synchronous concurrent algorithms: a case study of the Pixel Planes architecture ," in Parallel Processing for Computer Vision and Display, P.M. Dew, R.A. Earnshaw & T.R. Heywood, eds., Addison Wesley, 16–49.

E. Engeler [1993], Algorithmic Properties of Structures, World Scientific.

M. Fitting [1981], Fundamentals of Generalized Recursion Theory, North-Holland.

G. Gentzen [1969], "Investigations into logical deduction," in The Collected Papers of Gerhard Gentzen, M.E. Szabo, ed., North-Holland, 68–131.

S. Greibach [1975], Theory of Program Structures: Schemes, Semantics, Verification, Lecture Notes in Computer Science #36, Springer-Verlag.

N. A. Harman & J. V. Tucker [1988], "Clocks, retimings, and the formal specification of a UART," in The Fusion of Hardware Design and Verification (Proceedings of IFIP Working Group 10.2 Working Conference), G. Milne, ed., North-Holland, 375–396.

N. A. Harman & J. V. Tucker [1990], "The formal specification of a digital correlator I: User specification process," in Theoretical Foundations of VLSI Design, K. McEvoy & J.V. Tucker, eds., Cambridge University Press, 161–262.

N. A. Harman & J. V. Tucker [1993], "Algebraic methods and the correctness of microprocessors," in Correct Hardware Design and Verification Methods, G.J. Milne & L. Pierre, eds., Lecture Notes in Computer Science #683, Springer-Verlag, 92–108.

K. M. Hobley, B. C. Thompson & J. V. Tucker [1988], "Specification and verification of synchronous concurrent algorithms: a case study of a convoluted algorithm," in The Fusion of Hardware Design and Verification (Proceedings of IFIP Working Group 10.2 Working Conference), G. Milne, ed., North-Holland, 347–374.

A. V. Holden, M. Poole, J. V. Tucker & H. Zhang [1992], "Coupled map lattices as computational systems," American Institute of Physics — Chaos 2, 367–376.

A. V. Holden, B. C. Thompson & J. V. Tucker [1990], "The computational structure of neural systems," in Neurocomputers and Attention. I: Neurobiology, Synchronisation and Chaos, A.V. Holden & V.I. Kryukov, eds., Manchester University Press, 223–240.

A. V. Holden, B. C. Thompson & J. V. Tucker [1991], "Can excitable media be considered as computational systems," Physica D 49, 240–246.

C. A. Jervis [1988], A Theory of Program Correctness for Three Valued Logic, PhD Thesis, School of Computer Studies, University of Leeds.

K. Kaneko, ed. [1993], Coupled Map Lattices — Theory and Applications, J. Wiley.

S. C. Kleene [1952], Introduction to Metamathematics, North-Holland.

A. R. Martin & J. V. Tucker [1988], "The concurrent assignment representation of synchronous systems," Parallel Computing 9, 227–256.

B. McConnell & J. V. Tucker [1992], "Infinite synchronous concurrent algorithms: the specifiation and verification of a hardware stack," in NATO Advanced Study Institute, International Summer School at Marktoberdorf,, 1991, on Logic and Algebra for Specification, H. Schwichtenberg, ed., Springer-Verlag.

K. McEvoy & J. V. Tucker, eds. [1990], Theoretical Foundations of VLSI Design, Cambridge University Press.

K. Meinke [1992], "Universal algebra in higher types," Theoretical Computer Science 100, 385–417.

K. Meinke & J. V. Tucker [1992], "Universal algebra," in Handbook of Logic in Computer Science, S. Abramsky, D. Gabbay & T. Maibaum, eds., Oxford University Press, 189–411.

J. C. Shepherdson [1985], "Algorithmic procedures, generalized Turing algorithms, and elementary recursion theory," in Harvey Friedman's Research on the Foundations of Mathematics, L.A. Harrington, M.D. Morley, A. Ščedrov & S.G. Simpson, eds., North-Holland, 309–315.

R. Stephens & B. C. Thompson [1995], "Cartesian stream transformer composition," Fundamenta Informaticae, to appear.

B. C. Thompson [1987], A Mathematical Theory of Synchronous Concurrent Algorithms, PhD Thesis, School of Computer Studies, University of Leeds.

B. C. Thompson & J. V. Tucker [1985], "Theoretical considerations in algorithm design," in Fundamental Algorithms for Computer Graphics, R.A. Earnshaw, ed., Springer-Verlag, 855–878.

B. C. Thompson & J. V. Tucker [1991], "Algebraic specification of synchronous concurrent algorithms and architectures," Computer Science Department, University College of Swansea, Research Report CSR 9-91.

J. V. Tucker [1980], "Computing in algebraic systems," in Recursion Theory, its Generalisations and Applications, F.R. Drake & S.S. Wainer, eds., London Mathematical Society Lecture Note Series #45, Cambridge University Press, 215–235.

J. V. Tucker [1991], "Theory of computation and specification over abstract data types and its applications," in NATO Advanced Study Institute, International Summer School at Marktoberdorf,, 1989, on Logic, Algebra and Computation, F.L. Bauer, ed., Springer-Verlag, 1–39.

J. V. Tucker, S. S. Wainer & J. I. Zucker [1990], "Provable computable functions on abstract data types," in 17th International Colloquium on Automata, Languages and Programming, Warwick University, England, July 1990, M.S. Paterson, ed., Lecture Notes in Computer Science #443, Springer-Verlag, 660–673.

J. V. Tucker & J. I. Zucker [1988], Program Correctness over Abstract Data Types, with Error-State Semantics, CWI Monographs #6, North-Holland.

J. V. Tucker & J. I. Zucker [1989], "Horn programs and semicomputable relations on abstract structures," in 16th International Colloquium on Automata, Languages and Programming, Stresa, Italy, July 1989, G. Ausiello, M. Dezani-Ciancaglini & S. Ronchi della Rocca, eds., Lecture Notes in Computer Science #372, Springer-Verlag, 745–760.

J. V. Tucker & J. I. Zucker [1991], "Projections of semicomputable relations on abstract data types," International Journal of Foundations of Computer Science 2, 267–296.

J. V. Tucker & J. I. Zucker [1992a], "Theory of computation over stream algebras, and its applications," in Mathematical Foundations of Computer Science 1992: 17th International Symposium, Prague, I.M. Havel & V. Koubek, eds., Lecture Notes in Computer Science #629, Springer-Verlag, 62–80.

J. V. Tucker & J. I. Zucker [1992b], "Deterministic and nondeterministic computation, and Horn programs, on abstract data types," Journal of Logic Programming 13, 23–55.

J. V. Tucker & J. I. Zucker [1992c], "Examples of semicomputable sets of real and complex numbers," in Constructivity in Computer Science: Summer Symposium, San Antonio, Texas, June 1991, J.P. Myers, Jr. & M.J. O'Donnell, eds., Lecture Notes in Computer Science #613, Springer-Verlag, 179–198.

J. V. Tucker & J. I. Zucker [1993], "Provable computable selection functions on abstract structures," in Proof Theory, P. Aczel, H. Simmons & S.S. Wainer, eds., Cambridge University Press, 277–306.

J. V. Tucker & J. I. Zucker [1994], "Abstract computability, conditional equations and initial algebras," in preparation.

W. W. Wadge & E. A. Ashcroft [1985], Lucid, the Dataflow Language, Academic Press.

W. Wechler [1991], Universal Algebra for Computer Scientists, EATCS Monographs #25, Springer-Verlag.

S. Wolfram, ed. [1989], Theory of Cellular Automata, World Scientific.

J. I. Zucker & L. Pretorius [1993], "Introduction to Computability Theory," South African Computer Journal, No. 9, 3–30.

The Proof Theoretic Complexity of Recursive Programs

Stanley S. Wainer *

Dept. of Pure Mathematics, University of Leeds, Leeds LS2 9JT, UK

What does cut elimination have to do with program transformation ?

Abstract. We explore, from the point of view of mathematical logic, the structural relationships between recursive programs and their termination proofs, and between program transformation and proof transformation, particularly in the case of the recursive-to-tail-recursive transformation. Ideas from proof theory and recursion theory are applied, so as to measure in an abstract way, the logical cost of such transformations.

This paper continues the general theme of the author's two previous Marktoberdorf contributions Wainer (1991), Wainer (1993), the aim being to explore - from a mathematical/logical point of view - the deep structural connections between recursive programs and their termination proofs. In particular the main interest here lies in various relationships between program *transformations* and their corresponding proof *transformations* . Work of e.g. Goad

*Partially supported by an EC Science Plan Twinning Project between the mathematical logic groups in Leeds, Munich and Oslo, entitled "Proof Theory and Computation".

(1980), Schwichtenberg (1992 and in this volume), Murthy (1991), Madden (1992), Pfenning (1990), and Anderson (1993) already illustrates the potential applicability of proof-transformation as a means to synthesize and analyze useful program-transformations, but our present concern lies rather in the general proof-theoretic principles which underly such applications. Since most (though not all) interesting proof transformations can be viewed as cut-eliminations of one form or another, it seems likely that classical proof theory should supply appropriate methods and ideas.

The paper is split into three parts. Part One is a preliminary report on recent work with W. Sieg in Sieg-Wainer (1994), investigating in a very simple context, the proof-theoretic complexity of "recursive to tail-recursive" program transformation. Part Two takes a different look at this same transformation, but now from a quite general recursion theoretic point of view. Basic results about totally defined recursive operators show how to make natural ordinal assignments which measure the "sizes" of their computation trees, and from this we are able to read off the increased cost - in ordinal terms - of transforming an arbitrarily nested recursion into a tail recursion. These results are due, but with rather different proofs, to Tait (1961) and more recently Fairtlough-Wainer (1992). Part Three returns to our original question and shows how the general results of part 2, when viewed proof theoretically in the framework of Fairtlough (1991) and Handley-Wainer (1994), are actually cut elimination theorems at the equational level.

1 The Logic of Primitive Recursion

An old and well-known theorem of logic, due variously to Kreisel, Parsons, Mints, Takeuti and others, states that the primitive recursive functions are exactly those which can be proved to be total and computable in the fragment of arithmetic with induction restricted to existential (Σ_1) formulas only (see e.g. Wainer (1993) for a proof). This is an "extensional" result, characterizing a certain class of number-theoretic *functions* . What we are looking for here is something more "intensional", i.e. a logic which allows us to distinguish between different kinds of primitive recursive *programs* according to the structure of their respective termination proofs. Preferably it should provide a clear correspondence between proofs and programs, and also at the higher level between proof-transformations and program- transformations, so that "program-complexity" is measurable directly in terms of "proof complexity". See Feferman (1992) for another development of similar ideas.

The logic we arrive at below is a strictly "linear" version of Σ_1-arithmetic (no contraction, no weakening and no exchange !). It is obtained simply by analyzing just what one needs to prove the totality of primitive recursive definitions. The absence of exchange rules means that two cut rules are needed - an ordinary one and another one which we call "call-by-value cut" for reasons which will be obvious. It then turns out that, in the appropriate setting, the transformation from recursive to tail-recursive programs is precisely call-by-value-cut-elimination !

Definition. A *recursive program* over N, the natural numbers, is a finite sequence of function-definitions

$$f_i\,(\ldots) \;=\; T_i\,(f_0,\ldots,f_{i-1},f_i\,;x_1,\ldots,x_k)$$

where the arguments ... of f_i are the variables $x_1,\ldots,x_k$, or successors of them, or the constant 0; and T_i is a term built up from those variables and the constant 0 by arbitrary applications of the successor function, the previously defined functions $f_0,\ldots,f_{i-1}$ and possibly the function f_i itself.

Evaluation is by "substitution" or "call-by-value". That is, on given numerical inputs $\vec{n} = n_1,\ldots,n_k$ we are allowed successively to replace any subterm $f_j(\vec{n})$ of T_i by its previously-computed value until eventually a numerical value for $f_i(\vec{n})$ is obtained. However termination is not guaranteed in general.

A *(partial) recursive function* is then one which is defined by a recursive program, according to the usual least-fixed-point semantics.

Example. Given a program defining g, add to it the new equations :

$$\begin{array}{lcl} h(0,y,z) & = & y \\ h(x+1,y,z) & = & z \\ f(x) & = & h(g(x),x,f(x+1)) \end{array}$$

Then on input $x := n$, and assuming $g(n)$, $g(n+1)$, ... defined, we have

$$\begin{array}{lcll} f(n) & = & n & \text{if } g(n)=0 \\ & = & f(n+1) & \text{if } g(n)\neq 0 \\ f(n+1) & = & n+1 & \text{if } g(n+1)=0 \\ & = & f(n+2) & \text{if } g(n+1)\neq 0 \end{array}$$

etcetera, and therefore if g has a "zero" above n then f computes it :

$$f(n) \quad = \quad \text{least } m \geq n \text{ such that } g(m) = 0 \; .$$

Note on the other hand that if the third equation defining f is simply regarded as a left-to-right term rewrite then $f(n)$ can be expanded continually without ever giving a value.

Definition. Call a recursive (program defining a) function f *provably recursive* or *verifiable* in a given logic L if

$$L \vdash \forall \vec{x}\, \exists y \; (\, f(\vec{x}) = y \,) \; .$$

Obviously the more restrictive the logic, the more restricted will be the class of recursive programs we can prove to terminate in it. The aim here is to impose simple logics on the equation calculus in such a way that there is a clear and precise structural correspondence between termination proofs and known subclasses of recursive programs. We concentrate here on primitive recursive programs, though the ideas have a much wider range of application.

Definitions. A *primitive recursive* program is one in which every defining equation has one of the five forms "zero", "successor", "projection", "explicit definition" and "primitive recursion" as follows :

$$\begin{array}{llll}
(Z) & f_i(x) & = & 0 \\
(S) & f_i(x) & = & x+1 \\
(P) & f_i(\vec{x}) & = & x_j \\
(E) & f_i(\vec{x}) & = & T(f_0,\ldots,f_{i-1}\,;\vec{x}) \\
(PR_0) & f_i(0,\vec{x}) & = & f_{i_0}(\vec{x}) \\
(PR_1) & f_i(z+1,\vec{x}) & = & f_{i_1}(z,\vec{x},f_i(z,\vec{x}))
\end{array}$$

where in the (PR) scheme $i_0, i_1 < i$.

A *generalized primitive recursive* program is one in which the primitive recursion equation (PR_1) is generalized to allow substitution of terms for the parameters $\vec{x}$ in the recursive call $f_i(z,\vec{x})$ as follows :

$$(GPR_1) \quad f_i(z+1,\vec{x}) \;=\; f_{i_1}(z,\vec{x},f_i(z,f_{i_2}(z,\vec{x}),\ldots,f_{i_{k+1}}(z,\vec{x})))$$

where $i_0, i_1, i_2, \ldots, i_{k+1} < i$.

A primitive *tail recursive* program is one in which generalized primitive recursion is allowed, but only in the following restricted context, where the recursive call on $f_i(z,\ldots)$ is the final function-call made in the evaluation of $f_i(z+1,\vec{x})$:

$$(TR_1) \quad f_i(z+1,\vec{x}) \;=\; f_i(z,f_{i_1}(z,\vec{x}),\ldots,f_{i_k}(z,\vec{x}))\,.$$

Remark. Tail recursive programs

$$\begin{array}{lll}
f(0,x) & = & g(x) \\
f(z+1,x) & = & f(z,h(z,x))
\end{array}$$

are "efficient" since they can immediately be recast as while-loops in the following way :

while $z \neq 0$ **do** $z := z-1;\ x := h(z,x)$ **od** ; $f := g(x)$.

The following transformations are either explicit or implicit in R. Péter (1967). See also Wainer (1993) for further information on primitive recursion.

Theorem 1.1 *Every generalized primitive recursive program can be transformed into a primitive tail recursive program defining the same function. Every primitive tail recursive program can be transformed into an ordinary primitive recursive program defining the same function.*

Proof. (i) A generalized primitive recursion (i.e. with parameter substitution) such as

$$\begin{array}{lcl} f(0,x) & = & g(x) \\ f(z+1,x) & = & h(z,x,f(z,p(z,x))) \end{array}$$

can be transformed into a tail recursive program as follows (note however that three tail recursions seem to be needed - the two given here plus another one implicitly used in order to define the "modified minus" from the predecessor)

$$\begin{array}{llcl} (TR_0) & f_0(0,z,x) & = & x \\ (TR_1) & f_0(n+1,z,x) & = & f_0(n,z\dot{-}1,p(z\dot{-}1,x)) \\ (E) & f_1(n,z,x,y) & = & h(z\dot{-}(n+1),f_0(n,z,x),y) \\ (TR_0) & f_2(0,z,x,y) & = & y \\ (TR_1) & f_2(n+1,z,x,y) & = & f_2(n,z,x,f_1(n,z,x,y)) \\ (E) & f_3(z,x) & = & f_2(z,z,x,g(f_0(z,z,x))) \end{array}$$

The devoted (!) reader with a taste for intricate inductions might now like to verify that

$$\forall z\, \forall x\ (\ f_3(z,x) \ = \ f(z,x)\ )\ .$$

Hint : one needs first to check the following identities

$$f_1(n+1,z+1,x,y) \ = \ f_1(n,z,p(z,x),y)$$

$$f_2(n+1,z+1,x,y) \ = \ h(z,x,f_2(n,z,p(z,x),y))$$

and then a further induction on z yields the desired result.

(ii) A primitive tail recursion such as

$$\begin{array}{lcl} f(0,x) & = & g(x) \\ f(z+1,x) & = & f(z,p(z,x)) \end{array}$$

can be transformed into an ordinary primitive recursion as follows

$$\begin{array}{llcl} (PR_0) & f_0(0,z,x) & = & x \\ (PR_1) & f_0(n+1,z,x) & = & p(z\dot{-}n, f_0(n,z,x)) \\ (E) & f_1(z,x) & = & g(f_0(z,z\dot{-}1,x)) \,. \end{array}$$

The verification needs a preliminary induction on n to show

$$f_0(n+1,z,x) \;=\; f_0(n, z\dot{-}1, p(z,x))$$

and then by a further induction on z,

$$\forall z\,\forall x\,(\,f_1(z,x) \;=\; f(z,x)\,)\,.$$

We are now going to devise a logic exactly tailored to proofs about primitive recursive and generalized primitive recursive programs.

The Logic of Primitive Recursion (LPR).

The formulas $A, B, C, \ldots$ will be either atoms of the form $f(\vec{x}) \simeq y$ with y a variable, meaning $f(\vec{x})$ is defined with value y, or Σ_1-formulas $\exists y(f(\vec{x}) \simeq y)$ or Π_2-formulas $\forall \vec{x} \exists y(f(\vec{x}) \simeq y)$.

The axioms are of two kinds, the principal ones being purely relational sequents or "logic programs" describing the individual equations in a primitive recursive program, thus for example

$$f_0(\vec{x}) \simeq y_0, f_1(\vec{x}, y_0) \simeq y_1, \ldots, f_m(\vec{x}, y_0, \ldots, y_{m-1}) \simeq y_m \vdash f(\vec{x}) \simeq y_m$$

is the axiom for an explicit definition

$$f(\vec{x}) \;=\; f_m(\vec{x}, f_0(\vec{x}), f_1(\vec{x}, f_0(\vec{x})), \ldots)\,.$$

The other axioms simply express that the zero, successor and projection functions are defined :

$$(N-Ax) \quad \vdash \exists y(0 \simeq y)\,, \;\; \vdash \exists y(x+1 \simeq y)\,, \;\; \vdash \exists y(x \simeq y)\,.$$

The logic rules are the sequent rules for $\exists$ and $\forall$:

$$(\exists \vdash)\; \frac{\ldots\,,\, A(y)\,,\, \ldots \;\vdash\; B}{\ldots\,,\, \exists y A(y)\,,\, \ldots \;\vdash\; B} \qquad (\vdash \exists)\; \frac{\ldots \;\vdash\; B(y')}{\ldots \;\vdash\; \exists y B(y)}$$

$$(\forall \vdash)\; \frac{\ldots\,,\, A(x')\,,\, \ldots \;\vdash\; B}{\ldots\,,\, \forall x A(x)\,,\, \ldots \;\vdash\; B} \qquad (\vdash \forall)\; \frac{\ldots \;\vdash\; B(x)}{\ldots \;\vdash\; \forall x B(x)}$$

with the usual "eigenvariable" conditions on $(\exists \vdash)$ and $(\vdash \forall)$.

In addition there are two cut rules :

$$(C)\ \frac{\vdash C \quad C, \ldots \vdash B}{\ldots \vdash B} \qquad (CVC)\ \frac{\vdash C \qquad \ldots, C, \ldots \vdash B}{\ldots\ \ldots \vdash B}$$

and the induction rule :

$$(IND)\ \frac{\vdash B(0) \quad B(z) \vdash B(z+1)}{\vdash B(z)}\ .$$

Note. What you see is all there is ! The dots ... denote arbitrary finite sequences of assumptions and the logic is strictly linear in the sense that there are no hidden structural rules - no Contraction, no Weakening, and furthermore no Exchange ! Hence the need for two Cut rules, the second of which applies a cut "in context" and is called a "call by value" cut for reasons which will shortly become obvious. Note also that there are no other assumptions in the induction rule besides the induction hypothesis $B(z)$.

Definition. LPR($\exists$) and LPR($\forall\exists$) denote the logics restricted to Σ_1 and Π_2 formulas respectively. LPR($\forall\exists$)-(CVC) denotes the logic LPR($\forall\exists$) without call-by-value cuts.

Theorem 1.2 *We have the following equivalences :*

- *Primitive Recursive* $\equiv$ *LPR($\exists$) - verifiable.*
- *Generalized Primitive Recursive* $\equiv$ *LPR($\forall\exists$) - verifiable.*
- *Primitive Tail Recursive* $\equiv$ *LPR($\forall\exists$)-(CVC) - verifiable.*

Proof. We do not give a completely detailed proof here, but sufficient to display the basic relationships.

(i) That primitive recursive programs are LPR($\exists$) - verifiable is easily seen. Suppose for example that f is defined explicitly from g and h by

$$f(x) = g(h(x))$$

where g and h are already assumed to be LPR($\exists$) - verifiable. Then the starting axiom is

$$h(x) \simeq y \, , \; g(y) \simeq z \; \vdash \; f(x) \simeq z \, .$$

By ($\vdash \exists$) followed by ($\exists \vdash$) we then obtain

$$h(x) \simeq y \, , \; \exists z \, (g(y) \simeq z) \; \vdash \; \exists z \, (f(x) \simeq z) \, .$$

From this and the assumption $\exists z \, (g(y) \simeq z)$ we then have by a call by value cut (CVC),

$$h(x) \simeq y \; \vdash \; \exists z \, (f(x) \simeq z)$$

and then by ($\exists \vdash$),

$$\exists y \, (h(x) \simeq y) \; \vdash \; \exists z \, (f(x) \simeq z) \, .$$

Thus by the assumption $\exists y \, (h(x) \simeq y)$ and an ordinary cut (C),

$$\vdash \; \exists z \, (f(x) \simeq z) \, .$$

Note how the eigenvariable conditions on ($\exists \vdash$) rules completely determine the order of events in the above proof, so that the call by value cut was essential.

As a further example, suppose f is defined primitive recursively from g and h as follows ;

$$\begin{array}{lcl} f(0, x) & = & g(x) \\ f(z+1, x) & = & h(z, x, f(z, x)) \, . \end{array}$$

where $\vdash \; \exists y \; (g(x) \simeq y)$ and $\vdash \; \exists u \; (h(z, x, y) \simeq u)$ are assumed. Then the starting axioms are

$$g(x) \simeq y \; \vdash \; f(0, x) \simeq y$$

and

$$f(z, x) \simeq y \, , \; h(z, x, y) \simeq u \; \vdash \; f(z+1, x) \simeq u \, .$$

Concentrating on the induction step first, we have by ($\vdash \exists$) and ($\exists \vdash$),

$$f(z, x) \simeq y \, , \; \exists u \, (h(z, x, y) \simeq u) \; \vdash \; \exists y \, (f(z+1, x) \simeq y) \, .$$

Then by a call by value cut,

$$f(z,x) \simeq y \ \vdash \ \exists y\,(f(z+1,x) \simeq y)$$

and by $(\exists \vdash)$,

$$\exists y\,(f(z,x) \simeq y) \ \vdash \ \exists y\,(f(z+1,x) \simeq y)\,.$$

Applying $(\vdash \exists)$, $(\exists \vdash)$ and an ordinary cut to the first axiom we easily obtain $\vdash \ \exists y\,(f(0,x) \simeq y)$, and so by the induction rule we have

$$\vdash \ \exists y\,(f(z,x) \simeq y)$$

as required.

(ii) Next we show why LPR($\forall\exists$) - verifiable programs are generalized primitive recursive. Suppose we had a proof of

$$\vdash \ \forall x\,\exists y\,(f(z,x) \simeq y)$$

by induction on z. The induction step would therefore be

$$\forall x\,\exists y\,(f(z,x) \simeq y) \ \vdash \ \forall x\,\exists y\,(f(z+1,x) \simeq y)\,.$$

This deduction presumably used some recursive calls on "given" functions, so let us assume it came about by means of one ordinary cut on a function p and a call by value cut on a function h from :

$$\forall x\,\exists u\,(p(z,x) \simeq u), \ \ \forall x\,\exists y\,(f(z,x) \simeq y), \forall x\,\forall y\,\exists v\,(h(z,x,y) \simeq v) \\ \vdash \ \forall x\,\exists y\,(f(z+1,x) \simeq y)\,.$$

The eigenvariable conditions place heavy restrictions on how this could have been derived. Essentially it must have come about by applying $(\exists \vdash)$, $(\forall \vdash)$, $(\vdash \forall)$, in that order (!) to :

$$p(z,x) \simeq u\ , \ \ \forall x\,\exists y\,(f(z,x) \simeq y)\ , \ \forall x\,\forall y\,\exists v\,(h(z,x,y) \simeq v) \\ \vdash \ \exists y\,(f(z+1,x) \simeq y)\,.$$

Stripping away the quantifiers prefixing $f(z,x) \simeq y$ we now see that this would have come from

$$p(z,x) \simeq u, \ \ f(z,u) \simeq y, \forall x\,\forall y\,\exists v\,(h(z,x,y) \simeq v) \\ \vdash \ \forall x\,\exists y\,(f(z+1,x) \simeq y)$$

by applying ($\exists\vdash$) and then ($\forall\vdash$) with u as witnessing variable (the only other possible witnessing variables would have been z or x but these are less general). Now we can strip away the quantifier prefix on $h(z,x,y) \simeq v$ to see that this last line would have come about by applying ($\exists\vdash$) and ($\forall\vdash$) to :

$$p(z,x) \simeq u \ , \ f(z,x) \simeq y \ , \ h(z,x,y) \simeq v \ \vdash \ \exists y\,(f(z+1,x) \simeq y)\,.$$

Finally, this would have arisen by ($\vdash\exists$) from the axiom :

$$p(z,x) \simeq u \ , \ f(z,x) \simeq y \ , \ h(z,x,y) \simeq v \ \vdash \ f(z+1,x) \simeq v$$

describing a generalized primitive recursion :

$$\begin{array}{lcl} f(0,x) & = & g(x) \\ f(z+1,x) & = & h(z,x,f(z,p(z,x)))\,. \end{array}$$

By reversing the above we also obtain the converse, that every generalized primitive recursion is LPR($\forall\exists$) - verifiable. Note that if we took apart an LPR($\exists$) - inductive proof in a similar way then we would be prevented (by the absence of the $\forall\vdash$ rule) from substituting $p(z,x)$ for the variable x and so an ordinary primitive recursive program would be the only possible result. Hence the converse to part (i).

(iii) The only other crucial thing to note is that if call by value cuts were disallowed in the derivation in part (ii) above, then the h function could not appear and so the extracted program would have to be a tail recursion :

$$\begin{array}{lcl} f(0,x) & = & g(x) \\ f(z+1,x) & = & f(z,p(z,x))\,. \end{array}$$

This completes the proof.

Theorem 1.3 *Hence the transformation from generalized primitive recursive programs to primitive tail recursive programs corresponds exactly to the elimination of call by value cuts in LPR($\forall\exists$).*

2 Ordinal Analysis of Recursion

A recursive program, defining a function $f : N^k \to N$ from "given" functions $g_1, \ldots, g_m$, can be viewed as a monotone, continuous operator Φ on partial functions, which gives f as the least fixed point of the equation

$$f = \Phi(g_1, \ldots, g_m, f) .$$

For example the generalized primitive recursion

$$\begin{aligned} f(0, x) &= g(x) \\ f(z+1, x) &= h(z, x, f(z, p(z, x))) . \end{aligned}$$

defines f from g, h, p as the least fixed point of the operator

$$\begin{aligned} \Phi(g, h, p, f)(z, x) &= g(x) && \text{if } z = 0 \\ &= h(z-1, x, f(z-1, p(z-1, x))) && \text{else.} \end{aligned}$$

Henceforth, to simplify notation, we shall regard the given functions as having been coded into a single unary function $g(z) = \langle g_1(z), \ldots, g_m(z) \rangle$, and similarly the arguments of f as having been coded into a single argument $x = \langle x_1, \ldots, x_k \rangle$. Thus the fixed-point equation defining f is just

$$f(x) = \Phi(g, f)(x) .$$

The solution to this equation is

$$f = \bigcup_i f_i$$

where $f_i = \Phi(g)^i(\bot)$ so that $f_i \subseteq f_{i+1}$ for every $i = 0, 1, 2, \ldots$

Therefore for all $n, m \in N$,

$$f(n) = m \quad \Leftrightarrow \quad \exists i \, (f_i(n) = m)$$

and since the successful evaluation of $f_i(n) = \Phi(g)^i(\bot)(n)$ will require only finitely many calls on g, say $g(0), g(1), \ldots, g(j)$, we can go a step further and write

$$f(n) = m \quad \Leftrightarrow \quad \exists i \, \exists j \, (\Phi(\bar{g}(j))^i(\bot)(n) = m)$$

where $\bar{g}(j)$ denotes the finite subfunction of g from $g(0)$ up to $g(j-1)$. Furthermore since Φ is monotone we can contract the two existential quantifiers into one by taking any $y \geq \max(i,j)$ thus

$$f(n) = m \quad \Leftrightarrow \quad \exists y\, (\Phi(\bar{g}(y))^y(\bot)(n) = m)\,.$$

From this we can immediately derive Kleene's fundamental Normal Form for recursion, see e.g. Kleene (1952). Below we let U and V denote the left and right inverses of some standard bijective pairing function from N^2 to N. We also let σ range over the set of all finite sequences $\langle x_0, \ldots, x_{j-1}\rangle$ of natural numbers, so that $\sigma = \bar{g}(j)$ means $g(i) = x_i$ for each $i < j$.

Theorem 2.1 *With each operator Φ as above, we can associate a primitive recursive relation*

$$T_\Phi\ (n, \sigma, y) \quad :\equiv \quad \Phi(\sigma)^y(\bot)(n) = U(y)$$

such that if f is recursively defined from a given function g as the least fixed point of the equation

$$f \;=\; \Phi(g, f)$$

then f is expressible in the normal form

$$f(n) \;=\; U(\ \textit{least}\ y.\, T_\Phi(n, \bar{g}(y), y)\ )\,.$$

Definitions. With each operator Φ as above, associate the following *partial recursive functional* F_Φ from partial functions g to partial functions f,

$$F_\Phi(g) \;:=\; \text{the least fixed point } f \text{ of the equation } f = \Phi(g, f)\,.$$

Call Φ a *total operator* and F_Φ a *total recursive functional* if for each totally defined function $g : N \to N$ the output function $f = F_\Phi(g)$ is also totally defined.

Theorem 2.2 *The normal form for F_Φ is thus*

$$F_\Phi(g)(n) \;=\; U(\ \textit{least}\ y.\, T_\Phi(n, \bar{g}(y), y)\ )\,.$$

Furthermore

$$F_\Phi \textit{ is a total recursive functional} \quad \Leftrightarrow \quad \forall g\, \forall n\, \exists y\, T_\Phi(n, \bar{g}(y), y)\,.$$

Definition. With each operator Φ as above, associate the following *tree* of finite sequences of natural numbers :

$$\text{Tree}(\Phi) \;:=\; \{\, \langle x_0, \ldots, x_k \rangle \;:\; \forall y < k \, \neg T_\Phi(x_0, \langle x_1, \ldots, x_y \rangle, y) \,\} \,.$$

Tree(Φ) is called the *computation-tree* of Φ.

We imagine Tree(Φ) as growing downwards by extension, i.e. if σ and τ are any two nodes on the tree then σ comes below τ iff σ is an extension of τ. An infinite branch of the tree is thus determined by a number n and a total function $g : N \to N$ such that $\forall y \, \neg T_\Phi(n, \bar{g}(y), y)$. Therefore an infinite branch on Tree(Φ) is a witness to the fact that for some n and g, $F_\Phi(g)(n)$ is not defined. To say that the tree is "well-founded" is to say that there are no infinite branches, and hence :

Theorem 2.3

$$\Phi \textit{ is total} \;\Leftrightarrow\; \textit{Tree}(\Phi) \textit{ is well-founded.}$$

This equivalence is the basis for a natural theory of ordinal assignments. It allows us to assign to each total operator Φ a (recursive) ordinal $\|\Phi\|$ measuring the size of its computation-tree.

Definition. For each node σ on Tree(Φ) define

$$\begin{aligned} \|\sigma\| &= 0 && \text{if } \sigma \text{ is a leaf} \\ &= \sup{}^{+} \{\, \|\tau\| : \tau \text{ extends } \sigma \,\} && \text{otherwise.} \end{aligned}$$

where the $\sup^+$ of a set of ordinals is the least ordinal greater than all of them.

Then define

$$\|\Phi\| \;:=\; \overset{+}{\sup} \{\, \|\sigma\| \;:\; \sigma \in \text{Tree}(\Phi) \,\} \,.$$

Example. The iteration operator

$$\Phi(g, f)(n) \;:=\; \textbf{if } n = 0 \textbf{ then } 0 \textbf{ else } g(f(n-1))$$

computes the total recursive functional

$$F_\Phi(g)(n) \;=\; g^n(0) \,.$$

The terminal node or "leaf" on the branch through Tree(Φ) determined by n and g will be the shortest sequence

$$\langle n, g(0), \ldots, g^2(0), \ldots, g^3(0), \ldots, g^{n-1}(0), \ldots, g^n(0), \ldots, g(y-1) \rangle$$

such that $U(y) = g^n(0)$.

The value of $g^i(0)$, once chosen, thus determines the length of the following segment up to $g^{i+1}(0)$. Therefore as we take the supremum over all branches issuing from a given node

$$\langle n, g(0), \ldots, g(g^{i-1}(0) - 1) \rangle$$

the successive segments $g^i(0), \ldots, g^{i+1}(0)$ have unbounded length. So each such segment adds one more ω to the ordinal height of the tree. Since there are n such segments, the height of the subtree below node $\langle n \rangle$ must be $\omega \cdot n$. Therefore the height of the entire tree is

$$\|\Phi\| \;=\; \sup_n \,\omega \cdot n \;=\; \omega^2 .$$

Question. How does the ordinal $\|\Phi\|$ reflect the "complexity" of Φ ?

Definition. Associate with each total operator Φ the functional

$$H_\Phi \;:\; \text{Tree}(\Phi) \;\rightarrow\; (N \rightarrow N) \;\rightarrow\; (N \rightarrow N)$$

defined by the following tail recursion over nodes σ on Tree(Φ) :

$$\begin{aligned} H_\Phi\,(\sigma)(g)(m) \;&=\; m && \text{if } \sigma \text{ is a leaf} \\ &=\; H_\Phi\,(\sigma \frown g(m))(g)(m+1) && \text{otherwise} \end{aligned}$$

where $\sigma \frown z$ denotes the new sequence which arises by appending z to the end of σ.

Theorem 2.4 *For each total operator Φ we have*

$$H_\Phi\,(\langle n \rangle)(g)(0) \;=\; \textit{least } y.\; T_\Phi(n, \bar{g}(y), y)\;) \,.$$

Therefore

$$F_\Phi\,(g)(n) \;=\; U(\; H_\Phi\,(\langle n \rangle)(g)(0)\;) \,.$$

Therefore $m = F_\Phi(g)(n)$ *can be computed by the following abstract "while program" over Tree(*Φ*)*

$\sigma := \langle n \rangle$;
$m := 0$;
while σ *is not a leaf* **do** $\sigma := \sigma \frown g(m)$; $m := m+1$ **od** ;
$m := U(m)$.

Proof. Let $y_0 =$ least $y.\ T_\Phi(n, \bar{g}(y), y)$. Then from the definition above we compute

$$\begin{aligned} H_\Phi\ (\langle n \rangle)(g)(0) &= H_\Phi\ (\langle n, g(0)\rangle)(g)(1) \\ &= H_\Phi\ (\langle n, g(0), g(1)\rangle)(g)(2) \\ &= H_\Phi\ (\langle n, g(0), g(1), g(2)\rangle)(g)(3) \\ &= \ldots \\ &= H_\Phi\ (\langle n, g(0), \ldots, g(y_0 - 1)\rangle)(g)(y_0) \\ &= y_0 \ . \end{aligned}$$

The rest follows immediately from the Normal Form for F_Φ and from the direct translation of the tail-recursive definition of H_Φ into a while-loop.

Definition. Call a total operator Φ, and its associated recursive functional F_Φ, *α-recursive* where α is an ordinal, if there is a function $|.| : N \to \alpha$ such that for all g and n, the recursive calls on f made by Φ in evaluating the right hand side of

$$f(n) = \Phi(g, f)(n)$$

are all of the form $f(m)$ where $|m| < |n|$.

Example. The previous example of an iteration operator Φ is ω-recursive since we can simply take $|n| = n$.

Lemma 2.5 *If* Φ *is* α*-recursive then* $\|\Phi\| \leq \omega^\alpha$.

Proof. Suppose f is defined from a given function g by α-recursion thus

$$f(n) = \Phi(g, f)(n) \ .$$

For any fixed n let $f(n_0), f(n_1), \ldots, f(n_{k-1})$ be the recursive calls used in evaluating the right hand side, so $|n_i| < |n| < \alpha$ for each $i < k$. Note that in general there may be "nested" calls, meaning that the value of n_i might depend upon some other $f(n_j)$'s. We can always assume that the recursive calls are ordered so that whenever such a dependency occurs then $j < i$.

In the computation tree Tree(Φ), all of these recursive calls, and the successive calls needed for their evaluations etcetera, must be arranged linearly along the g-branch below node $\langle n \rangle$.

What we are looking for is a function ψ on ordinals such that

$$\|\langle n \rangle\| \leq \psi(|n|) .$$

But from the above comment we see that ψ must satisfy the following condition :

$$\psi(|n|) \leq \psi(|n_0|) + \psi(|n_1|) + \ldots + \psi(|n_{k-1}|) \leq \sup_i \psi(|n_i|) \cdot \omega$$

and so the obvious solution is

$$\psi(|n|) = \omega^{|n|} .$$

Therefore

$$\|\langle n \rangle\| \leq \omega^{|n|} < \omega^{\alpha}$$

and hence, taking the supremum over all nodes $\langle n \rangle$ in Tree(Φ), we obtain

$$\|\Phi\| \leq \omega^{\alpha} .$$

Note. If in the above Lemma, k is a fixed bound on the depth of nestings of recursive calls on f induced by Φ, then the ψ function need only satisfy :

$$\psi(|n|) \leq \psi(|n_0|) + \psi(|n_1|) + \ldots + \psi(|n_{k-1}|) \leq \sup_i \psi(|n_i|) \cdot k$$

and so we can then choose a better bound :

$$\psi(|n|) = k^{|n|} .$$

Therefore in this case we have

$$\|\Phi\| \leq k^{\alpha} .$$

Combining the Lemma with the previous Theorem, we thus have an "exponential trade-off" in the ordinal complexity of the transformation from general recursive programs to tail recursive ones. In fact the result below goes back originally to Tait(1961) who first showed how arbitrary nested recursions could be transformed into unnested ones. But it was re-examined more recently in Fairtlough-Wainer(1992) in the context of transformations to "while"-programs. The proofs there are completely different from Tait's, and different again from the ones presented here, although they all have the same germ.

Theorem 2.6 *Every α-recursive operator Φ can be transformed into an ω^α- tail-recursive definition of H_Φ such that for all $g : N \rightarrow N$ and all $n \in N$,*

$$F_\Phi\,(g)(n) \;=\; U(\,H_\Phi\,(\langle n\rangle)(g)(0)\,)\,.$$

Furthermore, if there is a fixed bound k on the depth of nestings of recursive calls in Φ, then ω^α can be strengthened to k^α.

Example. Let Φ be the operator defining a binary function $f = F_\Phi(g)$ as follows ;

$$\Phi(g,f)(n,m) \;=\; \textbf{if } n=0 \textbf{ then } g(m) \textbf{ else } f(n-1,f(n-1,m))\,.$$

Clearly Φ is ω-recursive under the assignment $|(n,m)| = n$ and, writing f_n to denote the function $\lambda m.f(n,m)$, we see that for all $n,m \in N$,

$$f_n\,(m) \;=\; f_{n-1}\circ f_{n-2}\circ\ldots\circ f_1\circ f_0\,(\,g(m)\,)\,.$$

Now for any strictly decreasing sequence of numbers $\sigma = \langle i_1,\ldots,i_r\rangle$ set

$$H(\sigma)(g) \;:=\; f_{i_1}\circ f_{i_2}\circ\ldots\circ f_{i_r}\,.$$

Then for all n, m we have

$$f_n\,(m) \;=\; H\,(\langle n\rangle)(g)(m)$$

and furthermore, H is definable by the following tail recursion :

$$\begin{array}{lll} H\,(\langle\rangle)(g)(m) & = & m \\ H\,(\sigma)(g)(m) & = & H\,(\tau)(g)(g(m)) \end{array}$$

where if $\sigma = \langle i_1, \ldots, i_r\rangle$ then

$$\tau = \langle i_1, \ldots, i_{r-1}, i_r - 1, i_r - 2, \ldots, 1, 0\rangle\ .$$

But notice that this tail recursive definition of H is a 2^ω-recursion under the assignment

$$|\sigma| = 2^{i_1} + 2^{i_2} + \ldots + 2^{i_r}\ .$$

Thus we have transformed the doubly-nested ω-recursion Φ into a 2^ω-tail- recursion.

Remark. Of course there is no difference *set theoretically* between the ordinals ω and 2^ω. However the structure of their respective well-orderings used in the above example is quite different. It is the increase in complexity of the well-ordering on 2^ω which reflects the cost in transforming the ω-recursive Φ to a tail recursion. In other words it is not really the set-theoretic notion of ordinal which matters here, but rather the "intensional" representation of it by means of a chosen well-ordering.

This distinction is brought out further in the next section, where we return to the original question : what does it all have to do with cut elimination ?

3 Ordinal Bounds on Equational Derivations

This section is based on work of Fairtlough (1991), Fairtlough-Wainer (1992) and Handley-Wainer (1994). The idea is to assign uniform ordinal bounds to derivations of recursive functions in Kleene's Equation Calculus, from which we can read off sub- recursive hierarchy classifications of their computational complexity. (Complexity is viewed "in the large" - we are not concerned here with questions about feasible or polynomially bounded computation, but rather with general methods for comparing the complexities of natural classes of recursive definitions.)

As noted at the end of section two, an ordinal for us is not simply a set. What matters is the way in which it is presented, for it is the presentation that provides a structure over which we can make recursive definitions.

Definitions. (i) By a *presentation* of a countable ordinal α we mean a chosen sequence of *finite* subsets $\alpha[i] \subset \alpha$ satisfying :

$$\alpha[0] \subset \alpha[1] \subset \ldots \subset \alpha[i] \subset \alpha[i+1] \subset \ldots$$

and

$$\alpha = \bigcup_i \alpha[i] .$$

We also insist that $0 \in \alpha[0]$ for all $\alpha > 0$.

(ii) Note that a presentation of α induces a presentation of each $\beta < \alpha$ by

$$\beta[i] = \alpha[i] \cap \beta .$$

(iii) Each presentation of α determines a sequence of predecessor functions $P_i : \alpha + 1 \to \alpha$ defined by

$$P_i(0) = 0 \text{ and } P_i(\beta) = \max \beta[i] \text{ if } 0 < \beta \leq \alpha .$$

Thus if k is the cardinality of $\beta[i]$ we have :

$$\beta[i] = \{0 = P_i^k(\beta) < P_i^{k-1}(\beta) < \ldots < P_i^2(\beta) < P_i(\beta)\} .$$

Definitions. Given fixed presentations of α and β, the predecessor functions P_i provide a natural way of recursively defining presentations of the sum, product and exponential as follows :

$$\begin{aligned}
\alpha + 0[i] &= \alpha[i] \\
\alpha + \beta[i] &= \alpha + P_i(\beta)[i] \cup \{\alpha + P_i(\beta)\} \\
\alpha \cdot 0[i] &= 0[i] \\
\alpha \cdot \beta[i] &= \alpha \cdot P_i(\beta) + \alpha[i] \\
\alpha^0[i] &= 1[i] \\
\alpha^\beta[i] &= \alpha^{P_i(\beta)} \cdot \alpha[i] .
\end{aligned}$$

Then if we give ω the presentation $\omega[i] = \{0, 1, \ldots, i\}$ the above definitions yield the so-called *standard* presentations for all ordinals below $\epsilon_0 = \sup_i\{1, \omega, \omega^\omega, \omega^{\omega^\omega}, \ldots\}$.

Definition. Let Φ be a total operator, let α be any fixed countable ordinal with a chosen presentation, and let h, g be given. Then the *hierarchy* generated by iterating Φ over α consists of the functions $f_\beta : N \to N$ where $\beta \leq \alpha$, defined as follows :

$$\begin{aligned} f_0(n) &= h(n) \\ f_\beta(n) &= \Phi(g, f_{P_n(\beta)})(n) \text{ if } 0 < \beta \leq \alpha \,. \end{aligned}$$

Lemma 3.1 *Suppose the total operator Φ satisfies the following "majorisation conditions" when restricted to strictly increasing, positive functions g and f :*

$$f(n) \;<\; \Phi(g, f)(n) \;<\; \Phi(g, f)(n+1)$$

and

$$\forall m \geq n\, (f(m) \leq f'(m)) \;\Rightarrow\; \Phi(g, f)(n) \;\leq\; \Phi(g, f')(n)\,.$$

Suppose also that h and g are strictly increasing and that g is positive, i.e. $g(n) > n$.

Then the hierarchy $\{f_\beta : \beta \leq \alpha\}$ generated by Φ satisfies :

- $\gamma \in \beta[n] \;\Rightarrow\; f_\gamma(n) < f_\beta(n)$.
- *f_β is strictly increasing.*

Proof. By inductions over $\beta \leq \alpha$.

If $\gamma \in \beta[n]$ then either $\gamma \in P_n(\beta)[n]$ or $\gamma = P_n(\beta)$ and so by the induction hypothesis and the first majorisation condition we have

$$f_\gamma(n) \;\leq\; f_{P_n(\beta)}(n) \;<\; \Phi(g, f_{P_n(\beta)})(n) \;=\; f_\beta(n)\,.$$

To see that f_β is strictly increasing, first notice that either $P_n(\beta) \in P_{n+1}(\beta)[n+1]$ or else $P_n(\beta) = P_{n+1}(\beta)$, because $\beta[n] \subset \beta[n+1]$. Therefore by the induction hypothesis we have

$$\forall m \geq n+1\, (f_{P_n(\beta)}(m) \leq f_{P_{n+1}(\beta)}(m))\,.$$

So by the induction hypothesis and the majorisation conditions,

$$\begin{aligned} f_\beta(n) &= \Phi(g, f_{P_n(\beta)})(n) \\ &< \Phi(g, f_{P_n(\beta)})(n+1) \\ &\leq \Phi(g, f_{P_{n+1}(\beta)})(n+1) \\ &= f_\beta(n+1) . \end{aligned}$$

This completes the proof.

Majorisation Hierarchies. The following are the main examples of hierarchies generated by the above Lemma - and we shall make use of them later.

- Choosing $\Phi(g, f) = g \circ f$ we obtain :

$$G_0(g)(n) \;=\; 0 \;; \;\; G_\beta(g)(n) \;=\; g(\; G_{P_n(\beta)}(g)(n)\;) \,.$$

- Choosing $\Phi(g, f) = f \circ g$ we obtain :

$$H_0(g)(n) \;=\; n \;; \;\; H_\beta(g)(n) \;=\; H_{P_n(\beta)}(g)(\; g(n)\;) \,.$$

- Choosing $\Phi(g, f) = f \circ f = f^2$ we obtain :

$$B_0(g)(n) \;=\; g(n) \;; \;\; B_\beta(g)(n) \;=\; B_{P_n(\beta)}(g)^2(n) \,.$$

These are called respectively the "Slow Growing", "Hardy" and (a version of the) "Fast Growing" hierarchies. When g is chosen to be the successor function we merely write $G_\beta(n)$, $H_\beta(n)$, $B_\beta(n)$ instead of $G_\beta(succ)(n)$, $H_\beta(succ)(n)$, $B_\beta(succ)(n)$. Although the definition of G does not quite satisfy the conditions of the Lemma, because we start with h the constant zero function, nevertheless the G functions will be strictly increasing after the first limit stage ω.

Lemma 3.2 $G_\beta(n) \;=\;$ *cardinality of* $\beta[n]$.

Proof. By a simple transfinite induction on β. If $\beta = 0$ then $\beta[n]$ is empty so its cardinality is $0 = G_0(n)$. If $\beta > 0$ then $\beta[n] = P_n(\beta)[n] \cup \{P_n(\beta)\}$ and so

$$\text{card } \beta[n] \;=\; 1 + \text{card } P_n(\beta)[n] \;=\; 1 + G_{P_n(\beta)}(n) \;=\; G_\beta(n) \,.$$

Lemma 3.3

$$H_{\gamma+\beta}(g) \;=\; H_\gamma(g) \,\circ\, H_\beta(g)$$

and

$$H_{2^\beta} \;=\; B_\beta\,.$$

Proof. (i) By induction on β. The first identitiy holds automatically if $\beta = 0$ since $H_0(g)$ is just the identity. If $\beta > 0$ then by the earlier definition of $\gamma + \beta[n]$ we see that $P_n(\gamma+\beta) = \gamma + P_n(\beta)$ and so

$$\begin{aligned} H_{\gamma+\beta}(g)(n) &= H_{P_n(\gamma+\beta)}(g)(g(n)) \\ &= H_{\gamma+P_n(\beta)}(g)(g(n)) \\ &= H_\gamma(g) \,\circ\, H_{P_n(\beta)}(g)\,(g(n)) \\ &= H_\gamma(g) \,\circ\, H_\beta(g)\,(n)\,. \end{aligned}$$

(ii) Again by induction on β. If $\beta = 0$ then

$$H_{2^0}(g)(n) \;=\; H_1(g)(n) \;=\; H_0(g)(g(n)) \;=\; g(n) \;=\; B_0(g)(n).$$

If $\beta > 0$ then by inspecting the definition of $2^\beta[n]$ given earlier we see that $P_n(2^\beta) = P_n(2^{P_n(\beta)} + 2^{P_n(\beta)})$, and therefore using the first part of the Lemma we have

$$\begin{aligned} H_{2^\beta}(g)(n) &= H_{P_n(2^{P_n(\beta)}+2^{P_n(\beta)})}(g)(g(n)) \\ &= H_{2^{P_n(\beta)}+2^{P_n(\beta)}}(g)(n) \\ &= H_{2^{P_n(\beta)}}(g) \circ H_{2^{P_n(\beta)}}(g)\,(n) \\ &= B_{P_n(\beta)}(g) \circ B_{P_n(\beta)}(g)\,(n) \\ &= B_\beta(g)(n) \quad \text{by definition of } B. \end{aligned}$$

Remark. As we shall see, the above relationship between the doubly nested definition of B and the tail recursive definition of H is in fact a cut elimination result.

Equational Derivations with Ordinal Bounds. Let E be a system of recursion equations (i.e. a recursive program) defining functions $f_0, \ldots, f_k$ from "given" functions $g = g_0, \ldots, g_l$ where g_0 denotes the constant 0 and g_1 the successor. Let t denote a variable-free term built up from numerals by repeated applications of the function symbols f_i and g_j. Write $t_0 \rightarrow_E t_1$ if t_0 rewrites to t_1 in one step by replacing any subterm $f_i(\vec{n})$ by its defining term T_i in E. Let ordinal α be given with a fixed presentation.

Then the relation

$$n : N \vdash_E^{\alpha} t = m$$

meaning "from inputs less than or equal to n we can derive that t has value m within ordinal bound α", is generated inductively according to the following rules (we omit the 'E' although the dependence on a given system of defining equations is to be understood) :

Ax $n : N \vdash^{\alpha} g(n') = m$ if $n' \leq n$ and $g(n')$ has value m.

E $$\frac{n : N \vdash^{\beta} t_1 = m}{n : N \vdash^{\alpha} t_0 = m} \quad \text{if } t_0 \rightarrow_E t_1.$$

C $$\frac{n : N \vdash^{\beta_0} t_0 = k \quad k : N \vdash^{\beta_1} t_1(k) = m}{n : N \vdash^{\alpha} t_1(t_0) = m}.$$

R In the E and C rules the restriction is that $\beta, \beta_0, \beta_1 \in \alpha[n]$.

Remark. Of course any derivation of $n : N \vdash_E^{\alpha} t = m$ is finite, so why need we consider possibly infinite ordinal bounds ? The reason is that if f is a function defined by a system of equations E, then what we want to do is find a *uniform* bound α such that

$$\forall n \in N \, (f(n) = m \Rightarrow n : N \vdash_E^{\alpha} f(n) = m).$$

It is the recursion rule R which allows us to do this. Then α will provide an abstract measure of the complexity of the program defining f.

Lemma 3.4 *Suppose* $n : N \vdash_E^{\alpha} t = m$. *Let* g *now denote any fixed positive, strictly increasing function which bounds all the givens* $g_0, \ldots, g_l$ *in E, and let* $g' = g + 1$. *Then* $m \leq B_\alpha(g)(n)$ *and the height of the derivation tree is bounded by* $B_\alpha(g')(n)$.

Proof. By induction over the derivation of $n : N \vdash^{\alpha}_{E} t = m$ according to the Ax, E, and C rules restricted by R. The Ax and E rules are trivial since if $n' \leq n$ then for each $j \leq l$ we have, for $\beta \in \alpha[n]$,

$$g_j(n') \leq g(n) \leq B_\beta(g)(n) < B_\alpha(g)(n) < B_\alpha(g')(n).$$

Now suppose $n : N \vdash^{\alpha} t = m$ arises by the C rule applied to premises

$$n : N \vdash^{\beta_0} t_0 = k \quad \text{and} \quad k : N \vdash^{\beta_1} t_1(k) = m\,.$$

Then by the induction hypothesis, and since $\beta_0, \beta_1 \in \alpha[n]$,

$$k \leq B_{\beta_0}(g)(n) \leq B_{P_n(\alpha)}(g)(n)$$

and

$$m \leq B_{\beta_1}(g)(k) \leq B_{P_n(\alpha)}(g)(k)\,.$$

Therefore

$$m \leq B_{P_n(\alpha)}(g)(\ B_{P_n(\alpha)}(g)(n)\ ) = B_\alpha(g)(n).$$

Furthermore the height of the derivation - viz. the least number greater than the heights of its two premises - is bounded by

$$\begin{aligned} \max\ &\{B_{\beta_0}(g')(n), B_{\beta_1}(g')(k)\} + 1 \\ &\leq \max\{B_{\beta_0}(g')(n), B_{\beta_1}(g')(B_{P_n(\alpha)}(g)(n))\} + 1 \\ &\leq \max\{B_{P_n(\alpha)}(g')(n) + 1, B_{P_n(\alpha)}(g')(B_{P_n(\alpha)}(g)(n) + 1)\} \\ &\leq B_{P_n(\alpha)}(g')(\ B_{P_n(\alpha)}(g)(n) + 1\ ) \\ &\leq B_{P_n(\alpha)}(g')(\ B_{P_n(\alpha)}(g')(n)\ ) \\ &= B_\alpha(g')(n)\,. \end{aligned}$$

Definitions. (i) Let α be an infinite countable ordinal with a fixed chosen presentation. Then there is a bijection $\pi : \alpha \to N$ and so we can represent each $\beta < \alpha$ as a number $b = \pi(\beta)$ and each predecessor function P_n as a number-theoretic function p_n such that $p_n(b) = \pi(P_n(\beta))$. We can assume without loss of generality that $0 = \pi(0)$. Then we say that a function f is defined by α - *recursion* from the givens h and g if for some term T

$$\begin{aligned} f(0,n) &= h(n) \\ f(b,n) &= T\,(g,f;n) \end{aligned}$$

wherein each recursive call on f inside T is of the form $f(p_t(b), t)$.

Lemma 3.5 *Suppose f is defined from the givens h, g by the α - recursion equations E as above. Then there is a fixed $k \in N$ such that for every $b = \pi(\beta)$ we have*

$$\forall n \in N \; (f(b,n) = m \;\Rightarrow\; n : N \;\vdash^{k.(\beta+1)}\; f(b,n) = m) \, .$$

Proof. The appropriate choice of k will emerge from the proof. Proceed by induction on $\beta < \alpha$. The case $b = 0$ follows from the axiom $n : N \;\; \vdash^0 \;\; h(n) = m$ by one application of the E rule, since $f(0,n) \rightarrow_E h(n)$ and $0 \in k[n]$ as long as $k > 0$. Now assume $b > 0$ and that the result holds for each recursive call $f(p_{n'}(b), n')$. Then since $k.(P_{n'}(\beta) + 1)[n'] \subset k.\beta[n']$ we can easily derive, for the correct value m,

$$n' : N \;\vdash^{k.\beta}\; f(p_{n'}(b), n') = m \, .$$

So if t is a subterm of T such that for some $i < k - 1$

$$n : N \;\vdash^{k.\beta+i}\; t = n'$$

we can use the C rule to combine these two in order to obtain

$$n : N \;\vdash^{k.\beta+i+1}\; f(p_t(b), t) = m \, .$$

Therefore by repeating this process over the build-up of the entire term T, and choosing $k - 1$ to be the height of the "formation tree" of T, we eventually get

$$n : N \;\vdash^{k.\beta+k-1}\; T = m$$

and then by one further application of the E rule, since $f(b,n)$ rewrites to T,

$$n : N \;\vdash^{k.(\beta+1)}\; f(b,n) = m.$$

This completes the proof.

Definitions. Let $REC(< \tau)$ denote the class of all functions definable from the constant, addition and (modified) subtraction functions, by repeated applications of composition and α - recursion with $\alpha < \tau$.

Let $DER(< \tau)$ denote the class of all functions f for which there is a recursive program E (with constants, addition and subtraction as the only givens) and an ordinal $\alpha < \tau$ such that

$$\forall \vec{n} \in N^l \ (f(\vec{n}) = m \ \Rightarrow \ \max \vec{n} : N \vdash^{\alpha}_{E} f(\vec{n}) = m) \, .$$

Let $COMP(B_{<\tau})$ denote the class of all functions computable (in any reasonable model of computation) with resource bound B_α for some $\alpha < \tau$. Since B_ω is exponential the size of B_α, even for relatively small infinite α, already blurs any fine distinctions between choices of computational models.

Theorem 3.6 *Suppose τ is a limit ordinal closed under addition and $> \omega$. Suppose also that we have a fixed presentation of τ and a bijection $\pi : \tau \rightarrow N$ such that for every $\beta < \tau$, $\beta \in \tau[\pi(\beta)]$. Then*

$$REC(< \tau) \ = \ \bigcup_k DER(< k \cdot \tau) \ = \ \bigcup_k COMP(B_{<k \cdot \tau}) \, .$$

Proof. (i) To show $REC(< \tau) \subset \bigcup_k DER(< k \cdot \tau)$ suppose for example that f is defined by an α - recursion from given functions g where $\alpha < \tau$ and where it is inductively assumed that $g \in DER(< k \cdot \gamma)$ for sufficiently large k. Then by Lemma 3.5 there is an appropriate k such that for all n and all $b = \pi(\beta)$ with $\beta < \alpha$,

$$n : N \ \vdash^{k.(\beta+1)} \ f(b, n) = m \, .$$

If we insert above the axioms introducing the givens g, their derivations with bound $k \cdot \gamma$, we obtain

$$n : N \ \vdash^{k.(\gamma+\beta+1)} \ f(b, n) = m$$

and hence

$$\max(b, n) : N \ \vdash^{k.(\gamma+\beta+1)} \ f(b, n) = m$$

where this is now an "outright" derivation from axioms defining only addition and subtraction. Now since $\beta \in \tau[b] \cap \alpha = \alpha[b]$ we have

$$k \cdot (\gamma + \beta) \in k \cdot (\gamma + \alpha)[b] \subset k \cdot (\gamma + \alpha)[\max(b, n)]$$

and hence

$$k \cdot (\gamma + \beta + 1)[\max(b, n)] \subset k \cdot (\gamma + \alpha)[\max(b, n)] .$$

Therefore we obtain for all $b, n \in N$,

$$\max(b, n) : N \vdash^{k.(\gamma+\alpha)} f(b, n) = m$$

and since τ is closed under addition, $\gamma + \alpha < \tau$, so f belongs to $DER(< k \cdot \tau)$.

(ii) To show $DER(< k \cdot \tau) \subset COMP(B_{<k \cdot \tau})$, suppose f is defined by a recursive program E (with only addition and subtraction as givens) in such a way that for some $\alpha < k \cdot \tau$,

$$\forall \vec{n}(f(\vec{n}) = m \Rightarrow \max \vec{n} : N \vdash^{\alpha} f(\vec{n}) = m) .$$

Then by Lemma 3.4, the finite height of the derivation tree for $f(\vec{n})$, and all the values m of subcomputations performed therein, are bounded by $B_\alpha(g)(\max \vec{n})$ for some linear function g. So for each input $\vec{n}$ the entire derivation tree for $f(\vec{n})$ can be encoded as a machine computation with resource bound exponential in $B_\alpha(g)$. Since $B_\omega(n) = B_\omega(succ)(n) = n + 2^{n+1}$, we have $g < B_{k \cdot \omega}$ provided $k > 0$, and hence $B_\alpha(g) < B_{k \cdot \omega + \alpha}$ and hence the resource bound in computing f will be less than $B_\omega \circ B_\alpha(g)$ which in turn is less than $B_{k \cdot \omega + \alpha} \circ B_{k \cdot \omega + \alpha} = B_{k \cdot \omega + \alpha + 1}$. But $k \cdot \omega + \alpha + 1 < k \cdot \tau$ and hence $f \in COMP(B_{<k \cdot \tau})$.

(iii) To show $COMP(B_{<k \cdot \tau}) \subset REC(< \tau)$ we only need to prove that $B_{k \cdot \alpha} \in REC(< \tau)$ for every $\alpha < \tau$, because if f is computable within resource bound $B_{k \cdot \alpha}$ then f is definable from $B_{k \cdot \alpha}$ by compositions and ω - recursions only, and so f will also be in $REC(< \tau)$. But notice that

$$B_{k \cdot \alpha}(n) = B_{k \cdot P_n(\alpha) + k}(n) = B^{2^k}_{k \cdot P_n(\alpha)}(n)$$

the 2^k - times iterate of $B_{k \cdot P_n(\alpha)}$. So $B_{k \cdot \alpha}$ is definable by a 2^k - times nested α - recursion, and therefore it belongs to $REC(< \tau)$.

This completes the proof.

Examples. $REC(<\omega^2)$ is the class of all primitive recursive functions, and $REC(<\epsilon_0)$ is the class of all provably recursive functions of formal arithmetic.

Derivations of Tail Recursions. A *tail* α - recursion from givens h and g is a definition of the form

$$\begin{aligned} f(0,n) &= h(n) \\ f(b,n) &= f(p_n(b), g(n)) \,. \end{aligned}$$

Following the idea of Lemma 3.5 we notice that the derivation of the recursion step uses only a simple form of the C rule where the left hand premise is an axiom :

$$\dfrac{\dfrac{n : N \vdash^0 g(n) = m' \quad m' : N \vdash^{2.P_n(\beta)} f(p_n(b), m') = m}{n : N \vdash^{2.P_n(\beta)+1} f(p_n(b), g(n)) = m}}{n : N \vdash^{2.\beta} f(b,n) = m} \,.$$

Thus if $\beta \in \alpha[\pi(\beta)]$ we can prove inductively that

$$\max(b,n) : N \vdash_0^{2.\alpha} f(b,n) = m$$

where $\vdash_0$ now denotes derivability using the Ax and E rules, but only these restricted forms of the C rule. /newline

This restricted form corresponds to the "ordinary" cuts of part one. The full C rule allows recursive "call by value", as is seen in the proof of Lemma 3.5.

Definitions. (i) Let $TAIL(<\tau)$ denote the class of functions definable from constants, addition and subtraction by repeated compositions and *tail* α - recursions where $\alpha < \tau$.

(ii) Let $DER_0(<\tau)$ denote the class of all functions f for which there is a recursive program E (with constants, addition and subtraction as the only givens) and an ordinal $\alpha < \tau$ such that

$$\forall \vec{n} \in N^l \; (f(\vec{n}) = m \;\Rightarrow\; \max \vec{n} : N \vdash_0^{\alpha} f(\vec{n}) = m) \,.$$

Theorem 3.7 *Let τ satisfy the same conditions as in the previous Theorem. Then*

$$TAIL(<\tau) \;=\; \bigcup_k DER_0(<k\cdot\tau) \;=\; \bigcup_k COMP(H_{<k\cdot\tau})\,.$$

Proof. The above comments are enough to show how the proof of the first part of Theorem 3.6 can be modified to yield $TAIL(<\tau) \subset DER_0(<k\cdot\tau)$, in fact with $k=2$.

The proof of $DER_0(<k\cdot\tau) \subset COMP(H_{<k\cdot\tau})$ follows exactly the same lines as the proof of the second part of Theorem 3.6. But instead of using Lemma 3.4 we now need a corresponding lemma giving the appropriate bounding functions for $\vdash_0$ derivations. Since only a restricted form of the C rule is allowed here, a careful inspection of the proof of 3.4 shows that it is the H - functions which now give the appropriate bounds. For if we have a derivation

$$\frac{n:N\ \vdash^0\ g(n)=m' \qquad m':N\ \vdash^\beta_0\ t(m')=m}{n:N\ \vdash^\alpha_0\ t(g(n))=m}$$

where inductively we assume $m \le H_\beta(g)(m')$ then since $\beta \in \alpha[n]$ we obtain

$$m \;\le\; H_\beta(g)(g(n)) \;\le\; H_{P_n(\alpha)}(g)(g(n)) \;=\; H_\alpha(g)(n)\,.$$

The proof of $COMP(H_{<k\cdot\tau}) \subset TAIL(<\tau)$ is again similar to the final part of the proof of 3.6. For any $\alpha<\tau$ it is easy to see that

$$H_{k\cdot\alpha}(g)(n) \;=\; H_{k\cdot P_n(\alpha)+k}(g)(n) \;=\; H_{k\cdot P_n(\alpha)}(g)(g^k(n))$$

and so $H_{k\cdot\alpha}(g)$ is definable by a tail α - recursion from the given function g^k. This completes the proof.

Theorem 3.8 *Let τ satisfy the same conditions as in the previous Theorems. Then for each k,*

$$DER(<k\cdot\tau) \;=\; DER_0(<2^{k\cdot\tau})$$

and hence

$$REC(<\tau) \;=\; \bigcup_k TAIL(<2^{k\cdot\tau})\,.$$

Proof. Simply recall that $B_\beta = H_{2^\beta}$ and apply the above Theorems.

Concluding remark. In terms of equational derivation, this is a cut elimination result. The first part shows that the "call by value" cut rule C can be reduced to the restricted form

$$\frac{g(n) : N \vdash t(g(n)) = m}{n : N \vdash f(n) = m} \quad \text{if } f(n) \to_E t(g(n))$$

at the cost of an exponential increase in the ordinal presentation. And the second part shows that this cut reduction amounts to the transformation from general recursive to tail recursive programs.

For further details and other results related to the work of this section, see Fairtlough-Wainer (1992) and Handley-Wainer (1994).

References

(1) P. Anderson, Program Derivation by Proof Transformation, Carnegie Mellon Technical Report CS - 93 - 206, 1993.

(2) M.V. Fairtlough, Ordinal Complexity of Recursive Programs and their Termination Proofs, Ph.D. dissertation Leeds University, 1991.

(3) M.V. Fairtlough and S.S. Wainer, Ordinal Complexity of Recursive Definitions, Information and Computation Vol. 99, 1992, pp 123 - 153.

(4) S. Feferman, Logics for Termination and Correctness of Functional Programs II, Logics of Strength PRA, in Proof Theory, Eds. P.Aczel, H. Simmons, S. Wainer; Cambridge 1992, pp 195 - 225.

(5) C. Goad, Computational Uses of the Manipulation of Formal Proofs, Stanford Technical Report CS - 80 - 819, 1980.

(6) W.G. Handley and S.S. Wainer, Equational Derivation versus Computation, Annals of Pure and Applied Logic 1994, to appear.

(7) P. Madden, Automatic Program Optimization through Proof Transformation, Proc. 11th. Intl. Conf. on Automatic Deduction, Ed. D. Kapur, Lecture Notes in AI 607, Springer Verlag 1992, pp 446 - 460.

(8) C. Murthy, An Evaluation Semantics for Classical Proofs, Logic in Computer Science, Amsterdam 1991.

(9) R. Péter, Recursive Functions, Academic Press 1967.

(10) F. Pfenning, Program Development through Proof Transformation, AMS Contemporary Mathematics Vol. 106, 1990, pp 251 - 262.

(11) H. Schwichtenberg, Proofs as Programs, in Proof Theory, Eds. P.Aczel, H. Simmons, S. Wainer; Cambridge 1992, pp 79 - 113.

(12) W. Sieg and S.S. Wainer, Program Transformation and Proof Transformation, in Computer Science Logic '93, Eds. E. Börger, Y. Gurevich, K. Meinke, Lecture Notes in Computer Science 832, Springer Verlag 1994, pp 305 - 317.

(13) W.W. Tait, Nested Recursion, Math. Annalen Vol. 143,1961, pp 236-250.

(14) S.S. Wainer, Computability - Logical and Recursive Complexity, in Logic, Algebra and Computation, Ed. F.L. Bauer, ASI Series F, Vol. 79, Springer Verlag 1991, pp 237 - 264.

(15) S.S. Wainer, Four Lectures on Primitive Recursion, in Logic and Algebra of Specification, Eds. F.L. Bauer, W. Brauer, H. Schwichtenberg, ASI Series F, Vol. 94, Springer Verlag 1993, pp 377 - 410.

NATO ASI Series F

Including Special Programmes on Sensory Systems for Robotic Control (ROB) and on Advanced Educational Technology (AET)

Vol. 71: Expert Systems and Robotics. Edited by T. Jordanides and B.Torby. XII, 744 pages. 1991.

Vol. 72: High-Capacity Local and Metropolitan Area Networks. Architecture and Performance Issues. Edited by G. Pujolle. X, 536 pages. 1991.

Vol. 73: Automation and Systems Issues in Air Traffic Control. Edited by J. A. Wise, V. D. Hopkin and M. L. Smith. XIX, 594 pages. 1991.

Vol. 74: Picture Archiving and Communication Systems (PACS) in Medicine. Edited by H. K. Huang, O. Ratib, A. R. Bakker and G. Witte. XI, 438 pages. 1991.

Vol. 75: Speech Recognition and Understanding. Recent Advances, Trends and Applications. Edited by P. Laface and Renato De Mori. XI, 559 pages. 1991.

Vol. 76: Multimedia Interface Design in Education. Edited by A. D. N. Edwards and S. Holland. XIV, 216 pages. 1992. *(AET)*

Vol. 77: Computer Algorithms for Solving Linear Algebraic Equations. The State of the Art. Edited by E. Spedicato. VIII, 352 pages. 1991.

Vol. 78: Integrating Advanced Technology into Technology Education. Edited by M. Hacker, A. Gordon and M. de Vries. VIII, 185 pages. 1991. *(AET)*

Vol. 79: Logic, Algebra, and Computation. Edited by F. L. Bauer. VII, 485 pages. 1991.

Vol. 80: Intelligent Tutoring Systems for Foreign Language Learning. Edited by M. L. Swartz and M. Yazdani. IX, 347 pages. 1992. *(AET)*

Vol. 81: Cognitive Tools for Learning. Edited by P. A. M. Kommers, D. H. Jonassen, and J. T. Mayes. X, 278 pages. 1992. *(AET)*

Vol. 82: Combinatorial Optimization. New Frontiers in Theory and Practice. Edited by M. Akgül, H. W. Hamacher, and S. Tüfekçi. XI, 334 pages. 1992.

Vol. 83: Active Perception and Robot Vision. Edited by A. K. Sood and H. Wechsler. IX, 756 pages. 1992.

Vol. 84: Computer-Based Learning Environments and Problem Solving. Edited by E. De Corte, M. C. Linn, H. Mandl, and L. Verschaffel. XVI, 488 pages. 1992. *(AET)*

Vol. 85: Adaptive Learning Environments. Foundations and Frontiers. Edited by M. Jones and P. H. Winne. VIII, 408 pages. 1992. *(AET)*

Vol. 86: Intelligent Learning Environments and Knowledge Acquisition in Physics. Edited by A. Tiberghien and H. Mandl. VIII, 285 pages. 1992. *(AET)*

Vol. 87: Cognitive Modelling and Interactive Environments. With demo diskettes (Apple and IBM compatible). Edited by F. L. Engel, D. G. Bouwhuis, T. Bösser, and G. d'Ydewalle. IX, 311 pages. 1992. *(AET)*

Vol. 88: Programming and Mathematical Method. Edited by M. Broy. VIII, 428 pages. 1992.

Vol. 89: Mathematical Problem Solving and New Information Technologies. Edited by J. P. Ponte, J. F. Matos, J. M. Matos, and D. Fernandes. XV, 346 pages. 1992. *(AET)*

Vol. 90: Collaborative Learning Through Computer Conferencing. Edited by A. R. Kaye. X, 260 pages. 1992. *(AET)*

Vol. 91: New Directions for Intelligent Tutoring Systems. Edited by E. Costa. X, 296 pages. 1992. *(AET)*

Vol. 92: Hypermedia Courseware: Structures of Communication and Intelligent Help. Edited by A. Oliveira. X, 241 pages. 1992. *(AET)*

Vol. 93: Interactive Multimedia Learning Environments. Human Factors and Technical Considerations on Design Issues. Edited by M. Giardina. VIII, 254 pages. 1992. *(AET)*

NATO ASI Series F

Including Special Programmes on Sensory Systems for Robotic Control (ROB) and on Advanced Educational Technology (AET)

Vol. 47: Advanced Computing Concepts and Techniques in Control Engineering. Edited by M. J. Denham and A. J. Laub. XI, 518 pages. 1988. *(out of print)*

Vol. 48: Mathematical Models for Decision Support. Edited by G. Mitra. IX, 762 pages. 1988.

Vol. 49: Computer Integrated Manufacturing. Edited by I. B. Turksen. VIII, 568 pages. 1988.

Vol. 50: CAD Based Programming for Sensory Robots. Edited by B. Ravani. IX, 565 pages. 1988. *(ROB)*

Vol. 51: Algorithms and Model Formulations in Mathematical Programming. Edited by S. W. Wallace. IX, 190 pages. 1989.

Vol. 52: Sensor Devices and Systems for Robotics. Edited by A. Casals. IX, 362 pages. 1989. *(ROB)*

Vol. 53: Advanced Information Technologies for Industrial Material Flow Systems. Edited by S. Y. Nof and C. L. Moodie. IX, 710 pages. 1989.

Vol. 54: A Reappraisal of the Efficiency of Financial Markets. Edited by R. M. C. Guimarães, B. G. Kingsman and S. J. Taylor. X, 804 pages. 1989.

Vol. 55: Constructive Methods in Computing Science. Edited by M. Broy. VII, 478 pages. 1989.

Vol. 56: Multiple Criteria Decision Making and Risk Analysis Using Microcomputers. Edited by B. Karpak and S. Zionts. VII, 399 pages. 1989.

Vol. 57: Kinematics and Dynamic Issues in Sensor Based Control. Edited by G. E. Taylor. XI, 456 pages. 1990. *(ROB)*

Vol. 58: Highly Redundant Sensing in Robotic Systems. Edited by J. T. Tou and J. G. Balchen. X, 322 pages. 1990. *(ROB)*

Vol. 59: Superconducting Electronics. Edited by H. Weinstock and M. Nisenoff. X, 441 pages. 1989.

Vol. 60: 3D Imaging in Medicine. Algorithms, Systems, Applications. Edited by K. H. Höhne, H. Fuchs and S. M. Pizer. IX, 460 pages. 1990. *(out of print)*

Vol. 61: Knowledge, Data and Computer-Assisted Decisions. Edited by M. Schader and W. Gaul. VIII, 421 pages. 1990.

Vol. 62: Supercomputing. Edited by J. S. Kowalik. X, 425 pages. 1990.

Vol. 63: Traditional and Non-Traditional Robotic Sensors. Edited by T. C. Henderson. VIII, 468 pages. 1990. *(ROB)*

Vol. 64: Sensory Robotics for the Handling of Limp Materials. Edited by P. M. Taylor. IX, 343 pages. 1990. *(ROB)*

Vol. 65: Mapping and Spatial Modelling for Navigation. Edited by L. F. Pau. VIII, 357 pages. 1990. *(ROB)*

Vol. 66: Sensor-Based Robots: Algorithms and Architectures. Edited by C. S. G. Lee. X, 285 pages. 1991. *(ROB)*

Vol. 67: Designing Hypermedia for Learning. Edited by D. H. Jonassen and H. Mandl. XXV, 457 pages. 1990. *(AET)*

Vol. 68: Neurocomputing. Algorithms, Architectures and Applications. Edited by F. Fogelman Soulié and J. Hérault. XI, 455 pages. 1990.

Vol. 69: Real-Time Integration Methods for Mechanical System Simulation. Edited by E. J. Haug and R. C. Deyo. VIII, 352 pages. 1991.

Vol. 70: Numerical Linear Algebra, Digital Signal Processing and Parallel Algorithms. Edited by G. H. Golub and P. Van Dooren. XIII, 729 pages. 1991.

NATO ASI Series F

Including Special Programmes on Sensory Systems for Robotic Control (ROB) and on Advanced Educational Technology (AET)

Vol. 25: Pyramidal Systems for Computer Vision. Edited by V. Cantoni and S. Levialdi. VIII, 392 pages. 1986. *(ROB)*

Vol. 26: Modelling and Analysis in Arms Control. Edited by R. Avenhaus, R. K. Huber and J. D. Kettelle. VIII, 488 pages. 1986. *(out of print)*

Vol. 27: Computer Aided Optimal Design: Structural and Mechanical Systems. Edited by C. A. Mota Soares. XIII, 1029 pages. 1987.

Vol. 28: Distributed Operating Systems. Theory und Practice. Edited by Y. Paker, J.-P. Banatre and M. Bozyiğit. X, 379 pages. 1987.

Vol. 29: Languages for Sensor-Based Control in Robotics. Edited by U. Rembold and K. Hörmann. IX, 625 pages. 1987. *(ROB)*

Vol. 30: Pattern Recognition Theory and Applications. Edited by P. A. Devijver and J. Kittler. XI, 543 pages. 1987.

Vol. 31: Decision Support Systems: Theory and Application. Edited by C. W. Holsapple and A. B. Whinston. X, 500 pages. 1987.

Vol. 32: Information Systems: Failure Analysis. Edited by J. A. Wise and A. Debons. XV, 338 pages. 1987.

Vol. 33: Machine Intelligence and Knowledge Engineering for Robotic Applications. Edited by A. K. C. Wong and A. Pugh. XIV, 486 pages. 1987. *(ROB)*

Vol. 34: Modelling, Robustness and Sensitivity Reduction in Control Systems. Edited by R.F. Curtain. IX, 492 pages. 1987.

Vol. 35: Expert Judgment and Expert Systems. Edited by J. L. Mumpower, L. D. Phillips, O. Renn and V. R. R. Uppuluri. VIII, 361 pages. 1987.

Vol. 36: Logic of Programming and Calculi of Discrete Design. Edited by M. Broy. VII, 415 pages. 1987.

Vol. 37: Dynamics of Infinite Dimensional Systems. Edited by S.-N. Chow and J. K. Hale. IX. 514 pages. 1987.

Vol. 38: Flow Control of Congested Networks. Edited by A. R. Odoni, L. Bianco and G. Szegö. XII, 355 pages. 1987.

Vol. 39: Mathematics and Computer Science in Medical Imaging. Edited by M. A. Viergever and A. Todd-Pokropek. VIII, 546 pages. 1988.

Vol. 40: Theoretical Foundations of Computer Graphics and CAD. Edited by R. A. Earnshaw. XX, 1246 pages. 1988. *(out of print)*

Vol. 41: Neural Computers. Edited by R. Eckmiller and Ch. v. d. Malsburg. XIII, 566 pages. 1988. *Reprinted as Springer Study Edition 1989, 1990.*

Vol. 42: Real-Time Object Measurement and Classification. Edited by A. K. Jain. VIII, 407 pages. 1988. *(ROB)*

Vol. 43: Sensors and Sensory Systems for Advanced Robots. Edited by P. Dario. XI, 597 pages. 1988. *(ROB)*

Vol. 44: Signal Processing and Pattern Recognition in Nondestructive Evaluation of Materials. Edited by C. H. Chen. VIII, 344 pages. 1988. *(ROB)*

Vol. 45: Syntactic and Structural Pattern Recognition. Edited by G. Ferraté, T. Pavlidis, A. Sanfeliu and H. Bunke. XVI, 467 pages. 1988. *(ROB)*

Vol. 46: Recent Advances in Speech Understanding and Dialog Systems. Edited by H. Niemann, M. Lang and G. Sagerer. X, 521 pages. 1988.

NATO ASI Series F

Including Special Programmes on Sensory Systems for Robotic Control (ROB) and on Advanced Educational Technology (AET)

Vol. 94: Logic and Algebra of Specification. Edited by F. L. Bauer, W. Brauer, and H. Schwichtenberg. VII, 442 pages. 1993.

Vol. 95: Comprehensive Systems Design: A New Educational Technology. Edited by C. M. Reigeluth, B. H. Banathy, and J. R. Olson. IX, 437 pages. 1993. *(AET)*

Vol. 96: New Directions in Educational Technology. Edited by E. Scanlon and T. O'Shea. VIII, 251 pages. 1992. *(AET)*

Vol. 97: Advanced Models of Cognition for Medical Training and Practice. Edited by D. A. Evans and V. L. Patel. XI, 372 pages. 1992. *(AET)*

Vol. 98: Medical Images: Formation, Handling and Evaluation. Edited by A. E. Todd-Pokropek and M. A. Viergever. IX, 700 pages. 1992.

Vol. 99: Multisensor Fusion for Computer Vision. Edited by J. K. Aggarwal. XI, 456 pages. 1993. *(ROB)*

Vol. 100: Communication from an Artificial Intelligence Perspective. Theoretical and Applied Issues. Edited by A. Ortony, J. Slack and O. Stock. XII, 260 pages. 1992.

Vol. 101: Recent Developments in Decision Support Systems. Edited by C. W. Holsapple and A. B. Whinston. XI, 618 pages. 1993.

Vol. 102: Robots and Biological Systems: Towards a New Bionics? Edited by P. Dario, G. Sandini and P. Aebischer. XII, 786 pages. 1993.

Vol. 103: Parallel Computing on Distributed Memory Multiprocessors. Edited by F. Özgüner and F. Erçal. VIII, 332 pages. 1993.

Vol. 104: Instructional Models in Computer-Based Learning Environments. Edited by S. Dijkstra, H. P. M. Krammer and J. J. G. van Merriënboer. X, 510 pages. 1993. *(AET)*

Vol. 105: Designing Environments for Constructive Learning. Edited by T. M. Duffy, J. Lowyck and D. H. Jonassen. VIII, 374 pages. 1993. *(AET)*

Vol. 106: Software for Parallel Computation. Edited by J. S. Kowalik and L. Grandinetti. IX, 363 pages. 1993.

Vol. 107: Advanced Educational Technologies for Mathematics and Science. Edited by D. L. Ferguson. XII, 749 pages. 1993. *(AET)*

Vol. 108: Concurrent Engineering: Tools and Technologies for Mechanical System Design. Edited by E. J. Haug. XIII, 998 pages. 1993.

Vol. 109: Advanced Educational Technology in Technology Education. Edited by A. Gordon, M. Hacker and M. de Vries. VIII, 253 pages. 1993. *(AET)*

Vol. 110: Verification and Validation of Complex Systems: Human Factors Issues. Edited by J. A. Wise, V. D. Hopkin and P. Stager. XIII, 704 pages. 1993.

Vol. 111: Cognitive Models and Intelligent Environments for Learning Programming. Edited by E. Lemut, B. du Boulay and G. Dettori. VIII, 305 pages. 1993. *(AET)*

Vol. 112: Item Banking: Interactive Testing and Self-Assessment. Edited by D. A. Leclercq and J. E. Bruno. VIII, 261 pages. 1993. *(AET)*

Vol. 113: Interactive Learning Technology for the Deaf. Edited by B. A. G. Elsendoorn and F. Coninx. XIII, 285 pages. 1993. *(AET)*

Vol. 114: Intelligent Systems: Safety, Reliability and Maintainability Issues. Edited by O. Kaynak, G. Honderd and E. Grant. XI, 340 pages. 1993.

Vol. 115: Learning Electricity and Electronics with Advanced Educational Technology. Edited by M. Caillot. VII, 329 pages. 1993. *(AET)*

NATO ASI Series F

Including Special Programmes on Sensory Systems for Robotic Control (ROB) and on Advanced Educational Technology (AET)

Vol. 116: Control Technology in Elementary Education. Edited by B. Denis. IX, 311 pages. 1993 *(AET)*

Vol. 118: Program Design Calculi. Edited by M. Broy. VIII, 409 pages. 1993.

Vol. 119: Automating Instructional Design, Development, and Delivery. Edited by. R. D. Tennyson. VIII, 266 pages. 1994 *(AET)*

Vol. 120: Reliability and Safety Assessment of Dynamic Process Systems. Edited by T. Aldemir, N. O. Siu, A. Mosleh, P. C. Cacciabue and B. G. Göktepe. X, 242 pages. 1994.

Vol. 121: Learning from Computers: Mathematics Education and Technology. Edited by C. Keitel and K. Ruthven. XIII, 332 pages. 1993. *(AET)*

Vol. 122: Simulation-Based Experiential Learning. Edited by D. M. Towne, T. de Jong and H. Spada. XIV, 274 pages. 1993. *(AET)*

Vol. 123: User-Centred Requirements for Software Engineering Environments. Edited by D. J. Gilmore, R. L. Winder and F. Détienne. VII, 377 pages. 1994.

Vol. 124: Fundamentals in Handwriting Recognition. Edited by S. Impedovo. IX, 496 pages. 1994.

Vol. 125: Student Modelling: The Key to Individualized Knowledge-Based Instruction. Edited by J. E. Greer and G. I. McCalla. X, 383 pages. 1994. *(AET)*

Vol. 126: Shape in Picture. Mathematical Description of Shape in Grey-level Images. Edited by Y.-L. O, A. Toet, D. Foster, H. J. A. M. Heijmans and P. Meer. XI, 676 pages. 1994.

Vol. 127: Real Time Computing. Edited by W. A. Halang and A. D. Stoyenko. XXII, 762 pages. 1994.

Vol. 128: Computer Supported Collaborative Learning. Edited by C. O'Malley. X, 303 pages. 1994. *(AET)*

Vol. 129: Human-Machine Communication for Educational Systems Design. Edited by M. D. Brouwer-Janse and T. L. Harrington. X, 342 pages. 1994. *(AET)*

Vol. 130: Advances in Object-Oriented Database Systems. Edited by A. Dogac, M. T. Özsu, A. Biliris and T. Sellis. XI, 515 pages. 1994.

Vol. 131: Constraint Programming. Edited by B. Mayoh, E. Tyugu and J. Penjam. VII, 452 pages. 1994.

Vol. 132: Mathematical Modelling Courses for Engineering Education. Edited by Y. Ersoy and A. O. Moscardini. X, 246 pages. 1994. *(AET)*

Vol. 133: Collaborative Dialogue Technologies in Distance Learning. Edited by M. F. Verdejo and S. A. Cerri. XIV, 296 pages. 1994. *(AET)*

Vol. 134: Computer Integrated Production Systems and Organizations. The Human-Centred Approach. Edited by F. Schmid, S. Evans, A. W. S. Ainger and R. J. Grieve. X, 347 pages. 1994.

Vol. 135: Technology Education in School and Industry. Emerging Didactics for Human Resource Development. Edited by D. Blandow and M. J. Dyrenfurth. XI, 367 pages. 1994. *(AET)*

Vol. 136: From Statistics to Neural Networks. Theory and Pattern Recognition Applications. Edited by V. Cherkassky, J. H. Friedman and H. Wechsler. XII, 394 pages. 1994.

Vol. 137: Technology-Based Learning Environments. Psychological and Educational Foundations. Edited by S. Vosniadou, E. De Corte and H. Mandl. X, 302 pages. 1994. *(AET)*

Vol. 138: Exploiting Mental Imagery with Computers in Mathematics Education. Edited by R. Sutherland and J. Mason. VIII, 326 pages. 1995. *(AET)*

Vol. 139: Proof and Computation. Edited by H. Schwichtenberg. VII, 470 pages. 1995.

Vol. 140: Automating Instructional Design: Computer-Based Development and Delivery Tools. Edited by R. D. Tennyson and A. E. Barron. IX, 618 pages. 1995. *(AET)*

NATO ASI Series F

Including Special Programmes on Sensory Systems for Robotic Control (ROB) and on Advanced Educational Technology (AET)

Vol. 142: Dialogue and Instruction. Modeling Interaction in Intelligent Tutoring Systems. Edited by R.-J. Beun, M. Baker and M. Reiner. IX, 368 pages. 1995. *(AET)*

Vol. 146: Computers and Exploratory Learning. Edited by A. A. diSessa, C. Hoyles and R. Noss. VIII, 482 pages. 1995. *(AET)*

MIX
Papier aus verantwortungsvollen Quellen
Paper from responsible sources
FSC® C105338

If you have any concerns about our products,
you can contact us on
ProductSafety@springernature.com

In case Publisher is established outside the EU,
the EU authorized representative is:
Springer Nature Customer Service Center GmbH
Europaplatz 3, 69115 Heidelberg, Germany

Printed by Libri Plureos GmbH
in Hamburg, Germany